Submarine Mass Movements and Their Consequences

Advances in Natural and Technological Hazards Research

Volume 27

Submarine Mass Movements and Their Consequences

3rd International Symposium

Edited by

VASILIOS LYKOUSIS
Institute of Oceanography
Athens, Greece

DIMITRIS SAKELLARIOU
Institute of Oceanography
Athens, Greece

JACQUES LOCAT
Laval University
Québec, Canada

Springer

A C.I.P. Catalogue record for this book is available from the Library of Congress.

Additional materials to this book can be downloaded from http://extras.springer.com

ISBN 978-1-4020-6511-8 ISBN 978-1-4020-6512-5 (eBook)
DOI 10.1007/978-1-4020-6512-5

Published by Springer,
P.O. Box 17, 3300 AA Dordrecht, The Netherlands.

www.springer.com

Printed on acid-free paper

TABLE OF CONTENTS

Section 2 - Mass waste evolution: From slump to distal turbidites

Section 3 - New techniques, approaches and challenges in submarine slope instability analysis

Section 4 - Monitoring stress on submarine slopes and sediment physical properties

CD-ROM enclosed, containing full colour images which are printed in black-and-white in the book.

FOREWORD

Submarine mass movements and their consequences are of major concern for coastal communities and infrastructures but also for the exploitation and the development of seafloor resources. A tragic example of the vulnerability of coastal communities has been provided by the Indonesian tsunami of December 2004. Since 2005, as part of the scientific community efforts to minimize the impact of such natural disasters, the International Union of Geological Science (IUGS) and the United Nation, Educational, Scientific, and Cultural Organization (UNESCO) have sponsored an International Geoscience Program on Submarine Mass Movements and Their Consequences (IGCP-511). One of the main objectives of IGCP-511 members is to hold bi-annual symposia on these types of marine and coastal geohazards. The first symposium of this series was held in Nice (2003) and the second in Oslo (2005).

This 3rd Symposium on submarine Mass Movements and Their consequences provides an opportunity to review the state of the art in risk evaluation from submarine landslides, deposit characterization and its implication for coastal and offshore development. By bringing together professionals from the industry and academia with a range of different expertise, these proceedings hope to cover the full spectrum of aspects related to subaqueous mass movements and related consequences. The interdisciplinary views gathered in this book, arising from the conference, help identify future challenges, mitigation strategies and better management of the seafloor. To that effect, the Santorini is quite a unique venue for scientists and engineers interested in marine and coastal geohazards.

The book is organized in 7 sections from environmental settings along margins to mass movements and tsunamis. It also brings together our recent knowledge on submarine failure and post-failure analysis and in situ monitoring of stress and geotechnical properties. It also presents recent techniques for either in situ or laboratory analysis. Over the recent years new areas along the coast, fjords and estuaries have been investigated and are reported on herein. Finally, the venue of the symposium at Santorini provided a unique incentive to present various case histories of submarine mass movements and consequences around volcanic islands.

We want to offer special thank to Petra Van Steenbergen of Springer for her cooperation during the preparation of this book.

This series of symposia on Submarine Mass Movements and Their Consequences shall continue in the future to maintain the necessary momentum to keep our community vibrant. We are strongly convinced that this is only by meeting and sharing our views that we can hope for a better understanding and mitigation of the consequences of these catastrophic geohazards. And, for the readers of this book, we only hope that the enthusiasm and dynamism of our scientific community will transpire from the various papers.

Vasilis Lykousis, Dimitris Sakellariou and Jacques Locat
May 15th 2007

Section 1 - Role of submarine slides in margin development

FRACTAL STATISTICS OF THE STOREGGA SLIDE

AARON MICALLEF *, CHRISTIAN BERNDT, DOUGLAS G. MASSON AND DORRIK A.V. STOW
National Oceanography Centre, European Way, Southampton, SO14 3ZH, UK.
Telephone: +44 23 80 596563; *Contact E-mail: *amicall@noc.soton.ac.uk*

Abstract

The statistics of submarine mass movement inventories are poorly characterised in comparison to those of subaerial mass movements. In this study we investigate the aggregate behaviour of the Storegga Slide by carrying out a statistical analysis of its constituent mass movements. By using area as a proxy for mass movement magnitude, we demonstrate that the non-cumulative frequency-magnitude distribution of mass movements within the Storegga Slide is a power law with an exponent of 1.52. The Storegga Slide has the characteristics of a dissipative system in a critical state, where the input of sediment is continuous in the form of hemipelagic sedimentation and glacial deposition, and the output is represented by mass movements that are spatially scale invariant. We conclude that the Storegga Slide may be modelled as a large-scale geomorphic system that exhibits self-organised critical (SOC) behaviour. In comparison to subaerial mass movements, the aggregate behaviour of submarine mass movements is more comparable to that of the theoretical 'sandpile' model. The origin of SOC may be linked to the retrogressive nature of the Storegga Slide. Since SOC is an emergent feature, the large-scale behaviour of the Storegga Slide should be autonomous of the smaller-scale elements. A power law distribution also implies that incomplete submarine mass movement inventories may be extrapolated within the limits of power law behaviour, which is important in terms of hazard management.

1. Introduction

Concepts of non-linear dynamic systems, such as scale invariance and the fractal model, provide a powerful approach to the representation of a wide range of geoscientific data (e.g. fluvial systems (e.g. Pelletier 1999), coastal profiles (Southgate and Möller 2000). Scale invariant properties of data inventories are identified by demonstrating a single power law exponent in a frequency-magnitude distribution (Mandelbrot 1983). A power law distribution implies that when we compare the number of events of size A or greater, with the number of events of size ηA or greater (η is an arbitrary factor), the number always differs by the same factor $\eta^{-\beta}$, regardless of the absolute size of the events (Hergarten 2003). A power law distribution can be replaced with other measures of the size of the event (e.g. area, volume and thickness of mass movements are strongly correlated with each other, and a distribution can be converted between variables (Hovius *et al.* 1997)); thus a power law distribution is free of a characteristic scale and can be described as fractal (Mandelbrot 1983).

The Storegga Slide, located 120 km offshore Norway, is a mega-scale geomorphic system (Figure 1). Like most other submarine slides, the Storegga Slide has been investigated using an engineering approach. In subaerial geomorphology, the statistical characteristics of landslide inventories have become a recent focus of study (e.g.

V. Lykousis, D. Sakellariou and J. Locat (eds.), Submarine Mass Movements and Their Consequences, 3–10.

Guzzetti *et al.* 2002). In comparison, the statistics of submarine mass movement data are still poorly characterised. The extensive coverage and the excellent quality of the acoustic imagery from the Storegga Slide allow us to investigate the aggregate behaviour of the Storegga Slide and carry out a statistical analysis of its constituent mass movements. The objectives of this study are to assess whether the size statistics of the Storegga Slide mass movements exhibit scale invariance, and to explain the origin and implications of such behaviour.

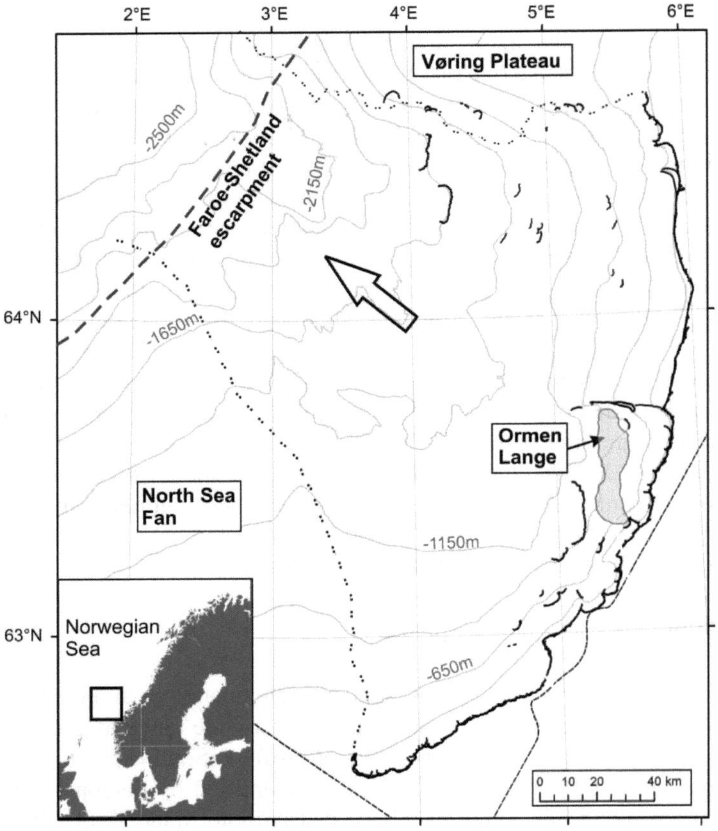

Figure 1. Bathymetric contour map of the Storegga Slide (contour interval of 250 m). The headwalls that were extracted from the bathymetric data set are represented by solid black lines. The arrow indicates the direction of sediment mobilisation. The location of the Storegga Slide is shown in the inset.

2. Method

The study is based on a high resolution multibeam bathymetry data set covering the slide scar from the main headwall down to a water depth of ca. 2700 m (Figure 1). Most of the data have a horizontal resolution of 25 m or better. A mass movement is defined as a single episode of slope failure where sediment moves downslope under the influence of gravity. The area of the mass movement is delineated by a steep scarp at the upslope limit (headwall) and the distal point of the depositional section at the downslope limit. We use mass movement area as a proxy for magnitude. The estimation of the slide area is hindered by the difficulty in defining the boundaries of quasi-simultaneous slides in a retrogressive slope failure. Thus we try to estimate mass movement area using the length of the associated headwalls, which constitute easily identifiable and prominent features located at the upslope limit of the mass movement. Previous studies of the Storegga Slide have estimated the dimensions of sixty-three mass movements (Haflidason *et al.* 2004). We plot the headwall lengths against the mass movement areas from these published data (Figure 2a). $R^2 = 0.91$ implies a strong statistical dependency between area and length in the form:

$$A = 0.87\, l^{\,1.98} \qquad (1)$$

where
A is the area of mass movement (in m^2)
l is the length of headwall (in m)

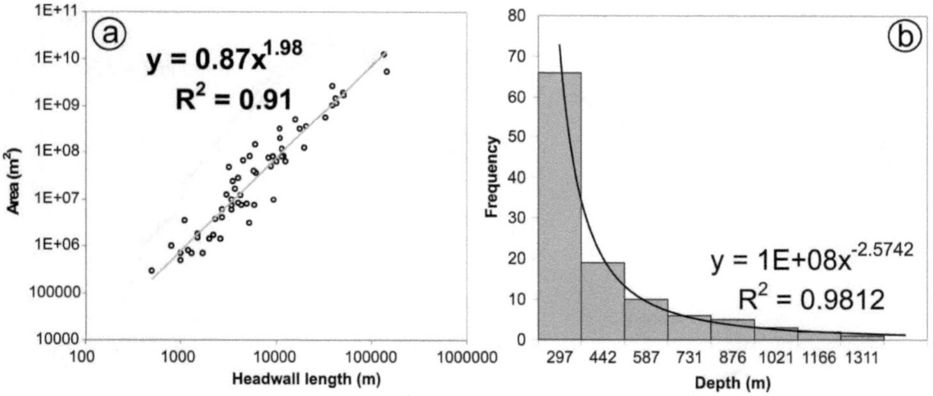

Figure 2. (a) Plot of mass movement area vs. headwall length for the mass movements identified in Haflidason *et al.* (2004). (b) Variation of the number of mass movements extracted from the bathymetric data set (frequency) with depth.

We used a suite of geomorphometric techniques to extract the headwalls automatically from the bathymetric data set. A geomorphmetric map, which is a parametric representation of a landscape decomposed into its elementary morphological units, was generated for the study area. The technique for producing a geomorphometric map is explained in more detail in Micallef *et al.* (2007). Headwalls are extracted as one-cell thick lineaments. Since the geomorphometric techniques delineate headwalls at the resolution of the bathymetric data, rather than at the scale at which the study area is being observed by an investigator, the techniques are more accurate than manual digitisation. Using geomorphometric mapping we were able to extract one hundred and five individual headwalls. The extent of a headwall is defined by the section of the headwall where sediment evacuation has occurred perpendicularly to the lineament. The length of each headwall was calculated using a Geographic Information System, and the area of the mass movement associated with each headwall was estimated using equation (1). A cumulative frequency-area graph was plotted for the mass movements. A non-cumulative distribution, defined in terms of the negative of the derivative of the cumulative distribution with respect to A, was then derived to enable comparison with previous studies (e.g. Guzzetti *et al.* 2002).

3. Results

The estimated areas of the mass movements range between 0.27 km^2 and 1174 km^2. The data in the non-cumulative distribution can be best correlated with an inverse power function (Figure 3a):

$$dN/dA = 3900\,A^{-1.52} \qquad (2)$$

where
N is the cumulative number of mass movements with an area > A

The exponent of this power function is 1.52. The range over which this function is valid is 0.3 – 100 km^2.

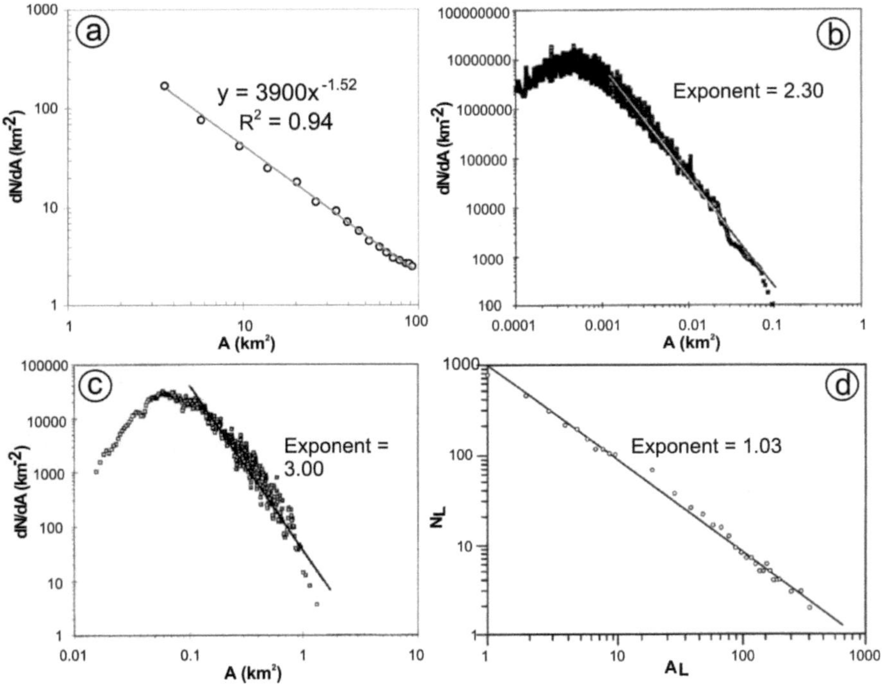

Figure 3. (a) Non-cumulative frequency-area distribution for mass movements within the Storegga Slide. Subaerial mass movements from (b) California (Harp and Jibson 1995) and (c) Akaishi ranges, central Japan (Ohmori and Sugai 1995) exhibit similar power law distributions, although the exponents are higher. (d) Non-cumulative frequency-area distribution for a 'sandpile' model based on a 50 × 50 grid (Kadanoff *et al.* 1989). The distribution is also power law with an exponent ~1. (Figures 3b-d are adapted from Turcotte (1999)).

4. Discussion

The inverse power law distribution of mass movement areas, observed over ~2.5 orders of magnitude of the area, is evidence of fractal spatial statistics within the Storegga Slide system. Similar power law distributions have been identified in numerous sub-aerial mass movements of different types and sizes, occurring in a range of environmental settings and triggered by a variety of mechanisms (Figure 3). A power law has also also been detected in other natural phenomena, such as earthquakes (Turcotte *et al.* 2006).

Explaining the origin of this fractal distribution in geological terms is difficult. The most prevalent explanation to this behaviour in subaerial environments has been self-organized criticality (SOC) (Bak *et al.* 1987). SOC is a property of complex systems whereby, in spite of heterogeneity at the small-scale of individual elements (e.g. sediment grains), the large-scale, aggregate behaviour of the system exhibits order in the form of a fractal distribution. This order is an emergent property of the system, which occurs through autogenic dynamics and feedback mechanisms (Phillips 1995). In a self-organised critical system, the "input" is nearly constant and the "output" is characterised by a series of events. Self-organised critical systems are characterised by three conditions (Bak *et al.* 1987): (i) the distribution of the 'outputs' is scale invariant; (ii) the system is in a quasi-stationary (critical) state and (iii) the temporal behaviour of the system is a 1/f (red) noise.

The Storegga Slide is a dissipative system, where sediment is mobilised or removed from the slide area in the form of mass movements. The driving force of this system has been the continuous deposition of glacially-derived material (during glacial maxima) and hemipelagic sedimentation (during interglacials), for at the least the last 3 million years (Rise *et al.* 2005). This deposition resulted in a progressive increase in sediment pore pressure, gravitationally-induced stress and surface slope gradient. Seismicity, associated to glacially-induced tectonic movements, may constitute another driving force as it enables the system to exceed thresholds. These are all characteristics of a system in a quasi-stationary state. The distribution of mass movements within the Storegga Slide is spatially scale invariant (Figure 3a). On the other hand, we are not able to demonstrate temporal scale invariance of the mass movements due to a low temporal resolution of the data. In consideration of the above, we conclude that the Storegga Slide may possibly exhibit SOC.

SOC can be theoretically modelled using the 'sandpile' model, which is a simple cellular automata model (Bak *et al.* 1988) (Figure 4). In this model, particles are dropped randomly and continuously into a square grid of boxes. When a box accumulates four particles, these are redistributed to the four adjacent boxes. This redistribution may result in further instability, creating an avalanche. The non-cumulative frequency-magnitude distribution of these avalanches was shown to satisfy a power law, with an exponent of ~1 (Kadanoff *et al.* 1989) (Figure 3d).

The elements of the 'sandpile' model can be related to the components of the Storegga Slide system. The dropping particles represent sediment deposition, the avalanches are the individual mass movements, and the thresholds are associated with changes in slope gradient, pore pressure and gravitationally-induced stress. The exponent associated with the power law distribution of subaerial mass movements is generally >2.2 (e.g. Dai and Lee 2002; Guzzetti *et al.* 2002; Malamud *et al.* 2004), whereas for the Storegga Slide, the exponent is 1.52. The value of the exponents for mass movements is higher than that of the theoretical 'sandpile' model.

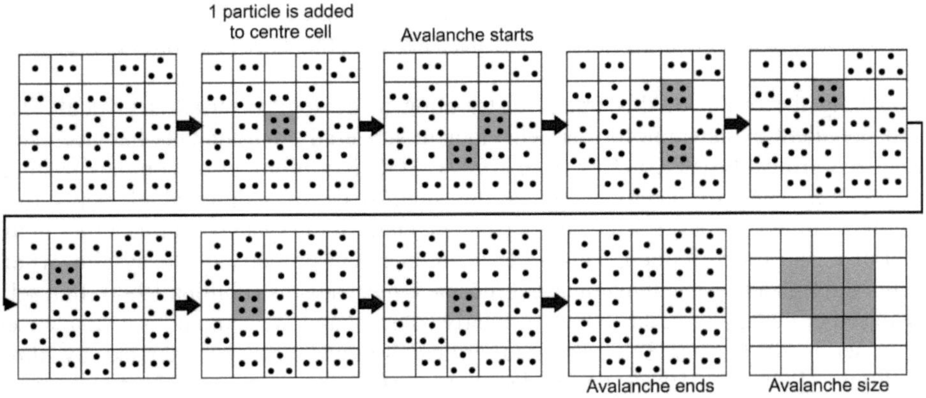

Figure 4. The theoretical 'sandpile' model based on a 5 × 5 grid. The dots indicate the number of particles within each cell of the grid. When a particle is added to the centre cell in this example, an avalanche of a size of 8 cells is triggered. In the 'sandpile' model, the frequency-magnitude distribution of these avalanches is power law.

The difference may be explained by the large number and variety of forces and controls associated with 3D 'real' mass movements, in comparison to the simpler 2D 'sandpile' model. The consideration of factors such as geological heterogeneity or soil moisture content tends to increase the exponent of frequency-size distributions in landslide models (e.g. Pelletier *et al.* 1997; Sugai *et al.* 1994). The fact that the exponent for mass movements within the Storegga Slide is considerably lower than that for subaerial slides could imply that, in comparison to subaerial mass movements, submarine mass movements are less complex and that the dynamics are more comparable to those of the 'sandpile' model. Submarine settings are characterised by gentler slopes, consistent geology and morphology over extensive areas (Shepard 1963), and therefore homogeneous boundary conditions. Subaerial settings, in contrast, consist of rougher landscapes where numerous driving forces, such as tectonic uplift and fluvial incision, interact with weathering and variable degrees of saturation, to generate a higher exponent for the power law distribution. An important role may also be played by cohesion. The sediments failing within the Storegga Slide are mainly clays. Mass movements in cohesive sediments were shown to exhibit lower exponents than those occurring in less cohesive material (Dussauge *et al.* 2003).

Some uncertainties do arise with the applicability of the 'sandpile' model to submarine mass movements, however. For example, the 'sandpile' model disregards aspects of inertia and cohesion, which are quite important in sliding within the Storegga Slide. SOC is not a sole property of cellular automata models. For example, Hergarten and Neugebauer (1998) developed a model of landsliding that exhibits SOC using partial differential equations. For our study area, another explanation of the fractal distribution of the mass movements and the potential SOC behaviour may be the fact that the Storegga Slide was a retrogressive slope failure (Haflidason *et al.* 2004). The slide was initiated close to the Faroe-Shetland Escarpment (Bryn *et al.* 2005). Large mass movements within this region destabilised neighbouring and upslope areas. The development of the Storegga Slide may be likened to a retrogressive cascade, because as the instability propagated upslope via the repeated collapse of the headwall, the mass movements became more numerous (Figure 2b) and smaller (Haflidason *et al.* 2004). The Storegga Slide extends over most of the continental slope, where topography is smooth, and boundary conditions are homogeneous (Shepard 1963). The extent of the Storegga Slide is in fact limited by changes in boundary conditions at its perimeter, in particular the decrease in slope gradient and the increase in the consolidation of sediments at the continental shelf (Gauer *et al.* 2005), as well as the presence of the North Sea Fan in the south and the Vøring Plateau in the north. The retrogressive cascade is also qualitatively similar to the activation of avalanches in the 'sandpile' model and may explain the fractal distribution of submarine mass movements. Other cascade models, such as the inverse cascade model, have been used to reproduce the self-organized critical behaviour of forest-fires (Turcotte *et al.* 1999).

The origin of the fractal distribution may also be attributed to factors that are unrelated to SOC. A power law distribution may be the signature of pre-defined geological structures (e.g. Hergarten 2003; Pelletier *et al.* 1997) or external mechanisms. Since we do not have detailed information about the spatial variation of geological structures within the Storegga Slide, we are unable to confirm the role of geological structures in relation to the observed fractal distribution.

5. Conclusions

Our results have direct implications relating to the modelling of submarine mass movements. SOC is put forward as the most likely origin of the observed power law distribution of submarine mass movements. The Storegga Slide may thus be modelled as a large scale geomorphic system in a critical state, incorporating dynamics of the 'sandpile' model. SOC is an emergent property of a system, and thus it is not built into the fundamental physical equations. This means that the aggregate behaviour of the Storegga Slide cannot be modelled using a reductionist approach based on the small-scale elements of the system. This also means, however, that limitations in data acquisition techniques can be circumvented when considering these emergent features. The retrogressive cascade, which is based on an open system where loss of support constitutes the threshold exceeding mechanism, fits the SOC behaviour well and emphasizes the importance of considering the interconnectivity of individual slides. The evolution of a retrogressive cascade on the continental slope, where boundary conditions are generally uniform, would explain the large size of the Storegga Slide. In fluvial systems, SOC has been associated with minimum energy dissipation (Rigon *et al.* 1994). We are not able to measure energy in a complex system such as the Storegga Slide, so we may only theorise that the Storegga Slide is a geomorphological system operating at the level of minimum energy dissipation, with SOC as an emergent feature. Another application of our results is in hazard management. Frequency-magnitude distribution of mass movements can be used to extrapolate incomplete inventories within the limits of power law behaviour (in our case, for mass movements ranging between 0.3 – 100 km^2 in area) and thus estimate event magnitude and total number of mass movements.

6. Acknowledgements

We would like to thank Norsk Hydro A.S. for providing the bathymetric data set.

7. References

Bak, P., Tang, C. and Wiesenfield, K., 1987. Self-organized criticality: An explanation of *1/f* noise. *Physical Review Letters*, 59: 381-384.

Bak, P., Tang, C. and Wiesenfield, K., 1988. Self-organized criticality. *Physical Review A*, 38: 364-374.

Bryn, P., Berg, K., Forberg, C. F., Solheim, A. and Kvalstad, T. J., 2005. Explaining the Storegga Slide. *Marine and Petroleum Geology*, 22(1-2): 11-19.

Dai, F. C. and Lee, C. F., 2002. Landslide characteristics and slope instability modeling using GIS, Lantau Island, Hong Kong. *Geomorphology*, 42: 213-228.

Dussauge, C., Grasso, J. R. and Helmstetter, A., 2003. Statistical analysis of rockfall volume distributions: Implications for rockfall dynamics. *Journal of Geophysical Research*, 108(B6): 2286.

Gauer, P., Kvalstad, T. J., Forsberg, C. F., Bryn, P. and Berg, K., 2005. The last phase of the Storegga Slide: Simulation of retrogressive slide dynamics and comparison with slide-scar morphology. *Marine and Petroleum Geology*, 22(1-2): 171-178.

Guzzetti, F., Malamud, B. D., Turcotte, D. L. and Reichenbach, P., 2002. Power-law correlations of landslide areas in central Italy. *Earth and Planetary Science Letters*, 195: 169-183.

Haflidason, H., Sejrup, H. P., Nygård, A., Bryn, P., Lien, R., Forsberg, C. F., Berg, K. and Masson, D. G., 2004. The Storegga Slide: Architecture, geometry and slide-development. *Marine Geology*, 231: 201-234.

Harp, E. L. and Jibson, R. L., 1995. Inventory of landslides triggered by the 1994 Northbridge, California earthquake. *In: US Geological Survey Open File 95-213*.

Hergarten, S., 2003. Landslides, sandpiles and self-organised criticality. *Natural Hazards and Earth System Sciences*, 3: 505-514.

Hergarten, S. and Neugebauer, H. J., 1998. Self-organized criticality in a landslide model. *Geophysical Research Letters*, 25(6): 801-804.

Hovius, N., Stark, C. P. and Allen, P. A., 1997. Sediment flux from a mountain belt derived by landslide mapping. *Geology*, 25(3): 231-234.

Kadanoff, L. P., Nagel, S. R., Wu, L. and Zhou, S. M., 1989. Scaling and universality in avalanches. *Physical Review A*, 39: 6524-6533.

Malamud, B. D., Turcotte, D. L., Guzzetti, F. and Reichenbach, P., 2004. Landslides inventories and their statistical properties. *Earth Surface Processes and Landforms*, 29: 687-711.

Mandelbrot, B. B., 1983. *The Fractal Geometry of Nature*. W.H. Freeman, New York

Micallef, A., Berndt, C., Masson, D. G. and Stow, D. A. V., 2007. A technique for the morphological characterization of submarine landscapes as exemplified by debris flows of the Storegga Slide. *Journal of Geophysical Research*, 112: F02001.

Ohmori, H. and Sugai, T., 1995. Toward geomorphometric models for estimating landslide dynamics and forecasting landslide occurrence in Japanese mountains. *Zeitschrift für Geomorphologie (Suppl. Bd.)*, 101: 149-164.

Pelletier, J. D., 1999. Self-organization and scaling relationships of evolving river networks. *Journal of Geophysical Research*, 104(B4): 7359-7375.

Pelletier, J. D., Malamud, B. D., Blodgett, T. and Turcotte, D. L., 1997. Scale-invariance of soil moisture variability and its implications for the frequency-size distribution of landslides. *Engineering Geology*, 48: 255-268.

Phillips, J. D., 1995. Self-organization and landscape evolution. *Progress in Physical Geography*, 19: 309-321.

Rigon, R., Rinaldo, A. and Rodríguez-Iturbe, I., 1994. On landscape self-organization. *Journal of Geophysical Research*, 99: 11971-11993.

Rise, L., Ottensen, D., Berg, K. and Lundin, E., 2005. Large-scale development of the mid-Norwegian margin during the last 3 million years. *Marine and Petroleum Geology*, 22(1-2): 33-44.

Shepard, F. P., 1963. *Submarine Geology (2nd edition)*. Harper and Row, London

Southgate, H. N. and Möller, N. K., 2000. Fractal properties of coastal profile evolution at Duck, North Carolina. *Journal of Geophysical Research*, 105(C5): 11489-11507.

Sugai, T., Ohmori, H. and Hirano, M., 1994. Rock control on magnitude-frequency distribution of landslides. *Transactions of the Japanese Geomorphological Union*, 15(3): 233-251.

Turcotte, D. L., 1999. Self-organized criticality. *Reports on Progress in Physics*, 62: 1377-1429.

Turcotte, D. L., Malamud, B. D., Guzzetti, F. and Reichenbach, P., 2006. Self-organization, the cascade model and natural hazards. *Proceedings of the National Academy of Sciences of the United States of America*, 99(1): 2530-2537.

Turcotte, D. L., Malamud, B. D., Morein, G. and Newman, W. I., 1999. An inverse-cascade model for self-organized critical behavior. *Physica A*, 268: 629-643.

SUBMARINE PALEO-FAILURE MORPHOLOGY ON A GLACIATED CONTINENTAL MARGIN FROM 3D SEISMIC DATA

J.S. LABERG AND K. ANDREASSEN
Department of Geology, University of Tromsø, N-9037 Tromsø, Norway

Abstract

The morphology of two lower – middle Pleistocene paleo-slope surfaces within a muddy glacigenic succession was studied on 3D seismic data. The lower surface is characterised by irregular relief. It terminates upslope by an escarpment that represents the upper part of a paleo-slide scar. The slide scar morphology is relatively similar to that of modern slide scars and failure is inferred to have occurred during or after a glacial maximum when the ice reached the paleo-shelf break. A large area immediately outside the paleo-slide scar was affected by sediment creep or sliding, thus the area of unstable sediments extends beyond the paleo-slide scar. The upper surface morphology is dominated by three straight to slightly meandering paleo-channels, at least one of them formed by mass wasting. Together, the two paleo-surfaces exemplify slope morphology that may result from sediment instability on glaciated margins.

Keywords: paleo-slide scar, paleo-channels, glaciated margin, Barents Sea

1. Introduction

Some of the largest submarine landslides have affected high-latitude glaciated continental margins, areas which now receive increased attention because of their hydrocarbon potential. On the Norwegian – Barents Sea – Svalbard margin submarine landslides have resulted in several slide scars including the Storegga (Bugge et al., 1987; Haflidason et al., 2004, 2005), Trænadjupet (Laberg and Vorren, 2000) and Hinlopen Slides (Vanneste et al., 2006; Winkelman et al., 2006). Paleo-slope records from these areas also reveal evidence of similar sized buried slide scars, indicating repeated large-scale sliding (Evans et al., 1996; Kuvaas & Kristoffersen, 1996; Laberg and Vorren, 1996; Solheim et al., 2005; Laberg et al., 2006). Another, less studied result of slope instability on glaciated margins is the development of large canyon – channel systems. Morphological studies of modern canyons have shown that sliding is an important process in their evolution and that they develop over a long period of time from the interaction of several processes which include repeated sliding (Laberg et al., in press).

In this study we have focused on the early – middle Pleistocene interval of a glacigenic depocentre in the south-western Barents Sea, the prograding wedge in front of the Bear Island Trough (Fig. 1). The objective of this study is to describe and discuss the morphology of two close-lying paleo-slope surfaces; the lower displays part of a paleo slide-scar, whilst the upper comprises several large paleo-channels, and to elucidate on the paleoenvironment during their formation.

V. Lykousis, D. Sakellariou and J. Locat (eds.), Submarine Mass Movements and Their Consequences, 11–18.

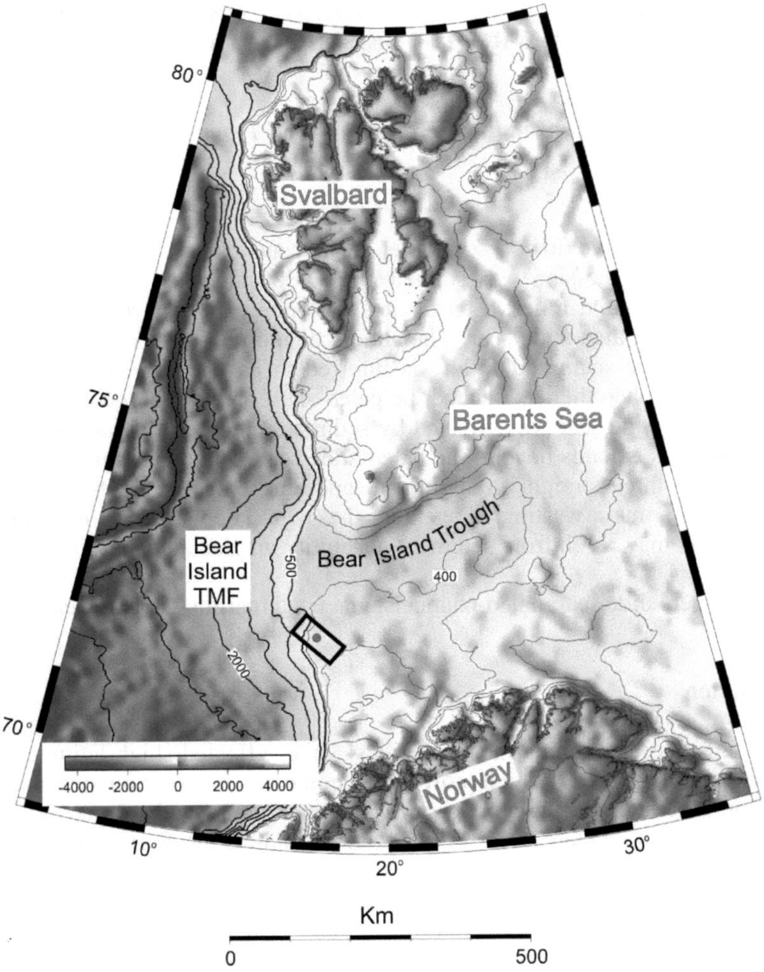

Fig. 1. Bathymetric map of the western Barents Sea continental margin. The box outlines the area of 3D seismic data and the red dot shows the location of well 7216/11-1. Contour interval is 100 m on the shelf and 500 m on the continental slope and in the deep sea.

2. Geological Setting

The Norwegian – Barents Sea – Svalbard continental shelf has a glacial morphology. Glacial erosion was most pronounced in the transverse troughs where ice streams were located during full glacial conditions when the Fennoscandian and the Barents Sea Ice Sheets extended to the shelf break along most of the Norwegian – Barents Sea shelf (e.g. Vorren, 2003). As a result of this spatially variable glacial erosion, some areas of the continental slope, i.e. the areas in front of the troughs received huge volumes of sediments. These depocenters have been called "Trough Mouth Fans" (Vorren et al., 1989). The largest trough mouth fan along the Norwegian – Barents Sea continental margin is the Bear Island Trough Mouth Fan (TMF) (Fig. 1) which developed due to deposition over repeated glacial episodes.

The onset of the Bear Island TMF development has been dated to about 2.75 - 2.3 Ma (Eidvin et al., 1993; Sættem et al., 1992; Mørk and Duncan, 1993). Based on shallow boreholes in the northern outer Bear Island Trough (c. 150 km north of our study area) Sættem et al. (1992) suggested that the lower part of the fan was deposited by “a high sediment input onto a shallow, sand-dominated continental shelf in front of a grounded ice margin”. Where sampled during commercial drilling (mainly cuttings and small sidewall cores), the upper part of the fan comprises clast-bearing muddy sediments inferred to be glacimarine deposits which include Ice-Rafted Material (IRD). Data available from well 7216/11-1 located within the study area indicate that the paleo-slope surfaces studied in the present paper (Fig. 2) are located within muddy glacimarine sediments (Ryseth et al., 2003). Studies of 3D seismic data from the paleo-shelf indicated that grounded ice reached the shelf break from the level of the lower surface that is investigated here and upwards (Andreassen et al., 2004, 2007).

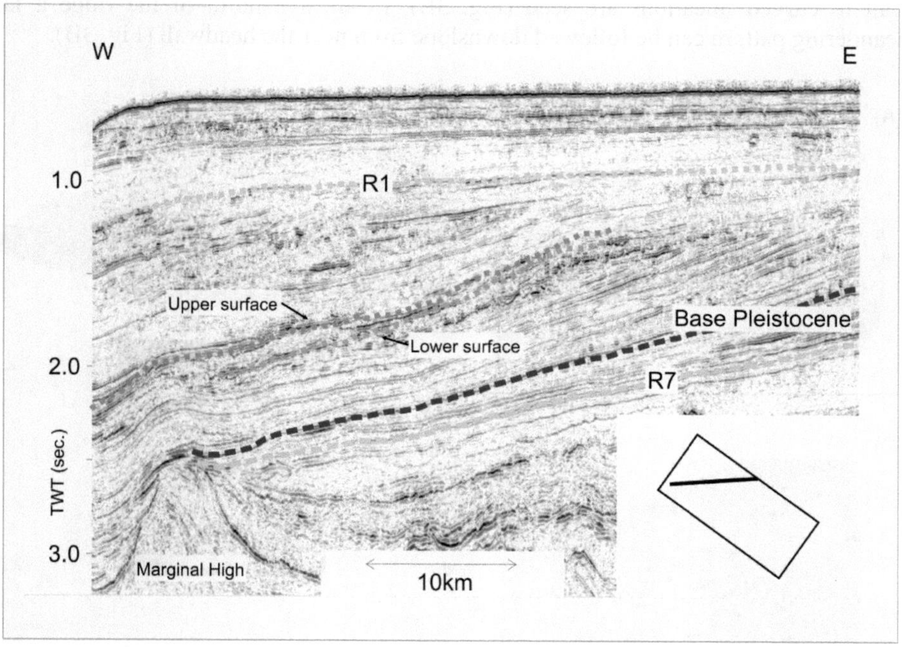

Fig. 2. Part of a seismic line showing the stratigraphic position of the lower and upper surface studied. Reflection R7 is the base of the late Pliocene – Pleistocene glacigenic sediments, Base Pleistocene the base of the Pleistocene sediments and R1 an upper regional unconformity of middle Pleistocene age.

3. Data Base

This study is based on a commercial 3D seismic data set covering an area of about 2900 km^2 (Fig. 1). The vertical resolution of the data is approximately 20-25 m. The theoretical limit for the horizontal resolution of 3D seismic data is ¼ of the seismic wavelet (Brown 2003), which here is ~ 20-30 m (using a seismic velocity of 2200 m/s). For the seismic interpretation the GeoFrame Charisma software was used.

4. Paleo-slope Morphology

4.1 THE LOWER SURFACE

The lower surface includes part of the upper paleoslope. It has been mapped from its upslope truncation by a semi-horizontal, slightly westward dipping "topset" reflection, downslope to the marginal high (Fig. 2). The central and northern part of this surface is dominated by irregular relief which terminates upslope by an escarpment (Fig. 3A). In some areas the escarpment is up to 50 ms (TWT) high and easily identified, in other areas it is more subdue probably because the height is below the vertical resolution of the seismic system. The area of irregular relief is separated into a northern and a southern part by a downslope oriented, steep-crested ridge which is up to 4 km wide and 100 ms (TWT) high (Figs. 3A-B). The ridge crest is irregular and dominated by amphitheatre-shaped depressions, except for in the lowermost part where a series of strait to curved lineations are seen (Fig. 3B). In an area north of the ridge a faint meandering pattern can be followed downslope from near the headwall (Fig. 3B).

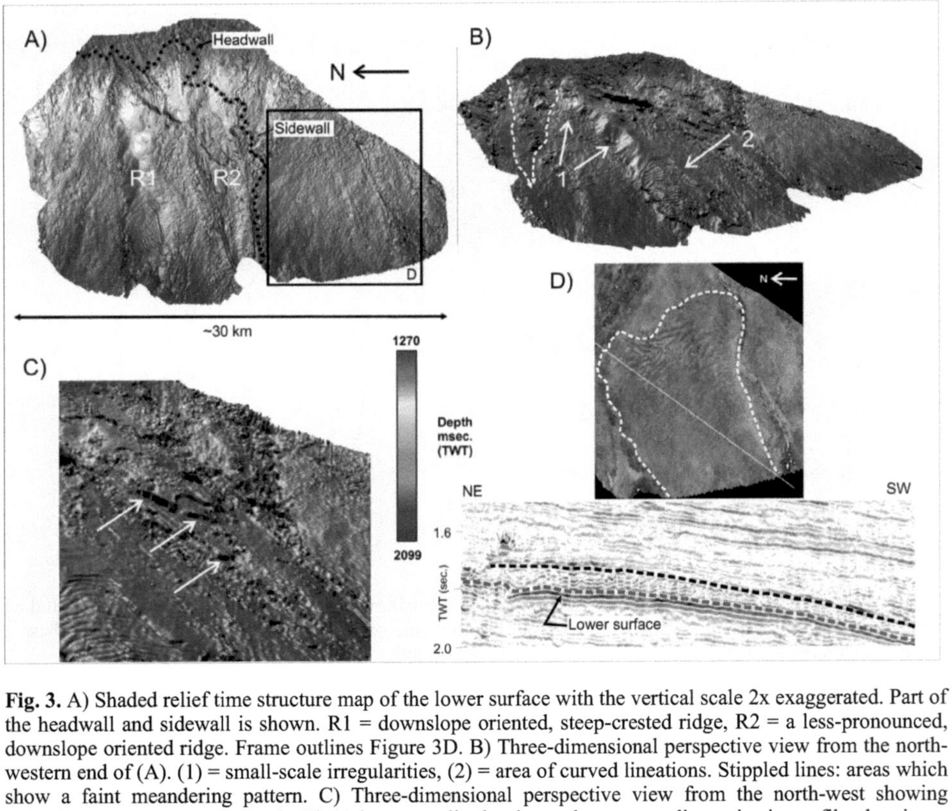

Fig. 3. A) Shaded relief time structure map of the lower surface with the vertical scale 2x exaggerated. Part of the headwall and sidewall is shown. R1 = downslope oriented, steep-crested ridge, R2 = a less-pronounced, downslope oriented ridge. Frame outlines Figure 3D. B) Three-dimensional perspective view from the north-western end of (A). (1) = small-scale irregularities, (2) = area of curved lineations. Stippled lines: areas which show a faint meandering pattern. C) Three-dimensional perspective view from the north-west showing secondary escarpments (arrows). D) Volume amplitude plot and corresponding seismic profile showing a large area of curved lineations immediately south of the slide scar. The area is outlined by the stippled line on the amplitude plot. On the corresponding seismic line it is located between the green (lower surface reflection) and the black stippled lines (see Fig A for location).

Small-scale irregularities are seen, most pronounced near the escarpment (Fig. 3B). Several secondary escarpments are seen within the upper, southern part of the irregular area (Fig. 3C). They occur within the upslope part of a less-pronounced, downslope oriented ridge (Fig. 3A). A small escarpment marks the southern limit of the irregular area. To the south of this escarpment the paleo-slope has a relatively smooth relief. Immediately above this surface the reflections are discontinuous and a volume amplitude plot shows a large area of curved lineations (Fig. 3D), similar to the lower part of the northern ridge (Fig. 3B).

The areas of irregular morphology are inferred to represent the upper part of a paleo-slide scar. The upper escarpment probably forms part of the paleo-headwall while the downslope oriented ridge is an erosional remnant. The amphitheatre-shaped depressions on this ridge are likely the result of smaller-scale mass wasting, whilst the curved lineations in the lowermost part are the result of sediment creep. South of the ridge secondary escarpments occur. Such features are not found north of the ridge, instead a faint meandering signature is seen, originating from near the headwall. This difference may be related to sediment physical properties variations, with the more consolidated sediments south of the ridge more difficult to mobilise into flows. A large area south of the slide scar was probably also affected by sediment creep or sliding, from the present data base it is not possible to discriminate between the two alternatives. This formed a pattern of curved fractures separating rafts or ridges of sediments and was possibly part of the same event that resulted in the formation of the slide scar.

4.2 THE UPPER SURFACE

Stratigrapically, the upper surface is located slightly above the lower in a similar physiographic setting (Fig. 2). Its morphology is dominated by three straight to slightly meandering channels (indicated 1, 2, 3 in Fig. 4A). The upslope part of the southern two channels (1-2) could not be mapped because this part of the paleo-slope has been removed by subsequent erosion. The southernmost channel (1) originates as two channels, then merges into one which keeps its identity as a straight channel downslope (Fig. 4B). The channel is V-shaped, 50 ms (TWT) deep and has a width of about 250 m. The channel is visible on the amplitude plot (Fig. 4D), showing an increased acoustic contrast downslope (Fig. 4D).

The middle channel (2) is slightly meandering (Fig. 4B), it has a depth and shoulder width of about 25 ms (TWT) and 250 m, respectively and has a U-shaped cross-section. It is also well displayed on the amplitude plot (Fig. 4D), indicating contrasting sediments at the bottom of the channel. This could be due to erosion and subsequent deposition of more coarse grained sediments brought downslope from the area of flow origin. The third channel (3), is U-shaped, c. 500 m wide and terminates upslope in a headwall area (Figs. 4A, C). The headwall is amphitheatre-shaped, about 3 km wide, and is incised by second-order channels (Fig. 4C). Downslope from the headwall channel 3 is seen on the amplitude plot, although the contrast is not as clear as for the other channels (Fig. 4D).

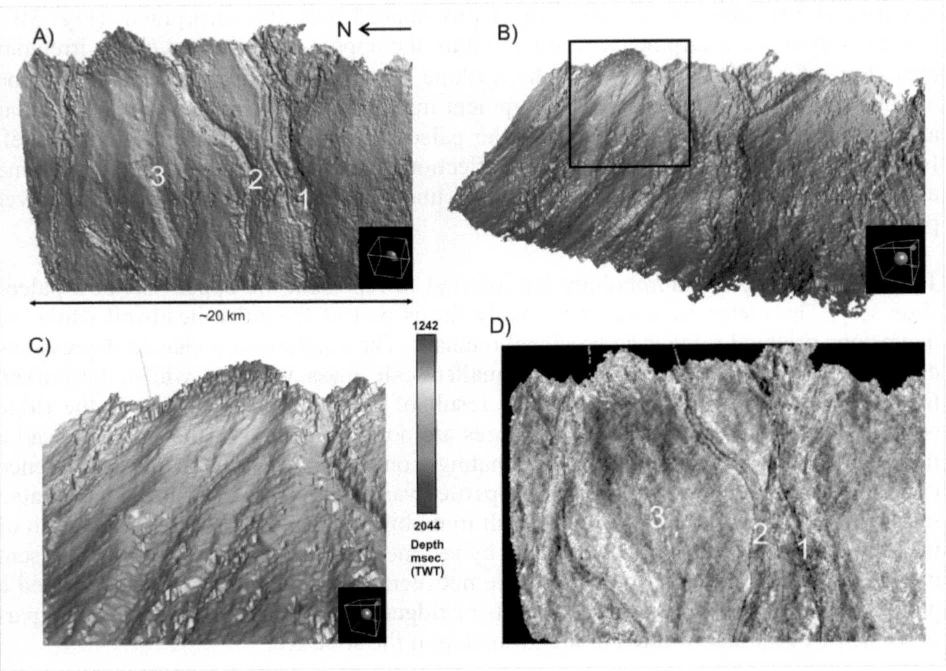

Fig. 4. A) Shaded relief time structure map of the upper surface with the vertical scale 2x exaggerated. Channels 1 – 3 are indicated. B) Three-dimensional perspective view from the south-west. Frame outlines Fig. 4C. C) Three-dimensional perspective view of the upper part of channel 3. D) Surface amplitude plot of area in (A). The location of channels 1 – 3 is indicated.

5. Discussion

5.1 PALEO-SLOPE PROCESSES

Although only part of the upper slide scar was identified on the lower surface, its morphology is relatively similar to that of modern slide scars. Secondary escarpments downslope of the headwall form the upper boundary of subparallel paleo-surfaces as seen for instance in the upper Trænadjupet (Laberg and Vorren, 2000) and Nyk (Lindberg et al., 2004) Slide scars. This indicates that sediments at different stratigraphic levels were affected by the failure, that the failure may have been initiated at specific stratigraphic levels forming layers of weakness, and that this may have occurred during one major event followed by smaller, secondary events.

The morphology of the upper surface differs from the lower surface, being dominated by paleo-channels of various sizes. One channel terminates upslope in a headwall area with second-order channels. Several second-order channels may indicate channel formation by sliding over a longer period, as seen in modern canyons (Laberg et al., in press).

5.2 FACTORS PROMOTING LARGE-SCALE SLIDING

Within the glacigenic sediments studied, the lower paleo-slope surface represents the oldest level of large-scale sliding. Below this surface, 2D seismic data display mainly acoustically laminated sediments where intervals of single channels, channel systems and small-scale sliding have been shown. So what caused large-scale sliding in this area at this time? Studies of the late Pleistocene succession on the Norwegian – Barents Sea continental margin have shown that sliding events tend to occur during, or immediately after, glacial maximum periods (e.g. Solheim et al., 2005; Laberg et al., 2006).

The advance of an ice sheet to the shelf break results in sediment erosion below the ice. This erosion is most intense beneath fast-flowing ice streams and large volumes of sediment will be deposited in front of these on the upper continental slope. This rapid sediment loading affects the physical properties of the underlying sediments, makes them more prone to failure (Bryn et al., 2005; Laberg et al., 2003; Kvalstad et al., 2005). We therefore suggest that the submarine landslide which resulted in the slide scar partly displayed on our lower surface was the result of increased sediment input to this part of the continental margin. This was probably due to the advance of an ice sheet to or near the shelf break, in accordance with Andreassen et al. (2004, 2007).

5.3 FACTORS PROMOTING CHANNEL DEVELOPMENT ON GLACIATED MARGINS

The upper surface morphology is also related to sediment reworking but why this resulted in channel features and not a slide scar morphology as the lower surface event is not known. The northern channel was probably formed by sliding over a longer period. However, the southern two may have been formed by a similar process as envisaged for the channels identified by Sættem et al. (1992) slightly north of our study area. These channels may have been formed at the margin of ice caps or ice sheets, possibly by meltwater erosion. This interpretation is supported by the fact that they have a more pronounced acoustic contrast (Fig. 4D) compared with the northern; i.e. that they were not formed by the reworking of slope sediments as the northern channel but related to large input of meltwater, introducing more coarse-grained sediments to the channels which caused the acoustic contrast. Thus the upper slope morphology was most likely a result of channel formation both by mass wasting and glacial meltwater erosion.

6. Acknowledgements

This work is a contribution to the project; *Ice ages: subsidence, uplift and tilting of traps: the influence on petroleum systems (GlaciPet)* funded by the Research Council of Norway and Norsk Hydro ASA. The University of Tromsø acknowledges Schlumberger for seismic interpretation software and Norsk Hydro ASA for providing the data. Figure 1 was created using Generic Mapping Tools (GMT) software (Wessel and Smith, 1998). The manuscript benefited from helpful comments by the reviewers Anders Solheim and Michele Rebesco, and Monica Winsborrow corrected the English language.

7. References

Andreassen, K.A., Nilssen, L.C., Rafaelsen, B., Kuilman, L. 2004. *Three-dimensional seismic data from the Barents Sea margin reveal evidence of past ice streams and their dynamics.* Geology 32, 729-732.

Andreassen, K., Ødegaard, C.M., Rafaelsen, B. 2007. *Imprints of former ice streams, imaged and interpreted using industry three-dimensional seismic data from the south-western Barents Sea.* Geological Society of London, Special Publication 277, pp. 151-169.

Brown, A.R. 2003. *Interpretation of Three Dimensional Seismic Data, 6th. Edition.* AAPG Memoir, vol. 42, 541 pp., Tulsa, OK.

Bryn, P., Berg, K., Forsberg, C.F., Solheim, A., Kvalstad, T.J. 2005. *Explaining the Storegga Slide.* Marine and Petroleum Geology 22, 11-19.

Bugge, T., Befring, S., Belderson, R.H., Eidvin, T., Jansen, E., Kenyon, N.H., Holtedahl, H., Sejrup, H.P. 1987. *A giant three-stage submarine slide off Norway.* Geo-Marine Letters 7, 191-198.

Eidvin, T., Jansen, E., Riis, F. 1993. *Chronology of Tertiary fan deposits off the western Barents Sea: implications for the uplift and erosion history of the Barents Shelf.* Marine Geology 112, 109-131.

Evans, D., King, E.L., Kenyon, N.H., Brett, C., Wallis, D. 1996. *Evidence for long-term instability in the Storegga Slide region off western Norway.* Marine Geology 130, 281-292.

Haflidason, H., Lien, R., Sejrup, H.P., Forsberg, C.F., Bryn, P. 2005. *The dating and morphometry of the Storegga Slide.* Marine and Petroleum Geology 22, 123-136.

Haflidason, H., Sejrup, H.P. et al. 2004. *The Storegga Slide: architecture, geometry and slide development.* Marine Geology 213, 201-234.

Kuvaas, B., Kristoffersen, Y. 1996. *Mass movements in glacimarine sediments on the Barents Sea continental slope.* Global and Planetary Change 12, 287-307.

Kvalstad, T.J., Nadim, F., Kaynia, A.M., Mokkelbost, K.H., Bryn, P. 2005. *Soil conditions and slope stability in the Ormen Lange area.* Marine and Petroleum Geology 22, 245-256.

Laberg, J.S., Vorren, T.O. 1996. *The Middle and late Pleistocene evolution of the Bear Island Trough Mouth Fan.* Global and Planetary Change 12, 309-330.

Laberg, J.S., Vorren, T.O. 2000. *The Trænadjupet Slide, offshore Norway - morphology, evacuation and triggering mechanisms.* Marine Geology 171, 95-114.

Laberg, J.S., Guidard, S., Mienert, J., Vorren, T.O., Haflidason, H., Nygård, A. In press. *Morphology and morphogenesis of a high-latitude canyon; the Andøya Canyon, Norwegian Sea.* Marine Geology.

Laberg, J.S., Vorren, T.O., Kenyon, N.H., Ivanov, M. 2006. *Frequency and triggering mechanisms of submarine landslides of the North Norwegian continental margin.* Norwegian Journal of Geology 86, 155-161.

Laberg, J.S., Vorren, T.O., Mienert, J., Haflidason, H., Bryn, P., Lien, R. 2003. *Preconditions leading to the Holocene Trænadjupet Slide offshore Norway.* In: Locat, J. Mienert, J. (Eds.), *Submarine mass movements and their consequences.* Kluwer Academic Publishers, Netherlands, pp. 247-254.

Lindberg, B., Laberg, J.S., Vorren, T.O. 2004. *The Nyk Slide – morphology, progression, and age of a partly buried submarine slide offshore mid-Norway.* Marine Geology 213, 277-289.

Mørk, M.B.E., Duncan, R.A. 1993. *Late Pliocene basaltic volcanism on the western Barents shelf margin: implications from petrology and ^{40}Ar – ^{39}Ar dating of volcaniclastic debris from a shallow drill core.* Norsk Geologisk Tidsskrift 73, 209-225.

Ryseth, A., Augustson, J.H. et al. 2003. *Cenozoic stratigraphy and evolution of the Sørvestsnaget Basin, southwestern Barents Sea.* Norwegian Journal of Geology 83, 107-130.

Solheim, A., Berg, K., Forsberg, CF., Bryn, P. 2005. *The Storegga Slide complex: repetitive large scale sliding with similar cause and development.* Marine and Petroleum Geology 22, 97-107.

Sættem, J., Poole, D.A.R., Ellingsen, L., Sejrup, H.P. 1992. *Glacial geology of outer Bjørnøyrenna, southwestern Barents Sea.* Marine Geology 103, 15-51.

Vanneste, M., Mienert, J., Bünz, S. 2006. *The Hinlopen Slide: a giant, submarine slope failure on the northern Svalbard margin, Arctic Ocean.* Earth and Planetary Science Letters 245, 373-388.

Vorren, T.O., 2003. *Subaquatic landsystems: Continental Margins.* In: D.J.A. Evans (Editor), *Glacial landsystems.* Arnold Publishers, London, pp. 289-312.

Vorren, T.O., Lebesbye, E., Andreassen, K. and Larsen, K.-B., 1989. *Glacigenic sediments on a passive continental margin as exemplified by the Barents Sea.* Marine Geology, 85, 251-272.

Wessel, P., Smith, W.H.F., 1998. *Improved version of the Generic Mapping Tools released.* EOS Transactions of the American Geophysical Union, 79, 579.

Winkelmann, D., W. Jokat, F. Niessen, R. Stein, and A. Winkler 2006. *Age and extent of the Yermak Slide north of Spitsbergen, Arctic Ocean.* Geochem. Geophys. Geosyst., 7, Q06007, doi:10.1029/2005GC001130.

SLOPE INSTABILITY AND MASS TRANSPORT DEPOSITS ON THE GODAVARI RIVER DELTA, EAST INDIAN MARGIN FROM A REGIONAL GEOLOGICAL PERSPECTIVE

C.F. FORSBERG, A. SOLHEIM, T.J. KVALSTAD
Norwegian Geotechnical Institute, P.O. Box 3930, NO-0806 Oslo, Norway

R.VAIDYA AND S. MOHANTY
Reliance Industries Ltd. Navi Mumbai, India

Abstract

The Krishna and Godavari rivers have formed the Krishna-Godavari delta where they enter the ocean. Both the rivers, which drain a significant part of the Indian peninsula, and the delta are influenced by a seasonal sediment supply controlled by the monsoon rains. The delta system receives nearly all its sediment during this annual flushing of the river system.

The slope instabilities on the delta are most likely driven by excess pore pressure induced by rapid sedimentation and therefore reflect sedimentation history and distribution. The overall geometry of delta shows that the Pleistocene subsidence on the inner part of the sub-aqueous delta is associated with toe-thrusts in more distal regions and that the morphology of the delta front is comprised of channel-levee systems with overbank and mass transport deposits. The climatically induced fluctuations in sea level have probably shifted the main depocentre through time with the more distal sedimentation occurring during sea-level lowstands.

This study is based on 2D and 3D-seismic data that has been combined with geotechnical information from boreholes. This investigation demonstrates the role regional geology has for the distribution and timing of mass movements.

Keywords: Submarine landslides, East Indian margin, delta deposits

1. Introduction

The eastern continental margin of India is an Atlantic type passive margin. It is characterized by a relatively narrow continental shelf, which widens toward the inner part of the Bengal Bay (Fig. 1), and a relatively wide area comprising the continental slope and rise. The present study area is located on the upper continental slope and consists two areas about 14 km apart. One with water depths between about 40 and 400 m, the other with water depths between about 300 to 1900 m (Fig. 2).

The Godavari River is the third largest river in India, after the Ganges and Brahmaputra. While the Krishna River drains a basin about two thirds of that of the Godavari, it only has half of the sediment flux. For both rivers a significant part of the annual discharge and therefore the sediment flux occurs during the relatively short monsoonal period. Together the Godavari and Krishna rivers deliver an extensive sediment load (255 million tonnes per year Sharma 2002, Sarin et al. 2002) to this part of the eastern Indian continental margin.

V. Lykousis, D. Sakellariou and J. Locat (eds.), Submarine Mass Movements and Their Consequences, 19–27.

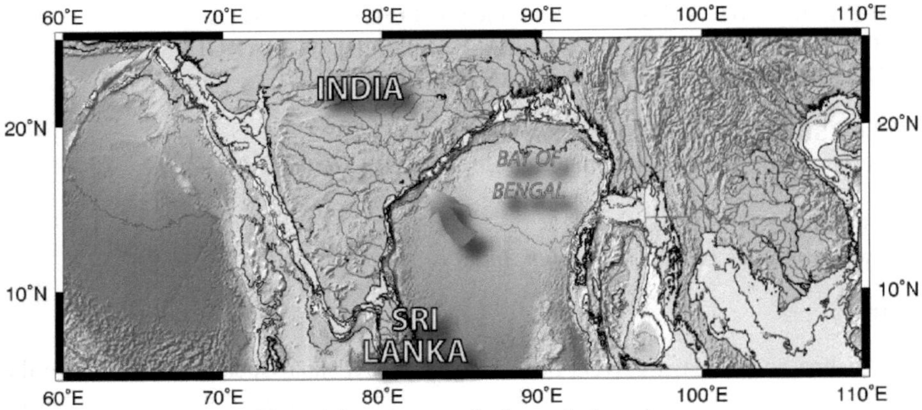

Figure 1. Location map. Study area is shown by arrow.

The present day sea level was established about 5000 years ago. Since that time, a near-shore sandy facies has formed bars, shoals and spits, whereas most of the shelf and slope has been covered by muddy deposits. A progradation of the Krishna-Godavari delta has also taken place since about 5000 years ago, with a seaward shift in the position of the shore line up until about 800 years ago (Bruckner, 1988). An important effect of sea level rise was a de-linking of the main rivers from slope valleys (Flood and Piper, 1997; Moscardelli et al. 2006; Piper et al. 1997; Posamentier and Kolla 2003), some of which terminate landward at around the 50 m isobath. Our main objective is therefore to demonstrate that there is a contrast between the history of shallow and deep water mass transport processes through our interpretations of the regional geology provided by 2D and 3D seismic data in combination with borehole information.

2. Seafloor Morphology

2.1. DEEP WATER AREA

The seafloor morphology of the deep water area exhibits several large channel systems and some smaller channels in regions shallower than about 750 m (Fig. 2 and Fig. 3). Between the channels in water depths greater than about 1000 m there are large areas with a hummocky seafloor. While these features shape the seafloor, seismic sections show that they are covered by a drape of acoustically laminated deposits (Fig. 3).

Numerous ENE-WSW trending synsedimentary listric faults offset the seabed in the northern part of the area. Evidence of toe thrust faulting can be seen in the south western and deepest parts (Fig. 2, right hand panel). The faults scarps on the seafloor provide slopes steep enough for slumps etc. to be triggered.

2.1.1. Hummocky seafloor and channel levee complex

The Channel B (Fig. 2) forms a channel levee complex that has developed through several avulsions. The adjacent hummocky seafloor is associated with units that have a chaotic seismic facies and are interpreted by us as overbank and mass transport deposits (Fig. 3). While the laminated levee deposits mostly underlie the mass transport deposits, there is also some inter-fingering on the levee flanks.

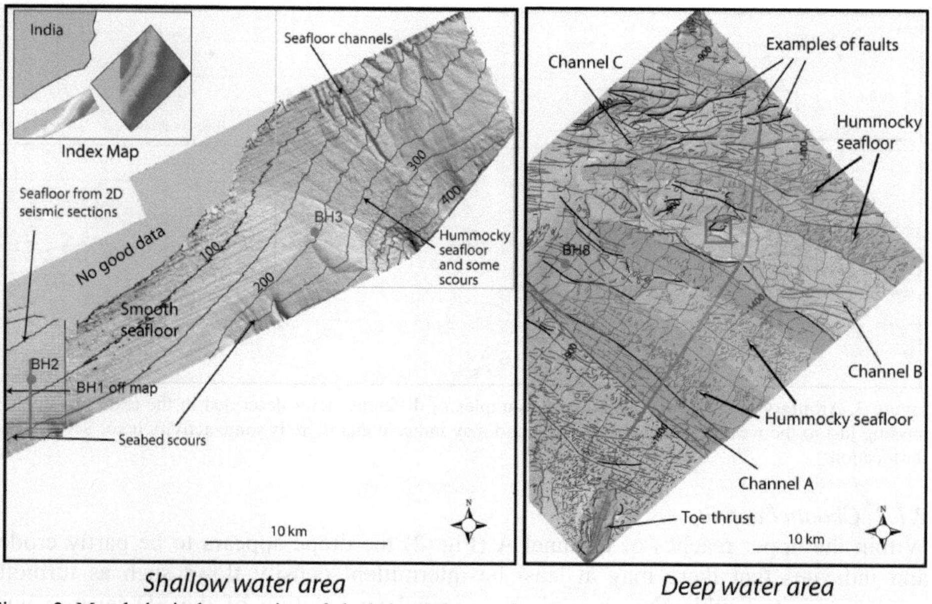

Figure 2. Morphological expression of shaded relief map of seafloor from 3D seismic surveys. On the right hand panel major channel systems are coloured red, yellow and pale green. The left hand panel shows the shallow water area, about 14 km to the West of the deep water area. Red line on right hand panel is location of Fig. 3 whereas red line on left hand panel is location of Fig. 5 and the red Box shows location of Figure 4a. The pale blue box on the left hand panel is the location of Figure 4b. Colours on Index Map indicate depths.

There is internal structuring within the mass transport units that demonstrate that they may be the result of several smaller episodes, perhaps inter-layered with or underlain by overbank deposits from the channel(s). The bases of the channels are sometimes associated with high amplitude reflectors (HARs) that probably consist of more coarse grained sediments (Flood and Piper, 1997; Piper et al. 1997). Because these partly underlie the levee deposits, they were probably formed before the channel levee system had matured and expanded into the area, perhaps after an avulsion, i.e. that they are analogous, but not identical to the turbidite deposits later formed at the mouth of the channels in the deep sea (Flood and Piper, 1997; Piper et al. 1997). The HARs having been formed on the slope whereas the deep sea turbidites deposits are on the relatively flat basin floor.

The hummocky seafloor terminates upslope along scarps that are interpreted to be the headwalls associated with the mass transport deposits (Fig. 2). In general there is a surface drape of acoustically laminated sediments across the whole deep water area. In several places this drape has erosional scars that only influence the uppermost sediment layers. These are therefore the youngest features in the area. They take the shape of elongated depressions with sharp upper and side boundaries and follow "valley" floors as would be expected for mass movements under the influence of gravity. The downslope termination is not as distinct and sometimes associated with a small mound. We interpret these features as having been formed by debris flows/slides/slumps and that can be seen to have originated along faults and other scarps or steeper areas on the sea floor. Their relief is about 2-5 m and they usually have quite a significant run-out (Fig. 4a).

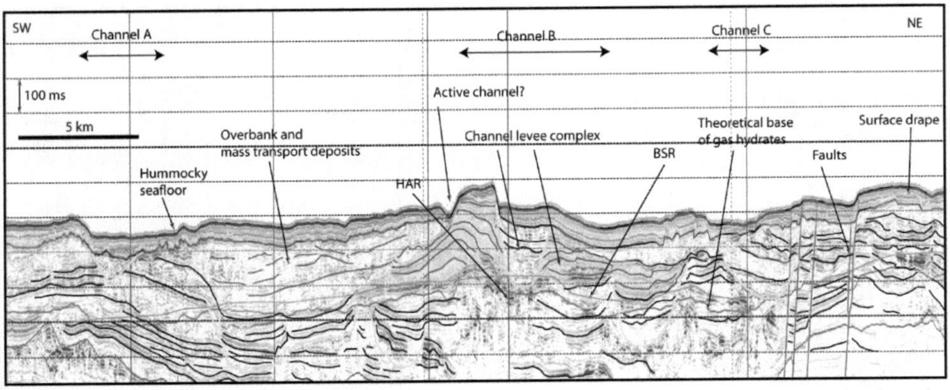

Figure 3. Arbitrary seismic section showing examples of different facies described in the text. The drape is missing just to the west of Channel B (Fig. 2) and may indicate that there is some activity here. See Figure 2 for location.

2.1.2. Channel activity

Within the upper reaches of Channel A (Fig. 2) the drape appears to be partly eroded and indicates that there may at least be intermittent density flows such as turbidity current activity within the channel. However the resolution of the 3D-seismic lines makes an unequivocal interpretation difficult. It should be noted that the channel is eroded into the underlying mass transport unit (Fig. 3) and has therefore been active after the deposition of this unit. Channel C is similar, but contains a drape that suggests that it is not active at present. Within this channel, the seafloor has been offset by several faults, but does nevertheless not show evidence of erosion etc. This channel is therefore interpreted to be inactive.

Channel B contains a couple of notable features. The first is a horseshoe shaped cut where it narrows sharply upslope (west of red line Fig. 2) that looks like a slump headwall. This impression is confirmed by seismic sections across the scarp, but also shows that it is an old feature that has later been covered by channel-levee deposits. The second is a V-shaped cut in the surface drape on the southern side of the levee (Fig. 3, middle of section) that indicates that there are still active processes that prevent deposition in this area.

2.2. SHALLOW WATER AREA

Four morphological zones have been recognized in the shallower area (Fig. 2), a channelled region, an area with hummocky seafloor, areas with seafloor scours and smooth seafloor. The scours are lineations that usually converge towards channel heads. The channelled area indicates active down slope transport. In the middle of the region there are signs both of depositional lobes in front of the channels and also a set of meandering shallow channels. These features influence the uppermost seafloor sediments showing that they are the result of recent processes.

Horizon A1 (A1-2D on Fig. 5) defines the slip plane of an old slide. The slide scar has later been filled in with stratified, but more acoustically transparent sediments (Fig. 5).

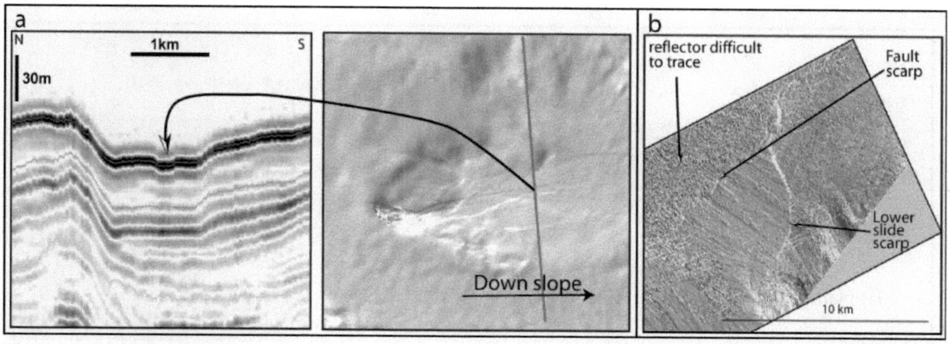

Figure 4. a: Elongated depression that follows "valley floor".The depression is only expressed in the uppermost sediments demonstrating that it is a young feature. b: Shaded relief map Horizon A1 in shallow water area showing the slip plane with striations indicative of the direction of slide movement. Illumination from the upper right. See Figure 2 for locations.

It is a very distinct feature in the shallow part of the study area and matches the slide identified by Hartevelt and van der Zwaag (1993) on 2D seismic data quite well.

The horizon displays at least two escarpments which are related to sliding (headwalls) indicating either that the sliding took place in distinct phases, or that it developed retrogressively and shifted the stratigraphic level of the failure plane as it developed (Fig. 4b). A similar development is well documented from e.g. the Storegga Slide offshore Norway (Bryn et al., 2005). The upper escarpment is defined by the limit of mapable horizon A1. In the eastern part of the block, the horizon is quite smooth on the lower side of the next escarpment, which coincides with one of the major faults in the area. The lower escarpment, however, is clearly created by sliding, as it cuts strata and does not have an underlying fault.

Whereas the upslope (NW) part of the horizon is difficult to trace and gives a rugged appearance (Fig. 4b), the lower parts (SE) are smooth and display striations evident of past sliding directions. There is also evidence for different phases in the slide although this does not mean that they are necessarily much separated in time.

2.3. RADIOCARBON AGES

Radiocarbon age determinations have been performed on foraminifera picked from samples from geotechnical boreholes within the upper 125 metres of the sediments (Table 1).

In BH8 from the deep water area, the two lowermost dates show ages (35.5 and 34.0 kyr) that are inverted, i.e. the oldest age is the shallower of the two. Both ages are relatively old with regards to the radiocarbon method, but are still about 10 kyr from the limit for radiocarbon dating. The estimated error range of around 500 years appears good and indicates that the ages are reliable. We therefore interpret the results as showing that the samples come from re-deposited sediments. Seismic sections through the borehole location show that these two samples are from basal layers of the surface drape.

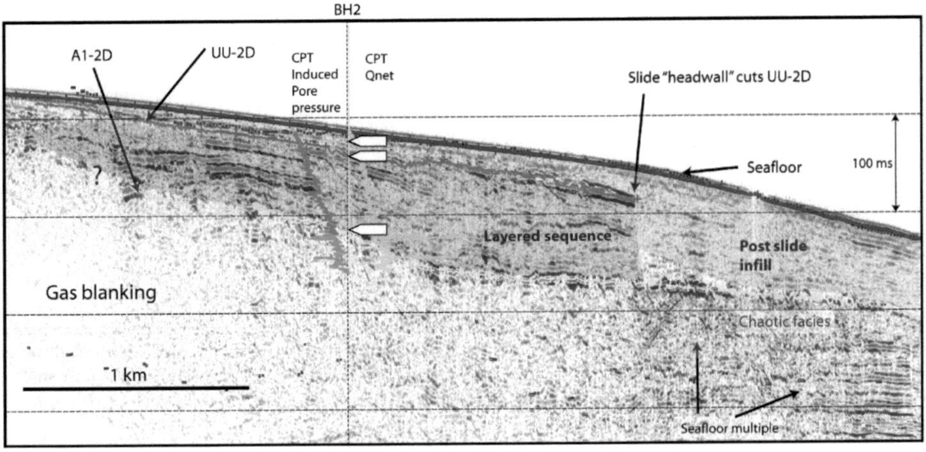

Figure 5. High resolution 2D line showing interpreted horizons and different seismic facies (coloured areas). The flags along BH2 indicate depths of dated samples. Upper: 440; Middle: 620 & Lower: 1620 Cal. years BP. The slide headwall cuts Horizon UU-2D (pink), and must be younger than this (~620 years). The CPT induced pore pressure depends on sediment type, and Qnet is the resistance to cone penetration. They indicate that the sediments are mostly clays although the peaks in Qnet show the presence of layers contain more sandy material. See Figure 2 for location.

2.4. RADIOCARBON AGES

Radiocarbon age determinations have been performed on foraminifera picked from samples from geotechnical boreholes within the upper 125 metres of the sediments (Table 1).

In BH8 from the deep water area, the two lowermost dates show ages (35.5 and 34.0 kyr) that are inverted, i.e. the oldest age is the shallower of the two. Both ages are relatively old with regards to the radiocarbon method, but are still about 10 kyr from the limit for radiocarbon dating. The estimated error range of around 500 years appears good and indicates that the ages are reliable. We therefore interpret the results as showing that the samples come from re-deposited sediments. Seismic sections through the borehole location show that these two samples are from basal layers of the surface drape.

Of the 12 samples selected for analysis from the shallow water area, six contained enough foraminifera for dating. These show that the sedimentation rate in the shallow area is significantly higher than in the deep water region. The deepest sample analyzed (123 m below seafloor from BH1, 40 m water depth) is only 6340 years old. The results from BH2 are the most significant because they allow the slide on Horizon A1 to be dated.

2.5. AGE OF MOVEMENTS

The timing of the mass movements on delta systems (Flood and Piper, 1997; Moscardelli et al 2006; Piper et al. 1997; Posamentier and Kolla 2003) can be related to the Pleistocene fluctuations in sea level and the consequent shift in the position of the main depocentres. Our results confirm such a relationship for the Krishna-Godavari delta. In general slope instabilities on delta systems are related to the supply of sediments to the delta front. This sediment supply is usually most rapid in conjunction with lowering of sea level and the exposure of the shallow shelf areas to wave and fluvial erosion and the consequent redeposition of sediments on the slope. The inverted ages from the bottom of borehole BH8 on the slope supports such timing and mechanism. They both pre-date the lowest sea level during the last glacial maximum that occurred about 22 kyr ago. Sediments that have been redeposited without addition of fresh radiocarbon will contain dateable material that reflects the original time of deposition. Furthermore, as erosion proceeds, progressively older sediments will be exposed. Thus the reworked deposits will produce old ages that may be oldest on top. The stratigraphic position at the base of the drape shows that the underlying mass transport deposits were formed first, probably in conjunction with falling sea level.

Table 1. The results of successful ^{14}C age determinations by Beta Analytic. Several other samples did not contain sufficient numbers of foraminifera for dating. BH1, BH2 and BH3 are form the shallow water area where the present sedimentation rate can be seen to much higher than in the deep water area (BH8). Conversion from ^{14}C years to calendar ages was performed using the INCAL98 calibration curve (Stuiver et al., 1998) with a tentative extension to 40k years BP (Beta Analytic). Data from Athersuth and Wonders (2005) and Kohl (2006).

Core, water depth (m)	Depth	^{14}C years	Calendar years	Lab. reference
BH1, 38.9	6.3	1740 +/- 40	1280 +/- 60	Beta 214825
	123.7	5940 +/- 40	6340 +/- 75	Beta 214828
BH2 , 84.06	8.4	790 +/- 40	440 +/- 90	Beta 214826
	20.0	1030 +/- 40	620 +/- 60	Beta 214827
	78.4	2060 +/- 40	1620 +/- 90	Beta 214829
BH3, 195.0	82.4	5060 +/- 40	5440 +/- 100	Beta 214830
BH8, 556.0	10.0	3660 +/- 40	3560 +/- 100	Beta 208053
	30.35	35500 +/- 560	n/a	Beta 208055
	50.36	33990 +/- 470	n/a	Beta 208054

The rapid sedimentation on the shallow water region demonstrated by BH2 shows that sediments are trapped on the shelf during sea level high stands such as the present. Numerical sedimentation models and geotechnical measurements (Kvalstad pers. comm.) show that excess pore pressures are probably developed due to this rapid sedimentation and have contributed to the observed Holocene slope instabilities near the shelf break. The dates in BH2 allow the ages of the upper horizons to be determined (Fig. 5). The horizon UU-2D is approximately 620 years old and can be seen to be cut by the headwall of the slide on reflector A1. The slide is therefore about 620 years old. The slide scar is however no longer apparent on the seafloor and demonstrates that the accommodation space provided by slide scars (and by inference other seafloor disturbances such as pockmarks or syn-sedimentary gravitational faulting) is quickly filled, contributing to local very high sedimentation rates. Numerical models (Kvalstad pers comm.) show that this sedimentation rate will cause excess porewater pressure to develop. There is therefore a positive feedback between both slope instability (and syn-sedimentary gravitational faulting) with sedimentation rate. The smooth sea floor (Fig. 2) is a result of this rapid filling of accommodation space.

3. Conclusions

Our conclusions are that:
The region's sedimentation history has provided the basis for the understanding the processes leading to slope instabilities with past mass movements having been observed on both the delta front (slope) and near the shelf break in shallower water.

The failures in deeper water occur during the lowering of global sea level in conjunction with glacial periods and are probably caused by elevated sediment supply due to both erosion from the shelf and a shift of the delta depocentre to the slope.

Mass movements in the shallow water area have occurred during the Holocene and reflect the present rapid deposition on the shelf because excess pore pressures and potentially unstable sediments probably develop as a result of the accumulation rate.

Both the deep water and shallow water mass movements are related to rapid sedimentation and, supported by numerical sedimentation models, we therefore suggest that this has induced excess pore pressures that have reduced slope stabilities.

Accommodation space on the shelf that is provided by both slide scars and syn-sedimentary gravitational faulting is quickly filled by fresh sediments and a positive feedback loops may therefore exist for both of these through the buildup of excess pore pressure and thick sediment packages respectively.

In the deep water area, evidence of quite young mass movements comes from elongated scars in the surface drape at the upper end of features resembling flow tracks.

4. Acknowledgements

This paper is based on the authors' interpretations of data kindly released by Reliance Industries for publication. We also thank Oddvar Longva and Anders Elverhøi for their constructive comments to the manuscript.

5. References

Athersuch J and Wonders A (2005) *StrataData report No.05/05 for Fugro Engineers* B.V. 27 September 2005.

Bruckner H (1988) *Indicators of formerly higher sea levels along the east coast of India and the Sandaman Islands.* Hamburger Geographic Studies. 44: 44-72.

Bryn P, Berg K, Forsberg CF, Solheim A, Kvalstad TJ (2005) *Explaining the Storegga Slide.* Marine and Petroleum Geology 22: 11-19.

Flood RD and Piper DJW (1997) *Amazon fan sedimentation: the relationship to equatorial climate change, continental denudation, and sea-level fluctuations.* In: Flood, RD, Piper DJW, Klaus A, and Peterson, LC (eds) College Station, TX, (Ocean Drilling Program). Proceedings of the Ocean Drilling Program, Scientific Results. 155: 653-675.

Hartevelt, J J A. and Zwaag van der, GL (1993) *Integrated geohazard study along the Krishna-Godavari Delta slope, east coast India.* In: Ardus DA, Clare D, Hill A, Hobbs R, Jardine RJ and Squire JM. Advances in Underwater Technology, Ocean Sciences and Offshore Engineering. ISBN 0-7923-2363-7, Kluwer Academic Publishers. Offshore Site Investigation and Foundation Behaviour. 28: 333-346.

Kohl B (2006) *Report on radiocarbon dating and biostratigraphy prepared for Fugro Geoservices*, Houston TX. March 18, 2006.

Moscardelli L, Wood L, Mann P (2006) *Mass-transport complexes and associated processes in the offshore area of Trinidad and Venezuela.* AAPG Bulletin 90: 1059-1088.

Piper DJW, Pirmez C, Manley PL, Long D, Flood RD, Normark WR and Showers W (1997). *Mass transport deposits of the Amazon Fan.* In: Flood RD, Piper DJW, Klaus A and Peterson LC (eds) College Station, TX, (Ocean Drilling Program). Proceedings of the Ocean Drilling Program, Scientific Results. 155: 109-146.

Posamentier HW and Kolla V (2003) *Seismic geomorphology and stratigraphy of depositional elements in deep-water settings.* Journal of Sedimentary Research 73: 367-388.

Sharma KP (2002) *A study of sediment fluxes to the coastal zone in south Asia and their relationship to human activities.* Project report on APN Project on a study of nutrient, sediment and carbon fluxes to the coastal zone in south Asia and their relationships to human activities. 20pp.

Sarin MM, Sudheer AK and Balakrishna K (2002) *Significance of riverine carbon transport: A case study of a large tropical river*, Godavari (India). Science in Chine (Series C). 45: 97-108.

Stuiver M, Reimer PJ, Bard E, Beck JW, Burr GS, Hughen KA, Kromer B, McCormac G, VanderPlicht J, Spurk M (1998) *INTCAL98 radiocarbon age calibration, 24,000-0 cal BP.* Radiocarbon 40: 1041-1083.

REPEATED INSTABILITY OF THE NW AFRICAN MARGIN RELATED TO BURIED LANDSLIDE SCARPS

AGGELIKI GEORGIOPOULOU[1*], SEBASTIAN KRASTEL[2], DOUGLAS G. MASSON[1], AND RUSSELL B. WYNN[1]

[1] *National Oceanography Centre, Southampton, European Way, Waterfront Campus, SO14 3ZH, UK,*

[2] *Department of Geosciences, University of Bremen, Klagenfurter Strasse, 28334 Bremen, Germany*

[*] *Corresponding author: Currently at 3DLab, School of Earth, Ocean and Planetary Sciences, Cardiff University, Main Building, Park Place, Cardiff, CF10 3YE, ag1000@cardiff.ac.uk*

Abstract

The Sahara Slide occurred approximately at 50-59 Ka offshore Western Sahara in a mid-slope setting (1900 m water-depth). The existence of several buried and stacked slide events, seen on high resolution seismic profiles, provide new insights into slide location and triggering mechanisms. Buried slide scarps coincide remarkably with scarps and boundaries of the Sahara Slide, presently exposed on the seafloor. The objectives of this work are to examine the long-term stability of this part of the margin and investigate the triggering mechanism(s) that led to these massive events.

Buried slide scarps occur in sediments of Miocene-Pliocene age. Multiple scarps becoming more closely spaced towards a larger scarp that may be the main headwall suggest that most of the buried slides developed as retrogressive slides. The seismic record shows that differential compaction across an area of depression bound by scarps generates compaction hinges (anticlines) leading to oversteepening and possible excess pore pressure. We propose that alignment of ancient and present scarps and vertically stacked slide deposits points towards differential compaction as being a key factor in landslide triggering.

1. Introduction

Submarine slides are an important architectural element of continental margins. Because of their widespread and often episodic nature, mass sediment movement events are important components of the modern stratigraphic record, and have been studied in connection with global climatic cycles, including sea level changes (e.g. McHugh et al., 2002). Several investigations address the geohazard potential of submarine slides (e.g. Locat & Mienert, 2003, and references therein) as submarine slides might destroy offshore infrastructure and trigger tsunamis.

The NW African continental margin is well known for the occurrence of large but infrequent slides (Weaver et al., 2000; Wynn et al., 2000; Krastel et al., 2006). The largest slides are the Mauritania Slide (Antobreh and Krastel, 2006) and the Sahara Slide (Gee et al., 1999; Georgiopoulou et al., in prep). These studies mainly discuss the dynamics of the slides but less is known about trigger mechanisms and why these slides occur at specific locations along the margin while other parts of the margin are stable. Early investigations of the Sahara Slide concentrated on the depositional part of the slide (see Gee et al., 1999 and references therein) but data of the head wall area were lacking.

V. Lykousis, D. Sakellariou and J. Locat (eds.), Submarine Mass Movements and Their Consequences, 29–36.

New acoustic and gravity core data from the headwall area of the Sahara Slide, off Western Sahara, were acquired during *RV Meteor* cruise M58/1 in 2003 (Fig. 1). These new data in combination with revisited older data allows analysis of the distribution and mechanics of the Sahara Slide form source to sink (Georgiopoulou et al., in prep.). Here we present a set of high-resolution seismic reflection profiles from the headwall area of the Sahara Slide, which document the presence of buried scarps and slide deposits which exhibit remarkable spatial coincidence with the present Sahara Slide headwall. We interpret the buried slide deposits as indicating a long term instability for this part of the margin and analyse the role of old slides for triggering subsequent slides at a similar location.

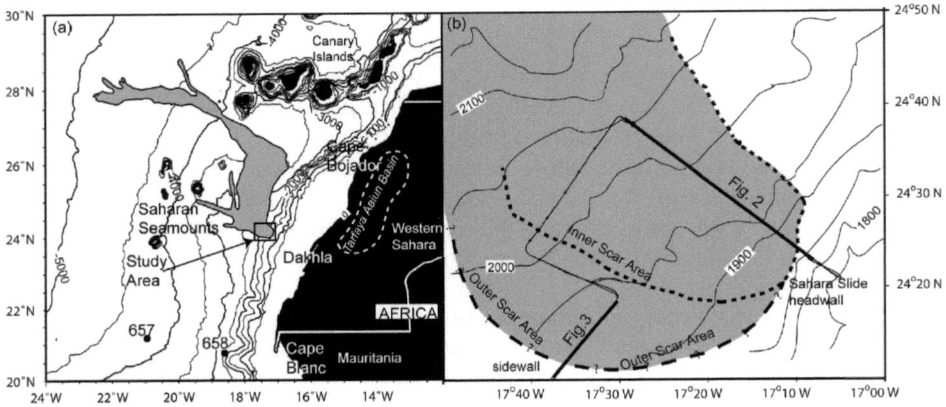

Fig. 1. Location of the study area on the NW African margin (box) and blow up with survey lines (b). (a) Map showing the outline of the Sahara Slide (grey-shaded area) on the NW African margin and the location of ODP sites 657 and 658. Contours are at 500 m. (b) Thin solid lines show the survey lines at the headwall of the Sahara Slide (grey-shaded area) and thick lines the location of Figs. 2 and 3.

2. Geological setting and stratigraphic framework

The continental shelf of the Northwest African margin is composed of seaward-dipping Cretaceous and Tertiary sediments (Summerhayes et al., 1976). Shelf widths are generally 40-60 km, although the maximum width is >100 km off the Western Saharan coast. The shelf-break is at around 110-150 m water depth (Summerhayes et al., 1976; Seibold, 1982; Wynn et al., 2000). Continental slope angles range from 1°-6°, while the continental rise displays gradients of <1° (Masson et al., 1992).

The study area lies in the Aaiun-Tarfaya basin, offshore NW Africa. The marine history of the Aaiun-Tarfaya Basin starts with a Jurassic transgression from the west. Maximum transgression occurred during the Cenomanian invading the Sahara and reducing terrigenous input to the Atlantic (Seibold, 1982). A strong regression and erosion started on the continental shelf and slope during the Oligocene/Miocene, which also marks the first development of submarine incision characterised by accentuated subsidence rates, averaging 6cm/ka off Cape Bojador, and possible sea-level lowering (Seibold, 1982). In post-Oligocene times canyons were filled and re-excavated and local slumping occurred (Seibold, 1982). The early Miocene is a phase of marked global warming at the same time as continuous sea-level rise (Sarnthein et al., 1982). Climate on the Northwest African margin was wet with hardly any aeolian dust input. Sea level continued to rise during the Middle Miocene, while major hiatuses occurred at the upper Early Miocene and end of Middle Miocene. During the Late Miocene (Messinian) there were episodes of cooling which were correlated with a short but marked sea level drop. During the Early Pliocene high sea-level was re-established (Sarnthein et al., 1982).

Quaternary sediments off NW Africa are essentially biogenic due to a continuous upwelling cell (Seibold and Hinz, 1974). Results from ODP site 658 show that the upper Pliocene-Holocene sediment section comprises three major hemipelagic lithologic units, the upper of which is divided into two subunits. Unit I spans the period Pleistocene to Holocene. Unit II represents the period between Upper Pliocene and Lower Pleistocene and Unit III deposited from Lower to Upper Pliocene (Ruddiman et al., 1988). Unit II correlates with an almost transparent unit which has velocities 900-950 m/s indicative of free gas in the sediment (Ruddiman et al., 1988).

3. The Sahara Slide

The Sahara Slide is a large submarine slide that took place retrogressively 50-59 ka on an open slope offshore the arid Western Sahara (Fig. 1) and involved approximately 600 km^3 of sediment. Recently acquired bathymetric data revealed that the scar area is shaped by two headwalls, each up to 100 m high, cut into mid-slope sediments. Between the two main escarpments the scar has a stepped profile consisting of a series of discrete glide planes at different stratigraphic levels separated by internal scarps, suggesting that the slide occurred retrogressively. The main headwall of the slide appears to have been reactivated as recently as ~2000 y.a. as indicated by detailed shallow seismic profiles and sediment cores. High resolution deep seismic data presented in this paper show that the headwall scarps of the Sahara Slide coincide remarkably with buried scarps of ancient slides that occurred at different times in the past.

4. Data and Methodology

High-resolution seismic data were collected using a 1.7L GI-Gun as source and 450 m long 72 channel Syntron streamer for signal recording. Processing included trace editing, static corrections, velocity analysis, normal moveout corrections, bandpass frequency filtering (frequency content: 55/110 – 600/800 Hz), stack, and time migration. A common midpoint (CMP) spacing of 10 m was applied throughout.

5. Ancient scarps and buried slides

Seismic Profile GeoB03-060 crosses the headwall area of the Sahara Slide (Fig. 2). The headwall is characterized by a ~80 m high step in morphology. The headwall cuts well-stratified sediments. A sidewall and an internal scarp each ~20m high were found on Profile GeoB06-057 (Fig. 3). The seismic sections are generally characterized by an interlayering of transparent and well stratified units. The upper most transparent unit represents the slide deposits of the Sahara Slide, which was the last major failure that took place retrogressively at 50-59 ka BP (Gee et al., 1999; Georgiopoulou et al., in prep). Four buried slides have been highlighted on the seismic sections (Fig. 2 and 3), although more can be identified, but either because they are very thin or because they cannot be associated with a prominent scarp, they are not described in detail in this paper. However, their presence contributes to demonstrating the degree of slope instability on this margin and therefore they will be considered in the discussion. The highlighted slides have been letter-coded and this letter is used to refer to them (A, B, C and D, from the deepest to the shallowest). A stratigraphic interpretation of the profile allows assigning approximate times to the major slide events. Seismic profiles published in von Rad and Wissmann (1982), as well as results from ODP sites 657 (Faugeres et al., 1989) and 658 (Ruddiman et al., 1988) helped establish the stratigraphy of the area. Sediment packages a-e correspond to the equivalent sediment packages in fig. 8 of von Rad and Wissmann (1982). Units a and b seem to match in thickness and seismic character with Units I and II at ODP site 658 (Ruddiman et al., 1988). The uncomformity marked on Fig. 2 between Middle and Early Miocene may be the Messinian sea level drop that caused the hiatus during 6.2-4.6 Ma reposted in Sarnthein et al. (1982).

Slide A appears to have taken place within Early Miocene sediments (Figs. 2, 3). The other three also took place in pre-Quaternary times, probably between Middle Miocene and Pliocene.

The headwall scarp of slide A, which is the most prominent feature on the seismic profiles, is characterised by complex morphology (Fig. 2). It is formed by multiple scarps, similar to the Sahara Slide headwall scarp, but those of slide A are higher. They range between 20 m and up to approximately 150 m (calculated using 1500 m/s sound velocity). The total height displacement of the headwall is 470 ms (~320 m). Blocks bounded by those scarps appear rotated and internally deformed (Fig. 2). This morphology indicates that this slide took place retrogressively. The deposit displays the typical chaotic internal seismic signature for slide deposits and also contains intact non *in situ* blocks. The sidewall of this slide can be seen on Fig. 3, where continuous strong reflectors interpreted as hemipelagic sediments can be seen interlayered in the slide deposits, suggesting that this slide is comprised of at least four events, separated by sediment packages of at least 10 m thickness. Considering an average sedimentation rate of 6 cm/ka (Oligocene/Miocene average sedimentation rate according to Seibold, 1982), this thickness indicates time periods of about 170 ka between events. Each of the events may have taken place retrogressively as suggested by the scarp morphology. The interpretation as retrogressive slides is suggested with caution, however, as the scarp morphology may have been the end result of these multiple events. However, we believe that the first event was dramatic enough to form this scarp and re-activation, perhaps because of over-steepening, generated the following events.

A period of general stability followed in the Middle Miocene, with only minor slide events taking place (outlined with green solid lines, Fig. 2), until slide B occurred. This slide does not have a prominent scarp associated with it, but it initiates approximately 5km downslope of what appears to be an anticline over one of the scarps of slide A.

Slide C appears to be the result of a re-activation of the major scarp of slide A as the slide A headwall area had not been fully infilled when slide C occurred and the scarp must have still been exposed on the seafloor. Slide C shows multiple scarps too that cut into slope stratified sediments in different stratigraphic levels.

The headwall of slide D that occurred in the Pliocene, is also situated approximately 5km forward of a forced fold that is formed on the edge of the major scarp.

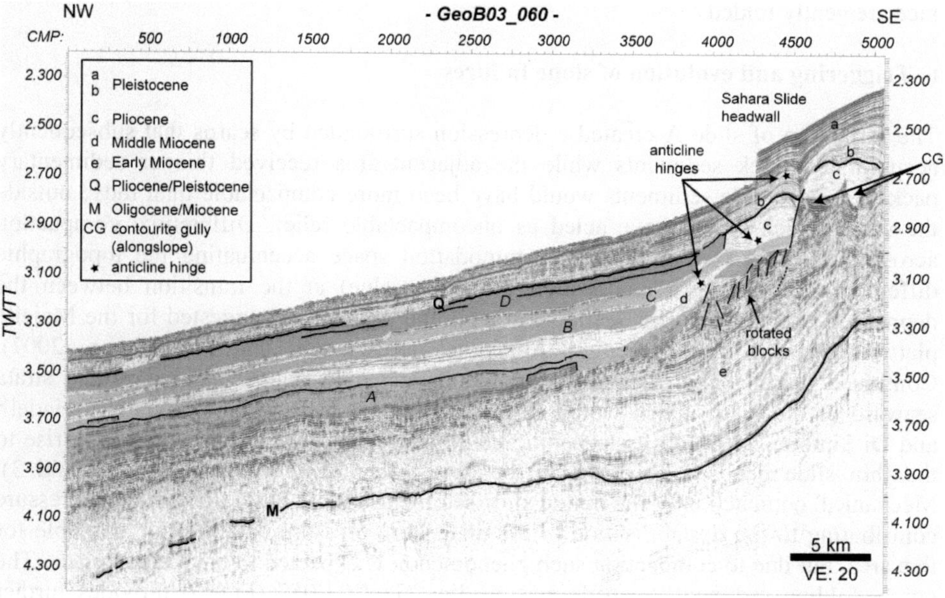

Fig. 2. Seismic reflection profile GeoB03-060 across the headwall area of the Sahara Slide (blue solid line) showing the stratigraphy and the distribution of slides *A*, *B*, *C* and *D*. The black stars indicate the location of anticline hinges. See Fig. 1 for location.

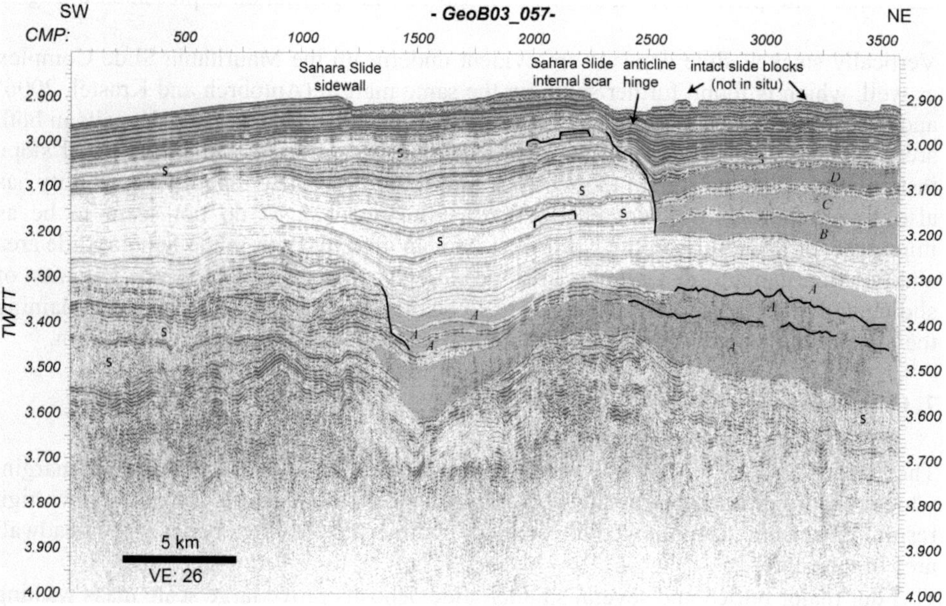

Fig. 3. Seismic reflection profile GeoB03-057 through the sidewall of the Sahara Slide (blue solid line) that shows the distribution of slides A, B, C and D. Note that slide A can be resolved into at least four events on this profile. Slides B, C and D have a common sidewall above which sediments fold gently forming an anticline hinge. More slide deposits are indicated with green solid lines or represented with the letter *s*. See Fig. 1 for location.

The three slides (B, C and D) share a common sidewall seen in Fig. 3, which appears to have propagated upwards through time. Sediments immediately adjacent to the scarp face are gently folded.

6. Triggering and evolution of slope failures

The formation of slide A created a depression surrounded by scarps that subsequently accumulated thick sediments while the adjacent area received thinner sedimentary packages. Post-slide sediments would have been more compactable than those outside the scarp which would have acted as uncompactable relief. Differential compaction across the scarps increased the accommodation space accentuating the topographic differences generating a compaction hinge (anticlne) at the transition between the depression and the adjacent seafloor, in a similar manner as suggested for the Maiella platform margin in the Central Appenines (Rusciadelli and Di Simone, 2007). Continued loading and compaction would have led to oversteepening of the strata seaward of the scarp which would further accentuate the anticline hinge (Rusciadelli and Di Simone, 2007). Oversteepening could cause intra-layer slumping giving rise to the thin slide deposits observed in the data (green solid lines in Figs. 2 and 3). Mechanical compaction of the buried slide sediments would build up pore fluid pressure contributing to the destabilisation of the area. Pore pressure data are not available for this area, but due to compaction such phenomenon is expected to have taken place. The gently folding sediments seen on seismic line GeoB03-057 (Fig. 3) provides further evidence that differential compaction takes place across the scarps. Subsequent sliding seems to regularly occur a few kilometres in front of anticline hinges suggesting that differential compaction across a dramatic scarp promotes repeated instability. A schematic conceptual model about the progress of slide generation is presented in Fig. 4.

Vertically stacked slide deposits are evident underneath the Mauritania Slide Complex as well, which is found further south on the same margin (Antobreh and Krastel, 2006) and has similar dimensions to the Sahara Slide. The vertical stacking of deposits in both areas suggests that the Northwest African margin has suffered a long history of slope instability. Unlike the area of the Storrega Slide complex where repetitive instability has also been reported (Solheim et al., 2005) climatic changes do not seem to be as important for instability on the Northwest African margin. Both in the Sahara Slide area and the Mauritania Slide Complex area the slope remains unstable after the initiation of sliding and differential compaction seems to be the main causal factor, hence explaining the long term instability of specific sections of the NW African continental Margin.

7. Conclusions

The Sahara Slide is one of the mega slides on the NW African continental margin, which occurred at approximately 50-59 Ka offshore Western Sahara. New high resolution seismic data allowed the study of sedimentary features beneath the headwall area of this slide.

- Four major buried and several smaller slide deposits prove large scale mass wasting at this section of the margin since Early Miocene times.

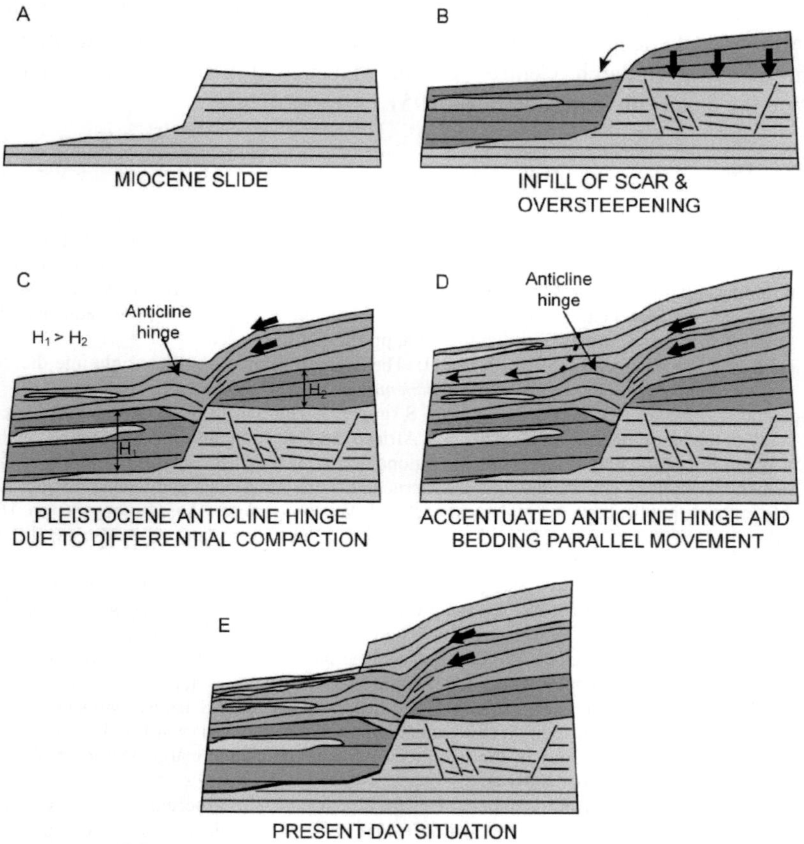

Fig. 4. Conceptual model of slide generation due to buried scarps (not to scale). (a) Slide A occurs in Miocene times creating a depression bound by scarps. (b) Infill of the scar with thick, underconsolidated sediments, differential compaction across the scarp begins. Oversteepening at the scarp produces minor failures (grey wedge in C). Lithostatic pressure on the uncompactable sediments outside the scar leads to fracturing (black lines indicating faults). (c) Loading and differential compaction across the escarpment generates an anticline hinge. (d) Continued loading and differential compaction accentuates the anticline hinge and leads to oversteepening and potential excess pore pressure. (e) Present day situation after failure of the Sahara Slide at 50-59 Ka.

- Several buried slide scarps coincide remarkably with scarps at shallower stratigraphic levels as well as the boundaries of the Sahara Slide that is presently exposed on the seafloor.
- Major slide events most likely occurred as retrogressive type failures.
- Loading and differential compaction across escarpments generates an anticline hinge and leads to oversteepening and potential excess pore pressure. Hence specific sections of the NW African continental Margin remain unstable after the initiation sliding.

8. Acknowledgements

The authors would like to acknowledge the invaluable assistance of fellow cruise participants as well as the Captain and crew members of RV Meteor Cruise M58/1 during data acquisition. A Georgiopoulou is thankful to the University of Southampton and the former Challenger Division of NOC for the provision of PhD funding. We are grateful to DFG-Research Centre "Ocean Margins" for the financial support for collecting the data (Publication RCOM0495).

9. References

Antobreh A, Krastel S (2006). Mauritania Slide Complex: morphology, seismic characterisation and processes of formation. *International Journal of Earth Science*, DOI: 10.1007/s00531-006-0112-8.

Faugeres JC, Legigan P, Maillet N, Sarnthein M, Stein R (1989). Characteristics and distribution of Neogene turbidites at Site 657 (Leg 108, Cap Blanc continental rise, Northwest Africa): Variations in turbidite source and contineta1 climate. In: W. Ruddiman, Sarnthein, M., et al. (Eds), Proceedings of the Ocean Drilling Program, Scientific Results, Vol. 108, pp. 329-348.

Gee MJR, Masson DG, Watts AB, Allen PA (1999). The Saharan debris flow: an insight into the mechanics of long runout submarine debris flows. *Sedimentology*, 46: 317-335.

Georgiopoulou A, Masson DG, Wynn RB, Krastel S (in prep). The Sahara Slide: A detailed source-to-sink study of a giant submarine slide on the NW African margin. To be submitted in Marine Geology.

Krastel S, Wynn RB, Hanebuth TJJ, Holz C, Georgiopoulou A, Henrich R, Schulz HD (2006). Mapping of seabed morphology and shallow sediment structure of the Mauritania continental margin, Northwest Africa: some implications for geohazard potential. *Norwegian Journal of Geology*, 86, pp. 163-176.

Locat, J, Mienert, J (2003). Submarine mass movements and their consequences. Advances in natural and technological hazards research 19, Kluwer, Dordrecht, 540pp.

Masson DG, Kidd RB, Gardner JV, Huggett QJ, Weaver PPE (1992). Saharan continental rise: Facies distribution and sediment slides. In: Poag CW, de Graciansky PC (Eds), Geologic evolution of the Atlantic continental rises. Van Nostrand Reinhold, New York, pp. 327-343.

McHugh MGC, Damuth JE, Mountain SG (2002). Cenozoic mass-transport facies and their correlation with sea level change, New Jersey continental margin. Marine Geology 184, pp. 295-334.

von Rad U, Wissmann G (1982). Cretaceous-Cenozoic history of the West Saharan continental margin (NW Africa): development, destruction and gravitationla sedmentation. In: von Rad U, Hinz K, Sarnthein M, Seibold E (Eds), Geology of the Northwest African continental margin. Springer-Verlag, Berlin Heidelberg, New York, pp. 106-131.

Ruddiman W, Sarnthein M, Baldauf J, shipboard scientific party (1988). Proceedings of the Ocean Drilling Program, Initial Reports. Vol. 108, Section 1, Part A. Eastern tropical Atlantic: covering Leg 108 of the cruises of the Drilling Vessel "Joides Resolution", Marseille, France to Dakar, Senegal, Sites 657-668, 18 February 1986-17 April 1986. College Station, TX: Texas A & M University Ocean Drilling Program, pp. 556.

Rusciadelli G, Di Simone S (2007). Differential compaction as control on depositional architectures across the Maiella carbonate platform margin (central Appenines, Italy). *Sedimentary Geology*, 196, pp. 133-155.

Sarnthein M, Thiede J, Pflaumann U, Erlenkeuser H, Fuetterer D, Koopmann B, Lange H, Seibold E (1982). Atmospheric and oceanic circulation patterns off Northwest Africa during the past 25 million years. In: von Rad U, Hinz K, Sarnthein M, Seibold E (Eds), Geology of the Northwest African continental margin. Springer-Verlag, Berlin Heideberg, New York, pp. 584-604.

Seibold E (1982). The Northwest African continental margin - An Introduction. In: von Rad U, Hinz K, Sarnthein M, Seibold E (Eds), Geology of the Northwest African continental margin. Springer-Verlag, Berlin Heidelberg, New York, pp. 3-20.

Seibold E, Hinz K (1974). Continental slope construction and deconstruction, West Africa. In: Burk CA, Drake CL (Eds), Geology of continental margins. Springer-Verlag, New York, pp. 179-196.

Solheim A, Berg K, Forsberg CF, Bryn P (2005). The Storegga Slide complex: repetitive large scale sliding with similar cause and development. *Marine and Petroleum Geology*, 22, pp. 97-107.

Summerhayes CP, Milliman JD, Briggs SR, Bee AG, Hogan C (1976). Northwest African shelf sediments: influence of climate and sedimentary processes. *Lournal of Geology*, 84, pp. 277-300.

Weaver, PPE, Wynn, RB, Kenyon, NH, Evans, J. 2000: Continental margin sedimentation with special reference to the Northeast Atlantic margin, *Sedimentology* 47, 239-256.

Wynn RB, Masson DG, Stow DAV, Weaver PPE (2000). The Northwest African slope apron: a modern analogue for deep water systems with complex seafloor topography. *Marine and Petroleum Geology*, 17, pp. 253-265.

ALONG SLOPE VARIATIONS IN MASS FAILURES AND RELATIONSHIPS TO MAJOR PLIO-PLEISTOCENE MORPHOLOGICAL ELEMENTS, SW LABRADOR SEA

M.E. DEPTUCK, D.C. MOSHER, D.C. CAMPBELL
Geological Survey of Canada (Atlantic), Bedford Institute of Oceanography, P.O. Box 1006, 1 Challenger Dr., Dartmouth, NS, B2Y 4A2, Canada

J.E. HUGHES-CLARKE
Ocean Mapping Group, Department of Geodesy and Geomatics Engineering, University of New Brunswick, Box 4400, Fredericton, NB, E3B 5A3, Canada

D. NOSEWORTHY
Canada Newfoundland and Labrador Offshore Petroleum Board, 140 Water St., St. John's, NL, A1C 6H6, Canada

Abstract

The slope along the SW Labrador Sea is a prospective exploration frontier with limited legacy data and geoscience knowledge. Newly acquired seismic reflection and multibeam bathymetry data provide a better understanding of mass failure processes. A semi-continuous seismic section along the upper slope from Flemish Pass to north of Hamilton Spur shows an alternation of major morphological elements that includes canyons and failure corridors, inter-canyon ridges, ice-outlet trough mouth fans (TMFs), and sedimentary spurs. Preliminary geohazard investigation shows a wide variety of Plio-Pleistocene mass failure products including shallow detachment faults, head-scarps, creep folds, decollement surfaces, and a preponderance of mass transport deposits (MTDs) of various origins. Particularly noteworthy are two newly identified fans outboard the Hawke Saddle and Notre Dame Channel (believed to be TMFs constructed of mass wasted material), and a large shallow buried failure complex north of the Hamilton Spur, which contains many km-scale slide blocks dispersed over thousands of square kilometers.

Keywords: SW Labrador Sea, slope morphology, mass failures, giant slide blocks

1. Introduction

Renewed exploration interest in the SW Labrador Sea includes deep water areas of the slope where there is poor data coverage and limited stratigraphic control. New high resolution GI gun seismic reflection profiles collected during two cruises (Hud2005-033b and Hud2006-040) combined with 5500 km^2 of 30 kHz multibeam bathymetry collected between 2003 and 2006 were used to study the regional Plio-Pleistocene geology and potential geohazards along the slope (Fig. 1). These data suggest that mass-transport deposits (MTDs) form a significant part of the sedimentary succession. They occur at a variety of scales ranging from local failures that modified deposits dominated by other processes (e.g. contourite drifts) to more regional failure complexes that significantly altered the morphology of the margin (e.g. off Makkovik Bank). Some of the failures identified are amongst the largest yet observed off eastern Canada. The purpose of this paper is to provide a preliminary account of the MTDs and associated features and their influence in shaping major morphological elements along the slope.

V. Lykousis, D. Sakellariou and J. Locat (eds.), Submarine Mass Movements and Their Consequences, 37–45.

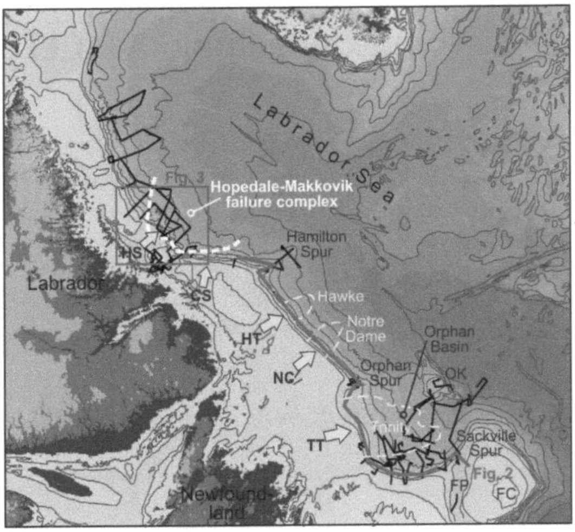

Figure 1. SW Labrador Sea showing profiles collected since 2003 and major morphological features. FC-Flemish Cap; FP-Flemish Pass; HT-Hawke Trough; HS-Hopedale Saddle; NC-Notre Dame Channel; TT-Trinity Trough; CS-Cartwright Saddle.

2. Slope between Sackville and Hamilton spurs

A semi-continuous 1330 km long composite seismic section was assembled from available high-resolution airgun profiles (Fig. 2). It shows dramatic changes in slope morphology between the Sackville and Hamilton spurs that are attributed to cross-slope ice-margin, along-slope contourite, and various mass wasting processes. Positive-relief fan-like deposits are found outboard the Trinity, Notre Dame, and Hawke shelf-crossing troughs (Fig. 1). They were deposited outboard major ice-outlets, probably during periods of maximum ice advance when large quantities of sediment were discharged onto the outer shelf (Hiscott & Aksu, 1996; Shaw et al., 2006; Tripsanas & Piper, submitted). Hence these features may be trough-mouth fans (TMFs; *sensu* O'Cofaigh et al. 2003). The most prominent bathymetric elements along the margin are the Sackville, Orphan, and Hamilton spurs. They are believed to be large contourite drifts (axial length >100 km; width > 50 km) deposited since at least the late Miocene, molded by a combination of the shallow south-flowing Labrador Current and the deeper Western Boundary Undercurrent (Myers & Piper, 1988; Kennard et al., 1990). Piston cores from their crests indicate late Pleistocene sediments consist predominantly of variably coloured clay to silty or sandy clay (e.g. Campbell et al., 2002; Goss, 2006; Tripsanas & Piper, submitted). The composition of older spur sediment is generally unknown. Canyons are also important morphological features. They are present on the upper slope south of the Trinity TMF, north of the Orphan Spur, between the Notre Dame and Hawke 'fans', and north and south of the Hamilton Spur (Fig. 2). In most places they are young, truncating shallow Pleistocene strata above thick largely canyon-less intervals of higher continuity Pliocene slope deposits. Hence, the prevalence of canyons at the seafloor is commonly a recent feature not widely observed in the subsurface, and like TMFs, is probably directly related to increased sediment supply associated with the growth and decay of the Laurentide ice sheet (Myers & Piper, 1988; Hesse et al., 1999). Canyons are commonly separated by inter-canyon ridges whose morphology ranges from narrow with sharp crests to wide with relatively flat-lying strata. Some consist of strata that aggraded contemporaneously with canyon incision, comprised of sediment that settles from meltwater surface plumes and spillover from fine-grained parts of turbidity flows (Hesse et al., 1999). Each of the above settings is, to varying degrees, influenced by mass failures, described in more detail below.

2.1 TROUGH-MOUTH FANS (TMFs)

Three prominent positive-relief (convex-upward) 'fans' are present on the upper slope outboard the Trinity, Notre Dame, and Hawke shelf-crossing troughs (Fig. 2). They are believed to be constructed predominantly of mass wasted material. The southern most of these is the Trinity TMF. It consists of a series of incoherent wedges, each with multiple lens-shaped bodies consisting of glaciogenic debris flow deposits (Hiscott & Aksu, 1996). On dip-oriented profiles the Trinity TMF progrades into the Orphan Basin with forsets inclined at 2 to 3°. Each incoherent wedge extends from the outer shelf, indicating debris flows were initiated near the ice-margin, perhaps from failure of till tongues during rapid delivery of glacial till to the upper slope (Piper & Brunt, 2006). The deepest incoherent interval may correspond to the initial excavation of the Trinity Trough (Campbell, 2005) during early shelf-crossing glaciations (beginning at about MIS 12; Piper, 2005). It is overlain by 4 other incoherent wedges, each separated by local erosion and intervals of stratified coherent reflections corresponding to hemipelagic drape deposited when the ice margin was far removed from the continental shelf-break (Hiscott & Aksu, 1996).

The newly identified deposits outboard the Notre Dame Channel and Hawke Saddle produce subtle seaward bulges on GEBCO bathymetric contours, which we used to approximate their perimeters (Fig. 1). The fan outboard the Notre Dame Channel is >100 km wide and 650 ms thick. The fan outboard the Hawke Saddle is 85 km wide and about 500 ms thick. They consist largely of incoherent higher amplitude seismic facies with lens-shaped 'patches' of low amplitude incoherent reflections. They were probably built through mass wasting processes, but lack the well-developed alternation of incoherent and coherent reflections seen in Trinity TMF. The more complex seismic facies may indicate a higher degree of gullying on the steeper slope in this area (2.5 to 4.0°), with poorer preservation of coherent hemipelagic intervals. Alternatively, they may have been supplied by a wider variety of mass failure deposits (e.g. glaciogenic debrites, slumps, slide blocks, turbidites). Sharply underlying both fans are higher-continuity slope deposits containing shingled reflections, sediment waves, and multiple prominent erosive surfaces indicative of marine deposition under the influence of ocean currents (Fig. 2e). Like the Trinity TMF, the abrupt change in deposition at their bases may correspond to the first shelf-crossing glaciations north of Trinity Trough. Canyon erosion south of the Notre Dame fan (Fig. 2a), however, prevents us from correlating seismic reflections to the south, and we are thus unable to determine the timing of the Trinity TMF relative to the deposits to the north. Their smaller dimensions might mean that the shelf-crossing troughs to the north were less important outlets for fast-flowing ice compared to the Trinity Trough, but more seismic profiles are required to better define their perimeters and seismic stratigraphy.

2.2 SPURS

The Sackville, Orphan, and Hamilton spurs are distinctly asymmetric and are dominated by highly continuous reflections, but also contain intervals of sediment waves and shingled and chaotic reflections. Oversteepening of their slopes through time, however, preconditioned the spurs to fail periodically either under their own weight, through ground shaking (Piper & McCall, 2003), or through gas hydrate dissociation (Mosher et al., in press). Consequently, their crests and flanks are variably truncated by head-scarps, bedding-plane detachments, and they locally contain MTDs.
The northern *upstream* sides (relative to prevailing ocean currents) of spurs are steeper and commonly erosive, where currents are most intense (Kennard et al., 1990). The up-stream side of the Sackville Spur (up to 4.6°) periodically shed sediment into the Orphan Basin, with multiple failure scarps (some up to 200 m high) creating a stepped morphology with sharply truncated reflections periodically draped by continuous reflections. Prominent scarps are also present on the upstream (northern) sides of the Orphan and Hamilton spurs. Spur *crests* are variably truncated by failure scarps. At least one Pleistocene failure was sourced from the southern crest of the Sackville Spur (Campbell et al., 2002), but for the most part its crest remained intact throughout the Plio-Pleistocene, shedding surprisingly little sediment to the SE. In contrast, the more northerly spurs experienced more crest and flank failures. A chaotic MTD is found along the crest of the Hamilton Spur in >2800 m of water, presumably sourced from a crest failure initiated up-slope, and failure scarps are present off the seaward nose of the Orphan Spur. The southern *downstream* sides of spurs are commonly less steep and more depositional, with coherent reflections thinning away from their crests and in the seaward direction. Continuous seismic reflections on the downstream flank on the Sackville Spur were variably truncated by, and interfinger with, several erosive-based MTDs (up to 130 m thick; Fig. 2b). They were likely sourced from scarps on the Flemish Pass slope, south and east of the spur (Piper & Campbell, 2005). Similarly, the downstream flanks of the Orphan Spur interfinger with erosive-based MTDs from the adjacent Trinity TMF (Fig. 2d), and the southern flank of the Hamilton Spur is onlapped by MTDs in > 3000 m of water.

2.3 CANYONS AND INTER-CANYON RIDGES

Canyons are widespread across the southern Labrador margin. Their axes, margins and the ridges that separate them may be influenced by a variety of mass failures described in numerous previous studies (e.g. Josenhans et al., 1987; Piper & McCall, 2003; Mosher et al., 2004; Jenner et al., 2007). The northern side of the Orphan Spur and the northern and southern sides of the Hamilton Spur are particularly heavily eroded by canyons, exposing deeper Pliocene strata at the seafloor. Up-arching of seismic reflections in these areas indicates an underlying tectonic control may have caused uplift, with increased erosion and generally thin preservation of Plio-Pleistocene strata (Fig. 2). These areas may have been subjected to increased mass failures, but the lack of data in deeper water precludes mapping associated deposits. The slope south of the Trinity Trough is also cut by a series of seafloor canyons that coalesce downslope (Tripsanas et al., in press). Below them the slope is highly complex and contains prominent erosive surfaces that could represent much larger canyons (e.g. Campbell, 2005) or submarine erosion associated with intensified ocean currents as described by Piper & Normark (1989), Kennard et al. (1990) and Deptuck (2003) (e.g. Figs. 2b-f).

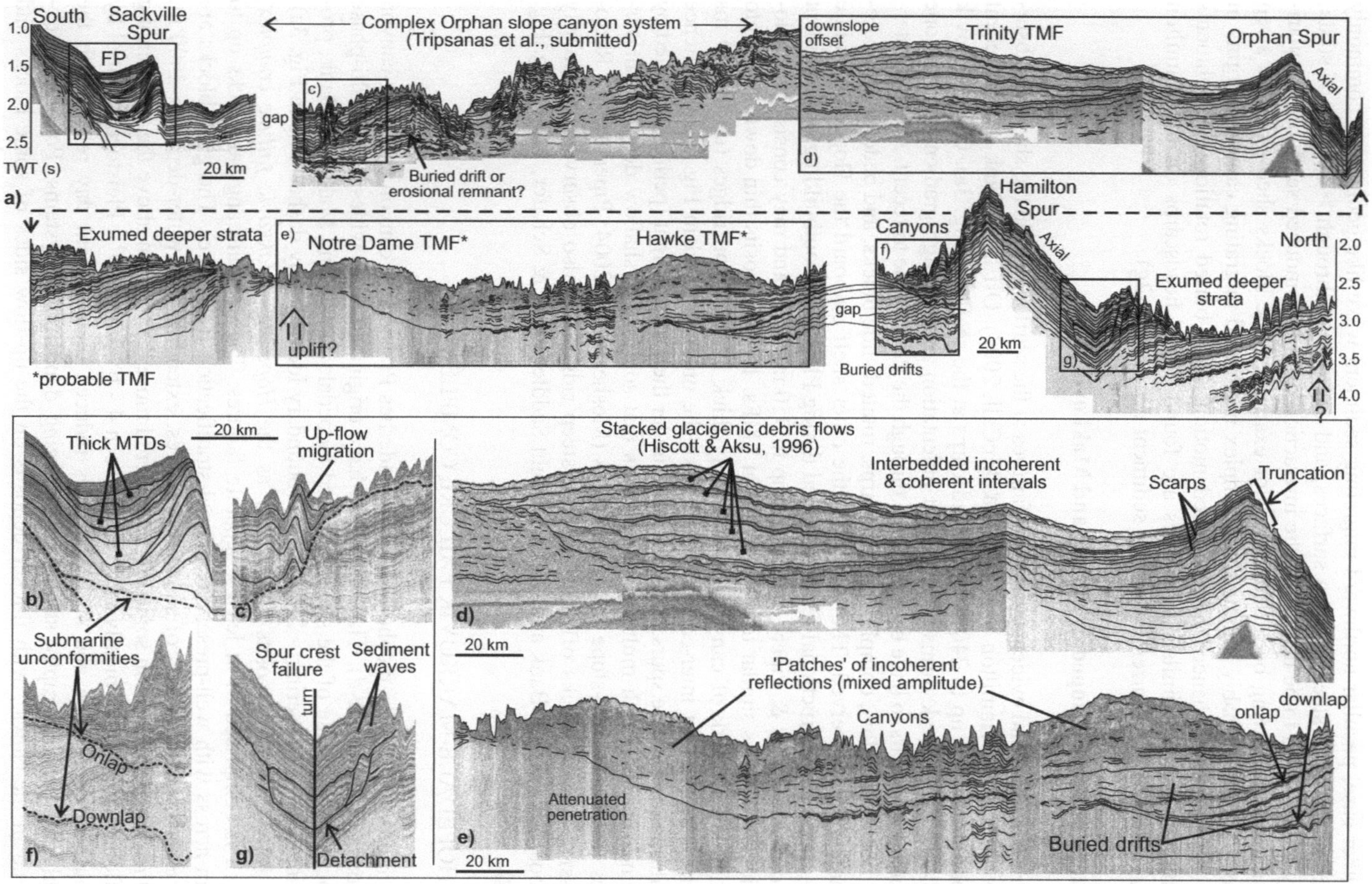

Figure 2. a) Along-slope profile extending from Flemish Pass (south) to north of Hamilton Spur (north), with close-ups (b-g). See text for details and figure 1 for location.

In either case, peculiar northward migrating sediment-wave-like geometries commonly develop along the steep incision surfaces, and can persist through > 600 ms of strata. They are believed to be caused by the interaction of failure-induced seafloor irregularities and south-flowing ocean currents. For example, slide blocks shed from the steep undercut slope probably initiated the complex northward migrating canyon margins in Fig. 2c, perhaps enhanced by continued motion along buried rotational detachments (Campbell, 2005). Similar geometries are found above headscarps on the Hamilton Spur, generating up-current migrating sediment waves (Fig. 2g).

3. Slope outboard Hopedale Saddle and Makkovik Bank

There is widespread evidence for mass failures at the seafloor and in the shallow subsurface north of the Hamilton Spur (Piper & McCall, 2003). Off Nain Bank several recent slope failures left up to 240 m high headscarps at the seafloor (Fig. 3a). Off Makkovik Bank the shelf-break forms a prominent indentation and the high gradient (> 5°) slope outboard it appears to have been unstable through the Pleistocene to recent. Amphitheater-like failure corridors, complex failure scarps, rotated slide blocks, and bedding plane detachments are observed. The 'bulge' in the outer shelf outboard the Hopedale Saddle (Fig. 3a) is a constructional feature formed during Pleistocene outbuilding of prominent clinoforms (Myers & Piper, 1988). It is up to 950 ms thick and may consist of glaciogenic debris flows similar to those found in TMFs. These deposits thin down-slope into a complex network of canyons and variably thick inter-canyon ridges (up to 400 ms thick). The thickest inter-canyon ridges, like the one identified in Fig. 3a, experience similar failures as the spurs. Some collapse under their own weight (perhaps initiated by ground-shaking), with multiple listric faults that sole out in shallow decollement surfaces that pass down-slope into creep folds (Mosher et al., 2004; Piper, 2005). Retrogressive rotated blocks off the flanks of steep ridges are also observed particularly outboard Makkovik Bank and the Cartwright Saddle (Praeg & Schafer, 1989) (e.g. Fig. 3f-g).

3.1 HOPEDALE-MAKKOVIK FAILURE COMPLEX

An interval of largely incoherent MTDs underlies Pleistocene canyon and inter-canyon deposits (e.g. Fig. 3c-d). They appear to have originated from widespread multi-phase Pliocene(?) collapse of the slope outboard Hopedale Saddle and Makkovik Bank, producing deposits covering > 28 000 km^2 (boundary identified by dashed line in Fig. 3a). We refer to these deposits collectively as the *Hopedale- Makkovic failure complex*, which is comprised of at least 4 separate failures. They contain angular blocks up to 6 km across with well-preserved internal stratification (Fig. 3d-e). The giant blocks are dispersed over thousands of square kilometres, extending into water-depths > 2500 m. At the time of deposition some blocks towered more than 350 m above the surrounding seafloor, with sides typically inclined between 4 - 6° (up to 20°). Pleistocene burial by turbidites, contourites and smaller MTDs reduced their present day relief, with the highly rugose topography strongly influencing depositional systems. Ponding is observed in the 'lows', with thicker deposits on the up-flow sides of some blocks and thinner deposits on their down-flow sides, in areas that were probably shadowed from sediment gravity flows. In some cases just the corners of angular blocks are exposed at the seafloor. In other cases long edges are exposed, producing 2-6 km lineations with up to 140 m of seafloor offset (Fig. 3c). Shallow detachment faults locally enhance their seafloor relief, with the Pleistocene overburden detaching along the steeply dipping faces of consolidated blocks (Fig. 3f).

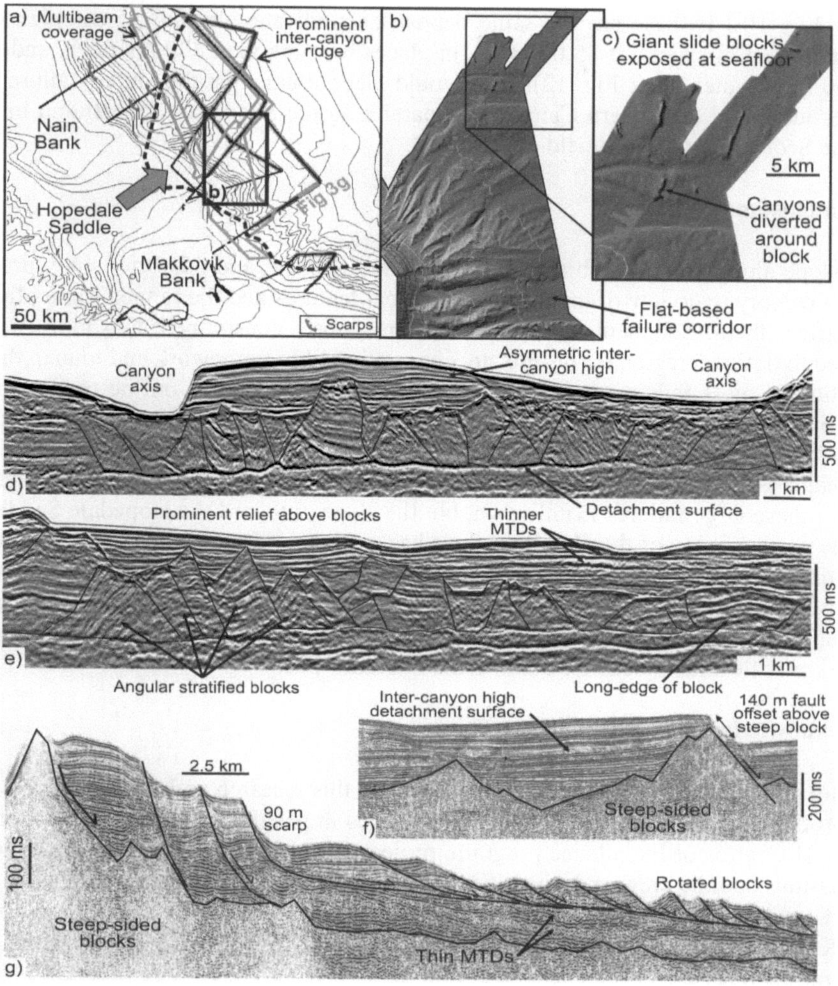

Figure 3. a) Location map; b-c) Multibeam bathymetry; d-e) Seismic profiles showing angular blocks within upper part of Hopedale-Makkovik failure complex (provided by GSI); f) High-resolution airgun profile above angular blocks below an inter-canyon high, onlapped by Pleistocene sediments; g) High-resolution airgun profile across recently collapsed inter-canyon ridge above blocky MTDs. See text for details.

The most recent failure in the complex forms above a bed-parallel detachment surface, constrained by lateral scarps as high as 280 m (Fig. 3d). The relatively low levels of deformation (subtle folding and tilting) and the angularity of blocks suggest strata were well-consolidated at the time of failure. We interpret the failure blocks to have formed through the break-up of a large detached slab consisting of relatively consolidated slope strata (perhaps above a weak layer). The dimensions of some blocks indicate the failed slab was more than 300 m thick. The failure probably originated near the Hopedale Saddle, where a 300 m high failure scarp is present below the more recent Pleistocene clinoforms. We speculate that the prominent indentation of the shelf-break outboard Makkovik Bank may also be a remnant scar associated with this failure complex. Stratigraphically similar chaotic deposits were mapped in vintage industry seismic profiles by Myers & Piper (1988) between their mid-Pliocene D and mid-Pleistocene A reflectors (their Fig. 10). If these are the same deposits, it indicates the failure complex could cover an area in excess of 85 000 km^2, in places be more than 700 ms thick, and reach >3500 m of water (their Fig. 12). This would place it amongst the largest failure complexes identified off eastern Canada, comparable to the Shelburne megaslump mapped off the Scotian margin (Shimeld et al., 2003).

4. Conclusions

The slope along the SW Labrador Sea is highly complex, and this brief study provides only a cursory overview of major morphological features and the range of mass failures that affect them. Sediment ages at present are poorly constrained, and more work is needed to define a regional Plio-Pleistocene stratigraphic framework and to map the distribution of mass failures and associated features. Mass failures on the predominantly depositional spurs, fans, and inter-canyon ridges are relatively local features. In contrast, the widespread Pliocene to early Pleistocene MTDs north of the Hamilton Spur are major failures that would have modified the seascape on a regional scale. The trigger for these large failures is unknown, but the slope outboard the Hopedale Saddle and on the northern side of the Hamilton Spur has experienced multiple magnitude 4 to 5.6 earthquakes over the past 50 years (USGS earthquake data-base). Hence earthquakes could have triggered these large failures (Piper et al., 2003), causing a multi-phase widespread collapse of a 200 km long segment of the Labrador slope, with transport of giant km-scale slide blocks far out into the Labrador Sea.

5. Acknowledgments

We are grateful to C-NLOPB for their support of this research and to GSI for providing the profiles in Figs 3d-e. The multibeam data was in part funded through the ArcticNet National Centres of Excellence program, project 1.6. We thank David Piper for helpful discussions and the crew and scientific staff aboard CCGS Hudson and the Ocean Mapping Group at UNB for their efforts in the field. We are grateful to Efthymios Tripsanas and Ken Skene for reviewing the manuscript and Vasilis Lykousis for organizing this volume.

6. References

Campbell, D.C., 2005. *Major Quaternary mass transport deposits in southern Orphan Basin, offshore New foundland and Labrador*. Geological Survey of Canada, Current Research, no. 2005-D3, 10 p.

Campbell, D.C., Piper, D.J.W., Douglas, E.V., Migeon, S., 2002. *Surficial geology of Flemish Pass: assessment of hazards and constraints to development.* Geological Survey of Canada, OF 1454, 60 p.

Deptuck, M.E., 2003. *Post-rift geology of the Jeanne d'Arc Basin, with a focus on early Paleogene submarine fans and insights from modern deep-water systems*. Unpublished Ph.D. thesis, Dalhousie University, 369 p.

Goss, S., 2006. *Quaternary seismic stratigraphy of the Hamilton Spur: A sediment drift on the Labrador continental margin*, Unpublished Honors Thesis, Dalhousie University, 56 p.

Hesse R., Klauke I., Khodabakhsh S., Piper D.J.W., 1999. *Continental slope sedimentation adjacent to an ice margin.* Marine Geology, 155, 249-276.

Hiscott, R.N., Aksu, A.E., 1996. *Quaternary sedimentary processes and budgets in Orphan Basin, south western Labrador Sea.* Quaternary Research, 45, 160-175.

Jenner, K.A., Piper, D.J.W., Cambell, D.C., Mosher, D.C., 2007. *Lithofacies and origin of late Quartenary MTDs in submarine canyons, central Scotian Slope, Canada*, Sedimentology, 54, 19-38.

Josenhans, H.W., Barrie, J.V., Kiely, L.A., 1987. *Mass wasting along the Labrador shelf margin: submersible observations.* Geo-Marine Letters, 7, 199-205.

Kennard, L., Schafer, C., Carter, L., 1990. *Late Cenozoic evolution of Sackville Spur: asediment drift on the Newfoundland continental slope*: Canadian Journal of Earth Sciences, 28, 863-878.

Mosher, D.C., Piper, D.J.W., Campbell, C., Jenner, K.A., 2004. *Near surface geology and sediment-failure geohazards of the central Scotian Slope.* AAPG Bull., 88, 703-723.

Mosher, D.C., Shimeld, J., Syhlonyk, G., in press. *Bottom simulating reflectors on Canada's east coast margin: evidence of gas hydrates.* Marine and Petroleum Geology.

Myers R.A., Piper D.J.W., 1988. *Seismic stratigraphy of late Cenozoic sediments in the northern Labrador Sea: a history of bottom circulation and glaciation.* CJES, 25, 2059-2074.

O'Cofaigh, C., Taylor, J., Dowdeswell, J.A., Pudsey, C.J., 2003. *Palaeo-ice streams, trough-mouth fans and high-latitude continental slope sedimentation,* Boreas, 32, 37-55.

Piper, D.J.W., 2005. *Late Cenozoic evolution of the continental margin of eastern Canada.* Norwegian Journal of Geology, 85, 305-318.

Piper, D.J.W., Brunt, R.A., 2006. *High-resolution seismic transects of the upper continental slope off south eastern Canada,* Geological Survey of Canada Open File 5310.

Piper, D.J.W., Campbell, D.C., 2005. *Quaternary geology of Flemish Pass and its application to geohazard evaluation for hydrocarbon development.* GAC Special Paper 43, 29-43.

Piper, D.J.W., McCall, C., 2003. *A synthesis of the distribution of submarine mass movements on the eastern Canadian margin.* In: Locat, J., Mienert, J. (eds), Submarine mass movements and their consequences. Kluwer, Dordrecht, 291-298.

Piper, D.J.W., Mosher, D.C., Gauley, B.J., Jenner, K., Campbell, D.C., 2003. *The chronology and recurrence of submarine mass movements on the continental slope off southeastern Canada.* In: Locat, J., Mienert, J. (eds), Submarine mass movements and their consequences. Dordrecht, 299-306.

Piper D.J.W., Normark W.R., 1989. *Late Cenozoic sea-level changes and the onset of glaciation: impact on continental slope progradation off eastern Canada.* Marine and Petroleum Geology 6, 336-348.

Praeg, D.B., Schafer, C.T., 1989. *Seabed features of the Labrador slope and rise near 55N revealed by SEA-MARC I sidescan sconar imagery*, Geological Survey of Canada Open File 2028. 28 p.

Shaw, J., Piper, D.J.W., Fader, G.B.I., et al., 2006. *A conceptual model of the deglaciation of Atlantic Canada,* Quaternary Science Reviews, 25, 2059-2081.

Shimeld, J.W., Warren, S.N., Mosher, D.C., MacRae, R.A., 2003. *Tertiary-aged megaslumps under the Scian Slope, south of the Lahave Platform, offshore Nova Scotia.* GSA Program with Abstracts.

Tripsanas, E.K., Piper, D.J.W., submitted. *Late Quaternary stratigraphy and sedimentology of Orphan Basin: implications for melt-water dispersal in the southern Labrador Sea.*

Tripsanas, E.K., Piper, D.J.W., Campbell, D.C., in press. *Evolution and depositional structure of earth quake-induced mass movements and gravity flows: southwest Orphan Basin, Labrador Sea*, Marine and Petroleum Geology.

SUBMARINE LANDSLIDES ALONG THE NORTH ECUADOR – SOUTH COLOMBIA CONVERGENT MARGIN: POSSIBLE TECTONIC CONTROL

G. RATZOV, M. SOSSON, J.-Y. COLLOT, S. MIGEON, F. MICHAUD, E. LOPEZ, AND Y. LE GONIDEC
UMR Géosciences Azur, BP 48, Port de la Darse, 06235 Villefranche-sur-Mer, France

Abstract

A morphometric analysis of submarine landslides on the continental slope of north Colombia – south Ecuador convergent margin provides insights into hazards, including mass movements locations, size of failures and location control. Bathymetric and seismic data acquired in 2005 revealed three distinctive types of locations with the occurrence of potentially tsunamogenic submarine landslides: 1) the erosive margin toe is characterized by three 5-6-km wide and up to 1500m high semi-circular scarps. The toe is also affected by a 35-km long area of linear scarps resulting from the imbrication of successive failures, 2) canyon walls are affected by five up to 500m high scarps, 3) the deformation front exhibits a 50x20 km potentially destabilized area characterized by intensive fracturation. All these features are controlled by active tectonics, Slope oversteepening is a key parameter facilitating the onset of slope failure for the two first types, and is associated with 1) seamount subduction, 2) subsidence related to basal erosion of upper plate, and 3) uplift along active tectonic structures. Regarding the third type, the destabilisation area is influenced by the intensive fracturing at the vicinity of a splay fault.

1. Introduction

Slope failures play a major role on sediment transport and sediment distribution on continental margins, and are largely responsible for shaping the seafloor of both deep sea and coastal environments. These mass-wasting events represent a hazard for e.g., offshore exploration safety, and communication cables. Coastal cities are exposed to a tsunami hazard if submarine slopes are affected by medium- to large-scale (over 5 km wide) collapses (McAdoo and Watts, 2004; von Huene et al., 1989). A good understanding of processes associated with submarine instabilities and their triggering factors is fundamental in order to quantify the hazard and mitigate the risk. To date, most studies have been carried out along passive margins (Canals et al., 2004; Evans et al., 1996) and volcanic islands (Masson, 1996; Masson et al., 2002). Triggering factors of submarine landslides include rapid accumulation of sediment, slope increase, sediment excess pore pressure, high physical stresses related to wave loading, tectonics and earthquakes (Hampton et al., 1996), eustatic sea-level variations , and gas hydrates dissociation (Maslin et al., 1998).

Regarding active margins, similar mechanisms were already reported, and tectonic activity is one of particular importance in triggering medium-scale (von Huene et al., 2000) to giant submarine landslides (Collot et al., 2001; Duperret et al., 1995). North Colombia – South Ecuador convergent margin has been explored in 2005 during

V. Lykousis, D. Sakellariou and J. Locat (eds.), Submarine Mass Movements and Their Consequences, 47–55.

AMADEUS cruise on board of R/V *L'Atalante* (Collot et al., 2005), providing original data on the seafloor.

The principal aim of this paper is to map the areas of the margin undergoing mass wasting. We concentrated our study on failures which upper scarps are larger than 5 km, that could possibly generate catastrophic tsunamis (McAdoo and Watts, 2004). The second aim is to better constrain what parameters are controlling localization and triggering of medium- to large-scale, submarine landslides on study area.

To achieve these objectives, this study is based on submarine landslides identification using: 1) Multibeam swath bathymetry (150m resolution except along Patia/Mira canyon where data was reprocessed to obtain a resolution of 60m), 2) Six channels seismic data from the Amadeus cruise (AMA profiles), and 24- channels seismic data provided by the *Agencia Nacional de Hidrocarburos* (ANH profiles), and 3) 10-m long gravity cores.

2. Geological setting

The north Ecuador - south Colombia active margin (from N0° to N3°30) is located along the subduction margin of the Nazca and South America plates (fig. 1). The eastward convergence rate of the Nazca Plate is 54 mm/year (Trenkamp et al., 2002). The margin is the scene of intense seismicity with the occurrence of four of the greatest earthquakes of the twentieth century ($M_W > 7.7$) in the region (Mendoza and Dewey, 1984). The northern part of the margin (N1°30 to N3°30) is characterized by an accretionnary wedge decreasing in width southward (Collot et al., 2006; Mountney and Westbrook, 1997). In contrast, the southern part of the margin shows a narrow slope (N0° to N1°30) related to tectonic erosion (Collot et al., 2002). The upper plate consists of an upper Cretaceous mafic basement overlain by Cenozoic sedimentary series (Jaillard et al., 1997). Offshore Colombia, the slope is incised by the Esmeraldas and Mira/Patia submarine canyons (Collot et al., 2005). (Fig. 1). Esmeraldas is V-shaped in cross section, sharply cutting the deformation front, thus indicating erosion is currently active. Patia and its tributary Mira are Z-shaped in map view and U shaped in cross-section, implying it has been less active than Esmeraldas, at least during the Holocene (Collot et al., 2005). The Ancon fault (Fig. 1) is a northeastward trending crustal scale fault which activity on the long term may reflect tectonics of the seismogenic zone (Collot et al., 2005).

3. Results

3.1 MARGIN TOE INSTABILITIES

3.1.1 Description

In the southern part of the margin (N0 to N1°30), the slope is 30-km wide between 200 and 3600 meter of water depth. The mean slope angle is about 6-7°. This narrow slope is affected by numerous sub-circular and linear scarps (Fig.1).

Sub-Circular scarps

Near 0°20'N of latitude (Fig. 2a) three semi-circular scarps, 5-6 km-wide, outline the top of a destabilized area covering about 300 km^2 and 22-km long. Scarps A and B are about 1500 m high, with a slope angle of 30°. Scarp C is up to 750 m high and dips 25° westward. Arcuate lineations, exhibiting a concavity toward the trench overhangs the scarps (Fig. 2a). They are interpreted as creeping features. Moreover, a topographic knoll is present upslope from the scarps. Gravity core KAMA01, collected at the base of scarp B (Fig. 2a), consists of decimetric blocks of consolidated clay in a muddy matrix that are interpreted as a debris flow deposits.

Linear scarps

At about 0°55'N of latitude, a 35-km long section of the slope is affected by linear scarps up to 2000 m high (Fig. 2b). They are characterized by a sharp slope break, with a dipping angle increasing from 6 to 25°. From the bathymetric map analysis (Fig. 2b), we interpret the linear shape of the scarps as the results of several imbricated small-scale scarps corresponding to successive sliding events. Lineations, parallel to the main direction of the scarps are also present on the upper part of the slope (Fig. 2b). They could correspond to secondary normal faults due to the collapse of the margin or to possible creeping structures.

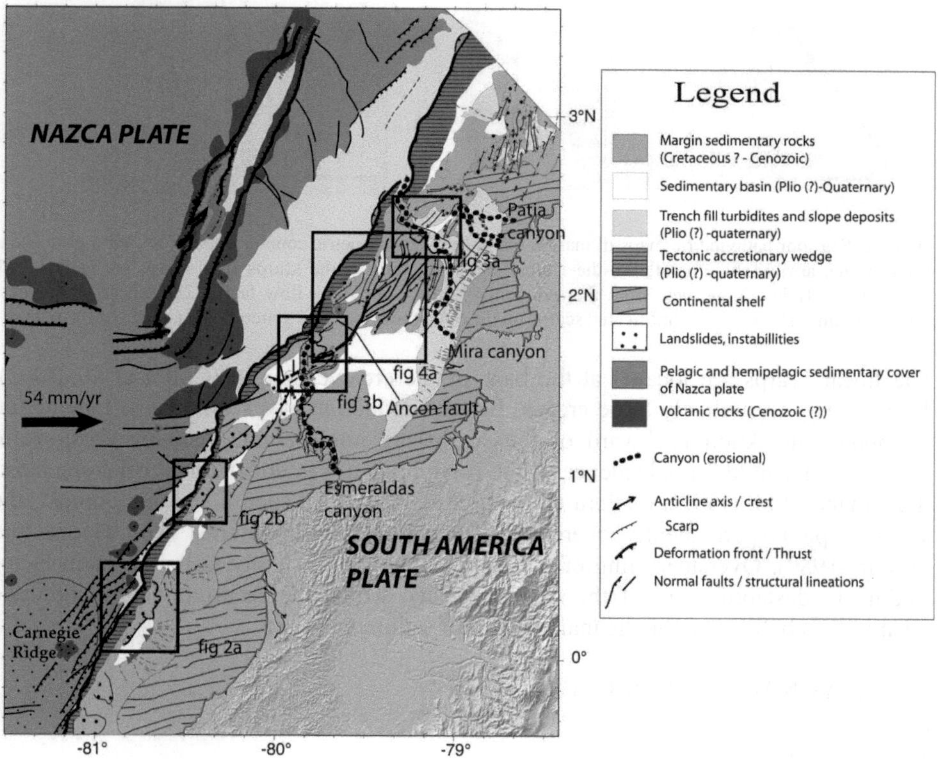

Figure 1. Structural map of North Ecuador - South Columbia margin. Black arrow is Nazca – South America plate convergence vector derived from (Trenkamp et al., 2002) GPS study. Blacks rectangles show localisation of bathymetric views of destabilization areas. Bold dotted lines show submarine canyons systems (modified from Collot et al., 2006).

3.1.2 Margin toe instabilities control

The origin of the sub-circular scarps is thought to result from the subduction of seamounts (Collot et al., 2005). The indentations shapes and the presence of bathymetric knoll landward of the scarps are similar to those described along the Costa Rica margin and unambiguously associated with seamount subduction (von Huene et al., 2000). Sandbox models also show similar morphological response of the seafloor to seamount subduction (Dominguez et al., 1998). This interpretation is supported by aligned seamounts located on the subducting plate a few kilometers in front of the scarps (Fig. 2a).

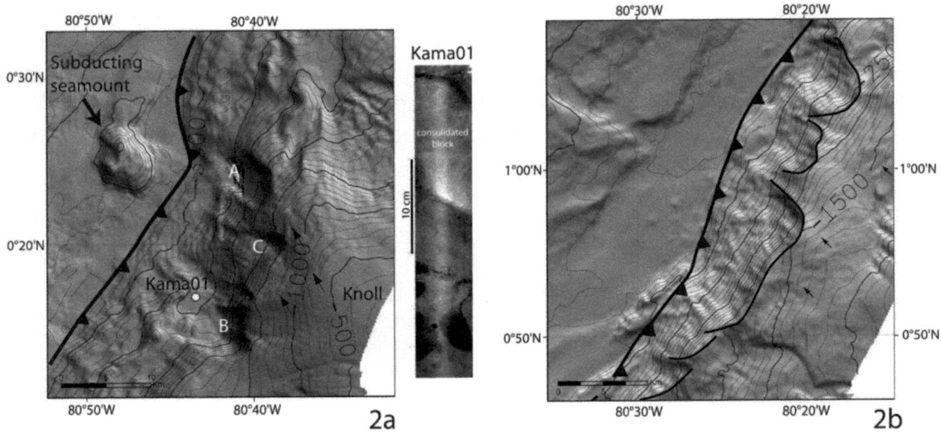

Figure 2. Seafloor bathymetric maps of margin toe failures. Bathymetric contour interval is 100m, grid size is 150m. Black arrows show parallel-to-the-scarps lineations. 2a) Circular scarps area: scarps are named A,B, and C. Kama01 is a Xray view from the sedimentary core with debris flow facies. 2b) Linear scarps area: black bold lines show scarps. Imbricated scarps on the SW corner lead to a general linear shape of the scarps.

The linear scarps are located at the base of the erosive margin (Collot et al., 2002). Their shapes are unlikely to be created by the subduction of a seamount as indentation or topographic knoll backward of the scarps are not present. Tectonic erosion are usually responsible for subsidence of the margin, as it was documented offshore Japan (Lallemand et al., 1992), or Peru (Sosson et al., 1994). In such a tectonic context, the lower slope is permanently oversteepened and undergoes failures (von Huene and Culotta, 1989). Oversteepening of the continental slope is likely to be the main factor leading to destabilization of the slope deposits, whilst the recurrent high-magnitude seismicity is believed to be the main trigger of sediment failure.

3.2 CANYON WALLS FAILURES

3.2.1 Description

The Patia/Mira canyons system

Three major slides (S1 to S3) (Fig.3a) are located close to the junction between the two canyons:

S1: A sub-rectangular scarp (8.5x5 km in map view and 500-m high) is located at the sharp bend of the Patia canyon (Fig. 3a). Hectometric blocks are located within the Patia canyon, downslope from the scarp. These deposits can be considered as a blocky avalanche (S1) in accordance with the classification of (Mulder and Cochonat, 1996). The length of rough seafloor morphology of the canyon axis beneath the scarp lead us estimate a 18 km run-out distance for the avalanche. The slid blocks in the canyon axis formed a dam that trapped the sediment supply and caused an infilling of the canyon (Fig. 3a-3c). The dam still overhangs the canyon floor by 30 m.

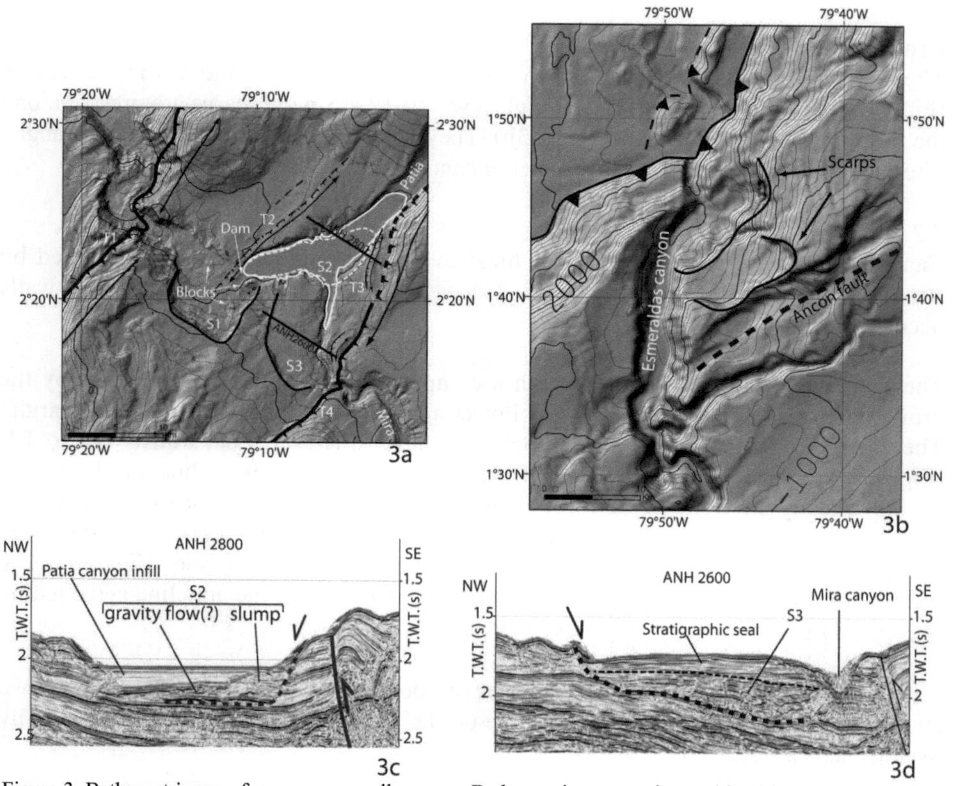

Figure 3. Bathymetric map from canyon walls scarps. Bathymetric contour interval is 100m, grid size is 60m for Fig. 3a, 150m for Fig. 3b. Black bold lines show scarps. 3a) Patia and Mira canyon walls failures. T1 to T4 are active thrusts. S1 to S3 are slides. White dashed line show buried S2 limits. White bold line show Patia canyon infill due to the dam. 3b) Esmeraldas canyon walls failures. Dashed line show location of Ancon fault. 3c) and 3d) Multichannel seismic profiles ANH2800 and ANH2800 across canyons junction area. Profiles are located on figure 3a).

S2: Seismic line ANH2800 shows tilted reflectors along a slip plane on the eastern wall of the Patia canyon (Fig. 3a). The reflectors are undisturbed at least close to the slip plane, but are getting disturbed and loose coherency further away. This structure (S2) is therefore interpreted as a rotational slump that possibly evolved into a gravity (mud?) flow. Its run-out distance is about 11 km, but must be greater as deposits in the canyon axis must have been eroded by the later overlapping S1 event. The scarp associated with slump S2 is buried beneath the canyon infill (Fig. 3c).

S3: A 12 km-wide, 100 to 200 m-high, amphitheater-like scarp is located at the junction of the Mira and Patia canyons (Fig. 3a). Seismic-reflection data (Fig. 3d) (ANH2600) shows weakly deformed material at the toe of the scarp. Reflectors are not tilted, and the Skempton ratio *h/l* between the depth *h* and the length *l* of the slide (0,04) is lower than 0,15. This allows us to interpret S3 as a translational slide (Mulder and Cochonat, 1996). The run-out distance of the slide is not clearly established, but it does not exceed a few hundred meters. The slide and scarp are buried under younger levee deposits that smoothed the pre-existing relief (Fig. 3d).

Esmeraldas Canyon
The distal part of the Esmeraldas canyon is linear and oriented obliquely to the margin (Fig. 3b). Its walls are up to 1000-m high. Two 10-12 km-wide scarps are present along the right-hand side of the canyon (Fig. 3b). They are 400 m high, with a 30° slope angle. The run-out distance of the sliding material cannot be estimated.

3.2.2 Canyon walls failures control
Canyon walls are destabilized by the local oversteepening of the slope generated by strong axial incision. However, the evidenced submarine landslides on canyon walls occur on specific locations:

The Patia and Mira canyons have both a complex channel pathway controlled by the growth of active thrusts and folds (Collot et al., 2005) mainly parallel to the margin. They merge in a slope basin located between these structures. Four thrusts (T1 to T4) (Fig. 3a) are responsible for the uplift of topographic highs bounding the basin. S2 occurs directly in front of active thrusts T3 and T4 where the seafloor is locally oversteepened. The Ancon Fault, interpreted as a splay fault (Collot et al., 2004) crosses perpendicularly the Esmeraldas canyon axis and forms an anticline (Fig. 3b). The location of the slump scarps correlates to the steepest flank of the anticline activated by the splay fault.

These series of scarps are located in front or above active structures. That is why we consider these structures could play a role on the localization of landslides as it locally oversteepens the seafloor.

A sub-rectangular scarp as S1 is unusual for a gravitational failure. Its north-west and south-east walls are parallel to the trend of active thrusts (Fig. 3a). Hence we believe that the direction of these walls is tectonically controlled, mainly by a structural fabric that developed parallel to the tectonic structures. The south-west wall of the scarp is parallel to the canyon. Strong erosion occurring at the base of the concave flank of the canyon bend has oversteepened the slope angle, and thus has facilitated the occurrence of S1. Therefore, in that area, the oversteepening is also accentuated by tectonic uplift.

3.3 ANCON FAULT FEATURES

3.3.1 Description
In the prolongation of the Ancon fault (Collot et al., 2004), a 50x20-km area located on the deformation front is characterized by the presence of arcuate-to-the-slope ridges (Fig. 4) that consist of east-facing and west-facing scarps and counterscarps. The scarps defined a staircase-like morphology descending to the west. Seismic-reflection data

show that scarps are associated to sub-vertical normal faults affecting the sedimentary cover (Fig. 4b-4c). Some of the flats are small sedimentary basins with infilling showing fan-like structures (Fig. 4c). Moreover, the size of the area encompasses the entire slope.

3.3.2 Structural control

The fault network suggests the area could be a horse-tail structure associated to the Ancon fault. However, the study area has geo-morphological features closed to Deep-Seated Gravitational-Slope Deformation (DSGSD) as documented onland in mountain belts (Agliardi et al., 2001; Di Luzio et al., 2004). These features are: the presence of scarps and counterscarps, and arcuate structures facing downslope. The steepness of the faults suggests they are deep-seated structures, but poor seismic penetration prevents an estimation of their extent. The Ancon fault is likely responsible to intense faulting of the area, allowing the gravitational deformation on the fractures associated to the fault.

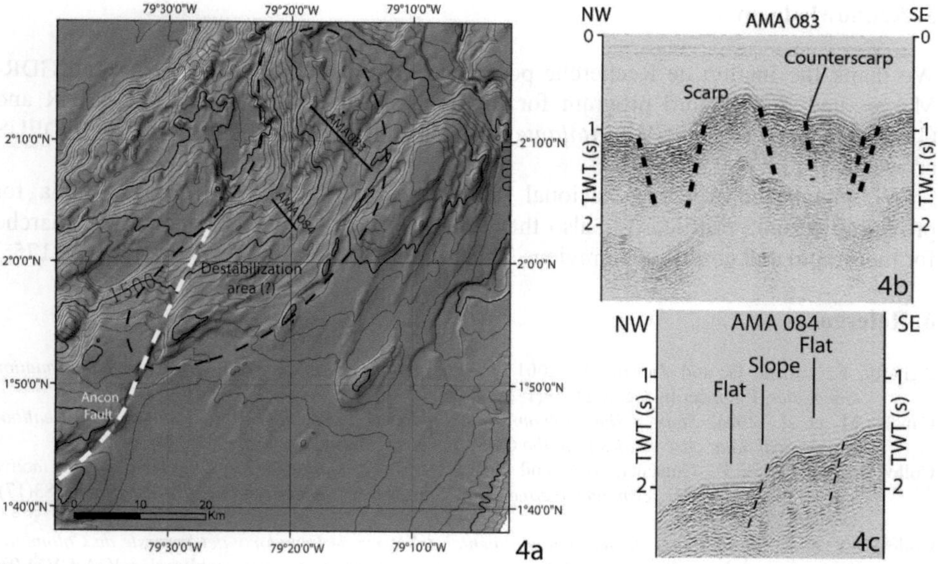

Figure 4. Bathymetric map and multichannel seismic reflection profiles across the possible giant destabilization area. Bathymetric contour interval is 100m, grid size is 150m. Black dashed line show the possible destabilistaion area. White dashed line show location of the Ancon fault.

4. Conclusions

On the North Ecuador – South Columbia active margin, medium (5-10 km) to large (50 km) - scale submarine failures have been observed 1) at the erosive margin toe (linear and sub-circular scarps), 2) along canyon walls (medium scale scarps), 3) in the vicinity of the Ancon fault (large scale margin destabilisation).

Destabilisation parameters:

- On the margin toe, the evidenced parameters controlling the localization of slope instabilities and the oversteepening of the margin are tectonically controlled. The oversteepening of the slope is associated to seamounts subduction, and strong subsidence of the margin caused by basal erosion of the upper plate.

- On canyon walls, oversteepening is associated to uplift along active thrusts. Strong canyon incision is a parameter that can also weaken the canyon walls and generate collapses. It is however not the only factor on canyon walls, as the failures are mainly located along active structures. The two parameters are considered act together.

- In the forearc, the 50-km wide area that could be undergoing a slow deep-seated gravitational deformation is located on the prolongation of a major crustal fault affected by intensive fracturation. In that case the structural framework controls the margin destabilisation.

The triggering parameters can not be doubtlessly established, but tectonic oversteepening of the seafloor is facilitating the destabilisation. Earthquake induced ground shaking is also suggested for failures sliding triggering.

5. Acknowledgements

We thank the Institut de Recherche pour le Développement (IRD), the French GDR-Marges and ECOS-Nord program for funding this research, and the IFREMER and GENAVIR crew of the R/V *L'Atalante* for their logistic support during the AMADEUS cruise.

We thank the Agencia Nacional de Hidrocarburos (ANH) from Colombia for providing seismic profiles. We also thank Dr Lies Loncke and Dr Geoffroy Lamarche for their rapid and constructive review. This is contribution from GDR-Marges n°2175.

6. References

Agliardi, F., Crosta, G. and Zanchi, A., 2001. *Structural constraints on deep-seated slope deformation kinematics.* Engineering Geology, 59(1-2): 83-102.

Canals, M. et al., 2004. *Slope failure dynamics and impacts from seafloor and shallow sub-seafloor geophysical data: case studies from the COSTA project.* Marine Geology, 213(1-4): 9-72.

Collot, J.-Y., Charvis, P., Gutscher, M.A. and Operto, S., 2002. *Exploring the Ecuador-Colombia active margin and interplate seismogenic zone.* EOS Transactions, American Geophysical Union, 83(17): 189-190.

Collot, J.-Y. et al., 2006. *Mapas del margen continental del Norte de Ecuador y del Suroeste de Colombia : Batimetría, Releive, Reflectividad Acústica e Interpretación Geológica.* publicación IOA-CVM-03-Post.

Collot, J.-Y. et al., 2005. *Seafloor margin map helps in understanding subduction earthquakes.* EOS Transactions, American Geophysical Union, 86(46): 464-466.

Collot, J.Y., Lewis, K., Lamarche, G. and Lallemand, S., 2001. *The giant Ruatoria debris avalanche on the northern Hikurangi margin, New Zealand: Result of oblique seamount subduction.* Journal of Geophysical Research-Solid Earth, 106(B9): 19271-19297.

Collot, J.Y. et al., 2004. *Are rupture zone limits of great subduction earthquakes controlled by upper plate structures? Evidence from multichannel seismic reflection data acquired across the northern Ecuador-southwest Colombia margin.* Journal of Geophysical Research-Solid Earth, 109(B11): B11103.

Di Luzio, E. et al., 2004. *Influence of structural framework on mountain slope deformation in the Maiella anticline (Central Apennines, Italy).* Geomorphology, 60(3-4): 417-432.

Dominguez, S., Lallemand, S., Malavieille, J. and von Huene, R., 1998. *Upper plate deformation associated with seamount subudction.* Tectonophysics, 293: 207-224.

Duperret, A., Bourgois, J., Lagabrielle, Y. and Suess, E., 1995. *Slope instabilities at an active continental margin: large-scale polyphase submarine slides along the northern Peruvian margin, between 5° S and 6° S.* Marine Geology, 122(4): 303-328.

Evans, D., King, E.L., Kenyon, N.H., Brett, C. and Wallis, D., 1996. *Evidence for long-term instability in the Storegga Slide region off western Norway.* Marine Geology, 130(3-4): 281-292.

Hampton, M.A., Lee, H.J. and Locat, J., 1996. *Submarine Landslides.* Review of Geophysics, 34: 33-59.

Jaillard, E., Benítez, S. and Mascle, G.H., 1997. *Les déformations paléogènes de la zone d'avant-arc sud-equatorienne en relation avec l'évolution géodynamique.* Bulletin de la Société Géologique de France, 168(4): 403-412.

Lallemand, S.E., Schnurle, P.S. and Manoussis., S., 1992. *Reconstruction of subduction zone paleogeometries and quantification of upper plate material caused by tectonic erosion.* Journal of Geophysical Research, 97(B1): 217-239.

Maslin, M., Mikkelsen, N., Vilela, C. and Haq, B., 1998. *Sea-level –and gas-hydrate–controlled catastrophic sediment failures of the Amazon Fan.* Geology, 26(12): 1107-1110.

Masson, D.G., 1996. *Catastrophic collapse of the volcanic island of Hierro 15 ka ago and the history of landslides in the Canary Islands.* Geology, 24: 231-234.

Masson, D.G. et al., 2002. *Slope failures on the flanks of the western Canary Islands.* Earth-Science Reviews, 57(1-2): 1-35.

McAdoo, B.G. and Watts, P., 2004. *Tsunami hazard from submarine landslides on the Oregon continental slope.* Marine Geology, 203(3-4): 235-245.

Mendoza, C. and Dewey, J.W., 1984. *Seismicity associated with the great Colombia-Ecuador earthquakes of 1942, 1958 and 1979: implications for barrier models of earthquake rupture.* Bulletin of the Seismological Society of America, 74(2): 577-593.

Mountney, N.P. and Westbrook, G.K., 1997. *Quantitative analysis of Miocene to recent forearc basin evolution along the Colombian margin.* Basin Research, 9: 177-196.

Mulder, T. and Cochonat, P., 1996. *Classification of offshore mass movements.* Journal of Sedimentary Research, 66(1): 43-57.

Sosson, M., Bourgois, J. and Mercier de Lépinay, B., 1994. *Seabeam and deep-sea submersible Nautile surveys in the Chiclayo canyon off Peru (7°S): subsidence and subduction-erosion of an Andean-type convergent margin since Pliocene times.* Marine Geology, 118: 237-256.

Trenkamp, R., Kellogg, J.N., Freymueller, J.T. and Mora, H.P., 2002. *Wide plate margin deformation, southern Central America and northwestern South America, CASA GPS observations.* Journal of South American Earth Sciences, 15(2): 157-171.

von Huene, R., Bourgois, J., Miller, J. and Pautot, G., 1989. *A large tsunamogenic landslide and debris flow along the Peru Trench.* Journal of Geophysical Research-Solid Earth, 94(B2): 1703-1714.

von Huene, R. and Culotta, R., 1989. *Tectonic erosion at the front of the Japan Trench convergent margin.* Tectonophysics, 160: 75-90.

von Huene, R., Ranero, C.R., Weinrebe, W. and Hinz, K., 2000. *Quaternary convergent margin tectonics of Costa Rica, segmentation of the Cocos Plate and Central American volcanism.* Tectonics, 19(2): 314-334.

THE SOUTHERN FLANK OF THE STOREGGA SLIDE: IMAGING AND GEOMORPHOLOGICAL ANALYSES USING 3D SEISMIC

J. GAFEIRA

University of Edinburgh, School of GeoSciences, West Mains Road, EH9 3JW UK

J. BULAT, D. EVANS

British Geological Survey, West Mains Road, Edinburgh, EH9 3LA, UK

Abstract

The Holocene Storegga Slide is one of the world's largest exposed slides and also the most studied of all the Norwegian slides. However due to its complexity it is far from being fully understood. Three-dimensional (3D) seismic combined with swath bathymetry data from the southern flank of the Storegga Slide have been used to study mass movement processes occurring in the region. The high spatial resolution provided by the 3D seismic data has allowed a detailed geomorphological analysis of sedimentary and deformational structures. The Holocene Storegga failure affected a significant part of the studied area. The predominant feature is a compression zone, comprising two lobes, where the seabed shows marked parallel ridges. Down slope it is possible to identify another compression zone. A relative chronology of events was established and it is proposed that these two compressions zones are the result of gravity-driven slope failures related to different stages of the Holocene Storegga Slide.

Keywords: 3D seismic, Storegga Slide, North Sea Fan, glacial debris flow, frontally confined submarine landslide

1. Introduction

The Holocene Storegga Slide event was the last of a large number of palaeoslides that have occurred on the mid-Norwegian Margin (Bugge *et al.*, 1987, 1988; King *et al.*, 1996; Evans *et al.*, 1996, 2005; Nygård *et al.*, 2005). This last slope failure, which took place 8200 years ago, displaced between 2500 and 3500 km^3 of Pleistocene sediments (Haflidason *et al.*, 2005). The Storegga Slide depression separates the North Sea Fan (NSF) from the Vøring Plateau (Fig. 1). With an area of approximately 142 000 km^2 (King *et al.*, 1996), the NSF is one of the largest trough-mouth fans on the Norwegian Margin.

The main aims of this paper are: (1) to document the geomorphology and underlying stratigraphy using seabed and subsurface acoustic data of an area on the edge of the NSF; (2) to highlight the value of detail analyses of 3D seismic data; (3) to assess the recent history of mass movements events in the study area, and (4) to relate it with the development of the adjacent Holocene Storegga Slide.

2. Data and methodology

This paper presents part of an extensive acoustic database gathered for interpretation of the mid-Norwegian continental margin (NDP, unpublished data 2004a,b): the seabed bathymetry provided by Norsk Hydro, covering the Storegga Slide scar, and Statoil's

V. Lykousis, D. Sakellariou and J. Locat (eds.), Submarine Mass Movements and Their Consequences, 57–65.

3D seismic volume (LS0105) from the boundary between the NSF and the Storegga Slide (Fig. 1). The focus of this study was the morphologic features observed on the seafloor covered by the 3D seismic data (1610 km^2, extending from 980 m to 1370 m water depth).

The methodology used was based on the interpretation of 3D seismic data with a seismic interpretation package and a Geographic Information System (GIS), along with the seabed bathymetry. This study has involved stratigraphic interpretation and detailed mapping of the southern flank of the Storegga slide using the different acoustic datasets with the objective of generating an improved geomorphological interpretation of the area. This was carried out by obtaining digital elevation models (DEM) of key reflectors and their respective seismic attributes (e.g. amplitude map and amplitude coherence). Geomorphological characterization of several surfaces was then carried out using spatial analysis tools and interpretation of seismic attribute 3D images (i.e. draping the map attribute on top of the related DEM). The incorporation of the seismic information into a GIS environment has improved greatly the geological interpretation, allowing extended geomorphological analyses.

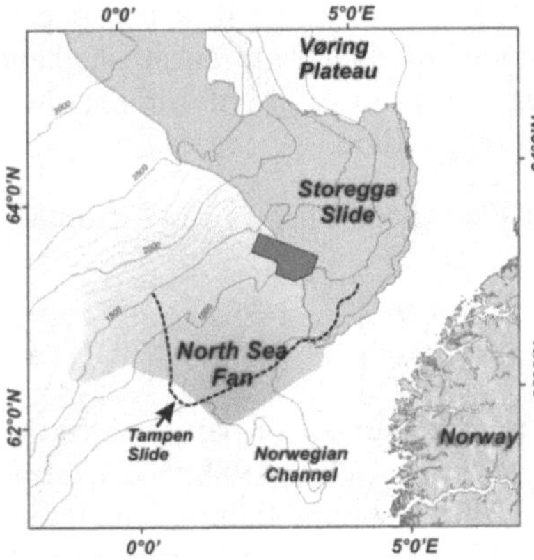

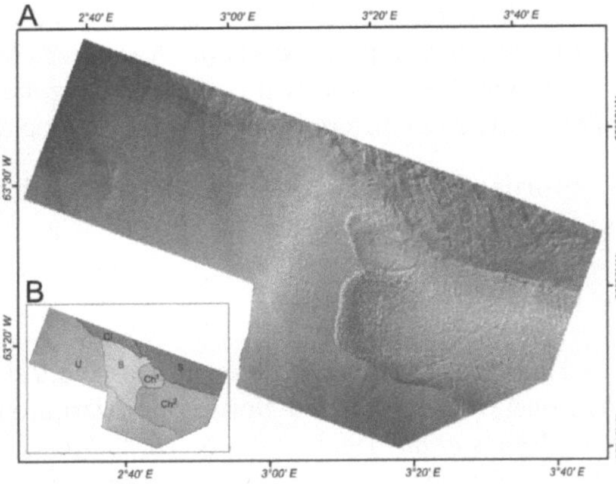

Figure 1 (left). Lower section of Mid-Norwegian Margin bathymetrical map, with the location of its main features. The study area is marked in dark grey.
Figure 2 (right) (A). Image of the seabed from the 3D seismic volume, obtained by tracking the first seismic reflection. (B) Distribution of the five zones identified on the detail study area.

3. Slope morphology

The study area seabed can be sub-divided into five morphological zones, according to their characteristics as topography, seismic signature and degree of disintegration (Fig. 2). These morphological zones reflect distinct erosional, compressional and depositional processes acting at the seabed. The five zones will be named in this paper as: *Zone S* (Storegga main failure); *Zone Ch* (higher compression zone) subdivided in two lobes: Ch^1 and Ch^2; *Zone Cl* (lower compression zone); *Zone B* (blocky debris) and *Zone U* (undisturbed North Sea Fan).

Zone S

Zone S corresponds to the seabed depression where the sediments appear largely remoulded by the Holocene Storegga deep failure. A progressive lost of coherence is observed heading upslope, from a seismic reflection character where it is still possible to recognize the original stratification to chaotic and transparent facies, through approximately 500 m (Fig. 3). The displaced Tampen Slide deposits provide a marker horizon within the remoulded material. They show that zone S comprises a succession of "pop-up" blocks close to its western edge demonstrating initial compression and a succession of graben and horst blocks to the east illustrating subsequent extension due to the escape of material to the north as part of the Holocene Storegga Slide (Fig. 3).

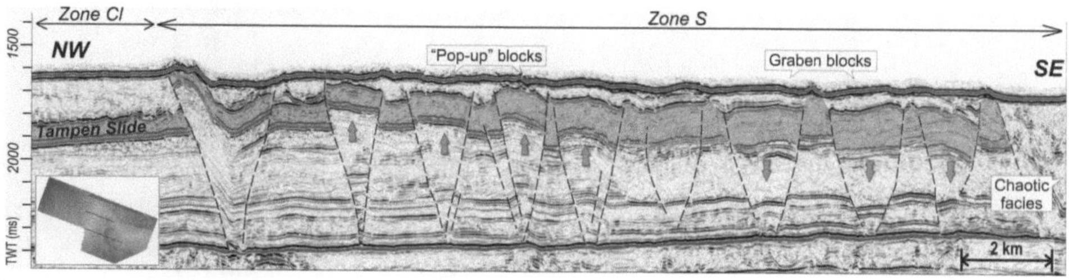

Figure 3. Section of the seismic profile across Zone S, showing a succession of "pop-up" blocks, a succession of graben blocks and chaotic facies within the material remoulded by the Holocene Storegga deep failure. It is marked on the map the presented section (red) and Fig. 4 and 6 profiles (blue).

Zone Ch

The most prominent feature in this area is a compression zone (Zone Ch), situated at water depths varying from less than 985 m to 1370 m. It is a mass transport deposit characterized by crenulation or destruction of most of the seismic reflectors throughout the top 200 ms (TWT). This zone comprises two lobes (Ch^1 and Ch^2) and appears well preserved at the present-day seabed as sub-parallel ridges of ~150 m width over an area of 380 km^2 (Fig. 4).

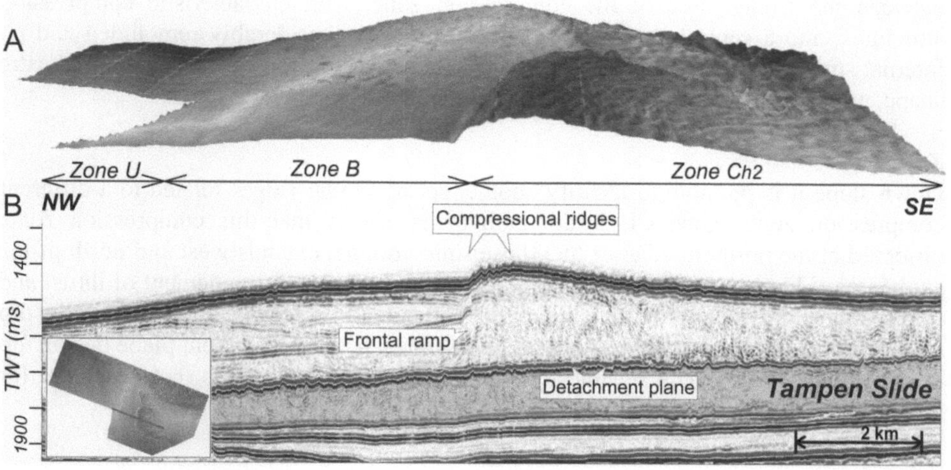

Figure 4. (A) 3D perspective view of the present-day seabed cutting through Ch^2 looking towards northeast Note the height difference between the front of both lobes of compression, Ch^1 and Ch^2. (B) Seismic profile across the toe region of the Ch^2 zone, showing compressional ridges, frontal ramp and well-defined detachment plane following the top of Tampen Slide deposits.

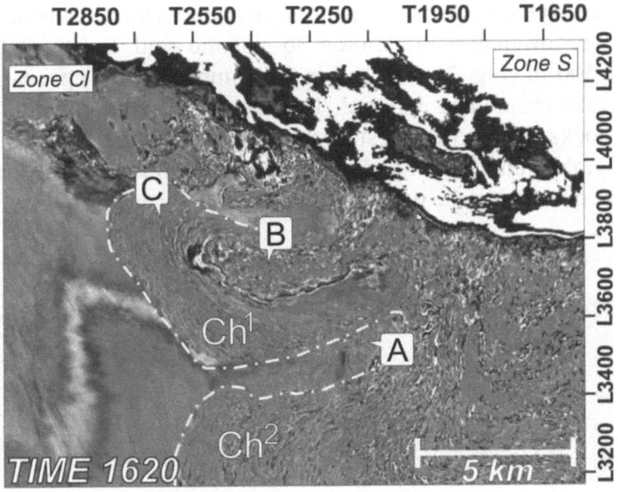

Figure 5. Detail of a time slice through the 3D seismic volume at 1620 ms depth showing the sub-zone Ch^1 in the centre of the image. It is possible to identify the undisturbed area (A) between the two lobes and the Ch^1 internal boundary the frontal area of the compressional facies (C) and the chaotic facies (B).

The deformation, by shortening and thickening of a succession of hemipelagic/ glacimarine sediment units, has resulted in a sequence of low angle thrust faults. The thrust faults can be traced down to a well-defined detachment plane on the top of the Tampen Slide deposits. Near the seabed surface the thrust faults are expressed as concentric fold traces oriented transverse to the flow direction.

An area, 6x1 km, remains totally undeformed separating the two lobes Ch^1 and Ch^2 (Fig. 5). Both lobes are cut by the Holocene Storegga event giving a minimum age to these compression features. The smaller of the two lobes shows an internal boundary between the frontal area of the compression zone, with characteristic compression structures, and a central area where the material was considerably remoulded and no internal structure was preserved. This marked boundary presents a curved amphitheatre shape, with steep sidewalls and a flat base at the top of Tampen deposits.

Zone Cl

Down slope it is possible to identify another set of similar ridges related to a different compression zone, Zone Cl. Swath bathymetry shows that this compression zone, observed at the northern edge of the 3D seismic volume, extends west and north of the study area. This compression zone corresponds to a complex arrangement of thrust and fold systems that are approximately parallel and extend for tens of kilometres in the dip direction and are inline with the zone S. As in zone Ch, the detachment plane lies on top of the Tampen Slide. The seismic internal structure also resembles that described for zone Ch.

Zone B

A blocky deposit dominates Zone B, a region of 93 km^2 in the central part of the study area. This deposit includes detached tabular blocks of ~100 to 200 m in width and ~5 m high and is associated with chaotic layer on the seismic data (Fig. 4 & 6). The thickness increases towards the north. In addition, a well-developed detachment plane can be observed in front of the Ch^1, ~600 m apart from it (Fig. 6).

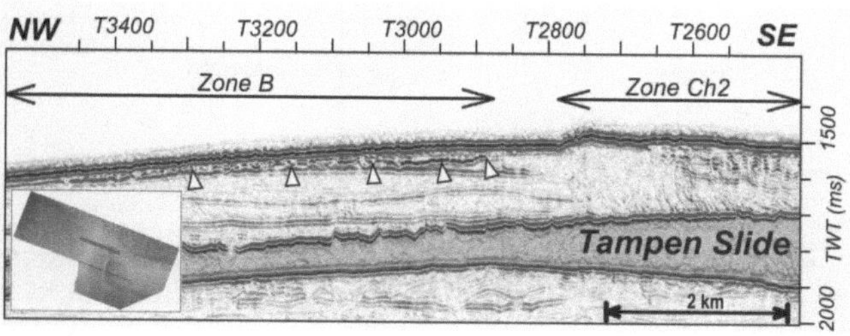

Figure 6. Seismic profile across Zone B and Zone Ch[1], showing the chaotic layer related to the blocky deposit identified at the seabed. Arrows point the base of the deposit. It is marked on the map the presented profile (red line) and profile on Fig. 4 (blue line).

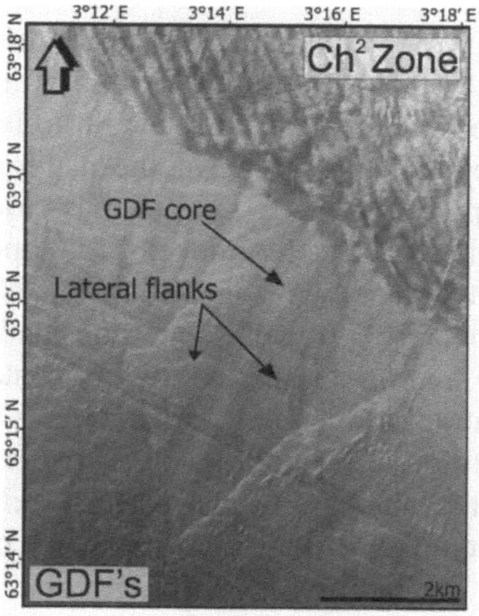

Figure 7. Detail from the sea-floor image (obtained from the 3D seismic data) showing GDF's deposits, on the area adjacent to the *Ch² zone*. The central GDF deposit, 2 km width, with strong chute-flank separation displays marked flow lines concentrated along the core suggesting a laminar movement within the GDF core. This GDF deposit is partially covered by more recent GDF deposits.

Zone U

In the areas of the North Sea Fan undisturbed by the Storegga Slide Complex (Zone U), it is possible to identify several glacial debris flow (GDF) deposits, from the last and uppermost sequence of glacigenic debris flows deposited during the Late Weichselian maximum (King *et al.*, 1998).

These deposits indicate long run-out transport towards the north, showing from a very well developed central core with strong chute-flank separation suggesting a laminar movement within the GDF core (Fig. 7).

4. Discussion

The higher lateral resolution and three-dimensional capacities provided by the 3D seismic data have allowed a greater understanding of the geometries (internal as well as external) and relative chronology of the main sedimentary features. In this section some of its aspects are discussed and a summary model of seabed development that incorporates these main features is presented in Conclusions.

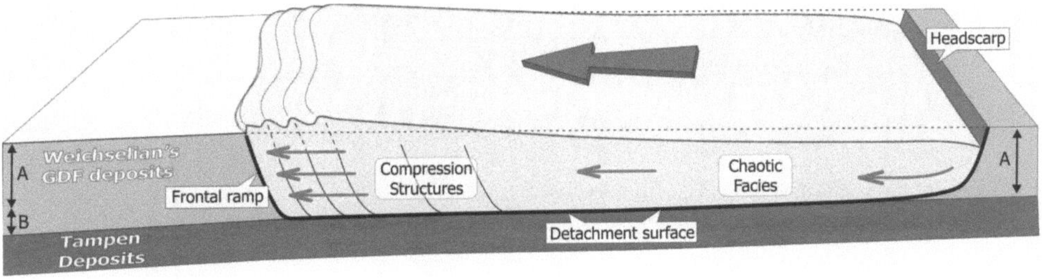

Figure 8. Schematic representation of the internal structure of zone Ch. The mass translation stops against frontral ramp without abandoning the detachment surface that lies on top of Tampen Slide deposits; development of compression structures at the toe region and gradual lost of internal structure upslope. Note: The thickness of Weichselian GDF deposits increases significantly downslope. No real scale.

Low angle thrust faults that originated at the flow base and extend through to the top of the flow deposit are common features near the termini of mass transport deposits (Frey-Martinez *et al.*, 2006). These thrust faults are thought to be oriented transversely to the flow direction; this implies that the source of the material is ESE of the study area. Although zone Ch is interpreted as the termini of mass transported deposit, it does not exhibit the conventional picture of a fully developed slope failure; this failure did not break through or overthrusted the down-slope sediments. Nevertheless it presents all the main characteristics of frontally confined submarine landslides (Frey-Martinez *et al.*, 2006): compressional toe region buttressed by a frontal ramp, small bathymetric expression compared to the total thickness and implying a relative modest downslope movement (Fig. 8). In addition, the compression zone Ch presents a gradual transition from a fragile environment, at the base of mass deposit (characterized by faulting), to a ductile environment, on the upper of the mass deposits (characterized by folding). This gradation must reflect the increase in shear strength and reduction in pore water with depth due to consolidation.

Downslope translation stops when the stress developed by the mass movement becomes lower than the strength of the foreland. That can happen by lost of mass potential energy and/or by the increase of the foreland strength. Considering that the thickness on the sediments above the detachment plane (the top of the Tampen Slide) increases into the NFS (Fig. 8) is reasonable to believe that the propagation of the mass movement downslope was prevented by the increase of foreland strength resultant from increase of thickness.

The zone Ch previously interpreted as result of lateral compression formed during the north-dipping mass-flow classified as Lobe 5 (Haflidason *et al.*, 2004), is now consider being associated to main failure on the Ormen Lange area (Bryn *et al.*, 2005). In both interpretations, the zone Ch is referred as being part a major compression zone that incorporates both Ch and Cl zones. However the Ch zone must had occurred in an early phase of sliding than Cl, as the Cl compression zone is directly related with the main Storegga failure and Ch compression zone is cut by its failure escarpment (Fig. 9 and 10B).

Lobe Ch^1 presents a retrogressive failure resulting of the removal of toe support and increased shear strain during the main failure event. During this small-scale failure, the central part of Ch^1 collapsed and the sediments that had been previously deformed by compression where remould filling part of zone S (Fig. 10C).

Figure 9. 3D perspective view of the southern flank of the Storegga Slide (Norsk Hydro bathymetry) looking eastward, with suggested chronology for the major events leading to the formation of the present-day morphology.

These observations corroborate that the compression zone Ch predates the main failure. Nevertheless, it is uncertain how early this compression took place. Whether this compression zone reflects an early phase of the main movement or a significantly earlier phase? If they were part of the main event, what is the implication for the retrogressive model proposed for Storegga? Further observations, outside of the main study area, are required to constrain the timing of Ch formation.

Other aspect of the evolution of the present-day seabed that is not totally understood and require further study is the timing and nature of Zone B. This blocky deposit is thought to be associated with sediment overbanking of the Storegga main failure, followed by localized slope instability in front of the compression zone.

5. Summary and Conclusions

3D seismic interpretation followed by geomorphologic analyses on the Southern Holocene Storegga Flank has shown the existence of five distinctive morphological zones in the study area. These zones have been given the informal names of: Zone S - where the slope sediments were remoulded; Zones Cl and Ch (1 and 2) - where the seabed has marked ridges resulting from compression of the sediments; Zone B - dominated by blocky deposits and Zone U - where the NSF sediments are undisturbed. The most prominent feature in this area is the compression zone Ch. This compression zone presents features described as characteristic of *frontally confined submarine landslides*. The base of the deformation lies on top of the Tampen Slide deposits. Although the exact timing of the formation of this compression zone and the other features described in this paper was not determined, a relative chronology has been established and can be summarized in the following way:

1) Tampen slide at 150 ka BP (0.15 Ma)
2) Deposition of Weichselian debris flow sediment from 24 to 13 ka BP; NSF build-up
3) Failure remote from study area to the east; some NSF sediments pushed by impact into compressive structures, formation of compression zone Ch (Fig. 10A)
4) Major failure continues downslope, originating Zone S that truncates Zone Ch; formation of compression zone Cl by response to the impact (Fig. 10B)
5) Due to lack of lateral support the northern lobe of the compression zone Ch, sub-zone Ch^2, collapsed into the major Storegga erosion area, zone S (Fig. 10C)

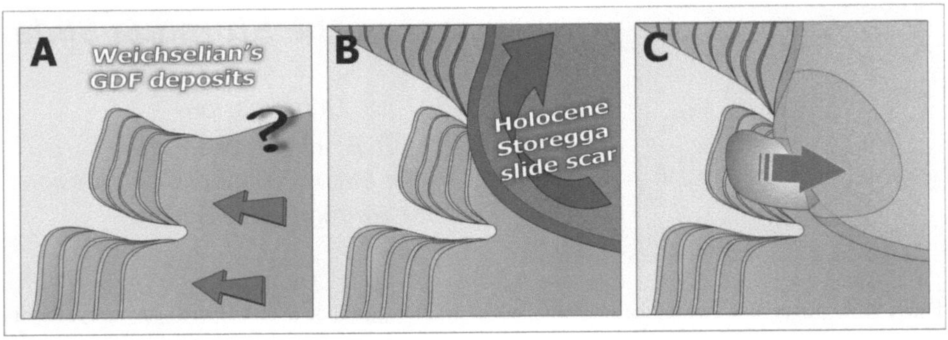

Figure 10. Sketch representation of the three main events shaping the seabed in the proximities of Zone Ch^1. (A) Formation of compression zone Ch. (B) Main slide, cut of zone Ch and formation of zones S and Cl (C) Collapse of the sub-zone Ch^2.

6. Acknowledgements

The authors are grateful to Statoil and Norsk Hydro for permission conceded to use and publish the data sets included in this paper. Special thanks go to Simon Ayers for enlightening discussions. The paper also benefited from the valuable comments from two reviewers, Dr. David Long and Dr. Jan Sverre Laberg. The Portuguese Foundation for Science and Technology (FCT) sponsored the contribution of the first author.

7. References

Bryn P, Berg K, Forsberg CF, Solheim A, Kvalstad TJ (2005) *Explaining the Storegga Slide.* Marine and Petroleum Geology 22: 11-19

Evans D, King EL, Kenyon NH, Brett C, Wallis D (1996) *Evidence for long-term instability in the Storegga Slide region off western Norway.* Marine Geology 130: 281-292

Evans D, Harrison Z, Shannon PM, Laberg JS, Nielsen T, Ayers S, Holmes R, Hoult RJ, Lindberg B, Haflidason H (2005) *Palaeoslides and other mass failures of Pliocene to Pleistocene age along the Atlantic continental margin of NW Europe.* Marine and Petroleum Geology 22: 1131

Frey-Martinez J, Cartwright J, James D (2006) *Frontally confined versus frontally emergent submarine landslides: A 3D seismic characterisation.* Marine and Petroleum Geology 23: 585-604

Haflidason H, Lien R, Sejrup HP, Forsberg CF, Bryn P (2005) *The dating and morphometry of the Storegga Slide.* Marine and Petroleum Geology 22: 123-136

Haflidason H, Sejrup HP, Nygard A, Mienert J, Bryn P, Lien R, Forsberg CF, Berg K, Masson D (2004) *The Storegga Slide: architecture, geometry and slide development.* Marine Geology 213: 201-234

King EL, Haflidason H, Sejrup HP, Lovlie R (1998) *Glacigenic debris flows on the North Sea Trough Mouth Fan during ice stream maxima.* Marine Geology 152: 217

King EL, Sejrup HP, Haflidason H, Elverhoi A, Aarseth I (1996) *Quaternary seismic stratigraphy of the North Sea Fan: glacially-fed gravity flow aprons, hemipelagic sediments, and large submarine slides.* Marine Geology 130: 293-315

Nygård A, Sejrup HP, Haflidason H, Bryn P (2005) *The glacial North Sea Fan, southern Norwegian Margin: architecture and evolution from the upper continental slope to the deep-sea basin.* Marine and Petroleum Geology 22: 71-84

Norwegian Deepwater programme (2004b) *Geomorphological study of the mid-Norwegian continental margin.* Seabed project phase II. Report No. SP2-02-FS-01R-00000-04 compiled by NGU/Fugro Survey Limited, 63pp

Nygård A, Sejrup HP, Haflidason H & Bryn P (2005) *The glacial North Sea Fan, southern Norwegian Margin: architecture and evolution from the upper continental slope to the deep-sea basin.* Marine and Petroleum Geology 22: 71-84

SUBMARINE MASS MOVEMENTS ON AN ACTIVE FAULT SYSTEM IN THE CENTRAL GULF OF CORINTH

M. CHARALAMPAKIS
Laboratory of Marine Geology & Physical Oceanography, Geology Department, University of Patras, Patras 26500, Greece (e-mail: cmarinos@upatras.gr)

A. STEFATOS
Rocksource ASA, Olav Kyrres gate 22, N-5808 Bergen, Norway

T. HASIOTIS
Department of Marine Sciences, University of the Aegean, University Hill, 81100 Mytilene, Lesvos, Greece

G. FERENTINOS
Laboratory of Marine Geology & Physical Oceanography, Geology Department, University of Patras, Patras 26500, Greece

Abstract

A very high-resolution shallow-seismic survey along the central part of the fault-bounded Corinth Gulf southern margin (offshore Xylocastro town) revealed that three morphological zones characterize the area: the shelf, the slope and the basin. Three E-W trending, right stepping basin bounding faults define the basin-slope contact, producing a step like configuration along the base of slope. Steep scarps, caused by mass failures, sculpt the fault plane surfaces, which act as part of the slope. The shelf and the slope are dissected by submarine canyons and numerous minor channels. The largest canyons are located immediately off the river mouths, run perpendicular to the slope and are linear. Seismic profiles across the canyons suggest that some of them are fault controlled. The head and the walls of the canyons are affected by mass failures. It is estimated that about 1.1 km^3 of mass failed sediments have been removed from the canyons and transported downslope to the basin floor.

Keywords: submarine failures, fault-bounded margin, escarpment, canyon, Xylocastro, Corinth Gulf

1. Introduction

Tectono-sedimentary evolution of active extensional basins is important for both scientific (preservation of fossil records of past changes in climate; growth, activity, decay and death of normal faults; record of extension; etc) and economic (huge economic reserves of hydrocarbon, water and minerals) reasons (Gawthorpe & Leeder, 2000). The purpose of this paper is to show the extensive presence of gravitative mass movements that are taking place on a fault controlled slope.

The Corinth Gulf is an active extensional basin that occupies the northern most part of the Plio-Quaternary Corinth rift (Stefatos et al., 2002), located in Central Greece (Figure 1). It is approximately 115 km long and 15 to 30 km wide, while reaches a depth of 900 m. The Corinth Gulf is bounded by E-W striking faults, situated both on- and offshore (Brooks & Ferentinos, 1984; Armijo et al., 1996; Sakellariou et al., 2001; Stefatos et al., 2002; Moretti et al., 2003; McNeill et al., 2005; Lykousis et al., 2007) (Figure 1).

V. Lykousis, D. Sakellariou and J. Locat (eds.), Submarine Mass Movements and Their Consequences, 67–75.

According to geological, geodetic and geophysical studies, extension takes place in a N-S to NNE-SSW direction with a rate of up to 15mm yr^{-1} across the gulf (Tselentis & Makropoulos, 1986; Billiris et al., 1991; Clarke et al., 1997; Davies et al., 1997), while the average rate of uplift of the southern bounding margin for the last 300,000 years is about 0.3mm yr^{-1} (Collier et al., 1992; Armijo et al., 1996; Dia et al., 1997).

2. Methodology

The area under investigation is part of the southern margin of the central Corinth Gulf, offshore Xylocastro. A dense grid of high-resolution, shallow-seismic, reflection profiles, combined with previously collected seismic and sonar data were used for the study of the seafloor morphology, the structural setting and the gravitational mass movements that affect the fault-bounded basin margin of the central part of the Gulf (Figure 1). More than 500 km of Sparker (1.1 kJ) and 3.5 kHz sub-bottom reflection profiles were acquired during a single cruise in 2004. The data were recorded in a Triton Elics digital acquisition system. Filters and gain controls were applied during the processing to enhance the signal. The average penetration achieved was varying from 30 to 120 m for the Sparker and between 7.5 and 40 m for the 3.5 KHz sub-bottom profiler. The lowest penetration achieved over the slope due to the steep gradient. The survey vessel was navigated by conventional GPS, with an accuracy of approximately 20 m. Air-gun seismic reflection and side-scan-sonar records, collected in previous surveys, were used to complement our data, in order to comprehend the structural setting of the basin margin.

Identification of the fault traces was very difficult due to the very steep morphology of the slope. Despite the reduced speed of the vessel and the preferred direction of travelling upslope during the seismic profiling, it was very difficult to penetrate the slope for more than a few meters. For this reason fault mapping was carried out taking into account the contact between the slope and the sedimentary cover in the basin.

The results from the processing of the data were analysed and presented within a Geographic Information System (GIS). A geographical database was developed, integrating various offshore morphological data from different sources. A series of thematic maps were compiled illustrating the spatial distribution of the major morphological features of the seafloor, such as canyons, faults and scarps (Figure 1b, c).

3. Presentation of data – Results

A detailed digital elevation model (DEM) for the study area is compiled using bathymetric data taken from 3.5 KHz sub-bottom profiler records, complemented with multi-beam swath bathymetry data collected by Brian Taylor of the University of Hawaii (Fig 1b). The study of the bathymetric map shows that the basin's margin is characterized by three morphological zones: the shelf, the slope and the basin floor. The shelf extends up to a water depth of 100 m and is approximately 1500m wide, locally narrowing to less than 20m. The slope extends from a water depth of 100 to 800 m, with an average NNE dip. The slope gradient ranges between 17° and 22° in the upper slope, whilst in the lower slope gradient is much steeper, exceeding 30°. The basin floor is flat; reaching a depth of 830 m. Cone-shaped fans have been formed at the base of the slope (Ferentinos et al., 1988).

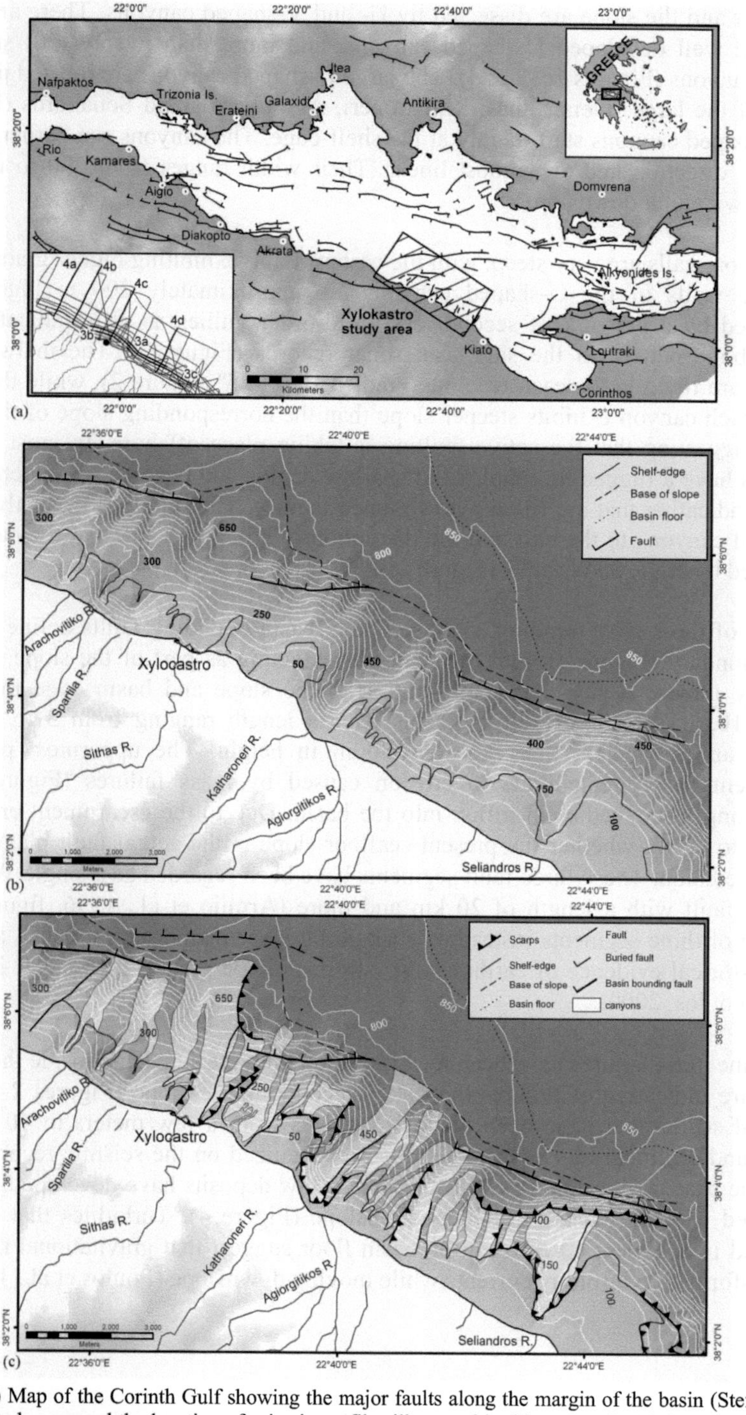

Figure 1. (a) Map of the Corinth Gulf showing the major faults along the margin of the basin (Stefatos et al., 2002), the study area and the location of seismic profiles illustrated in this paper. (b) Digital elevation model (DEM) showing the morphological (shelf, slope, basin) and the major structural features. (c) Bathymetric map showing the spatial distribution of the major morphological and structural features described in the text.

The shelf and the slope are dissected by U- and V-shaped canyons. There are four (4) large and well developed U-shaped canyons, and more than twenty (20) smaller V-shaped canyons (Figures 1c, 2 & 3). The large U-shaped canyons are located just off the mouth of the large rivers Sithas, Katharoneri, Agiorgitikos and Seliandros (Figure 3). The V-shaped canyons start mainly at the shelf-edge. The canyons run perpendicular to the slope direction and are almost linear. Their width ranges from 200 to more than 1400 m, widening downslope.

The canyon walls are very steep, with the eastern flank exhibiting higher gradient, up to 24°. The walls of the U-shaped canyons are approximately 200 m high and are intersected by a network of second- and third-order gullies, as it is indicated by the herring bone pattern in the side scan sonar records (Figure 3). The thalweg of the canyons are tilted to the east, reaching gradients up to 5° (Figure 2), while the thalweg axis of each canyon exhibits steeper slope than the corresponding slope of the onshore rivers, suggesting that the active faulting is taking place offshore. In most cases, the thalwegs have a rugged morphology at shallow depths and become smoother at greater depths, indicating that are filled with slumped masses. The tilting of the thalweg of the U-shaped canyons to the east and the disruption of strata indicate that the canyons are controlled by NNE-SSW faults (Figure 2 & 3)

A series of three E-W trending, right stepping, basin bounding faults define the basin-slope boundary (Figure 4) with the fault planes acting as part of the slope. This fault geometry impose a step-like configuration to the slope and basin edge morphology (Figure 1b,c). The three fault segments have a length ranging from 3 to 5 km and produce an escarpment that exceeds 580 m in height. The uppermost part of the escarpment has retreated due to erosion caused by mass failures (Figure 4), thus contributing to the sediment influx into the basin. Due to the escarpment erosion it is difficult to verify whether the present seafloor slope is the actual fault plane or is the result of erosion. These three fault segments have been regarded as a single WNW-ESE trending fault with a length of 20 km and more (Armijo et al., 1996, figure 2). The presence of three segments instead of a single fault seems more reasonable, since there is no historical evidence of earthquakes with magnitude greater than 6.5 R in the area (Papadopoulos, 2000).

Submarine mass failures have been recognized on both the slope and inside the canyons. Scarps are indicative of failures that take place along the slope (Figures 2 & 4). The height of scarps that have been identified ranges from a few meters to 40 m. Slided masses and sediment blocks have also been recognized on the seismic records (Figure 2). At the base of the slope, overlapping mass flow deposits have developed, which are associated with the unstable sediments upslope (Figure 4). Turbidites that have been identified in sediment cores from the basin floor suggest that gravitational mass flows are transformed to turbidity currents while moving downslope (Poulos et al., 1996).

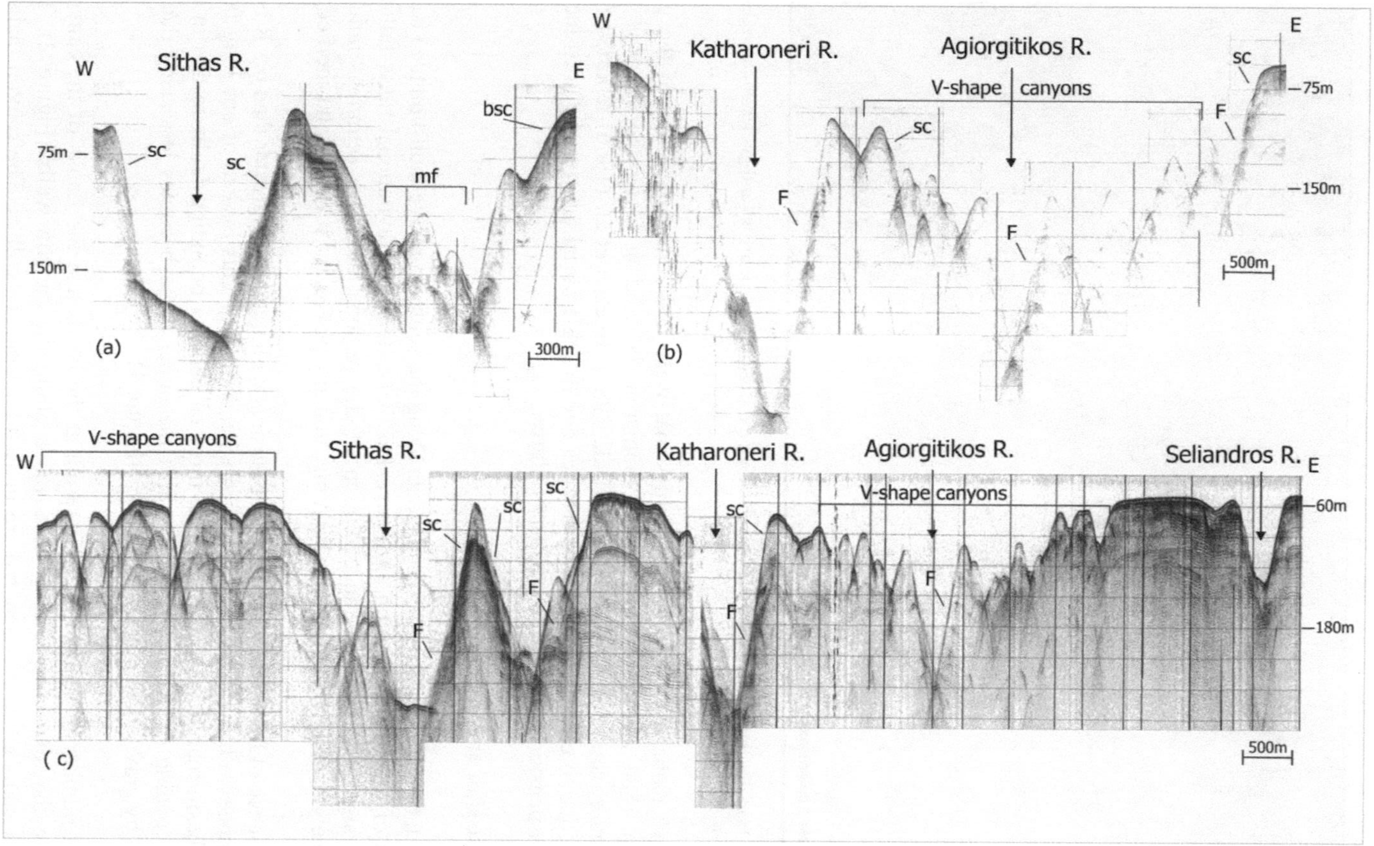

Figure 2. 3.5 kHz (a, b) and Sparker (c) sub-bottom profiles across the slope showing the U- and V-shaped canyons. sc: scarp, bsc: buried scarp, ms: mass flow, F: fault (see figure 1a for location).

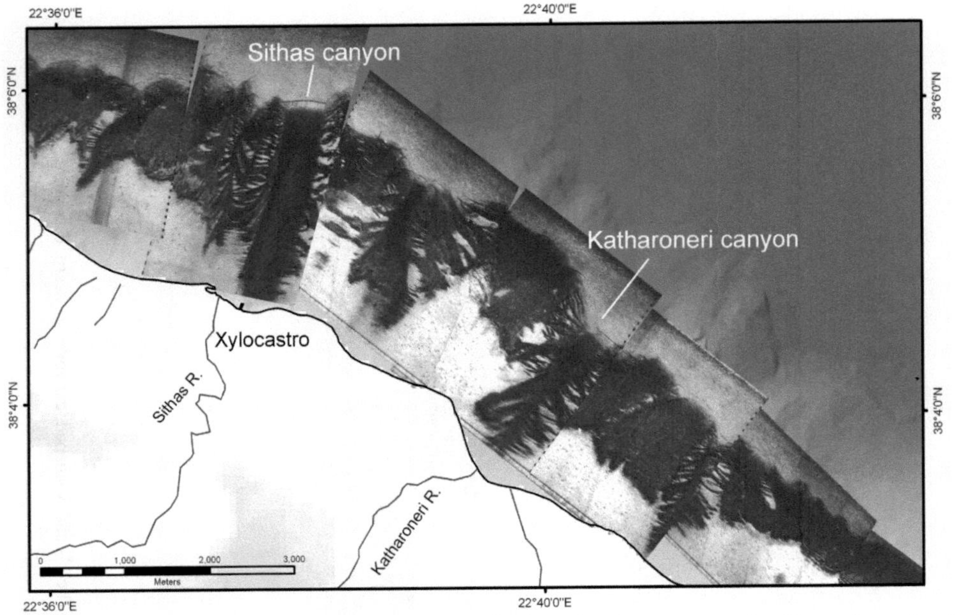

Figure 3. Map illustrating side scan sonar images along the slope showing canyons and gullies on the walls of the canyons.

Extensive damage to 3 submarine cables that were laid across the slope of the gulf between 1884 and 1957 proves that submarine landsliding along the slope is the dominant erosional factor (Ferentinos et al., 1988). Heezen et al. (1966) reported five failures due to cable breakage, two failures due to cable burial and two failures due to cable suspension, all attributed to mass failures.

4. Discussion and conclusions

Interpretation and analysis of seismic data along the southern active fault-bounded basin margin of the central Corinth Gulf revealed that extensive gravitational mass failures play a significant role in the evolution of the region. Extensive sediment failures effectively degrade the upper slope forcing, in many places, the retreat of the shelf edge.

The absence of historical earthquakes in the area with magnitude greater than 6.5 R is in accordance with the existence of three smaller fault segments instead of one long fault at the basin margin. Activation of the three en echelon bounding faults results in the oversteepening of the lower slope, whilst the upper slope becomes less steep due to erosion by gravitational mass movements (Figure 4).

A great number of scarps are also present at the heads and the flanks of the canyons showing that extensive sliding takes place along the canyon walls (Figure 2). The canyons walls are very steep and are incised by tributary gullies. The canyon thalweg receives material both from the gullies and from the flank failures. Buried scarps have also been observed, covered by recent sediments of at least 6-8 m (Figure 2).

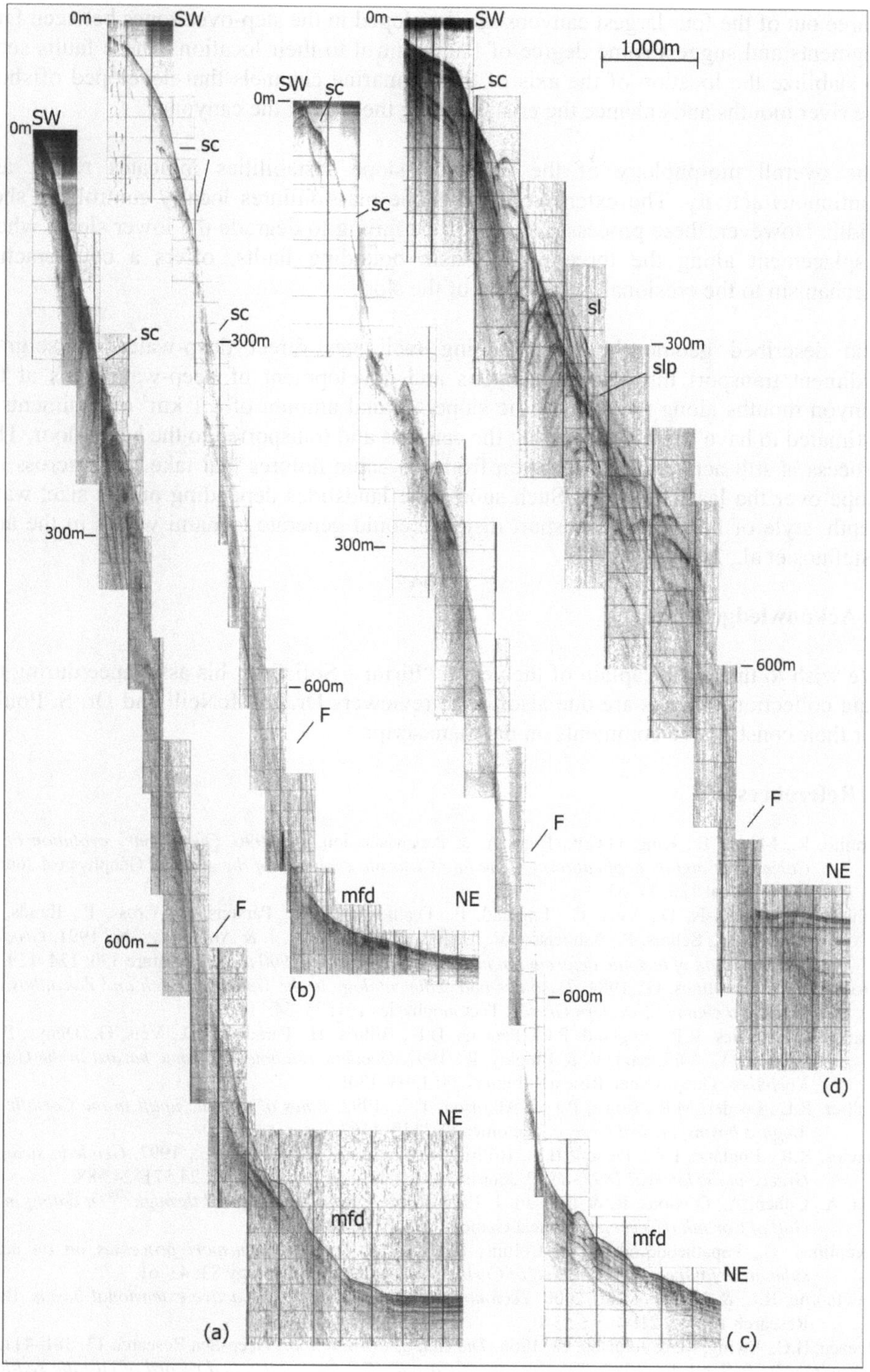

Figure 4. 3.5 kHz (a, b, c) and Sparker (d) sub-bottom profiles along the slope showing submarine failures and the location of the basin bounded faults. sc: scarp, sl: slide, slp: slide plane, mfd: mass flow deposits, F: fault (see figure 1a for location).

Three out of the four largest canyons are developed in the step-over zones between fault segments and suggest some degree of fault control to their location. These faults seem to stabilize the location of the axis of the submarine channels that developed offshore the river mouths and enhance the erosion along the axis of the canyons.

The overall morphology of the observed slope instabilities indicates recent and continuous activity. The extensive upper slope mass failures locally control the shelf width. However, these processes appear to be failing to degrade the lower slope, where displacement along the three active basin bounding faults, offers a counteractive mechanism to the erosional degradation of the slope.

The described geomorphological setting facilitates direct deep-water coarse-grain sediment transport through the canyons and development of deep-water fans at the canyon mouths along the base of the slope. A total amount of 1.1 km^3 of sediments is estimated to have been eroded along the canyons and transported to the basin floor. This process is still active, as can be seen from the cable failures that take place across the slope over the last 100 years. Such submarine landslides depending on the size, water depth, style of failure and transport distance could generate tsunami waves in the area (Stefatos et al., 2006).

5. Acknowledgements

We wish to thank the captain of the vessel "Eirini – Sofia" for his assistance during the data collection. Thanks are due also to the reviewers Dr. L. McNeill and Dr. S. Poulos for their constructive comments on the manuscript.

6. References

Armijo, R., Meyer, B., King, G.C.P., Rigo, A. & Papanastassiou, D., 1996. *Quarternary evolution of the Corinth rift and its implications for the late Cenozoic evolution of the Aegean*. Geophysical Journal International 126: 11–53.

Billiris, H., Paradissis, D., Veis, G., England, P., Featherstone, W., Parsons, B., Cross, P., Rands, P., Rayson, M., Sellers, P., Ashkenazi, V., Daavison, M., Jackson, J. & Ambraseys, N., 1991. *Geodetic determination of tectonic deformation in Central Greece from1900 to 1988*. Nature 350: 124–129.

Brooks, N. & Ferentinos, G., 1984. *Tectonics and sedimentology in the Gulf of Corinth and Zakynthos and Kefallinia channels, western Greece*. Tectonophysics 101: 25–54.

Clarke, P.J., Davies, R.R., England, P.C., Parsons, B.E., Billiris, H., Paradissis, D., Veis, G., Denys, P.H., Cross, P.A., Ashkenazi, V. & Bingley, R., 1997. *Geodetic estimate of seismic hazard in the Gulf of Korinthos*. Geophysical Research Letters 24: 1303–1306.

Collier, R.E., Leeder, M.R., Rowe, P.J. & Atkinson, T.C., 1992. *Rates of tectonic uplift in the Corinth and Megara basins, central Greece*. Tectonics 11:1159–1167.

Davies, R.R., England, P.C., Parson, B.E., Billiris, H., Paradissis, D. & Veis, G., 1997. *Geodetic strain of Greece in the interval 1892– 1992*. Journal of Geophysical Research 102: 24,571–24,588.

Dia, A., Cohen, A., O'Nions, R. & Jackson, J. 1997. *Rates of uplift investigated through* ^{230}*Th dating in the Gulf of Corinth (Greece)*. Chemical Geology 138:171–184.

Ferentinos, G., Papatheodorou, G. & Collins, M.B., 1988. *Sediment transport processes on an active submarine fault escarpment: Gulf of Corinth, Greece*. Marine Geology 83: 43–61.

Gawthorpe, R.L. & Leeder, M.R., 2000. *Tectono-sedimentary evolution of active extensional basins*. Basin Research 12: 195-218.

Heezen, B.C., Ewing, M. & Johnson, G., 1966. *The Gulf of Corinth floor*. Deep-Sea Research 13: 381–411.

Keraudren, B. & Sorel, D., 1987. *The terraces of Corinth (Greece)—a detailed record of eustatic sea-level variations during the last 500,000 years*. Marine Geology 77: 99–107.

Lykousis, V. et al. *Late Quaternary basin evolution of the Gulf of Corinth: Sequence stratigraphy, sedimentation, fault– slip and subsidence rates.* Tectonophysics (2007), doi:10.1016/j.tecto.2006.11.007

McNeill, L., Cotterill, C., Stefatos, A., Henstock, T., Bull, J., Collier, R., Papatheoderou, G., Ferentinos G., & Hicks, S., 2005. *Active faulting within the offshore western Gulf of Corinth, Greece: implications for models of continental rift deformation.* Geology 33/4: 241–244.

Moretti, I., Sakellariou, D., Lykousis, V. & Micarelli, L., 2003. *The Gulf of Corinth: an active half graben?* Journal of Geodynamics 36: 323–340.

Papadopoulos, G.A., 2000. *A new tsunami catalogue of the Corinth Rift: 373 B.C.–A.D. 2000.* In: Papadopoulos, G.A. (Ed.), Historical Earthquakes and Tsunamis in the Corinth Rift, Central Greece. Inst. Geodynamics, Natural Observatory, Athens, pp. 121–126.

Poulos, S., Collins, M., Pattiaratchi, C., Cramp, A., Gull, W., Tsimplis, M. & Papatheodorou, G., 1996. *Oceanography and sedimentation in the semi-enclosed, deep-water Gulf of Corinth (Greece).* Marine Geology 134: 213–235.

Sakellariou, D., Lykousis, V. & Papanikolaou, D., 2001. *Active faulting in the Gulf of Corinth, Greece.* 36th CIESM Congress Proceedings 36: 43.

Stefatos, A., Papatheodorou, G., Ferentinos, G., Leeder, M. & Collier, R., 2002. *Seismic reflection imaging of active offshore faults in the gulf of Corinth, their seismotectonic significance.* Basin Research 14/4: 487–502.

Stefatos, A., Charalambakis, M., Papatheodorou, G. & Ferentinos, G., 2006. *Tsunamigenic sources in an active European half-graben (Gulf of Corinth, Central Greece).* Marine Geology, 232: 35–47.

Tselentis, G.A. & Makropoulos, K., 1986. *Rates of crustal deformation in the Gulf of Corinth (Central Greece) as determined from seismicity.* Tectonophysics 124: 55–66.

ANALYSIS OF MULTIBEAM SEAFLOOR IMAGERY OF THE LAURENTIAN FAN AND THE 1929 GRAND BANKS LANDSLIDE AREA

DAVID C. MOSHER AND DAVID J.W. PIPER
Geological Survey of Canada, Natural Resources Canada, Bedford Institute of Oceanography, 1 Challenger Dr., Dartmouth, NS, B2Y 4A2, dmosher@nrcan.gc.ca

Abstract

The 1929 Grand Banks earthquake, landslide and tsunami were pivotal in geologic history as they led to the first unequivocal recognition of a landslide-triggered tsunami and turbidity current. The event is well constrained in terms of trigger, timing, sequence of events and impact. The landslide site was surveyed in September of 2006 with a 12 kHz multibeam echosounder. Regionally, these bathymetric data show canyons, valleys and gullies, somewhat typical of the continental slope in the region. No major headscarp related to the event is recognized (cf. the Storegga Slide). Most significant are a series of shallow gullies with small headwalls about mid-slope. Upslope from these is a series of shallow escarpments that probably represent upslope retrogression of the failure. The landslide appears to have been relatively shallow (top 5-100 m) and laterally extensive. There is no evidence of a single massive submarine landslide with major headscarp and debris lobe. The landslide presumably evolved rapidly into turbidity currents that flowed along existing canyon and valley corridors. In the case of the 1929 Grand Banks event, a damaging tsunami was generated following a landslide for which the bathymetric signature is not clearly identifiable from most of the regional sea floor of the Canadian East Coast margin. The tsunami was generated either by widely distributed, shallow sediment failure, or by the ensuing turbidity current. In either case, remnant morphology is difficult to distinguish. This fact suggests that assessment of tsunami hazard based on recognizable morphologic evidence alone may underestimate the landslide and tsunami risk.

Keywords: Submarine landslide, mass-failure, tsunami, multibeam, geohazard, seafloor geomorphology, submarine canyon, submarine valley, submarine fan

1. Introduction

On November 18, 1929, a M7.2 earthquake under the Laurentian Channel was felt widely around Maritime Canada and on the southern coast of the Island of Newfoundland (Gregory, 1929, 1931) (Fig. 1). Nearly simultaneously, 12 undersea trans-Atlantic communication cables were severed (Doxsee, 1948). Within two hours of the earthquake, a devastating tsunami struck the Burin Peninsula of the south coast of Newfoundland, claiming 28 lives and wreaking havoc on coastal communities in the region. That there was some sort of connection between the earthquake, cable breakages and tsunami was recognized (Keith, 1930; Hodgson and Doxsee, 1930). It was not until understanding of undersea geologic processes had vastly improved; however, before it was realized that an earthquake-generated submarine landslide precipitated these events. In a fundamental suite of investigative work, Heezen and others (Heezen and Ewing,1952; Heezen et al., 1954; Heezen and Drake,1964) pieced together a story that a slump-generated turbidity current with flow velocities in excess of 69 km/h caused the

V. Lykousis, D. Sakellariou and J. Locat (eds.), Submarine Mass Movements and Their Consequences, 77–88.

sequential severance of the undersea cables. This story and this region have been revisited numerous times over the decades as survey technologies evolved to understand more fully the processes of submarine sediment mass failure, turbidity current formation and propagation, and tsunami generation.

In this paper, new results from multibeam bathymetric data, acquired in September 2006, are presented. On first impression, these data show a seafloor that appears much like the remainder of the continental slope in this region, with numerous canyons, valleys and seafloor escarpments, but with no evidence of a single major headwall scarp and massive slump region. In combination with earlier interpretations of the seafloor geology of the region, it is therefore hypothesized that the 1929 Grand Banks landslide was relatively thin (< 80 m) but distributed over a large area, and that the resulting failed sediment quickly entered and was confined by canyons and gullies and evolved rapidly into sustained turbidity currents. This hypothesis has significant consequences with respect to understanding tsunamigenic processes.

Figure 1. Location of the study area (white box) situated within the East Coast of Canada.

1.1 REGIONAL GEOLOGY

The Laurentian Channel is a major Pleistocene ice outlet corridor that extends across the continental shelf from the Gulf of St. Lawrence (Shaw et al., 2006) (Fig. 1). It was fed by a series of tributary troughs draining ice originating in the Appalachian Ice Complex. At the channel's termination is the Laurentian Fan; a major ice margin depo-centre throughout the Pleistocene (Stow, 1981; Skene and Piper, 2003; Piper et al., 2005). It is the largest deep-sea fan on the Atlantic margin of Canada and merges seaward with the Sohm Abyssal Plain. The shelf break lies in about 400 m water depth and the transition to the abyssal plain is in about 5000 m water depth. A Mesozoic transform fault (the Cobequid-Chedabucto fault) is a significant tectonic element that forms the southwestern edge of the Grand Banks and sweeps beneath the Laurentian Channel (Pe-Piper and Piper, 2004). It is probably on this tectonic structure that ongoing low-grade seismicity occurs (Bent, 1995; Adams and Halchuk, 2003).

1.2 METHODS

Bathymetry Data

A multibeam bathymetric data set covering 32,150 km^2 of the upper Laurentian Fan was acquired in September, 2006 with the vessel *Kommandor Jack.* Data were collected by Fugro Jacques GeoSurveys Inc. of St. John's, Newfoundland. The vessel was equipped with a Kongsberg Simrad EM120 multibeam system that operates at a nominal frequency of 12 kHz, with 191 receive beams covering an ideal swath width of 150° (de Moustier, 2001). The sounder's specifications for sounding accuracy are 0.2% of water depth across the swath.

In the environment of the Laurentian Fan, maximum swath widths of about 10 km were achieved before data quality suffered beyond acceptance. Mosher et al. (2006a) showed data density is the greatest restriction on multibeam sonar resolution in deep water. In this study, data density is on the order of 2300 soundings per km^2 in 1000 m of water and 84 soundings per km^2 in 3000 m. These values imply that horizontal resolution is nominally about 40 m at the shallower depth and 400 m in the deep. In addition to this latest survey, to maximize data coverage, existing digital bathymetric data from a variety of sources were integrated before final grid production and seafloor rendering. These data sources include a 12 kHz Seabeam data set from 1986 over a portion of the study region (Hughes Clarke et al. 1990), single beam sounding data digitized from previous Geological Survey of Canada expeditions, bathymetric field sheet data from the Canadian Hydrographic Service, and first arrival picks from 2D and 3D seismic data in the region.

2. Results

The recently acquired multibeam data in the Laurentian Fan region, merged with other bathymetric data of the area (Fig. 2) reveal the following general aspects of this part of the continental margin: (1) the overall slope angle on the Laurentian Fan is two degrees, being steepest (up to six degrees) near the shelf break (Fig. 2C), and 2) the most prominent features are the numerous canyons and valleys with complex upslope tributary systems. On the Laurentian Fan specifically, these erosional systems include the Eastern, Western and Central Valleys (Fig. 2). The largest, Eastern Valley is about 20 km-wide at its upslope limit, but reaches over 30 km width by the 4000 m isobath (Fig. 3). It is locally as much as 500 m deeper than adjacent levees. Western Valley forms at 3400 m water depth at the confluence of four valleys emanating from the western portion of the mouth of Laurentian Channel (Fig. 4). At this point of confluence, its valley floor is 11 km wide, nearly 30 km-wide from levee to levee and 590 m deep from levee crest to valley floor (Fig. 2). Central Valley appears as a narrow (<8 km wide) corridor branching from Eastern Valley at about the 2800 m isobath and merging with Western Valley at about the 3800 m contour (Fig. 2). St. Pierre Valley cuts the western part of St. Pierre Slope and enters Eastern Valley near the 2600 m isobath (Figs. 2 and 3). Farther east, significant canyon systems seaward of Halibut and Haddock Channels (Fig. 5) converge to form the Grand Banks Valley that enters Eastern Valley near the 4200 m isobath. The upper reaches of Grand Banks Valley are 10-15 km wide and 300 m deep.

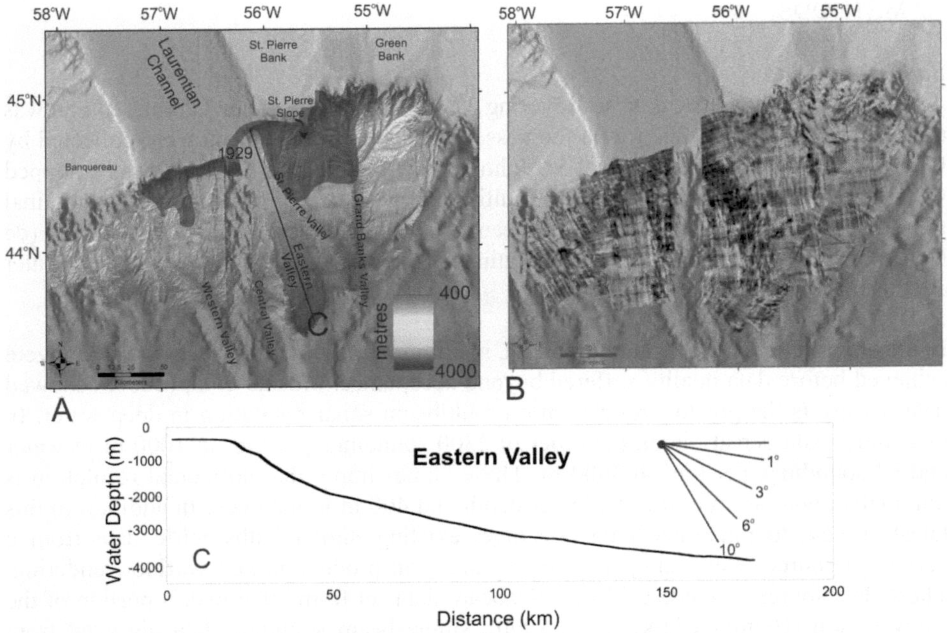

Figure 2. (**A**) Sun-illuminated (360° azimuth) sea floor bathymetric image of the EM120 data acquired in September, 2006 of the Laurentian Fan region, compiled with existing bathymetric data available from the region. Vertical exaggeration is 10x. The shaded ploygon represents the zone where 100% of the seafloor failed in 1929 (Piper et al., 1999). (**B**) EM120 backscatter reflectivity data; darker colours indicating higher reflectivity, and (**C**) Profile down Eastern Valley at a 10:1 vertical exaggeration, showing typical slope angles of 2-4°, with the steepest angles near the shelf break.

The canyon and valley walls demonstrate pinnate ridges, gullies, terraces, and escarpments (Figs. 2-5). Slope angles of the walls of these canyons are typically 10-20° but can reach 40°. The floors of these valleys and tributaries are broadly flat at coarse scale, but on close inspection their features are numerous. Many small channels and talwegs are apparent, particularly in the upper slope portion of Eastern Valley (Fig. 3). Sediment waves and scours (flute marks) are apparent on valley floors, well-known from previous sidescan sonar investigations (Hughes Clarke et al., 1990; Shor et al. 1990; Piper et al., 2007). These features as well as other anomalous pits, scours, buttes, channels and talwegs are notable on the valley floors (Figs 3-5). Debris lobes are also apparent on the valley floors, such as within Central Valley. Acoustic backscatter values from multibeam data show stronger reflectivity within channels and tributaries than on adjacent ridges; particularly notable within the Eastern Valley (Fig. 2b).

In the central study area, off St. Pierre Bank from the shelf break to the 2500 m isobath, is a broad (32 km-wide) relatively flat area known as the St. Pierre Slope (Figs. 2 and 6). The area is bounded east and west by canyons. On the lower slope, four box-shaped valleys (several km wide and 200 m deep) cut the slope. The mid slope (800 – 2000 mbsl) has a terraced morphology with several sinuous escarpments extending across it. These escarpments range from 2 to over 100 m in height (Fig. 6). The shallowest of these escarpments is in 730 m water depth. Similar features are visible elsewhere on flat inter-ridge regions within this data set.

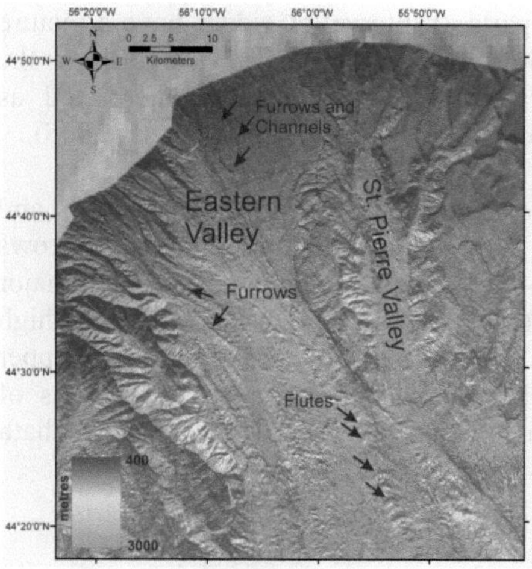

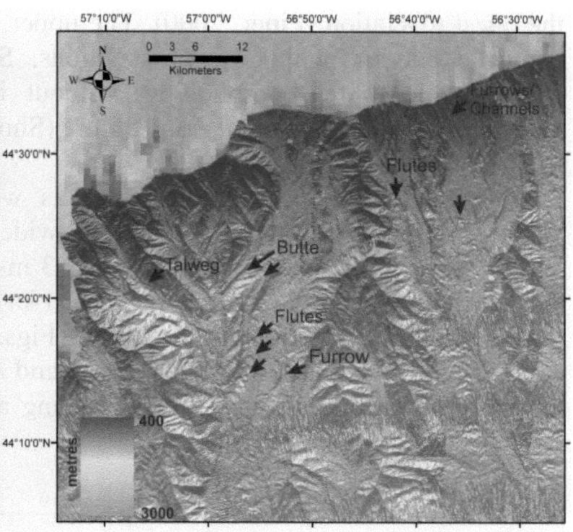

Figure 3 (left). Sun-illuminated (360° azimuth) bathymetry of Eastern Valley of the Laurentian Fan, showing the erosion characteristics of the valley, including flutes, furrows, scours and channels within the valley floor. Note the great width of Eastern Valley, uncharacteristic for most of the canyon and valley systems of the Canadian east coast margin. Vertical exaggeration is 10x.

Figure 4 (right). Sun-illuminated (360° azimuth) bathymetry of the western Laurentian Channel and Fan region, including the tributary system leading into Western Valley. This image demonstrates a number of the characteristic features of the region, including dendritic gully patterns of the valley walls, erosional remnant buttes, channel talwegs, and flute marks and furrows on the valley floors. Vertical exaggeration is 10x.

3. Discussion

As the Laurentian Fan was a major sediment depo-center on the eastern Canadian margin at least since early Tertiary and was the major ice-outlet conduit throughout Pleistocene glaciations (Piper et al., 2005; Shaw et al., 2006; Piper et al., 2007), high sedimentation rates involving highly mixed sediment textures resulted. Depositional processes included sub-glacial on the outer shelf and uppermost slope, channelized turbidity currents with associated levees, unconfined turbidity currents, hyperpycnal flows, hemipelagic deposition from nepheloid plumes, contour currents and ice-rafting (e.g. Stow, 1981; Piper and Normark, 1982; Piper et al., 2005; Piper et al., 2007). A diverse range of sedimentary environments and sedimentary deposits is apparent, as a consequence. This diversity is demonstrated in the seafloor morphology.

The new multibeam imagery where sediment failure from the 1929 Grand Bank's earthquake is widely observed (Figs. 2 and 4) is divided into five distinct slope sectors. From west to east, these are:

1. Seaward of the western part of Laurentian Channel are several broad relatively straight valleys that coalesce near the 3400 m isobath to form Western Valley of Laurentian Fan. Erosional furrows in these valleys extend all the way to the upper limit of the survey near the 600 m isobath (Fig. 4). Consistent valley width downslope, continuation to the uppermost slope, and erosional furrows on the flat valley floor are similar to slope valleys elsewhere, interpreted to have been cut by hyperpycnal flows (Normark et al. in press; Piper et al. 2007; Tripsanas and Piper, pers. comm.). The most recent event in these valleys was probably in the late Wisconsinan during recession of

the latest glaciation (Piper, 2000). The upper parts of intervalley ridges have a mature dendritic pattern, with a few exceptions. Several sites on these ridges show the characteristic muted morphology, without intense gullying, that is interpreted as evidence for recent sediment mass-failure (Shor and Piper, 1989; McCall et al., 2005).

2. Eastern Valley of Laurentian Fan is well-known from previous sidescan and submersible observations. It is unusually wide for a slope valley, has erosional furrows and residual buttes, and is floored by a 1-3 m-thick conglomerate deposited by a major sub-glacial meltwater discharge at 19.5 cal ka (Piper et al., 2007), as shown by the high backscatter values from the valley floor (Figs. 2 and 3). Retrogressive failure of upper slope muds took place in 1929 (Piper and MacDonald, 2002) and large blocks of indurated mudstone are observed overlying a broken cable near the 1200 m isobath (Hughes Clarke et al., 1989).

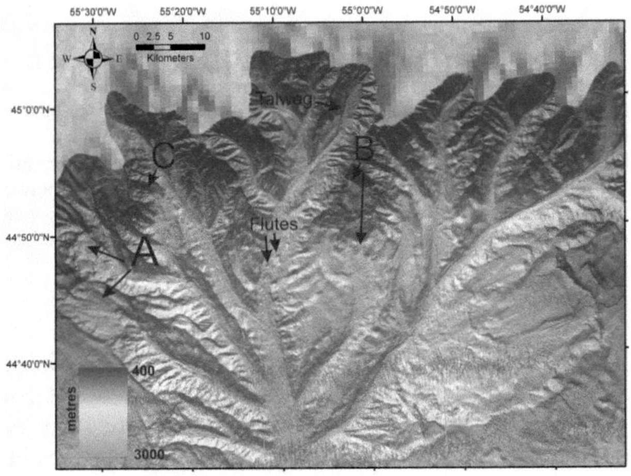

Figure 5. Sun-illuminated (360° azimuth) bathymetry of the tributary system for Grand Banks Valley. The regions without characteristic dendritic gully patterns (labelled A, B and C) on the valley walls are interpreted to be relatively "fresh" surfaces, possibly failed during the 1929 event. Vertical exaggeration is 10x.

3. St. Pierre Valley and its tributary canyons lead from the southwestern part of St. Pierre Bank (Figs. 2 and 6). The canyons/valleys are old features that predate the 1929 failures, but sidescan sonar and seismic profiles show that there was widespread failure on the intercanyon ridges (Piper et al., 1985a,b; Piper et al., 1999; McCall, 2006) which retrogressed upslope to the limit of glacial till near the 500 m isobath (Bonifay and Piper, 1988). Piper et al. (1999) inferred that debris-flows that flowed down steep valley walls went through hydraulic jumps, entraining water and becoming transformed into turbulent turbidity currents.

4. Eastern St. Pierre Slope is a broad flat area of relatively low gradients (3-6°) seaward of the eastern part of St. Pierre Bank (Fig. 6). It broadens seaward with a regional increase in gradient between the 2000 and 2500 m isobaths, related to salt tectonism (Shimeld, 2004). Between 500 and 2000 m water depth, the predominant morphological features are sinuous escarpments presumably formed by retrogressive failure (Piper et al. 1999). The steeper area between the 2000 and 2500 m isobaths is

incised by a series of 100-150 m deep sub-parallel gullies with few tributaries. These resemble valleys interpreted as retrogressive headwall failure on the Scotian Slope (Mosher et al. 2004). Failure in this area is interpreted to have been widespread (Fig. 4). McCall (2006) estimated that the total amount of failed sediment in the area of St Pierre Valley and St Pierre Slope, between the 500 and 2000 m isobaths, was about 93.5 km^3, of which about half was evacuated and half remained in place.

5. A dendritic series of slope canyons is developed seaward of Halibut Channel, Green Bank and Haddock Channel, which lead downslope to the Grand Banks Valley (Fig. 5). Seismic profiles and cores indicate widespread failure in 1929 at the western edge of this drainage system (e.g. Figs. 2 and 4). Elsewhere, cores suggest that there were local failures on some canyon walls and a turbidity current flow in 1929, confirmed by the distribution of cable breaks. Again, areas where the dendritic pattern is muted probably represent areas of failure (e.g. A, B, C in Fig. 5).

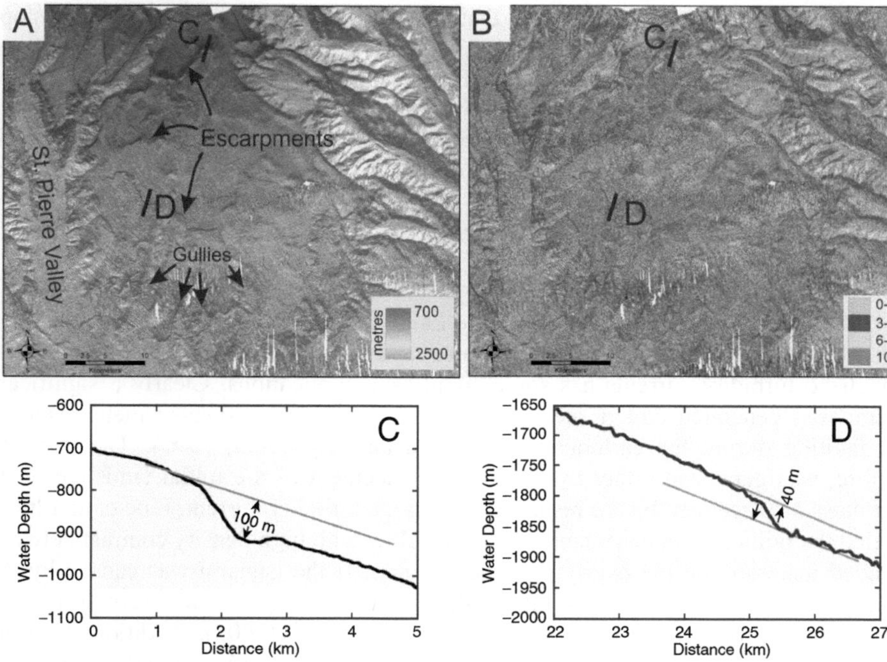

Figure 6. (**A**) Sun-illuminated (360° azimuth) bathymetry of the St. Pierre Slope, with associated slope angle map **(B)** of the same area. These maps show the relatively broad, flat nature of the slope in this region, bounded by canyon systems east, west and south. Average slope angles are between 3° and 4° on the St. Pierre Slope. Several escarpments are noted with slopes up to 10°. Profiles across two of these escarpments are shown in C and D, showing up to 120 m relief on the shallowest one. Vertical exaggeration in **A** is 10x.

Piper et al. (1999) interpreted the area where 100% of the seafloor failed in 1929 (Fig. 2) using a variety of sidescan sonar, subbottom profile, and sediment core data. The results of this study, incorporated into those of previous investigations, clearly demonstrate that the greatest amount of sediment failure occurred on the St. Pierre Slope, in area (4) of the above description (Fig. 6). The bathymetric compilation shows no evidence of a single large sediment mass failure; no single major headwall scarp and subsequent slump and debris lobe, as observed at Storegga, for example (Haflidason et al., 2004). These data show a series of escarpments, but horizontal resolution is insufficient in deep water to map failure deposits below these escarpments (Piper et al., 1985a; Piper et al., 1999; McCall et al., 2005). Based on escarpment heights, an average thickness of failure of 20 m for the entire region is assumed. This translates to a total volume of failed material of 144 km^3 (see Piper and Aksu, 2097).

Masson et al. (1985) and Piper et al. (1985b) also found no evidence for a single major slump, but rather widespread shallow failure on the St. Pierre Slope at the head of Eastern Valley. Sediment mass-failure likely began on areas like the St. Pierre Slope, where thick deposits of unconsolidated possibly gas-charged sediment reside. The escarpments with associated debris lobes are the remaining evidence of these failures. The uppermost prominent escarpment is in 730 m water depth. Piper et al. (1999) concluded that all failure occurred in water depths greater than 650 m. It is quite possible that failure commenced in mid-slope areas, perhaps in the region demarcated by the gullies shown in Figure 6, and that failure retrogressed upslope as downslope support was removed. Failure appears to have occurred on specific bedding planes; perhaps beds of lower shear strength (cf. Haflidason et al., 2002). Much of the failed sediment rapidly fell into the existing network of canyons and valleys to evolve into turbidity currents. High backscatter intensities on valley floors (Fig. 2) supports this interpretation, suggesting gravel and sand on the floor resulting from erosional lag or deposited from strong turbidity currents or hyperpycnal flows. By studying levee regions, Hughes Clarke et al. 1989) estimate the 1929 turbidity current in Eastern Valley was at least 400 m thick.

The fact that the submarine landslide on the Laurentian Fan occurred in relatively deep water, affecting the shallow sediment section over a large area and rapidly evolving into channelized turbidity currents has some significant implications. Clearly a significant tsunami was generated and it seems conclusive that seafloor displacement resulting from faulting during the earthquake itself was unlikely (Bent, 1995). The tsunami, therefore, was generated either by seafloor displacement of the initial landslide or by water mass displacement by the resulting rapid large turbidity current. Fine et al. (2005) modeled the failure as an instantaneous viscous flow with high density contrast between the failed material and the overlying water column. If the tsunami was caused by the turbidity current, refinements to the model would be required to account for the different fluid properties, as acknowledged by Fine et al. (2005). Additionally, time would need to be allotted for generation of the current after initial failure and subsequent retrogression. This lag between the earthquake and tsunami generation may account for the time differential observed by Fine et al. (2005) between their model and recorded tsunami arrival times.

The features recognized throughout this modern bathymetric compilation, such as shallow escarpments and modern debris within canyons, are also recognized throughout the Scotian Slope where multibeam data exist (Mosher et al., 1989; Mosher et al., 1994; Jenner et al., 2006). Although many of these features elsewhere on the Scotian Slope are draped with Holocene sediment, the lateral extent of escarpments is significant and of the same vertical magnitude as those on St. Pierre Slope (Mosher et al., 2004). This fact raises the issue as to whether each of these features represents an event that may have generated tsunamis of similar magnitude to the 1929 event. If thin-skinned failure and turbidity currents can cause tsunamis, then the tsunami threat to the Canadian East Coast margin, and in fact most margins with continental slopes of unconsolidated sediment, is significantly increased. Additionally, these thin failures and turbidity current records are difficult to recognize without high resolution data, and are increasingly difficult to recognize in the subsurface as seismic resolution deteriorates. Recurrence intervals of mass failures with tsunami generation capability on Canada's East Coast, as suggested by Piper et al. (2003), therefore, may be significantly underestimated.

4. Conclusions

Newly acquired 12 kHz multibeam bathymetric sonar data from the Laurentian Fan and St. Pierre Slope region of the east coast of Canada lend significant new detail to existing knowledge of the terrain in this region. Specific to the 1929 Grand Banks earthquake landslide and tsunami, these data show sites on canyon walls that appear to have failed recently, based on geomorphologic criteria, and they show numerous fresh escarpments ranging from 5 to 100 m in height on the St. Pierre Slope. They show no evidence of a single large slump, with headscarp, slump and debris lobe. These data, therefore, support earlier interpretations that the landslide was relatively thin-skinned (5-100 m thick but averages about 20 m) and dispersed over a relatively large area (~7,200 km^2) (Masson et al., 1985; Piper et al., 1988, 1992, 1999). The series of escarpments suggests a retrogressive style of failure. This terrain is similar to the morphology seen everywhere on the Scotian Slope, particularly on broad flat intercanyon regions (Mosher et al., 2004). This fact raises the issue as to whether even thin-skinned mass-transport events including large turbidity currents in deep water are tsunamigenic. If so, then the tsunami-hazard potential of the Canadian East Coast margin and likely most continental margins throughout the globe increases significantly.

5. Acknowledgements

The authors would like to express their appreciation to the officers and crew of the Kommandor Jack, including the field acquisition parties of Fugro Jacques Geosurveys of St. John's, NF and Mr. Kevin Desroches of GSC-Atlantic. This project would not have taken place without the support of Dr. R. Pickrill and Mr. J. Shimeld. Funding for data acquisition for this project was supplied by Canada's Ocean Action Plan and Natural Resources Canada. The work was conducted in part under a collaborative agreement with ConocoPhillips and Murphy Oil. Mr. John Preston of Quester-Tangent Corp. conducted the backscatter processing on a difficult data set. We thank Mr. C. Campbell for his much-needed help with GIS. We also thank the reviewers for critiquing and improving this manuscript, Drs. R. Courtney and B. Bornhold, and Mr. C. Campbell. Natural Resources Canada Earth Sciences Sector Contribution No. 2007085.

6. References

Adams J. and Halchuk S., 2003. *Fourth generation seismic hazard maps of Canada: Values for over 650 Canadian localities intended for the 2005 National Building Code of Canada.* Geological Survey of Canada Open File 4459: 1-155.

Bent, A.L., 1995. *A complex double-couple source mechanism for the Ms 7.2 1929 Grand Banks earthquake.* Bulletin of the Seismological Society of America, Vol. 85, p. 1003-1020.

Bonifay, D. and Piper, D.J.W., 1988. *Probable Late Wisconsinan ice margin on the upper continental slope off St. Pierre Bank, eastern Canada;* Canadian Journal of Earth Sciences, Vol. 25, p. 853-865.

de Moustier, C., 2001. *Field Evaluation of Sounding Accuracy in Deep Water Multibeam Swath Bathymetry.* IEEE Oceans Conference and Exhibition, Honolulu, Hawaii, USA, Vol. 3, p.1761-1765.

Doxsee, W.W., 1948. *The Grand Banks Earthquake of November 18, 1929.* Publications of the Dominion Observatory, Vol. 7.

Fine, I.V., Rabinovich, A.B., Bornhold, B.D., Thomson, R.E., and Kulikov, E.A., 2005. *The Grand Banks landslide-generated tsunami of November 18, 1929: preliminary analysis and numerical modeling.* Marine Geology, Vol. 215, p. 45-57.

Gregory, J.W., 1929, *The earthquake south of Newfoundland and submarine canyons.* Nature, Vol. 124, p. 945.

Gregory, J.W., 1931. *The earthquake off the Newfoundland banks of 18 November 1929.* Geographical Journal, Vol. 77, p. 123-129.

Haflidason, H., Sejrup, H.P., Berstad, I.M., Nygård, A., Richter, T., Bryn, P., Lien, R. and Berg, K., 2002. *Weak layer features on the northern Storegga Slide escarpment,* in J. Mienert and P. Weaver, eds., European margin sediment dynamics: Berlin, Springer, p. 55-63.

Haflidason, H., Sejrup, H.P., Nygård, A., Mienert, J., Bryn, P., Lien, R., Forsberg, C.F., Berg, K. and Masson, D., 2004. *The Storegga Slide: architecture, geometry and slide development.* Marine Geology, Vol. 213, p. 201-234.

Heezen, B.C. and Drake, G.L., 1964. *Grand Banks slump.* American Association of Petroleum Geologist Bulletin, Vol 48, p. 44-108.

Heezen, B.C. and Ewing, M., 1952. *Turbidity currents and submarine slumps and the 1929 Grand Banks earthquake.* American Journal of Science, Vol. 250, p. 849-873.

Heezen, B.C., Ericson, D.B. and Ewing, M., 1954. *Further evidence for a turbidity current following the 1929 Grand Banks earthquake.* Deep Sea Research, Vol. 1, p. 193-202.

Hodgson, E.A. and Doxsee, W.W., 1930. *The Grand Banks earthquake, November 18, 1929.* Eastern Section of the Seismological Society of America, Proceedings of the 1930 meeting, Washington D.C., p. 72-79.

Hughes Clarke, J.E., 1990. *Late stage slope failure in the wake of the 1929 Grand Banks earthquake.* Geo-Marine Letters, Vol. 10, p. 69-79.

Hughes Clarke, J.E., Mayer, L.A., Piper, D.J.W. and Shor, A.N., 1989. *Pisces IV submersible observations in the epicentral region of the 1929 Grand Banks earthquake.* In: Piper, D.J.W. (ed), Submersible observations off the east coast of Canada, Geological Survey of Canada Paper 88-20, p. 57-69.

Hughes Clarke, J.E., Shor, A.N., Piper, D.J.W. and Mayer, L.A., 1990. *Large-scale current-induced erosion and deposition in the path of the 1929 Grand Banks Turbidity Current,* Sedimentology, Vol. 37, p. 613-629.

Jenner, K.A., Piper, D.J.W.P., Campbell, D.C. and Mosher, D.C., 2006. *Lithofacies and origin of late Quaternary mass transport deposits in submarine canyons, central Scotian Slope, Canada.* Sedimentology, Vol. 53., 20 pp.

Keith, A., 1930. *The Grand Banks earthquake.* Eastern Section of the Seismological Society of America, Supplement to the Proceedings of the 1930 Meeting, Washington D.C., p. 1-9.

Masson, D.G., Gardner, J.V., Parson, L.M., and Field, M.E., 1985. *Morphology of Upper Laurentian Fan using GLORIA Long-range side-scan sonar.* American Association of Petroleum Geologists Bulletin, Vol. 69, p. 950-959.

McCall, C., 2006. *Character and distribution of the 1929 landslides.* Unpublished M.Sc (applied), St. Marys University, Halifax, NS.

McCall, C., Morrison, M.L. and Piper, D.J.W., 2005. *Geological data from the St. Pierre Slope around the epicentre of the 1929 Grand Banks earthquake,* Geological Survey of Canada, Open File 4879, 15 pp (+CD-Rom).

Mosher, D.C., Piper, D.J.W., Vilks, G., Aksu, A.E., and Fader, G.B., 1989. *Evidence for Wisconsinan glaciations in the Verrill Canyon area, Scotian Slope.* Quaternary Research, 31: 27-40.

Mosher, D.C., Moran, K. and Hiscott, R.N, 1994. *Late Quaternary sediment, sediment mass-flow processes and slope stability on the Scotian Slope:* Sedimentology, v. 41, p. 1039– 1061.

Mosher, D.C., Piper, D.J.W.P., Campbell, D.C., and Jenner, K.A., 2004. *Near surface geology and sediment failure geohazards of the central Scotian Slope.* American Association of Petroleum Geologists Bulletin, 88, p. 703-723.

Mosher, D.C., Bigg, S. and LaPierre, A. 2006a. *3D seismic versus multibeam sonar seafloor surface renderings for geohazard assessment: Case examples from the central Scotian Slope.* The Leading Edge, V. 25, p. 1484-1494.

Normark, W.R., Piper, D.J.W., Romans, B.W., Covault, J.A., Dartnell, P. and Sliter, R.W., 2007. *Submarine canyon and fan systems of the California Continental Borderland.* GSA Special Paper (in press)

Pe-Piper, G. and Piper, D J W. 2004. *The effects of strike-slip motion along the Cobequid - Chedabucto - southwest Grand Banks fault system on the Cretaceous-Tertiary evolution of Atlantic Canada.* Canadian Journal of Earth Sciences, Vol. 41, p. 799-808.

Piper, D.J.W., 2000. *Pleistocene ice outlets on the central Scotian Slope, Offshore Nova Scotia.* Geological Survey of Canada Current Research 2000-D7, 8 pp.

Piper, D.J.W. and Aksu, A.E., 1987. *The Source and origin of the 1929 Grand Banks turbidity current inferred from sediment budgets,* Geo-Marine Letters, Vol. 7, p. 177-182.

Piper, D.J.W., and MacDonald, A. 2002. *Timing and position of late Wisconsinan ice margins on the upper slope seaward of the Laurentian Channel.* Géographie physique et Quaternaire, Vol. 55, p. 131–140.

Piper, D.J.W. and Normark, W.R., 1982. *Effects of the 1929 Grand Banks earthquake on the continental slope off eastern Canada.* Geological Survey of Canada Current Research, Part B, Paper 82-01B, p. 147-151.

Piper, D.J.W., Sparkes, R., Mosher, D.C., Shor, A.N. and Farre, J.A., 1985a. *Seabed instability near the epicentre of the 1929 Grand Banks earthquake.* Geological Survey of Canada, Open File 1131, 29 pp.

Piper, D.J.W., Farre, J.A. and Shor, A.N., 1985b. *Late Quaternary slumps and debris flows on the Scotian Slope.* Geological Society of America Bulletin, 96: 1508-1517.

Piper, D.J.W., Shor, A.N. Clarke, J.E.H., 1988. *The 1929 "Grand Banks" Earthquake, Slump, and Turbidity Current.* In: Clifton, H.E. (ed), Sedimentologic Consequences of Convulsive Geologic Events, Geological Society of America, Special Paper , no. 229, p. 77-92.

Piper, D.J.W., Cochonat, P., Ollier. G., Le Drezen, E., Morrison, M., Baltzer, A., 1992. *Evolution progressive d'un glissement rotationnel en un courant de turbidite : cas du seisme de 1929 des Grands Bancs [Terre - Neuve].* Comptes rendus de l'academie des sciences, Serie II, Mechanique, Physique, Chimie, sciences de l'univers, sciences de la terre, tome 314, p. 1057-1064.

Piper, D.J.W., Cochonat, P. and Morrison, M.L., 1999. *The sequence of events around the epicentre of the 1929 Grand Banks earthquake: initiation of debris flows and turbidity current inferred from sidescan sonar.* Sedimentology, Vol. 46, p. 79-97.

Piper, D.J.W., Mosher, D.C., Gauley, B-J., Jenner, K. and Campbell, D.C., 2003. *The chronology and recurrence of submarine mass movements on the continental slope off southeastern Canada.* In: Locat, J. and Mienert, J. (eds.), Submarine Mass Movements and Their Consequences. Kluwer, Dordrecht, The Netherlands, p. 299-306.

Piper, D.J.W., Macdonald, A.W.A., Ingram, S., Williams, G.L. and McCall, C., 2005. *Late Cenozoic architecture of the St. Pierre Slope.* Canadian Journal of Earth Sciences, Vol. 42, p. 1987-2000, doi: 10.1139/E05-059.

Piper, D.J.W., Shaw, J. and Skene, K.I., 2007. *Stratigraphic and sedimentological evidence for late Wisconsinan sub-glacial outburst floors to Laurentian Fan.* Palaeogeography, Palaeoclimatorlogy, Palaeoecology, Vol. 246, p. 101-119, doi: 10.1016/ j.palaeo.2006.10.029.

Shaw, J., Piper, D.J.W., Fader, G.B., King, E.L., Todd, B.J., Bell, T., Batterson, M.J., and Liverman, D.J.E., 2006. *A conceptual model of the deglaciation of Atlantic Canada.* Quaternary Science Reviews, Vol. 25, p. 2055-2081.

Shimeld, J. 2004. *A comparison of salt tectonic subprovinces beneath the Scotian Slope and Laurentian Fan.* In: P.J. Post, D.L. Olson, K.T. Lyons, S.L. Palmes, P.F. Harrison, and N. Rosen (eds). Salt–sediment interactions and hydrocarbon prospectivity: concepts, applications, and case studies for the 21st century. 24th Annual Gulf Coast Section of the Society of Economic Paleontologists and Mineralogists Foundation Bob F. Perkins Research Conference, Houston, Texas, Dec. 5-8, 2004.

Shor, A.N. and Piper, D.J.W., 1989. *A large Late Pleistocene blocky debris flow on the central Scotian Slope.* Geomarine Letters, 9, 153-160.

Shor, A.N., Piper, D.J.W., Hughes Clarke, J.E., and Mayer, L.A., 1990. *Giant flute - like scour and other erosional features formed by the 1929 Grand Banks Turbidity Current.* Sedimentology, Vol. 37, p. 631-645.

Skene, K.I. and Piper, D.J.W. 2003, *Late Quaternary stratigraphy of Laurentian Fan: a record of events off the eastern Canadian continental margin during the last deglacial period*; Quaternary International, Vol. 99-100, p. 135-152.

Stow, D.A.V., 1981. *Laurentian Fan: Morphology, sediments, processes and growth pattern.* American Association of Petroleum Geologists Bulletin, Vol. 65, p. 375-393.

LANDSLIDE AND GRAVITY FLOW FEATURES AND PROCESSES OF THE NAZARÉ AND SETÚBAL CANYONS, WEST IBERIAN MARGIN

R.G. ARZOLA, R.B. WYNN, D.G. MASSON, P.P.E. WEAVER
National Oceanography Centre Southampton, European Way, Southampton, SO14 3ZH, UK

G. LASTRAS
GRC Geociències Marines, University of Barcelona, c/Martí i Franquès s/n, 08028 Barcelona, Spain

Abstract

The Nazaré and Setúbal Canyons of the west Iberian margin are highly complex seafloor environments, displaying a range of sedimentary features and processes that reflect the transition from erosive upper to depositional lower canyon. Upper sections are characterised by a deeply incised, narrow, V-shaped thalweg, and frequent localised intra-canyon slope failures. Lower sections have a U-shaped floor with heterogeneous sediment distribution. Two types of gravity flow are observed: thin-bedded, fine-grained deposits that may be the result of frequent turbidity currents generated by high sediment supply to canyon heads, and thicker, siliciclastic coarse sandy turbidites, probably generated by larger earthquake-triggered slope failures on much longer timescales. Our results highlight the complex interplay of sedimentary processes operating within major canyon systems.

Keywords: Nazaré, Setúbal, submarine canyon, continental margin, mass wasting, turbidity current

1. Introduction and aims

Previous work on submarine canyons has established that they are major pathways for the transport of sediment from land to the deep ocean (Berner, 1982; van Weering *et al.*, 2002; Canals *et al.*, 2006). Sediment accumulates at the canyon head from fluvial and/or shelf sources, producing a temporary sediment reservoir (Mastbergen & van den Berg, 2003). Sediment instability, due to higher supply rates, faster progradation and intense resuspension during storms and floods, is enhanced by the presence of rough topography and steep slopes in the upper canyon (Mulder *et al.*, 2001; Mullenbach *et al.*, 2004; Puig *et al.*, 2004). Failure can also be induced by earthquake-triggered deformation of sediment (Jones & Omoto, 2000). The failed sediment is then transported mainly by low frequency, high-energy gravity flows (Normark & Piper, 1991). Thus, the main sedimentary processes in canyons seem to be failure by mass wasting and subsequent transport by gravity flows. However, the dynamics of these processes in most modern canyons are poorly constrained, as is their overall role in the offshore export of sediment, due to difficulties in direct monitoring of sediment transport in canyons and in sampling in such rugged canyon topography.

Submarine canyons along European continental margins have recently been extensively studied as part of the EC-EUROSTRATAFORM project and the currently ongoing

V. Lykousis, D. Sakellariou and J. Locat (eds.), Submarine Mass Movements and Their Consequences, 89–98.

HERMES project (Weaver *et al.*, 2004). In particular, the two projects have generated a significant amount of new data from Nazaré and Setúbal Canyons, offshore west Iberia, some of which are presented here. The principal aim of this study is to highlight the key gravity flow and mass wasting (i.e. slope failure) processes in the Nazaré and Setúbal Canyons. De Stigter *et al.* (in press) have recently suggested that gravity flows in Nazaré Canyon vary between yearly and centennial timescales, depending on the nature of the material being transported. Although the timing of gravity flow events has also been analysed as part of this study, it will be covered in detail in a future contribution and is only briefly discussed here.

2. Methodology and database

This study is based upon geophysical and sedimentological data collected during EC-EUROSTRATAFORM cruise CD157 (2004) and HERMES cruises D297 and CD179 (2005-2006). The data include multibeam bathymetry surveys, medium-resolution (30 kHz) deep-towed sidescan sonar mapping, and 3.5 kHz shallow seismic profiles. Over 40 shallow piston cores were recovered from accurately targeted sites along the two canyon systems in order to ground-truth the geophysical data. Piston cores were visually inspected, including estimates of grain size, and photographed.

3. Regional setting

The Nazaré and Setúbal Canyons (Fig. 1) are the two largest canyons of the west Iberian margin and are located on the central part of the margin, oriented roughly perpendicular to the coast in an east-west direction. Nazaré Canyon, ~100 km north of Lisbon, cuts across the shelf almost to the beach but is not connected to a major river system. Distally the canyon leads into the Iberian Abyssal Plain, some 210 km from the coast at a water depth (WD) of ~5000 m. About 30 km south of Lisbon, the Lisbon and Setúbal Canyons extend landwards across the shelf towards the mouths of the Tagus and Sado Rivers, respectively. Lisbon Canyon connects to Setúbal Canyon as a tributary at ~1500 m WD, and Setúbal Canyon then continues downslope until it reaches the Tagus Abyssal Plain at ~4840 m WD, some 175 km from the canyon head.

4. Observations of mass wasting and gravity flow features

4.1. NAZARE CANYON

4.1.1. Terraces and gullies in the upper section

The upper Nazaré Canyon (< ~4000 m WD) is characterised by rugged topography and steep slopes (Fig. 2), with abundant gullies incising into semi-circular erosional scarps (Figs. 3A,B). Intra-canyon terraces, which are generally long and narrow and oriented roughly parallel to the canyon axis, are observed only in the steepest locations, where the narrow V-shaped canyon has undergone multiple phases of incision into surrounding bedrock and semi-lithified sediments.

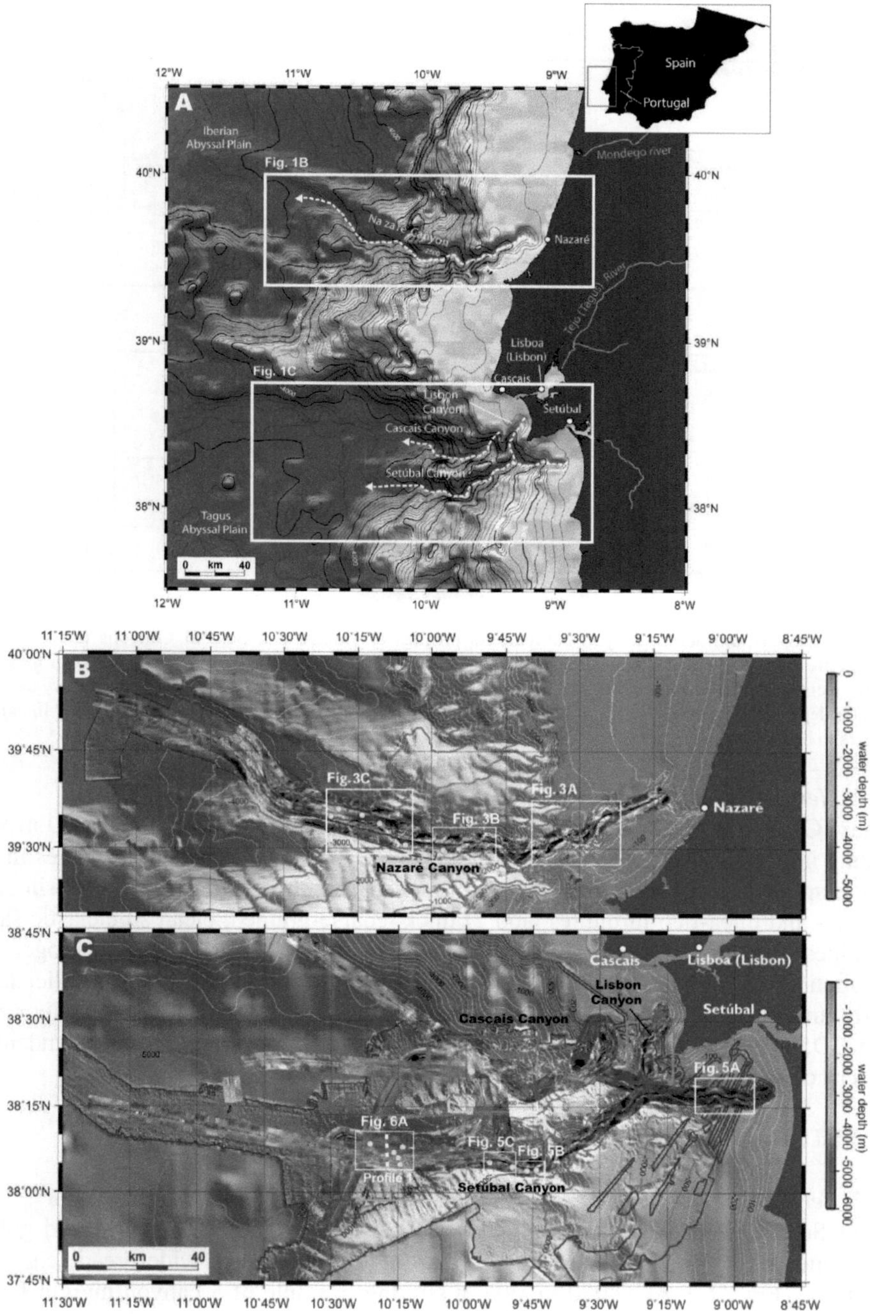

Figure 1. (A) Bathymetry map of the west Iberian margin showing the locations of Nazaré and Setúbal Canyons. (B,C) Coverage of sidescan sonar data for each canyon. Yellow dots represent the positions of piston cores. The dashed white line is shallow seismic profile 1 (Fig. 6B). Contours are every 25 m down to 200 m, then every 100 m.

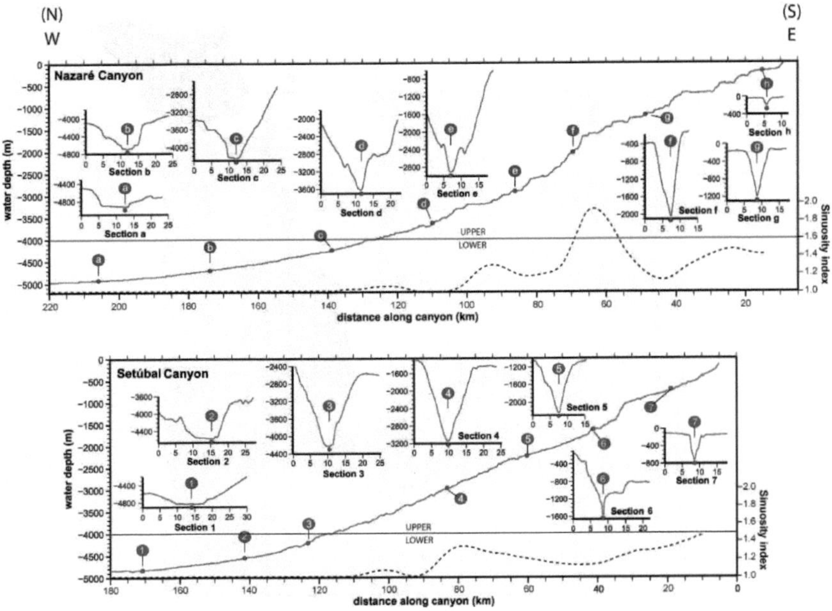

Figure 2. Down- and across-canyon bathymetric profiles of Nazaré and Setúbal Canyons (across-canyon profiles are taken facing up-slope, i.e. north is on left). A clear distinction can be made between the steep, deeply incised upper section, and the much flatter lower section in both canyons, with the boundary at ~4000 m water depth (WD). The sinuosity index is the sinuous distance over the straight distance (dashed lines), and appears to be highest in the upper sections and decreases considerably towards the lower sections.

4.1.2. Heterogeneous turbidite deposition in the lower section

Nazaré Canyon widens out abruptly ~130 km from the shelf edge and at ~4000 m WD, leading into the lower section with a 4-5 km-wide flat floor (Figs. 2, 3C). Cores in this area display two distinct types of turbidite: 1) thin-bedded, fine-grained silt-sand, organic- and mica-rich turbidites, and 2) thick, coarse-grained, clean siliciclastic sandy turbidites (Fig. 4). Backscatter variations on sidescan sonar data indicate heterogeneous distribution of sediment across the canyon floor, with the coarser-grained, siliciclastic sandy turbidites dominating on the thalweg and deeper parts of the canyon floor (e.g. core CD56419), and the finer-grained, thin-bedded turbidites being more abundant on the terraces and canyon margins (e.g. cores CD56420 and D15756).

4.2. SETUBAL CANYON

4.2.1. Terraces, small-scale mass wasting and gravity flows in the upper section

Upper Setúbal Canyon (< 4000 m WD) is V-shaped with steep terraced and gullied walls and a narrow thalweg (Figs. 2, 5A-C). Turbidites are widespread, e.g. core CD56416 (Fig. 5D). At ~4000 m WD, a striking example of a canyon margin failure can be seen (Fig. 5C), with several 100 m-wide blocks scattered across the canyon floor adjacent to an area of 1 km-long erosional lineations. Core CD56826 was recovered from the opposite side of the canyon to this rockfall, on a 140-160 m-high terrace. It displays a spectacular polymict debrite containing a variety of sub-rounded, semi-lithified silt and mud clasts up to 25 cm in diameter (Fig. 5D).

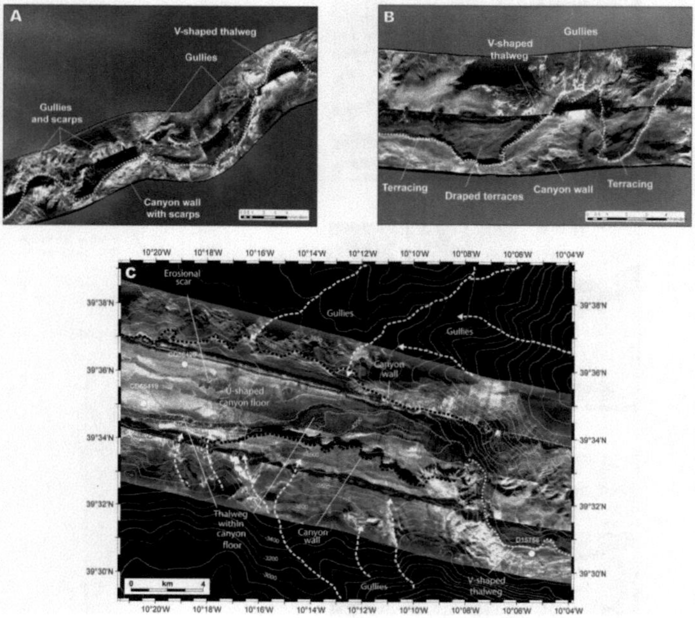

Figure 3. (A-B Sidescan sonar images of upper Nazaré Canyon features down to 3500 m WD. Note the narrow and sinuous V-shaped thalweg (white dotted line), and steep gullied and terraced walls. (C) The abrupt transition to the wider lower canyon occurs at ~4000 m WD. 100 m contours; cores described in Fig. 4.

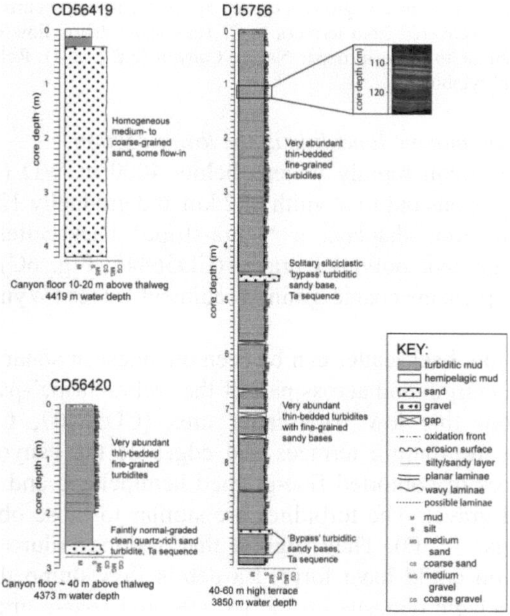

Figure 4. Logs of cores CD56419, CD56420 and D15756, recovered from Fig. 3C, show two types of turbidite: thick siliciclastic sandy and thin-bedded, fine-grained organic- and mica-rich (photographed).

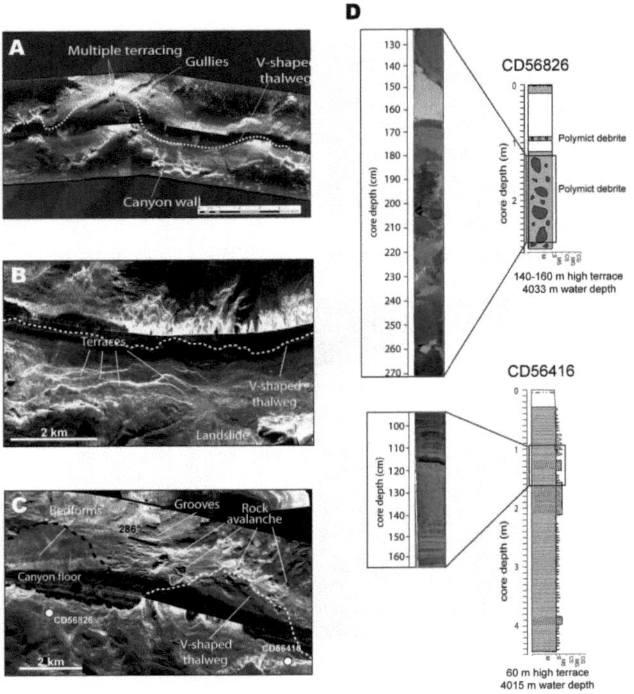

Figure 5. (A-C) Sidescan sonar images of upper Setúbal Canyon (~1000-1500 m, ~3000-3600 m and ~3700-4200 m WD respectively). Landslides and rockfalls are common, as is terracing due to incision of the narrow thalweg into the steep canyon margin, and grooves caused by erosive turbidity currents. (D) Sedimentary logs and photographs of cores, recovered from terraces in C. They show debris flow deposits and thin-bedded, fine-grained turbidites similar to those in upper Nazaré Canyon (c.f. Fig. 4). Refer to Fig. 4 for a key of sedimentary structures and symbols.

4.2.2. A large-scale submarine landslide in the lower section

The lower Setúbal Canyon rapidly widens below 4000 m WD (Figs. 1, 2). The flat canyon floor initially opens out to a width of 2 km and gradually 12 km near the canyon mouth. The canyon floor displays a 'zebra-stripe' backscatter pattern and cores recovered normally-graded, polymict gravel (CD56845, Fig. 6C), indicating that the backscatter pattern represents coarse-grained sediment waves (Wynn and Stow, 2002).

An area of uniform low backscatter can be seen on sidescan sonar data extending from the northern canyon margin and across part of the 'zebra-stripe'-patterned canyon floor (Fig. 6). Cores from this low backscatter area (CD56407, CD56408, CD56414, CD56415), both on the margin terraces and edges of the canyon floor, show thick remobilised sequences of contorted fine-grained hemipelagic and turbiditic sediments, as well as ungraded gravel. The turbidites are similar to those observed on the upper canyon terraces (Figs. 4, 5D). They suggest that a slope failure on the N margin of lower Setúbal Canyon could have formed a debris flow/slump that remobilised both turbiditic and hemipelagic deposits on the terraces and coarse gravel bedforms on the canyon floor, but bypassing the area around un-remobilised core CD56845, possibly because it is slightly elevated (Fig. 6B).

5. Mass wasting and gravity flow processes in Nazaré and Setúbal Canyons

5.1. MASS WASTING PROCESSES

Several examples of intra-canyon mass wasting events are observed in the upper (< ~4000 m WD) Nazaré and Setúbal Canyons. These events are localised and small-scale (< 10 km^2), and include submarine landslides, debris flows and rockfalls; they are especially numerous in the upper Setúbal Canyon (Fig. 5). Such slope failures can be initiated by two types of factor (e.g. Masson *et al.*, 2006); external factors include short-period (minutes to hours) ground shaking by earthquakes, and longer-lasting (hundreds to thousands of years) effects of sea level change (Weaver and Kuijpers, 1983). Internal factors include sediment overpressure caused by rapid deposition, overloading, oversteepening, etc. (e.g. Terrinha *et al.*, 2003; Puig *et al.*, 2004).

The slope failures in the Portuguese Canyons are most likely caused by under-cutting and oversteepening of upper canyon margins by erosive turbidity currents (evidence is in the large number of terraces and high wall steepness), and/or ground shaking during regional earthquakes (e.g. Mulder *et al.*, 1998). The latter is believed to be particularly important here since the Iberian Peninsula is located just north of the present-day Africa/Eurasia plate boundary. Seismic activity along this fault zone is believed to be associated with historical large earthquakes in the area (such as the 1755 Lisbon earthquake, e.g. Weaver *et al.*, 2000). Only one instance of mass wasting beyond 4000 m WD is observed (Fig. 6). It is unknown what caused this slump/debris flow; however, a possible cause might be seismic-induced failure rather than overloading of sediment by storms or floods, as the latter are more influential in the shallower reaches of the canyons rather than at these depths > 4000 m.

5.2. TURBIDITY CURRENT PROCESSES

There appear to be two main types of turbidity current in the Nazaré and Setúbal Canyons: those that form the stacked, thin-bedded, fine-grained sand turbidites that are rich in mica and organic material, and those that produce the thicker, siliciclastic, coarser-grained sand turbidites. The former deposits have been successfully cored both on upper and lower canyon terraces ~40-60 m above the floor in both canyons (Figs. 4, 5D, 6C). They are also inferred to be present in the upper canyon floor along with the coarse siliciclastic turbidites; however, coring has been unsuccessful in this area due to poor penetration. The thin-bedded, closely-stacked deposits of the fine-grained turbidites indicate that they were regular flows that probably entered the canyon semi-continuously as small pulses of sediment that remained mainly in the upper canyon. Based on the unvarying small size of the deposits throughout the canyons (1-2 cm-thick bases, < 5 cm-thick mud caps; Figs. 4, 5D, 6C), it seems unlikely that the turbidity currents transported much sediment beyond the canyon mouths. These turbidity currents are thus inferred to be small-scale, relatively regular events (possibly on an annual scale, as proposed by de Stigter *et al.*, in press) that are probably the result of failures generated by overloading and over-steepening of sediment during floods and/or storms (e.g. Mulder *et al.*, 1998).

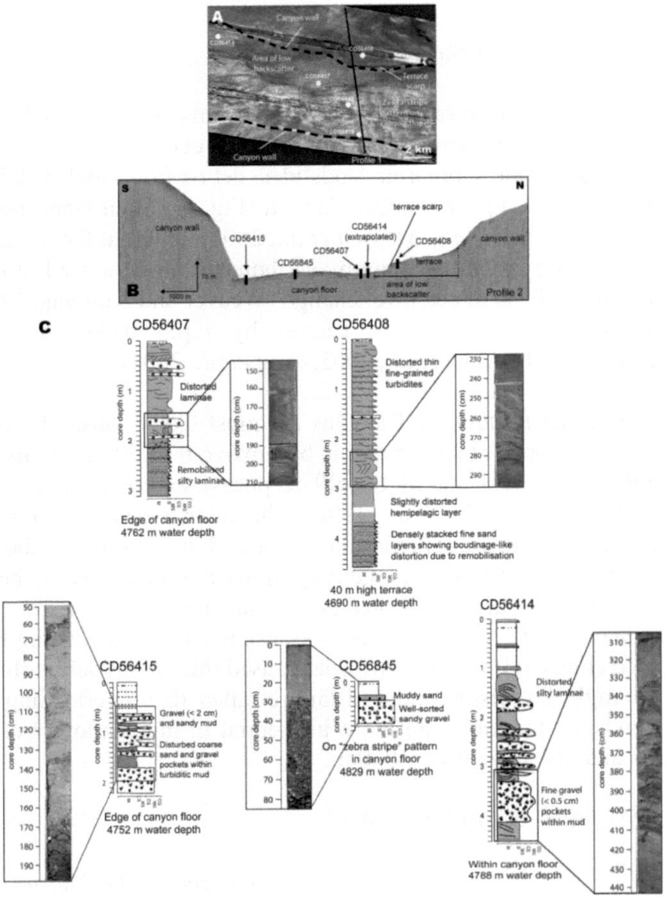

Figure 6. (A) Low backscatter area across lower Setúbal Canyon (canyon floor edges in dashed black lines). A suite of cores was recovered across this area (shown in C). (B) 3.5 kHz profile 1 taken across this area, with labelled location of cores. (C) Sedimentary logs and photographs of the cores shown in A and B. They imply that the low backscatter area is a debris flow/slump deposit that remobilised both terrace turbidites and canyon floor gravel waves, but that its trajectory did not affect core CD56845, possibly because of its slightly raised location (see B). Refer to Fig. 4 for a key of sedimentary structures and symbols.

The other type of turbidite, the clean siliciclastic sand, is generally observed on the terraces and floor of the lower sections, beyond ~4000 m WD (Fig. 4). The reason for this is probably varying settling velocities within different parts of the flow causing them to deposit in separate areas. The occasional occurrence of siliciclastic sandy bases on upper canyon terraces is probably due to centrifugal forces causing basal part of the flows to be deflected towards the outside (apex) of large bends, where they undergo superelevation and spill over the canyon margins (Keevil *et al.*, 2006). These flows are inferred to flush through the entire canyons, as their deposits are found in the lower canyon floor, the abyssal plains (Thomson and Weaver, 1994) and are inferred to be present in the upper canyon floor (thalweg) due to repetitive failed coring. They are erosive flows, evidenced by considerable erosion in the lower canyon and mouth floor of both canyons, including large-scale scours and grooves that incise recent sediment deposits (Fig. 3C). They are therefore inferred to have completely eroded any earlier deposits of the stacked, thin-bedded, fine-grained turbidite type in the canyon floor, which is why none have been preserved here. A likely cause for these large and less frequent turbidity currents is therefore a larger-scale, less regular trigger such as earthquakes (e.g. Mulder *et al.*, 1998), and which are also relatively commonplace in the west Iberian margin. A likely centennial or longer timescale is suggested for this type of turbidity current in the Portuguese Canyons (de Stigter *et al.*, in press).

6. Conclusions

This new dataset comprises data at very different scales, from hundreds of km-scale imaging (multibeam bathymetry) to cm-scale detail (sediment cores). The integration of all these data show the high level of complexity that exists throughout all scales in the canyons of the west Iberian margin. The processes of gravity flow and mass wasting can, however, be simplified and the main similarities and differences observed in the Nazaré and Setúbal Canyons are listed below.

1. There are two main types of gravity flow that occur in Nazaré and Setúbal Canyons: 1) small, regular, organic- and mica-rich turbidity currents that deposit mainly on the shallower intra-canyon terraces, and 2) large, less regular, canyon-flushing turbidity currents that deposit mainly in the deeper parts of the canyon and the abyssal plains.
2. Turbidity currents seem to be the dominant mode of sediment transport and erosion taking place in the canyons, evidenced by the dominance of gravity flow deposits and erosive scours throughout both canyons.
3. Mass wasting events tend to be small-scale (< 10 km^2), localised and in the steep upper canyon sections, especially in Setúbal Canyon. Failure events in the lower sections are rare, as these areas are dominantly depositional and have more gentle gradients; however, one large failure deposit has been identified in lower Setúbal Canyon. These results suggest that Setúbal Canyon may have been more recently unstable than Nazaré Canyon, although there is no evidence for any significant difference in the gravity flow activity between the two canyons.

7. Acknowledgements

This research was supported by the EUROSTRATAFORM (EVK3-CT-2002-00079) and HERMES projects (GOCE-CT-2005-511234), both funded by the European Commission's Fifth and Sixth Framework Programmes respectively. The authors wish to thank the crew of cruises CD157, D297 and CD179, and V. Huhnerbach, A. Akhmetzhanov and V. Huvenne for varied data processing and assistance. H. Haflidiason and T. van Weering are gratefully acknowledged for providing helpful comments on an earlier manuscript.

8. References

Berner, R.A., 1982. *Burial of organic carbon and pyrite sulphur in the modern ocean: its geological and environmental significance.* Am. J. of Science 282, 451-473.

Canals, M., Puig, P., Durrieu de Madron, X., Heussner, S., Palanques, A., Fabres, J., 2006. *Flushing submarine canyons.* Nature 444, 354-357.

de Stigter, H., Boer, W., de Jesús Mendes, P.A., Jesús, C.C., Thomsen, L., van den Berg, G.D., van Weering, T.C.E., in press. *Recent sediment transport and deposition in Nazaré Canyon, Portuguese continental margin.* Marine Geology.

Jones, A.P., Omoto, K., 2000. *Towards establishing criteria for identifying trigger mechanisms for soft-sediment deformation: a case study of Late Pleistocene lacustrine sands and clays, Onikobe and Nakayamadaira Basins, northeastern Japan.* Sedimentology 47, 1211-1226.

Keevil, G.M., Peakall, J., Best, J.L., Amos, K.J., 2006. *Flow structure in sinuous submarine channels: Velocity and turbulence structure of an experimental submarine channel.* Marine Geology 229, 241-257.

Masson, D.G., Harbitz, C.B., Wynn, R.B., Pedersen, G., Lovholt, F., 2006. *Submarine landslides: processes, triggers and hazard prediction.* Phil. Trans. R. Soc. A 364, 2009-2039.

Mastbergen, D.R., van den Berg, J.H., 2003. *Breaching in fine sands and the generation of sustained turbidity currents in submarine canyons.* Sedimentology 50, 625-637.

Mulder, T., Savoye, B., Piper, D.J.W., Syvitski, J.P.M., 1998. *The Var submarine sedimentary system: understanding Holocene sediment delivery processes and their importance to the geological record.* In: Stoker, M.S., Evans, D., Cramp, A. (eds.) Geological processes on continental margins: Sedimentation, mass-wasting and stability. Geological Society, London, Spec. Pub. 129, 145-166.

Mulder, T., Migeon, S., Savoye, B., Faugères, J.-C., 2001. *Inveresly graded turbidite sequences in the deep Mediterranean: a record of deposits from flood-generated turbidity currents?* Geo-Mar. Lett. 21, 86-93.

Mullenbach, B.L., Nittrouer, C.A., Puig, P., Orange, D.L., 2004. *Sediment deposition in a modern submarine canyon: Eel Canyon, northern California.* Mar. Geol. 211, 101-119.

Normark, W.R., Piper, D.J.W., 1991. *Initiation processes and flow evolution of turbidity currents: implications for the depositional record.* SEPM Spec. Publ. 46, 207-230.

Puig, P., Ogston, A.S., Mullenbach, B.L., Nittrouer, C.A., Parsons, J.D., Sternberg, R.W., 2004. *Storm-induced sediment gravity flows at the head of the Eel submarine canyon, northern California margin.* J. of Geophys. Res. 109, doi: 10.1029/2003JC001918.

Terrinha, P., Pinheiro, L.M., Henriet, J.-P., Matias, L., Ivanov, M.K., Monteiro, J.H., Akhmetzhanov, A., Volkonskaya, A., Cunha, T., Shaskin, P., Rovere, M., 2003. *Tsunamogenic-seismogenic structures, neotectonics, sedimentary processes and slope instability on the southwest Portuguese Margin.* Mar. Geol. 195, 55-73.

Thomson, J., Weaver, P.P.E., 1994. *An AMS radiocarbon method to determine the emplacement time of recent deep-sea turbidites.* Sedimentary Geology 89, 1-7.

Weaver, P.P.E., Kuijpers, A., 1983. *Climate control of turbidite deposition on the Madeira Abyssal plain.* Nature 306, 360-363.

Weaver, P.P.E., Wynn, R.B., Kenyon, N.H., Evans, J., 2000. *Continental margin sedimentation, with special reference to the north-east Atlantic margin.* Sedimentology 47, 239-256.

Weaver, P.P.E., Billett, D., Boetius, A., Danovaro, R., Freiwald, A., Sibuet, M., 2004. *Hotspot Ecosystem Research on Europe's Deep-Ocean Margin.* Oceanography 17 (4), 132-143.

van Weering, T.C.E., de Stigter, H.C., Boer, W., de Haas, H., 2002. *Recent sediment transport and accumulation on the NW Iberian margin.* Prog. in Oceanography 52, 349-371.

Wynn, R.B., Stow, D.A.V., 2002. *Classification and characterisation of deep-water sediment waves.* Mar. Geol. 192, 7-22.

Section 2 - Mass waste evolution: From slump to distal turbidites

EXPERIMENTAL STUDIES OF SUBAQUEOUS VS. SUBAERIAL DEBRIS FLOWS – VELOCITY CHARACTERISTICS AS A FUNCTION OF THE AMBIENT FLUID

H. BREIEN, M. PAGLIARDI, F.V. DE BLASIO, D. ISSLER
Department of Geosciences/International Centre for Geohazards (ICG), University of Oslo, Post box 1047 Blindern, N-0316 Oslo, Norway, email: hedda.breien@ngi.no

A. ELVERHØI
Department of Hydraulic and Environmental engineering, University of Pavia, Italy

Abstract

A series of comparable subaerial and subaqueous debris flow experiments of sand-clay-water mixtures has been performed at the St Anthony Falls Laboratory (SAFL) at University of Minnesota. Different compositions were tested and velocities measured in detail using PIV (Particle Image Velocimetry) techniques. The experimental series provides a unique data set highlighting the effects of the ambient and interstitial fluid in comparable subaerial and subaqueous debris flows. Based on our experimental data we emphasize the differences in the dynamical behaviour associated with the two environments and suggest important mechanisms to be included in numerical models.

Keywords: Subaerial debris flows, subaqueous debris flows, experiments, velocity, PIV

1. Introduction

Decades of ongoing research has shown that gravity mass flows exhibit a wide range of complex dynamics. To study mass flow phenomena three lines of investigation are possible: traditional field analysis, numerical modelling and experimental work. A reason why experimental studies are needed is that mass flows are not always directly observable and are usually difficult to monitor in the field. Nor can observations of the deposits yield complete information on the flow dynamics.

Several researchers have performed experiments in order to understand the dynamics of gravity mass flows. Concerning subaqueous mass flows, Kuenen and Migliorini (1950) and Hampton (1972) were among the first to perform laboratory flume experiments. The interest in this field significantly increased in more recent years in conjunction with improved mapping of the seafloor and further studies of old deposits on land, and has resulted in numerous experiments that for brevity cannot be summarized here (Talling et al. (2002), Marr et al. (2001), Ilstad et al. (2004) and Felix and Peakall (2006) among others). A much larger number of experimental studies at various scales have been performed for subaerial debris flows. As key literature we refer to Iverson (1997) and Pudasaini and Hutter (2007) and references therein.

Despite the same rheology, subaerial and subaqueous debris flows exhibit very different transport dynamics. Subaqueous slides stand out in their ability to run out long distances, their mismatch between static shear strength and dynamical shear stress measured in the lab (Locat and Lee 2001; De Blasio et al. 2006), hydroplaning (Mohrig et al. 1998; Harbitz et al. 2003; De Blasio et al. 2004a) and the transport of fines into the

V. Lykousis, D. Sakellariou and J. Locat (eds.), Submarine Mass Movements and Their Consequences, 101–110.

overriding turbidity currents (Mohrig and Marr 2003; Felix and Peakall 2006). The differences in dynamics between the two settings throw light upon the strong dependence on a slide's ambient fluid. Our data provide a unique opportunity to study this effect.

It is uncommon to study both subaerial and subaqueous flows in a similar setting and with the same slurry material. This paper addresses both types of debris flows, focusing on the analysis of the velocity field as a function of ambient medium as well as slurry composition. One of the main objectives of our experimental work is to provide better information on mass flow dynamics, which in the end can be used in improving numerical models for various types of flow. While several models have been suggested for the behaviour of subaerial debris flows, at present there is no comparably realistic model for subaqueous flows. This is especially true for non-cohesive multi-phase debris flows which are extremely complex. Cohesive debris flows, on the other hand, have occasionally been simulated in subaqueous settings, but even in these cases the role of ambient water is introduced with a simple drag force model, assuming no sediment disruption (Norem et al. 1990; De Blasio et al. 2004b; Gauer et al. 2005). Through velocity distribution analysis we try to pinpoint physical phenomena which distinguish subaqueous from subaerial debris flows and which need to be taken into account in a dynamical model.

It should be kept in mind that a direct extrapolation to the field scale is probably flawed because a complete scaling does not yet exist. In addition, the effect of the sidewalls needs to be taken into account. Despite this, experiments provide information that cannot be obtained from field observations and will further our physical understanding of gravity mass flow.

2. Experimental setup

The experiments were performed in a 10 m long, 3 m high and 0.6 m wide tank (Fig. 1). The tank can be filled completely with water and one sidewall built out of glass for observation. A 0.2 m broad Plexiglas flume is submerged in the tank, inclined at an angle of 6°. In our experiments the flume bed is coated with granular roof shingle with a roughness of around 1 mm to prevent undesired slip.

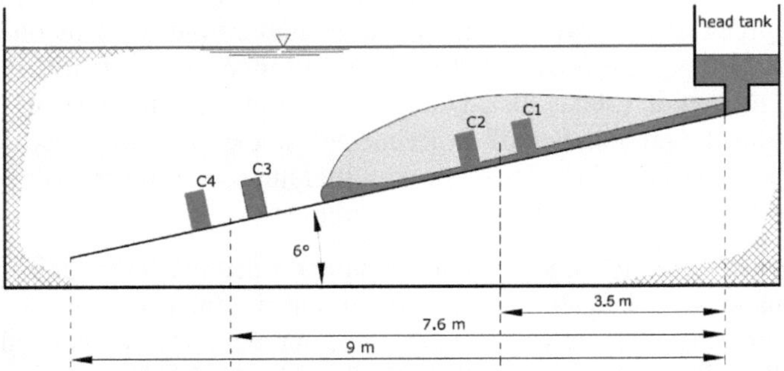

Figure 1. Flume set-up. C1-C4 indicate the view frames of the four high-speed cameras (not to scale). A total amount of 100 litres of slurry is released from the head tank into the flume during each run.

A total of 47 debris flow experiments (15 subaqueous and 32 subaerial) were performed. The slurries were composed of sand, kaolin, tap water and coal slag (for visualisation), and were prepared with 5 different compositions by varying clay and sand content (Table 1) while water and coal slag were held constant at 28 % and 5 % by weight. The density of the slurries was around 1.8 g/cm^3 and the total volume used in each run was 100 litres. All slurries were run both subaerially and subaqueously and all experiments were repeated in order to confirm the observed behaviours.

Table 1 Concentration of the different slurry components in percentage of weight of total mass.

	5 % slurry	10 % slurry	15 % slurry	20 % slurry	25 % slurry
Sand + coal slag	67 %	62 %	57 %	52 %	48 %
Clay	5 %	10 %	15 %	20 %	25 %
Water	28 %	28 %	28 %	28 %	28 %

Four high-speed cameras recording 240 frames per second were placed 3.6 m, 4.1 m, 7.3 m and 7.8 m downstream of the gate. A PIV algorithm (Sveen and Cowen 2004), applied to the high-speed camera images, permitted us to calculate the velocity automatically. In particular we calculated the velocity field (u and v component) for each image, i.e. every 0.004 s, and with a spatial resolution of about 1 mm. Contrary to usual PIV applications, no laser sheet was used and the velocity was calculated very close to the side wall. This introduces a wall effect that needs to be taken into account in the data analysis.

In addition to the high-speed cameras, two standard digital-video cameras were used to capture a wider view of the phenomena.

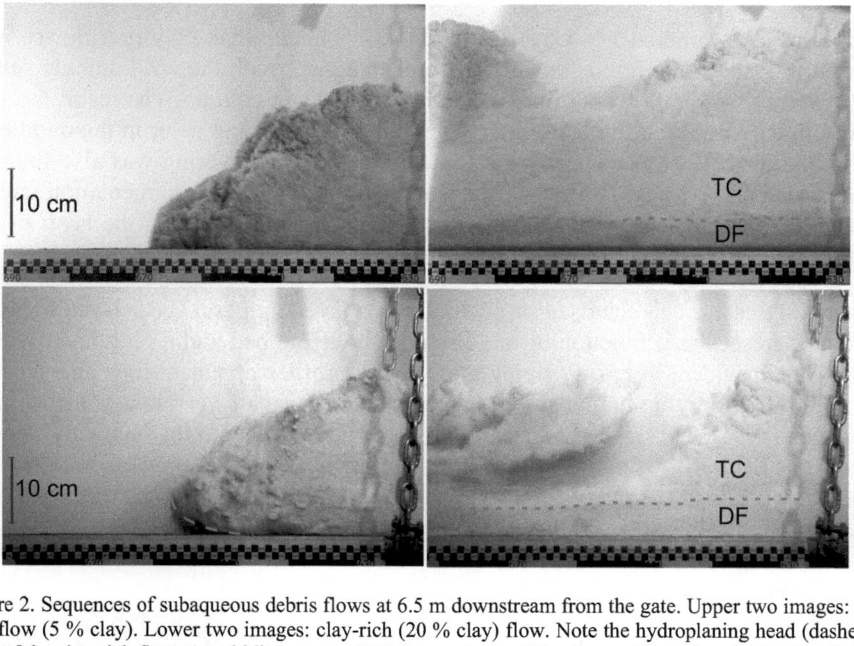

Figure 2. Sequences of subaqueous debris flows at 6.5 m downstream from the gate. Upper two images: sand-rich flow (5 % clay). Lower two images: clay-rich (20 % clay) flow. Note the hydroplaning head (dashed red line) of the clay-rich flow. A turbidity current (TC) forms on top of the denser debris flow (DF) when fines are ejected into the water column. Note the clearly visible difference between debris flow and turbidity current in the case of the sand-rich flow.

3. Results

3.1 SUBAERIAL AND SUBAQUEOUS DEBRIS FLOWS – FLOW CHARACTERISTICS

As seen from Figs. 2 and 3 the most visible differences between debris flows in water compared to those moving through air are their diffuse shape and the generation of turbidity currents on top of the denser subaqueous flows. As evident in Fig. 4, the subaerial flows are constantly decelerating during movement downstream, whilst subaqueous debris flows show larger velocity fluctuations. Turbulence is pronounced throughout a large part of the subaqueous sand-rich flows and several high-velocity cells can be identified (Fig. 3).

Fig. 4 shows the averaged velocity over time for the four different high speed cameras. It is seen that for subaerial flows, after the passage of a high-velocity front, the velocity is the same in all four locations, but decreasing exponentially in time. In our set-up the movement ceases after roughly 8 seconds.

The behaviour of the front and average velocity for the submarine debris flows differs from the subaerial in several ways. Average velocity over time is strongly dependent on clay concentration (Fig. 4), as it decreases with growing clay content. For low clay content mixtures the decrease is smooth and almost linear, while for high clay contents a head of higher velocity develops. In subaqueous environments, the highest velocities are obtained with clay-rich slurries, and their fronts seem to accelerate or stay constant in velocity.

Sand-rich flows (5 and 10 % clay) are much less coherent than clay-rich flows. When released in water, they immediately start to disintegrate, with the sand quickly settling out to the base of the flow and fines ejected into the water column. The result is a three layered flow, with a settling layer near the bed, a fluidized sand layer in the middle and a turbidity current composed of fine sediment on top. This layering was also found by Ilstad et al. (2004). There is a clear connection between a higher sedimentation rate and a lower the clay content for the slurry. Sedimentation can be seen as the layer of zero velocity (growing with time) in Fig. 3.

Subaerial cases are characterised basically by two layers: a basal shear layer where the velocity increases as a function of the distance above the bed and a plug layer with no shear. This is evident in Fig. 5, giving the velocity profiles of subaqueous and subaerial debris flows of varying composition. We also see a layer of very low velocity in the subaqueous cases. This is also found in Fig. 3 and coincides with the boundary between subaqueous debris flow and turbidity current. It is also seen that the velocity profile in subaerial and subaqueous cases to a larger degree are comparable in the sand-rich cases than in the clay-rich cases.

4. Discussion

Studying slurries of identical composition in subaerial and subaqueous environments emphasizes the importance of the ambient and interstitial fluid on debris flow dynamics. These differences are not only qualitatively observed but have also been measured quantitatively using PIV techniques. This allows us to accurately resolve the velocity field. We note however, the measurements are collected close to the sidewall, and assume that this near-wall velocity is representative of the interior velocity field. Fig. 3 underlines the differences between subaerial debris flows, which show a well-defined shape and a smooth velocity profile, and subaqueous debris flows which exhibit a poorly defined shape with large velocity fluctuations.

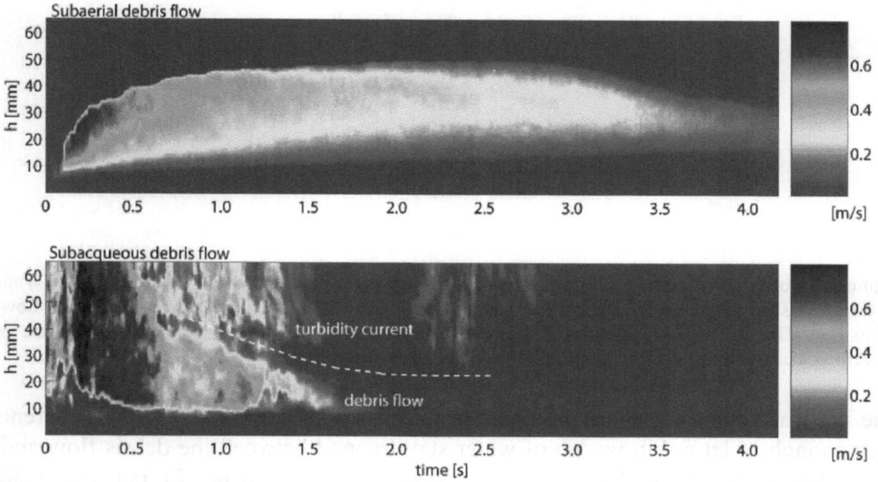

Figure 3. An example of subaerial (5 % clay) and subaqueous (20 % clay) velocity distributions (averaged velocity in x direction). The zero velocity close to the bed represents deposition. The subaerial debris flow is sharp and well-defined, with a velocity field constantly decelerating towards the tail. The subaqueous flow is divided into a debris flow and a turbidity current and exhibits complex behaviour with vortexes and high-velocity cells.

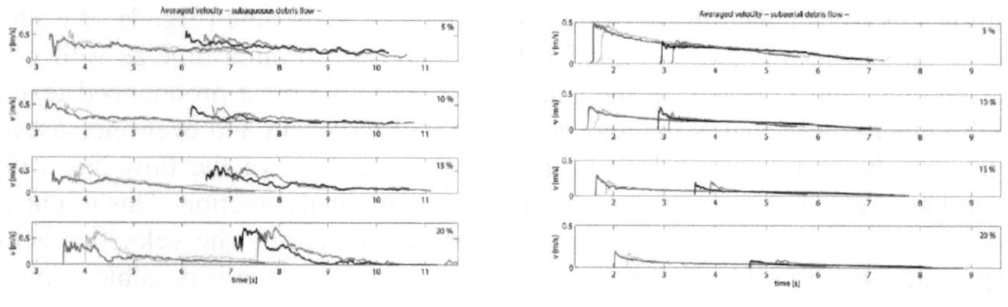

Figure 4. Average velocity over time for different compositions in subaerial and subaqueous flows. Blue line represents the velocity as the flow passes the first upstream camera (at 3.6 m), green line represents second camera (4.1 m), black line represents third camera (7.3 m) and red line represents the most downstream camera (7.8 m). Note the decelerating behaviour of subaerial flows (with time and clay content) and the constant-to-weakly accelerating behaviour of the subaqueous flows. The large velocity differences between head and debris flow body in the subaqueous environment (stretching) is also well demonstrated.

4.1 HYDROPLANING AND STRETCHING

The subaqueous debris flow front exhibits a high vertical velocity component. Generation of the turbidity current (depleting the debris flow in fines) on the top of the debris flow is probably related to this high velocity.

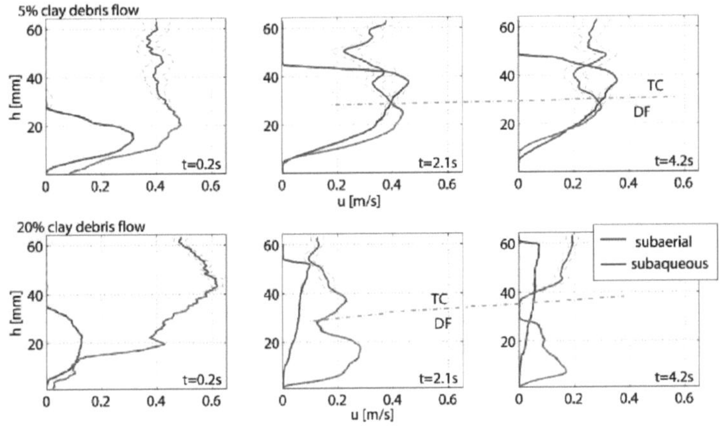

Figure 5. Velocity profiles for subaerial and subaqueous debris flows for front (t=0.2 s), body (t=2.1 s) and tail (t=4.2 s). Dashed green lines represent the standard deviation, and dashed black line divides the flow into debris flow (DF) and turbidity current (TC). Flume bed is located at zero in the beginning of the flow. Where the zero velocity is found at a higher position deposition has occurred.

The vertical velocity is most pronounced for clay-rich slurries where the coherence is high enough to let a thin wedge of water stay trapped between the debris flow and the bed. This effect of hydroplaning is shown in photos (Fig. 2) but our velocity measurements also resolve this behaviour (Fig. 3). A completely different behaviour is observed in the subaerial cases. The friction at the bed is much higher here than at the interface with air, where it is negligible. This introduces a counter-clockwise rotation similar to a conveyor belt.

The PIV technique gives us the velocity field (u(x,y,t), v(x,y,t)) at each camera position. This allows us to calculate the average flow velocity versus time (Fig. 4). We see that the subaerial slurries exhibit a gradual velocity decrease through time as well as with increasing clay concentration. In the subaqueous cases the most pronounced feature is the high-velocity head that is maintained throughout the event and even increases with time for the clay-rich slurries. Due to the fact that the head of the flow has a higher velocity than the body, the mass stretches in the longitudinal direction. This might cause intrusion of water and changes in slurry properties. Comparing the velocities between two adjacent camera stations (C1 and C2, C3 and C4 in Fig. 4), it is possible to measure the stretching of the flow: $\varepsilon_{xx} = \dfrac{\partial v}{\partial x} \cong \dfrac{v_1 - v_2}{\Delta x}$

Where v_1 and v_2 are the average velocity of the flow at two different positions and □x is the distance between the cameras (50 cm). Using Fig. 4 we find that in the case of subaerial debris flows the stretching decreases with increasing clay content, from a value of 0,18 s^{-1} (for 5 % clay) to 0,1 s^{-1} (for 20 % clay). In the subaqueous case the behaviour is opposite; the stretching increases with increasing clay content, from 0,46 s^{-1} to 0,86 s^{-1}. For subaqueous debris flows the stretching is almost one order of magnitude higher than for subaerial flows. This is most pronounced in the high clay content debris flow where the head moves with an almost constant velocity, like a gliding block. The explanation for this is the hydroplaning behaviour of the frontal part of clay-rich, subaqueous debris flows, lowering the friction at the base and increasing the velocity of the head compared to the body.

4.2 VELOCITY PROFILES

Fig. 5 shows velocity profiles from head, body and tail in sand-rich and clay-rich subaerial and subaqueous debris flows. Also here this large difference between head and body velocity is clear in the clay-rich subaqueous cases. The low velocity found in the middle of the velocity profiles in the clay-rich flows coincides with the boundary between debris flow and the turbidity current. This layer is most pronounced in the clay-rich subaqueous flows and does not exist for subaerial cases, which do not produce turbidity currents. This gives a velocity profile with two maxima, one in the middle of the debris flow, and one in the overriding turbidity current, as opposed to the proposed parabolic velocity profile of Mohrig and Marr (2003). They draw the highest velocity in the border stratum between debris flow and turbidity current. However, our velocity profiles have common features with Felix and Peakall's (2006) velocity profiles of high kaolinite concentration flows.

Note the fairly comparable behaviour of sand-rich subaerial and subaqueous flows (upper part of Fig. 5). The bigger differences in velocity profile in the clay-rich flows point towards a much larger dependence on the ambient fluid in the clay-rich flows (lower part of figure) than in more sandy debris flows.

4.3 FRONT VELOCITY

In Fig. 6 we present the travelled distance of the flow front versus time. The velocities of the subaerial debris flows are decreasing with increasing clay content, whilst the velocity of the subaqueous debris flows seems largely rheology independent. This is in agreement with the findings of Mohrig et al. (1997). When varying the viscosity of the slurry (but using only one clay content), they found that rheology played only a secondary role for subaqueous flows compared to subaerial flows. However, calculating the velocity between the two camera couples (C1 and C2, C3 and C4) reveals a more complex behaviour of the subaqueous flows (Fig. 4). The velocity is almost constant for clay contents up to 15 %, whilst for the 20 % clay slurry the velocity increases by 10 % moving from the upstream to the downstream stations. This is due to the hydroplaning frontal part. On the contrary, in the subaerial environment, the front velocity decreases significantly (up to 50 %) in function of the clay content.

It must be noted that the front velocities in Fig. 6 differ from the average velocities calculated at the wall using PIV (Fig. 4). This is a result of the wall effect, which is much larger in the subaerial cases than in the subaqueous ones.

5. Conclusions and outlook

In this paper we have presented a set of experiments carried out with novel technologies allowing us to measure accurate velocities. This gives us a unique opportunity to better understand the physics of subaerial and subaqueous debris flows. Also, our observations and analysis provide important findings in the development of new and/or better models for submarine debris flows.

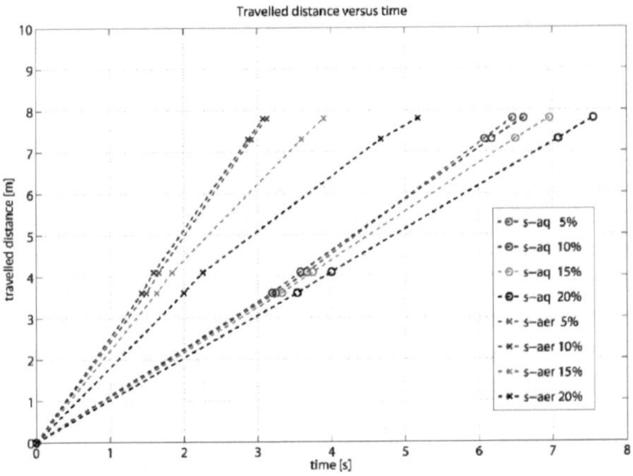

Figure 6. Comparison of travelled distance versus time for the different clay fractions and ambient fluids. Note that the velocities of the subaqueous flows are largely rheology independent.

Numerous attempts have been made to model subaerial mass flow. Based on the composition of the flow, mainly two approaches are in use. Granular models (Savage and Hutter 1989; Campbell et al. 1995; Denlinger and Iverson 2001; Takahashi 1991) put emphasis on the grain-grain interactions, whilst rheological approaches, usually with a Herschel-Bulkley rheology (Huang and Garcia 1999; Imran et al. 2001; Gauer et al. 2005) treat the debris as a non-Newtonian fluid with a somewhat fixed rheology. Hybrid models like NIS are constructed that account for both cohesion and granular behaviour (Norem et al. 1987; Norem et al. 1990). Some of them may be helpful for understanding the dynamics of subaerial rock and snow avalanches as well as debris flows, but their applicability to subaqueous slides is limited because they lack the affect of ambient water. Only cohesive debris has been occasionally simulated in subaqueous settings, and even in this case the role of ambient water is introduced with a simple drag force model, assuming no sediment disruption (Norem et al. 1990; De Blasio et al. 2004b; Gauer et al. 2005).

From the velocity profiles shown in Fig. 5 it is evident that sand-rich debris flows exhibit a quite similar behaviour in air and water. This might point towards using a granular debris flow approach in the modelling of these types of flow.

Huang and Garcia (1999) tested a model with Bingham rheology based on data from Mohrig et al. (1997). For subaerial cases they reproduce the overall flow and run-out distance. However, the model fails to reproduce subaqueous flows. Fig. 5 shows that the ambient fluid influences the dynamics of the clay-rich slurries in such a way that a model working in subaerial conditions becomes inappropriate to use in subaqueous environment. This is mainly due to hydroplaning, stretching and resulting change of properties of the flow. Our work is able to contribute to quantifying these characteristics in a more accurate way. Hydroplaning has been implemented in the BING model in a simplified way (De Blasio et al. 2004b), but modelling stretching and break-up and the influence on slurry properties is just beginning (Elverhøi et al. 2005).

In conclusion, from our systematic experiments we see that the affect of the ambient fluid on the dynamics of debris flows is fundamental. Going from a subaerial to a subaqueous environment changes the behaviour of the slurry to a great extent. This paper shows that turbulence, hydroplaning, stretching and dynamically changing slurry properties that result from the settling of sand, intrusion of water into the debris flow body and the depletion of fine particles due to turbidity current generation are all related to the ambient fluid being water. Also, to a large degree, subaqueous debris flow velocities are seen to be largely rheology and composition independent (in the composition range we tested) as opposed to subaerial flows.

A challenge in the coming years will be to develop a dynamical model that captures the dynamical properties of subaqueous debris flows and accurately predicts their run-out distances and velocities. This paper highlights some of the features that distinguish subaqueous flows from subaerial flows and that need to be taken into account in such a model.

6. Acknowledgement

This work was carried out as part of a project funded by the VISTA, the cooperation between The Norwegian Academy of Science and Statoil, project 6249. Additional funds were also provided by International Centre for Geohazards (ICG). The authors would like to thank the staff at the St. Anthony Falls Laboratory, University of Minnesota, for their technical as well as their scientifical support. We are also grateful to Kaare Høeg, University of Oslo and Jose Cepeda from ICG for help and discussions. Reviewers J. G. Marr and D. Mohrig are acknowledged for valuable comments.

This is the International Centre for Geohazards (ICG) contribution number 151.

7. References

Campbell CS, Cleary PW and Hopkins M (1995) *Large scale landslide simulations: Global deformations, velocities and basal friction.* J. Geophys. Res. 100: 8267-8283

De Blasio FV, Elverhøi A, Engvik L, Issler D, Gauer P and Harbitz CB (2006) *Understanding the high mobility of subaqueous debris flows.* Norwegian Journal of Geology, Special issue: Second International Symposium on Submarine Mass Movements and their Consequences, Solheim A (editor), 86: 275-284

De Blasio FV, Engvik L, Harbitz CB and Elverhøi A (2004a) *Hydroplaning and submarine debris flows.* J. Geophys. Res., 109, C01002, doi:10.1029/2002JC001714

De Blasio FV, Elverhøi A, Issler D, Harbitz CB, Bryn P and Lien R (2004b) *Flow models of natural debris flows originating from overconsolidated clay materials.* Marine Geology. 213: 439-455

Denlinger RP and Iverson RM (2001) *Flow of variably fluidized granular masses across three-dimensional terrain 2. Numerical predictions and experimental tests.* J. Geophys. Res – Solid Earth,106 (B1), jan 10: 553-566

Elverhøi A, Issler D, De Blasio FV, Ilstad T, Harbitz CB and Gauer P (2005) *Emerging insights into dynamics of submarine debris flows.* Natural Hazards and Earth System Sciences, 5: 633-648

Felix M and Peakall J (2006) *Transformation of debris flows into turbidity currents: mechanisms inferred from laboratory experiments.* Sedimentology, 53: 107-123

Gauer P, Kvalstad TJ, Forsberg CF, Bryn P and Berg K (2005) *The last phase of the Storegga slide: simulation of retrogressive slide dynamics and comparison with slide-scar morphology.* Marine Petrol. Geol., 22: 171-178

Hampton MA (1972) *The role of subaqueous debris flows in generating turbidity currents.* Journal of Sedimentary petrology, 42: 775-793

Harbitz CB, Parker G, Elverhøi A, Marr JG, Mohrig D and Harff P (2003) *Hydroplaning of subaqueous debris flows and glide blocks: Analytical solutions and discussion.* J. Geoph. Res. 108, B7, 2349, doi: 10.1029/2001JB001454

Huang X and Garcia MH (1999) *Modelling of non-hydroplaning mudflows on continental slopes.* Marine Geology, 154: 131-142

Ilstad T, Elverhøi A, Issler D and Marr J (2004a) *Subaqueous debris flow behaviour and its dependence on the sand/clay ratio: a laboratory study using particle tracking.* Marine Geology, 213: 415-438

Imran J, Harff P and Parker G (2001) *A numerical model of submarine debris flows with graphical user interface.* Computers and Geosciences, 27 (6): 721-733

Iverson R (1997) *The physics of debris flows.* Rev. Geophys. Res., 35: 245-296

Locat J and Lee HJ (2001) *Submarine landslides: advances and challenges.* Can. Geotech. J., 39: 193-212

Kuenen PHH and Migliorini CI (1950) *Turbidity currents as a cause of graded bedding.* The Journal of Geology, 58: 91-107

Marr JG, Harff PA, Shanmugam G and Parker G (2001) *Experiments on subaqueous sandy gravity flows: the role of clay and water content in flow dynamics and depositional structures.* Geological Society of America Bulletin, 113: 1377-1386

Mohrig D, Elverhøi A and Parker G (1997) *Experiments on the relative mobility of muddy subaqueous and subaerial debris flows, and their capacity to remobilize antecedent deposits.* Marine Geology, 154: 117-129

Mohrig D and Marr JG (2003) *Constraining the efficiency of turbidity current generation from submarine debris flows and slides using laboratory experiments.* Marine and Petroleum Geology 20: 883-899

Mohrig D, Whipple KX, Hondzo M, Ellis C, and Parker G (1998) *Hydroplaning of subaqueous debris flow.* Geological Society of America Bulletin, 110: 387-394

Norem H, Irgens F and Schieldrop B (1987) *A continuum model for calculating snow avalanche velocities.* In: Avalanche formation, movement and effects, Proceedings of the Davos symposium, September 1986. IAHS publication no. 162: 363-379

Norem H, Locat J and Schieldrop B (1990) *An approach to the physics and the modelling of submarine flowslides.* Marine Geotechnology, 9: 93-111

Pudasaini SP and Hutter K (2007) *Avalanche dynamics.* Springer, Heidelberg

Savage SB and Hutter K (1989) *The motion of a finite mass of granular material down a rough incline.* J. Fluid Mech., 199: 177-215

Sveen JK and Cowen EA (2004) *Quantitative imaging techniques and their application to wavy flows.* In Grue J, Liu PLF and Pedersen G (eds) PIV and Water Waves, World Scientific, New Jersey

Takahashi T (1991) *Debris flow.* IAHR Monograph series, Balkema, Rotterdam

Talling PJ, Peakall J, Sparks RS, Cofaigh CO, Dowdeswell JA, Felix M, Wynn RB, Baas JH, Hogg AJ, Masson DG, Taylor J and Weaver PPE (2002) *Experimental constraints on shear mixing rates and processes: implications for the dilution of submarine debris flows.* In: Dowdeswell JA and Cofaigh CÓ (eds.) Glacier-Influenced Sedimentation on High-Latitude Continental Margins, Geological Society of London, Special Publication, 203: 89-103

THE GENERAL BEHAVIOR OF MASS GRAVITY FLOWS IN THE MARINE ENVIRONMENT

A.W. NIEDORODA, C.W. REED, H. DAS AND L. HATCHETT
URS Corporation, 1625 Summit Lake Drive, Suite 200, Tallahassee, FL, 32317, USA

Abstract

Deep sea turbidity currents, mud flows, and debris flows have been the subject of a number of industry and government studies over the past two decades. While evidence of these flow events are common in a wide variety of continental slope and rise locations, the mode, scale, and frequency of these events have been shown to vary widely from place to place. Based on over more than a dozen field and modeling projects, we present an overview of the controls, scale, flow type, and flow behavior. The most general controlling factors are the type and scale of the triggering event, the slope and morphology of the seafloor, and the material properties of the flow. In this overview we focus on details of the evolving flows that need to be included in quantitative analyses with numerical models.

1. Introduction

Mass gravity flows include a variety of natural phenomena that are characterized by the near-bottom downslope flow of sediment and water. In the marine environment these take on a number of specific characterizations ranging from thin suspensions of sediment to flows of fluid mud, some with relatively high values of density.

In this paper we consider the interplay between the form of the mass gravity flows and the seafloor conditions where they occur. This includes the type and scale of the triggering event, the relevant seafloor morphology, and the physical properties of the materials associated with the flows. Many of the flow types demonstrate a remarkable degree of scale independence. The appearance and gross behavior of the flows are similar over about five orders of magnitude ($\sim 10^{-2}$ to $\sim 10^{3}$ km). However, the specific behaviors, including flow type, flow speed, erosiveness, run-out distance, and deposit thickness and width, as well as the ability to disrupt antecedent sediment, vary greatly.

This is a synthesis of information learned from a large number of studies of marine mass gravity flows in a wide range of deep sea environments. It is incorporates published research findings. We intend to provide an overview of the subject as a framework for guiding evaluations of these geohazards and for future studies.

2. Flow Types in the Marine Environment

Mass gravity flows consist of more or less rapid movement of fluid sediment masses that are driven downslope by gravity. Marine debris flows are a subclass. Middleton and Hampton (1973) discuss the whole range of mass gravity flows, including uncommon examples such as topples and grain flows. The most common forms encountered by the offshore industry are turbidity currents, debris flows, and mud flows. Gani (2004) showed that these flows are distinguished by four characteristics: sediment concentration,

V. Lykousis, D. Sakellariou and J. Locat (eds.), Submarine Mass Movements and Their Consequences, 111–118.

sediment-support mechanism, flow rate, and rheology. Of these, rheology is the most diagnostic.

3. Event Triggers

In almost all cases, marine mass gravity flows are started abruptly by a triggering event. These control the flow type. The most common trigger is some form of gravitational soil mass failure. These can originate in several different ways. Although steep parts of the continental slope often exhibit signs of mass gravity flows, steep slopes are not necessary. Mass gravity flows have been triggered on slopes as little as 1 or 2 degrees and continue to flow on slopes well less than 1 degree. Deposits can even indicate upslope flow, because the slope of the top surface is controlling.

Turbidity currents are often triggered by gravitational soil mass failures, but there are a number of other mechanisms that can start these flows. These flows tend to persist for much greater distances than other forms of marine mass gravity flows.

3.1 GRAVITATIONAL SLOPE FAILURES

Rotational and slab displacements are the two most common forms of gravitational soil mass failures. These events result from: 1) an increase in the force above a stability limit, 2) a decrease in soil material strength, or 3) a change in the slope geometry due to outside events. The failure envelope in a rotational collapse is a concave-upward curved surface. After a failure occurs, a distinctive scar is left. With proper measurements this can be used to estimate the volume of material that moved. Slab failures commonly occur when a layer of sediment resides on a stronger underlying soil and there is a distinct plane boundary between them. The plane boundary provides a sloped surface where stress concentrations develop. After a failure, mapping of this boundary permits an estimate of the sediment volume that moved during the event.

3.2 CAUSES OF SLOPE FAILURES

Whether submarine slopes are steep or gentle, they can remain stable for very long periods of time unless disturbed. Earthquake accelerations and sediment accumulations are examples of rapid and slow disturbances. During earthquakes the ground accelerations add to the destabilizing forces and, when combined with the existing internal stress field, can cause the slope stability criteria to be exceeded.

Ongoing sedimentation can slowly add to the total thickness of a weak sediment layer. As the thickness increases, the magnitude of the shear force across the plane boundary at its bottom can increase to above the point of stability.

Either slab or rotational failures can be caused by increased pore water pressure. This comes about in a number of ways. Collapse of grain-to-grain support is an example. Agitation by earthquake motions or fluctuating pore pressure due to steep storm waves causes loosely packed sediment grains to jostle and move into a tighter packing. During this process the overburden load is transferred from the grain framework to the pore water. The sediment mass suddenly loses its resistance to shear. The overburden is supported by the excess pore pressure. In these circumstances gravitational collapse can occur under small loads and low slopes. Venting of gas or pore water from underlying sediments can also create excess pore pressure.

Changes in the geometry of submarine slopes are the third general cause of gravitational soil mass failures. This can occur in several ways. The down-cutting of a submarine canyon or erosion in an adjacent tributary canyon can undercut a slope and cause it to become unstable. Some submarine slopes are actively growing. For example, the lower portion of the continental slope off Texas and Louisiana is characterized by the steep 900-m high Sigsbee Escarpment. This slope is deforming as a result of underlying creep in salt deposits. In the Caspian Sea, deep-seated compression due to converging continental plate movements expel liquid mud, forcing it upward where it can inflate isolated strata or erupt at the seafloor.

3.3 ORIGINS OF MUD FLOW EVENTS

Mud flows represent the same flow behavior as debris flows but tend to be more fluid. Outstanding examples occur in the relatively shallow water depths of marine deltas. Prior and Coleman (1977), and others, have shown the association of these features with wave-induced gravitational soil mass failures.

Mud volcanoes are remarkably similar in appearance to their igneous equivalents. These form both above and below the sea. They are common along the eastern margins of the Caspian Sea and are known in many ocean locations. Deep-seated tectonic processes squeeze fluid mud upward along passages that vent at the seafloor. The resulting discharges are mud flows that can last for hours or days.

3.4 ORIGINS OF TURBIDITY CURRENT EVENTS

There are several important triggers for turbidity currents. Turbidity currents triggered by sudden events such as gravitational soil mass failures, their resulting debris flows, or mud flows are considered short events. The duration of the turbidity current can be far longer than its triggering event because the current becomes elongated as it travels.

Very persistent turbidity currents also develop. Imran and Syvitski (2000) have described conditions at river mouths where the suspended sediment load is so high that the discharge plume sinks to the bottom as it enters the ocean (hyperpycnal flow). Only a few of the rivers of the world carry such high sediment loads, and then only for a few hours or days. When these flows reach the shelf edge they tend to continue as turbidity currents, often contained within a submarine canyon.

P. Traykovski et al. (2001) have found that low frequency storm waves resuspend muddy river plume deposits on the inner Eel River shelf off Northern California. In storms the wave orbital boundary layer becomes saturated with suspended sediment. A strong density contrast develops at the top of the wave boundary layer, which inhibits further upward dispersion of the suspended sediment. The thick sediment load, suspended by the strong fluid shear in the wave boundary layer, is drawn downslope by gravity. The result is an offshore transport of sediment which either reaches the shelf edge or dissipates on the outer shelf where the wave orbital activity diminishes.

4. General Flow Behavior

Debris and mud flows can be rapid (10s m/sec) or slow (1/10th m/sec). The whole mass of sediment and entrained water acts as a single fluid much like ketchup. The densities of these fluids vary from a little more than the surrounding water to values approaching those of the parent materials from which they derive. Elverhoi et al. (2000) and Harbitz et al. (2003) have identified four stages of debris flow events. These are: 1) initial failure (trigger event), 2) transition, 3) flow, and 4) deposition. The transitional stage follows the trigger event. The internal particle-to-particle structure is deranged and the material strength drops due to remolding as the soil mass begins to accelerate downslope. Submarine events also have an opportunity to uptake water. The details of these processes are poorly known (Harbitz et al. 2003).

Turbidity currents are different. These are suspensions of sediment grains in a turbulent flow. The suspension is most dense near the seabed and decreases to the value of the surrounding water at the top of the flow. Averaged over the whole height of the flow, these suspensions are on the order of 3 to 5 % by volume. A downslope flow of turbid water is said to “ignite” when it erodes as much, or more, sediment than is settling out. Under these conditions the mass of the turbid water increases and further acceleration occurs. The dynamics of stable turbidity currents necessitate that sediment particles are eroded at nearly the same rate as they settle to the bed. However, where the flow is accelerating, erosion tends to dominate and deposition occurs where the flow is slowing.

4.1 COMPOUND EVENTS

Although turbidity currents can form without associated debris or mud flows, the converse is rare. The triggering event and subsequent rapid downslope flow of the mud-rich debris flows cause high fluid shear at the upper boundary where the flow passes beneath the ambient water. The sediment in the debris flow is often weak and easily eroded. The resulting “cloud” of turbid water is accelerated by the boundary shear. This can reach the turbidity current ignition condition. Ilstad et al. (2004) have shown that a mud-poor debris flow can evolve directly into a turbidity current as excess ambient water penetrates the head. In other cases, the cloud of turbid water does not accelerate enough to reach the ignition condition, so that it slowly deposits its load and dissipates.

4.2 COMPOUND SCALES

Many of the subsea environments where mass gravity flows are of concern have complex histories. There are typically several cycles of sedimentation and erosion portrayed in the sculpted morphologies of the seafloor. Often some of these cycles are related to variations in the sea level and the supply of sediment brought about by the waxing and waning of the huge Pleistocene glaciers. A common result is that deposits of significantly different strength and water content alternate on the seafloor. Steep slopes on strong material often have systems of valleys. These can be active sedimentation sites collecting weak sediments. As the recent sediment deposits become thicker they can reach a point of instability. It is important to recognize this compound arrangement of potential trigger events because we have often found that strong and rare events are needed to cause the underlying slopes to fail. However, the weaker surficial deposits can be a much greater hazard because they can reach failure conditions more frequently.

5. Material Properties

Mass gravity flows are controlled, in large part, by the material of which they are comprised. However, this holds true in different ways for debris flows and turbidity currents.

Like their terrestrial counterparts, marine debris flows have a wide range of compositions; silt and mud are common constituents. The composition of a debris flow is determined by the relative amount of cohesive clay and granular particles, the size of the grains and clasts, the clay mineralogy, and the water content. In coarse-grained debris flows where the clay content is relatively low, the flow is characterized by both an internal shearing and grain-to-grain dispersive pressures (Huang and Garcia 1998). However, in our experience the most common marine debris flows are not coarse-grained, have considerable clay content (> 25 %), and deform plastically. Most studies have found that a Bingham Fluid representation is adequate to represent all but coarse granular debris flows.

Sensitivity is the ratio of the undisturbed and remolded shear strengths. This comes about because a soil loses its strength when internal shearing disrupts the grain-to-grain soil structure. Strength decreases by factors of two to three are common (Locat and Lee 2002) and in extreme cases may be an order of magnitude or more (Locat and Demers 1988). Carbonate sediment tend to have high values because grains shatter and collapse.

The relevant material properties of debris flows are notoriously difficult to measure. These properties are transient, changing from the initial values as the soil mass fails, to the reduced strength during the flow, and then converting to yet other values as the resulting deposits dewater and age. For this reason, most attempts to study debris flows with numerical models are forced to treat the Bingham shear strength and viscosity as parameters to be fitted during model calibration.

The sediment properties associated with turbidity currents are generally those related to most sediment transport analyses. These define the erosion and deposition properties. Here again, the role of mud is important because as little as 8 to 10 % generally causes cohesive behavior in the seabed. Additionally, similar amounts of fine clay particles in the suspension settle extremely slowly, so that they serve to help perpetuate the current once started.

The erodibility of the seafloor sediment can be determined from a Shields Curve (Middleton and Hampton 1973) provided that it is entirely granular. Although there have been various attempts to develop equations to express cohesive sediment erosion parameters in terms of traditional geotechnical measurements, these are not widely accepted. Instead, the fluid stress threshold for erosion and the rate of erosion are determined is special apparatus (Briaud et al. 2001)

Once a turbidity current has accelerated to the ignition condition its stability is determined by a balance between the rate that new sediment is entrained in the flow and the rate that sediment is deposited. The deposition is controlled, in turn, by a balance between turbulent diffusion, which acts to diffuse the grains upward, and grain settling due to gravity. Thus, a quantitative analysis of a turbidity current depends on an accurate knowledge of the particle settling speed. This can be accomplished in any of a variety of grain size analysis methods. The caution is that the sediment grains in the bed are not a true representation of what the current is carrying, because fine and very fine grains travel much further than silt and fine sand. Therefore, samples need to be taken along a considerable length of the flow path before the average grain size distribution of the suspended sediment can be estimated.

6. Role of Antecedent Conditions

As mentioned earlier, it has been commonly observed that marine debris flows run out further than their terrestrial counterparts. One mechanism to explain this is hydroplaning at the nose. However, the flowing portions of debris flows can be 10s of kilometers long. An alternate explanation of this far-traveling behavior can be found in the low strength and high water content in the upper 10s of centimeters of marine sediments. Figure 1 illustrates this process. The stronger, more competent soil makes up the steep slope. When a gravitational soil mass failure occurs, a debris flow forms, and it accelerates downslope to emerge onto the more flat-lying marine sediments. The high water content in the upper sediments (~0.1 to 1.0 m thick) provides lubrication. The shear shifts from within the debris flow to the lubricating layer.

Evidence for this process is found in the common observation that soft mud clasts suspended in a mud matrix (i.e. a soft sediment breccia) often characterize debris flow layers in marine cores, even 10s of kilometers from their source. These clasts have been fractured and sheared in the initial downslope run of the debris flow. However, once the flow passes onto the nearly flat-lying sediments, the shearing ceases. In this way these delicate textured layers can be transported 10s of kilometers from their sources.

7. Erosion and Deposition

Both turbidity currents and debris flows are known to erode. This behavior in turbidity currents is much like that in rivers, where erosion is associated with accelerating conditions and deposition occurs as the flows slow down.

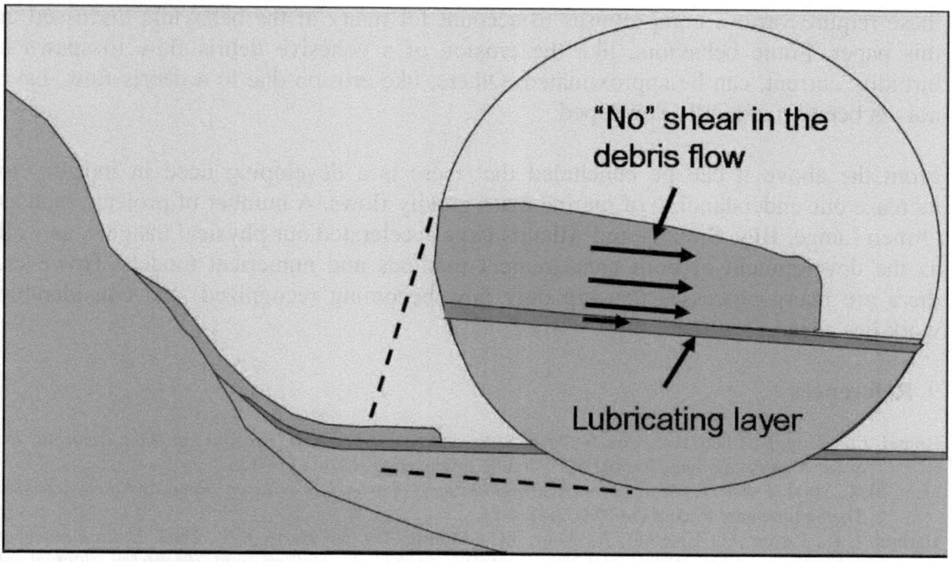

Figure 1. Schematic of a debris flow lubricated by the high-water-content sediment zone.

Not all debris flows erode. Usually, they simply pass over the seafloor sediment. When the surface of the antecedent sediments is weak, with a density close to that of the debris flow, it is possible for the basal shear to propagate downward into the underlying sediments. A thought experiment to illustrate this consists of envisioning pouring a small volume of ketchup onto a sloped board. This will run down and become thinner until it stops. Then, if another small volume is poured on top of the initial "deposit," both layers will move downslope together until they again reach a limiting thickness. Behavior of this type has been observed in natural debris flows (Schellmann et al. 2005).

8. Discussion and Conclusions

A geohazards investigation where ongoing mass gravity flow activity is suspected has four parts. Background data are used in an initial site evaluation and for the design of a survey. The survey leads to the creation of a geological model. Finally, numerical models are applied to both replicate past activity and to forecast future events.

Numerical models of turbidity currents and debris/mud flows play an important role in these investigations. Here it is assumed that the seafloor features and deposits are a record of past events. When good models demonstrate the capability of reproducing the physical features and deposits of the existing seafloor, they are thought to be trustworthy as indicators of the speeds, dimensions, and run-out lengths of potential future flows.

Models today are powerful but still incomplete. None follow the full sequence from the trigger event to the final run-out. Instead, sequences of models are typically applied, and these require various compromises to account for many of the behaviors discussed in this paper. Some behaviors, like the erosion of a cohesive debris flow to spawn a turbidity current, can be approximated. Others, like erosion due to a debris flow, have not yet been successfully developed.

From the above it can be concluded that there is a developing need in industry to increase our understanding of marine mass gravity flows. A number of projects such as Ormen Lange, Blue Stream, and Atlantis have accelerated our physical insights, as well as the development of both measurement methods and numerical models. However, there are many processes that are only now becoming recognized, and considerable work lies ahead.

9. References

Briaud, J.L., Ting, F., Chen, H.C, Coa, Y., Han, S.W., and Kwak, K., 2001. *Erosion Function Apparatus for Scour Rate Predictions*, Journal of Hydraulic Engineering, 127(2) 114-125.

Gani, M.R., 2004. *From Turbid to Lucid, A straightforward approach to sediment flows and their deposits*, The Sedimentary Record (SEPM), 2(3), 4-11.

Harbitz, C.B., Parker, G., Elverhoi, A., Marr, J.G., Mohrig, D., and Harff, P.A., 2003. *Hydroplaning of subaqueous debris flows and glide blocks: Analytical solutions and discussion.* Journal of Geophysical Research 108 (B7).

Huang, X. and Garcia, M.H., 1998. *A Herschel-Bulkley model for mud flow down a slope*, Journal of Fluid Mechanics, 374, 305-333.

Elverhoi, A., Harbitz, C.B., Dimakis, P., Mohrig, D., Marr, J, and Parker, G., 2000. *On the dynamics of subaqueous debris flows.* Oceanography 13(3), 109-125.

Ilstad, T., Elverhoi, A., Issler, D., and Marr, J.G., 2004. *Subaqueous debris flow behavior and its dependence on the sand/clay ratio: a laboratory study using particle tracking.* Marine Geology 213, 415-438.

Imran, J. and Syvitski, J.P.M., 2000. *Impact of extreme river events on the coastal ocean*, Oceanography, 13(3):85-92.

Locat, J. and Demers, D., *Viscosity, yield strength, remolded strength and liquidity index relationships for sensitive clays,* Canadian Geotechnical Journal, 25, 799-805.

Locat, J. and Lee, H.J., 2002, *Submarine landslides: advances and challenges*, Canadian Geotechnical Journal, 39, 193-212.

Middleton, G.V. and Hampton, M.A., 1973. *Sediment gravity flows: Mechanics of flow and deposition*, in Middleton G.V. and Bouma, A.H. (eds.), Turbidites and Deep-water Sedimentation, S.E.P.M., Los Angeles, 1- 38.

Prior, D.B. and Coleman, J.M., 1977. *Disintegrating retrogressive landslides on very-low-angle subaqueous slopes, Mississippi Delta.* Marine Geotechnology, 3(1), 37-61.

Schnellmann, M., Anselmentti, F.S., Giardini, D. and McKenzie, J.A., 2005. *Mass movement-induced fold-and-thrust belt structures in unconsolidated sediment in Lake Lucerne (Switzerland),* Sedimentology, 52, 271-289.

Traykovski, P., Reed, C., and Geyer, W.R., 2001. "*Observations and Modeling of Gravity Flows of Fluid Mud Trapped Within the Wave Boundary Layer*," AGU Chapman Conference, Ponce Puerto Rico, June 17-19, 2001.

SUBMARINE SPREADING: DYNAMICS AND DEVELOPMENT

A. MICALLEF *, D.G. MASSON, C. BERNDT AND D.A.V. STOW
National Oceanography Centre, European Way, Southampton, SO14 3ZH, UK.
Telephone: +44 23 80 596563; *Contact E-mail: *amicall@noc.soton.ac.uk*

Abstract

Spreading is a pervasive type of ground failure in subaerial environments, but its occurrence has hardly been documented in submarine settings. However, recent advances in seafloor imaging techniques show that repetitive extensional patterns of parallel ridges and troughs, oriented perpendicular to the direction of mass movement and typical of spreading, are widespread offshore. A spread develops via the failure of a surficial sediment unit into coherent blocks. These blocks are displaced downslope along a quasi-planar slip surface. Two modes of failure can be identified: retrogressive failure of the headwall, and slab failure and extension. Mechanical modelling indicates that loss of support and seismic loading are the main triggering mechanisms. The extent of displacement of the spreading sediment is controlled by gravitationally-induced stress, angle of internal friction of sediment, pore pressure escape and friction. The resulting block movement patterns entail an exponential increase of displacement and thinning of the failing sediment with distance downslope. A deeper insight into submarine spreading is important because of the widespread occurrence of ridge and trough morphology in numerous submarine slides, particularly in the vicinity of submarine infrastructures.

1. Introduction

Spreading involves the fracturing of a thin surficial layer of rock or soil into coherent blocks and their finite lateral displacement on gently sloping ground (Varnes 1978). The associated ground deformation is generally characterised by extensional fissures, resulting in a series of parallel ridges and troughs at the surface (Bartlett and Youd 1995). Recent advances in acoustic data acquisition techniques allow the identification of ridge and trough morphology within the Storegga Slide. Located on the mid-Norwegian margin, the Storegga Slide occurred 8100±250 cal. yrs BP as sixty-three quasi-simultaneous slide events (Haflidason *et al.* 2005). It is one of the largest known submarine mass movements, and it has been thoroughly surveyed by state-of-the-art acoustic imagery. Spreading has seldom been observed in the submarine realm, and the Storegga Slide provides an ideal setting for a deeper investigation of this type of mass movement. The objectives of this study are to identify the mechanisms that give rise to submarine spreading, and to characterise the associated morphology.

2. Data sets

There are three acoustic data sets available for this study (Figure 1a). The first consists of high resolution multibeam bathymetry that covers the Storegga Slide scar with a horizontal resolution of 25 m or better. The second comprises Towed Ocean Bottom Instrument (TOBI) sidescan imagery of 60% of the slide scar. The nominal resolution of these data is 6 m. The third data set consists of high resolution 2D seismic lines located in the northeast of the slide scar.

V. Lykousis, D. Sakellariou and J. Locat (eds.), Submarine Mass Movements and Their Consequences, 119–127.

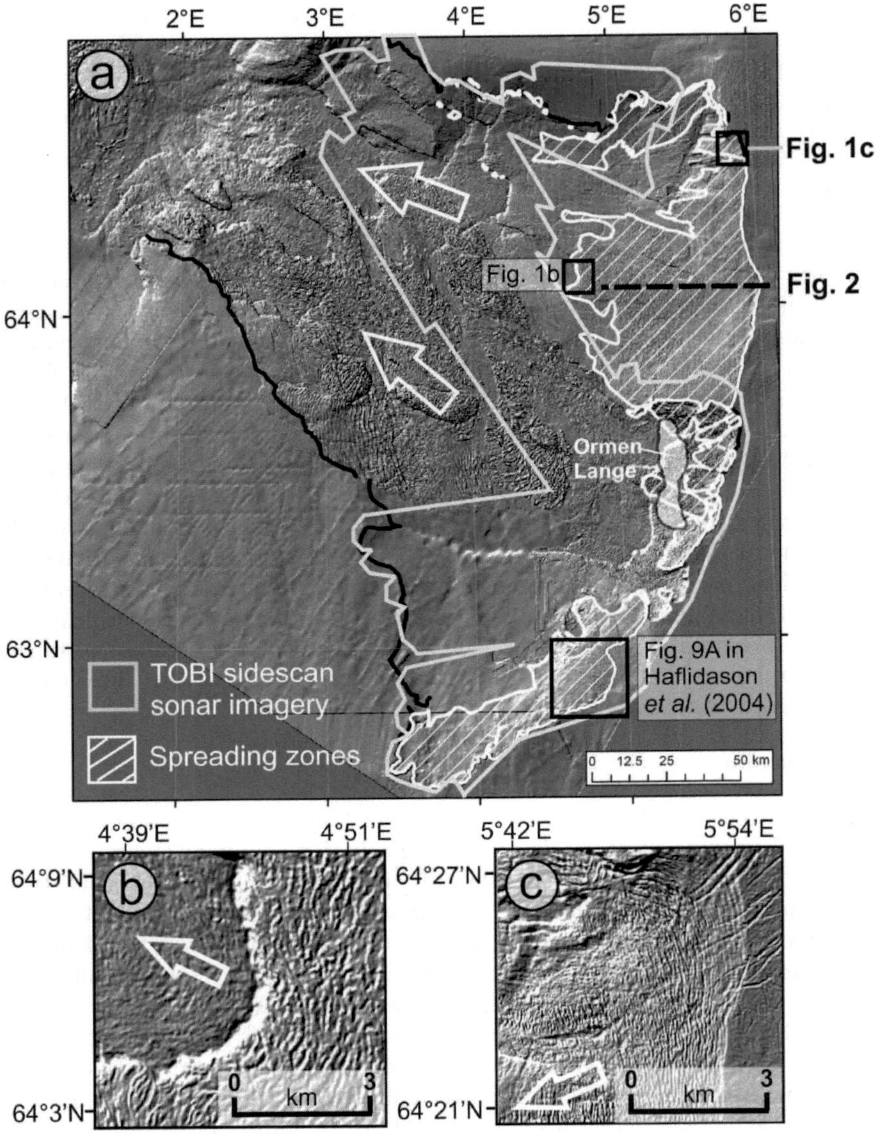

Figure 1. (a) Shaded relief map of the Storegga Slide (illumination from NW, 3× exaggeration), showing the zones of spreading and the coverage of TOBI sidescan sonar imagery. (b) Enlarged section of the shaded relief map from the distal part of a spread, where a steep and high escarpment is located. Upslope of the escarpment, the ridges are clearly visible and closely spaced, whereas downslope of the escarpment the morphology is more subdued and blocky. (c) Enlarged section of the shaded relief map from the upper part of a spread, close to the main headwall. The ridge pattern, which is more closely spaced in this region, is intersected by a number of iceberg ploughmarks. These are older than the spread and demonstrate the limited extension that has taken place near the headwall. The white arrows indicate the direction of sediment mobilisation. Also shown is the location of Figure 9a in Haflidason *et al.* (2004), which is a good example of sidescan sonar imagery of the ridge and trough morphology from the southern part of the Storegga Slide.

3. Results

Ridge and trough morphology is widespread within the Storegga Slide (Figure 1). The ridges and troughs are generally aligned parallel to a headwall. The plan pattern of the ridges and troughs is concave-downslope in the upper parts of the slope, changing to linear or convex-downslope with increasing distance from the headwall.

A vertical profile of the ridge and trough morphology in one of the high resolution seismic lines is shown in Figure 2. The seismic section consists of an upper sequence of downslope dipping reflectors, located at the top of reflector packets dipping at a similar downslope gradient. The reflector packets are separated by a series of upslope dipping reflectors (Figures 2b and 2d). At the bottom of the seismic section are four quasi-planar continuous reflectors. The upper seismic reflectors and the reflector packets are slightly steeper dipping than the bottom reflectors (Figures 2b and 2d). The reflectors in this seismic section are interpreted as layers in a stratified sediment package. This package is undeformed in the deeper parts, whereas it is broken up into blocks in the shallower parts. The lengths of these blocks were estimated at ~130 m. Using an acoustic velocity of 1700 ms^{-1} for the sediments, the mean slope gradient of the surface of the undeformed section was estimated at 1°. The estimated thickness of the failing sediment decreases gradually from 80 m in the upslope part of the seismic line, to 25 m in the downslope part (Figure 2c). The dips of the interfaces between the blocks have an average angle of ~25°. We interpret the blocks as having translated along a quasi-planar slip surface as the sediment unit was extended. During this displacement, the blocks have tilted anticlockwise and downslope, and in the process the upper part of the blocks has been exposed. This has resulted in a step-like morphology that is responsible for the ridges and troughs at the surface. The block pattern is best preserved close to the headwall, and there is progressive deformation of the blocks with distance downslope (Figure 2a).

Having established that the ridge and trough morphology is representative of spreading, we mapped the spatial extent of this mass movement across the Storegga Slide (Figure 1a). Ridge and trough morphology can be observed over a total area of 6000 km^2, or ~25% of the slide scar. It is mainly located along the main Storegga headwall, in the northeastern and southern parts of the slide scar. The spreading regions are generally bounded by a gentle, shallow headwall at their upslope limit (Figure 1c), and a steep, high escarpment at their downslope limit (Figure 1b). In the upslope sections, a number of iceberg ploughmarks, older than the Storegga Slide, can be identified, which are indicative of the limited displacement of the sediment (Figure 1c). Downslope of the distal escarpment, the surface is characterised by either blocks or very subtle ridges (Figure 1b).

Ridge and trough morphology varies across the spread in Figure 2a. The depth of the troughs shows a general increase from 1.5 m near the headwall to 5.5 m towards the toe. The length of individual ridges along the crest decreases from 360 m upslope to 250 m downslope. The number of ridges per unit area increases away from the headwall, reaching values of up to 12 km^{-2}. Spread morphology also varies across the Storegga Slide, with ridges in the Ormen Lange region being longer, more widely spaced, and having troughs up to 3× deeper than elsewhere. In the Ormen Lange region, the downslope face of the ridges is steeper than the upslope face. Failure takes place in the deep and thick O4-O7 sediments (200-130 ka) of the Naust formation, which consist of glacial till and debris flow deposits (Berg *et al.* 2005). The ridges in the rest of the spreading zones are shorter, more closely spaced and have shallower troughs. The downslope face of these ridges is gentler than the upslope face, and failure takes place in the shallow and thin O1-O2 sediments (30-15 ka), consisting of basal and deformation till, and O3 sediments (130-30 ka), which are fine-grained hemipelagic and glacial marine clays (Bryn *et al.* 2005) (for more information on the Storegga Slide stratigraphy refer to Berg *et al.* (2005)).

4. Discussion

4.1 MODE OF FAILURE

We propose two models for the failure dynamics of spreading based on the morphology and internal structure of the ridges within the Storegga Slide (Figure 3). The first model (model 1) entails failure in a thin slab of sediment underlain by a failure plane or 'weak layer'. The slab, which is under tension, breaks up into a series of coherent blocks, with shear planes dipping upslope (Figure 3a). This occurs because of extensional forces that are higher downslope and basal resisting forces that are stronger upslope (due to, for example, changes in the thickness of the failing layer (Berg *et al.* 2005) and/or a decrease in excess pore pressure with distance upslope (Strout and Tjelta 2005)). Failure can initiate anywhere along the slope. In the Ormen Lange region, the ridge morphology is different. Failure in this region can be explained by the model of Kvalstad *et al.* (2005) (model 2) where spreading propagates upslope via the repeated unloading of the headwall and the fracturing of the sediment into blocks (Figure 3b). The shear planes dip downslope and the blocks are rotated in a clockwise direction. The difference in the mode of failure is explained by the different properties of the failing sediment, with the failing sediment in the Ormen Lange being thicker and having lower clay content (Berg *et al.* 2005).

4.2 TRIGGERS OF A SPREAD

Both models of failure can be modelled using a limit-equilibrium model (Figure 3a). We considered the driving and resisting forces acting on a series of equidimensional blocks resting on a planar slip surface:

Driving forces:

$$Sin\theta \, (W_T) \tag{1}$$

Resisting forces:

$$Tan \, \o \, [W_T(Cos \, \theta) - u] + c + P \tag{2}$$

W_T total weight of sediment upslope of a block = γSl
γ submerged unit weight (in 2D)
S sediment unit thickness prior to failure

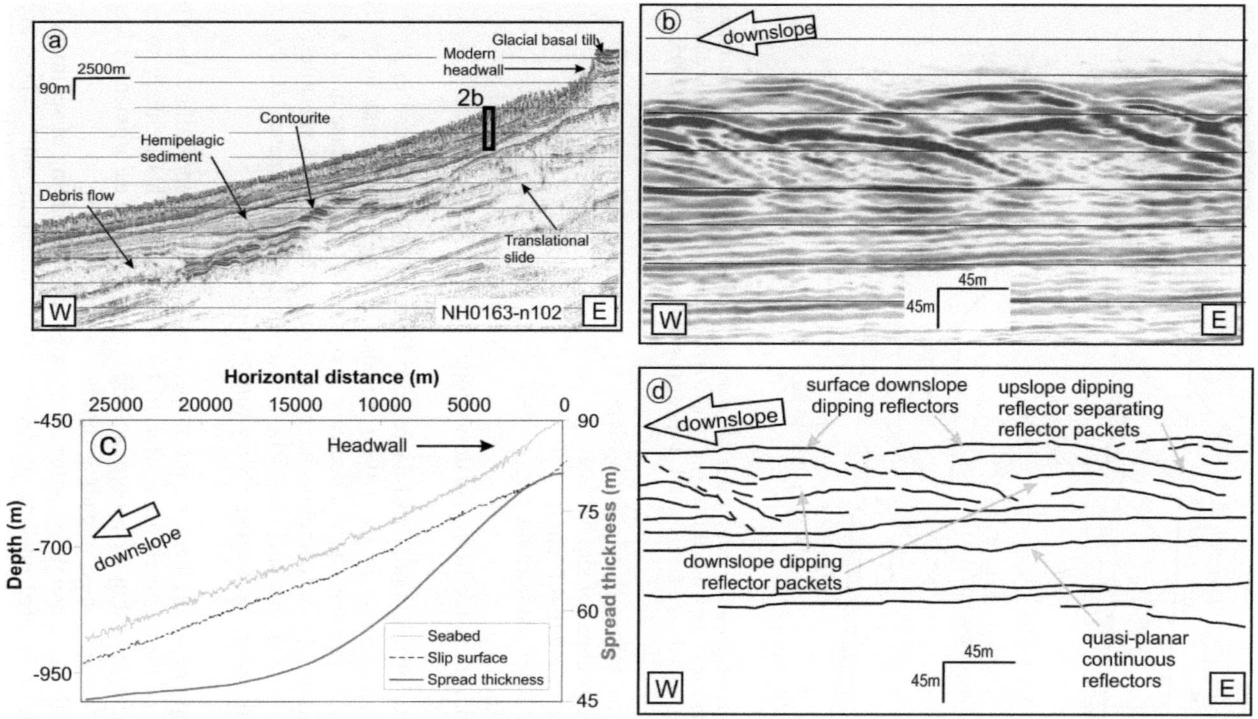

Figure 2: (a) Annotated profile of seismic line NH0163-n102; location is shown in Figure 1a. (b) Enlarged section of a part of the seismic seismic profile, showing the internal structure of the ridge and trough morphology. (c) Interpreted seabed and slip surface of the seismic profile in Figure (a), and the calculated thickness of the spreading sediment. (d) Labelled interpretation of the enlarged seismic section in (b). See section 3 for a more detailed explanation.

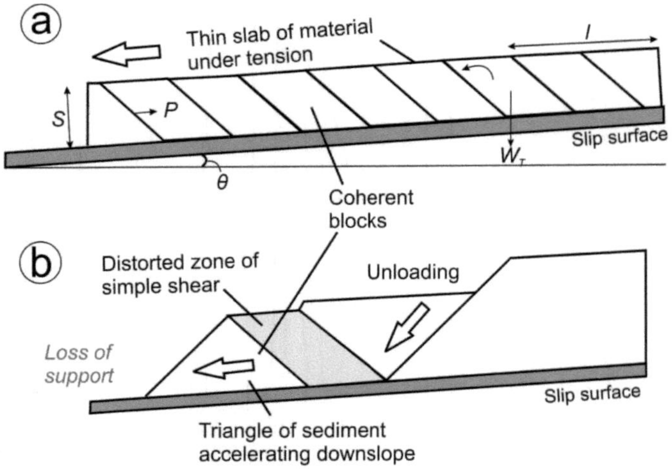

Figure 3. Schematic illustration of the two models of failure: (a) model 1 (slab extension and rupturing); (b) model 2 (repeated failure of the headwall). Figure (a) also shows some of the static forces and dimensional attributes considered in the limit-equilibrium model. Figure (b) is adapted from Kvalstad *et al.* (2005).

l	distance from a fixed point upslope
θ	slope gradient of slip surface
$\o$	angle of internal friction
u	pore water pressure (in 2D)
c	cohesion
P	supporting force from slab downslope

The factor of safety of this model will decrease if there is: (a) a decrease in P; (b) an increase in u; (c) an increase in W_T; (d) an increase in γ; (e) an increase in S; (f) an increase in l; (g) an increase in θ; (h) a decrease in $\o$; and/or (i) a decrease in c. A slope failure can be initiated by a temporal change in one/many of these variables, if the magnitude of the change is such that the factor of safety decreases below zero. Therefore, a spread within the Storegga Slide can be triggered by:

(a) Loss of support at the base of the slope, caused by a slope failure occurring downslope of the sediment unit affected by spreading. This is likely to be an important trigger within the Storegga Slide because a steep escarpment is located in the distal edge of most spreading zones.

(b) Increase in the total weight of sediment upslope, due to loading of sediment from a slope failure located upslope of the spreading sediment unit. There are no indications that this is a significant trigger within the Storegga Slide.

(c) Seismic loading, caused by the glacio-isostatic rebound following the deglaciation of Scandinavia (Atakan and Ojeda 2005), may have induced downslope shear stresses, leading to short term failure and strength loss, and/or increase in pore pressure.

4.3 PATTERN OF BLOCK DISPLACEMENT

We apply equations of motion to the blocks in Figure 3a to infer their behaviour during spreading. We can calculate the acceleration, velocity and distance travelled by individual blocks if we assume that the failure was instantaneous, and that fluid resistance and friction at the base are constant along the entire slip surface. We estimate the values for the different variables in the model from a 35 km section of the seismic line NH0163-n102. The blocks are 130 m wide, and θ and S are constant at 1° and 80 m, respectively. γ and $\o$ increase upslope from 9 kN m^{-2} to 10 kN m^{-2}, and from 25° to 27.5°, respectively. The reason for this variation is that the sediments become more consolidated upslope towards the shelf edge due to compaction by glacial advance during glacial maxima. u increases from 1000 kN m^{-2} downslope to 600 kN m^{-2} upslope, due to unloading of the sediment in the upslope region after the Last Glacial Maximum (Strout and Tjelta 2005). c is constant at 7 kPa (Sultan *et al.* 2004).

Changes in the velocity and distance covered by individual blocks along a section of the slope, after a spread is triggered, are shown in Figure 4a. The first block to be released (block 1) attains the highest velocity and covers the longest distance. This is because in the downslope section of the slope, W_T, l and u have high values, and γ and $\o$ have low values, creating the conditions for the highest resultant force. The values of W_T, l, u, γ and $\o$ change with distance upslope, resulting in a lower velocity attained and lower distance covered by successive blocks. The variation of block displacement with distance downslope is an exponential increase (Figure 4b).

According to this model, the velocity of the blocks and the distance they cover should increase continuously with distance. This is not the case because the distal remnants of spreads can be identified downslope of the distal escarpment. As the spreading sediment collapses over the escarpment, the blocks fragment. This allows the escape of excess pore pressure from the base of the spread, which reduces the velocity of the blocks. The sediment remaining upslope of the distal escarpment is likely to be slowed down by loss of excess pore pressure via the remoulding of sediment due to friction. Furthermore, as the spreading unit extends and breaks up, it also becomes thinner (Figure 2c). The decrease in S reduces the gravitationally-induced stress and retards block displacement further.

5. Conclusions

Spreads within the Storegga Slide are most probably triggered by either a loss of support due to mass movements taking place downslope of a spreading unit, or by seismic loading. When a spread is triggered, shear planes and coherent blocks form within the sediment unit. The mode of failure involves either the retrogressive failure of the headwall, as for the Ormen Lange region, or slab extension and rupturing, as for the rest of the spreading zones. The blocks in the distal edge of the spread are displaced the most and generally collapse over a pre-existing escarpment. The sediment either preserves the ridge and trough morphology or develops into a debris flow.

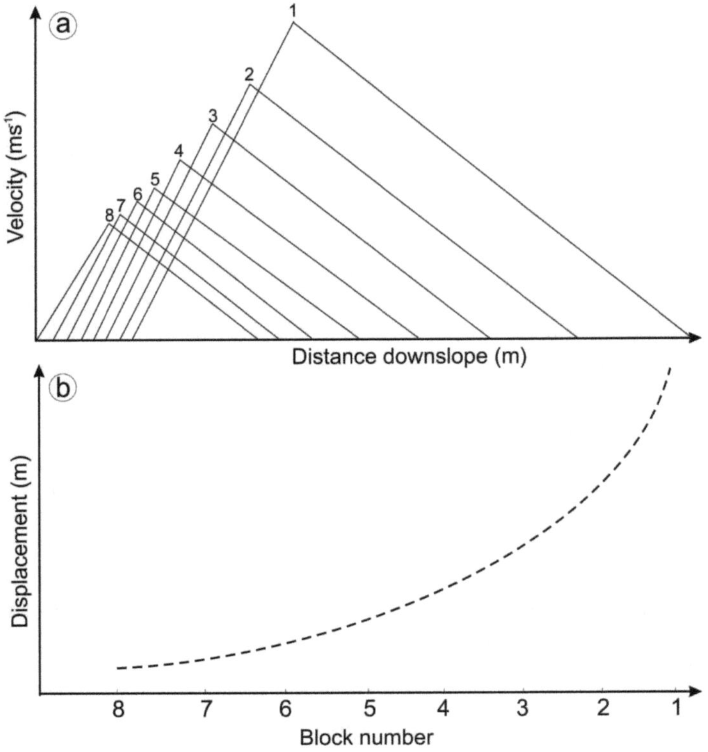

Figure 4. (a) Plot of velocity vs. distance downslope for a number of blocks in a theoretical spread, using values from the Storegga Slide. Block number indicates the order in which the blocks are displaced. (b) Exponential variation of block displacement with block number.

In the remaining part of the spread located upslope of the distal escarpment, extension, friction and water resistance combine to fragment and remould the sediment blocks as they are displaced downslope. Loss of excess pore pressure and thinning of the sediment unit retard the blocks and finally bring them to a halt. Block disintegration, displacement and velocity decrease exponentially upslope until a gentle headwall is formed. Within Storegga, this pattern is controlled by changes in pore pressure, gravitationally-induced stress, and the angle of internal friction of the sediment.

A recurrent extensional pattern of parallel ridges and troughs, oriented perpendicular to the direction of movement, can be observed in numerous well-known submarine slides around the world (e.g. Trænadjupet Slide in Laberg *et al.* (2002); Nyk Slide in Lindberg *et al.* (2004); BIG'95 Slide in Lastras *et al.* (2003); Hinlopen Slide in Vanneste *et al.* (2006)). This may indicate that spreading is a widespread type of submarine mass movement. Spreading tends to occur on gentle terrain over extensive regions, some of which are located in the vicinity of oil and gas exploitation infrastructure (e.g. Ormen Lange). A better understanding of spreading as a potential geohazard, and its role in the evolution of submarine slides, is thus pertinent.

6. Acknowledgements

We are grateful to Norsk Hydro AS for providing the bathymetric data and the high-resolution 2D seismic lines. European North Atlantic Margin (ENAM) II Programme is acknowledged for making the TOBI sidescan sonar imagery available.

7. References

Atakan, K. and Ojeda, A., 2005. *Stress transfer in the Storegga area, offshore mid-Norway.* Marine and Petroleum Geology, 22(1-2): 161-170.

Bartlett, S. F. and Youd, T. L., 1995. *Empirical prediction of liquefaction-induced lateral spread.* Journal of Geotechnical Engineering, 121(4): 316-329.

Berg, K., Solheim, A. and Bryn, P., 2005. *The Pleistocene to recent geological development of the Ormen Lange area.* Marine and Petroleum Geology, 22(1-2): 45-56.

Bryn, P., Berg, K., Forberg, C. F., Solheim, A. and Kvalstad, T. J., 2005. *Explaining the Storegga Slide.* Marine and Petroleum Geology, 22(1-2): 11-19.

Haflidason, H., Lien, R., Sejrup, H. P., Forsberg, C. F. and Bryn, P., 2005. *The dating and morphometry of the Storegga Slide.* Marine and Petroleum Geology, 22(1-2): 123-136.

Haflidason, H., Sejrup, H. P., Nygård, A., Bryn, P., Lien, R., Forsberg, C. F., Berg, K. and Masson, D. G., 2004. *The Storegga Slide: Architecture, geometry and slide-development.* Marine Geology, 231: 201-234.

Kvalstad, T. J., Andersen, L., Forsberg, C. F., Berg, K., Bryn, P. and Wangen, M., 2005. *The Storegga slide: evaluation of triggering sources and slide mechanisms.* Marine and Petroleum Geology, 22(1-2): 245-256.

Laberg, J. S., Vorren, T. O., Mienert, J., Evans, D., Lindberg, B., Ottesen, D., Kenyon, N. H. and Henriksen, S., 2002. *Late Quaternary palaeoenvironment and chronology in the Trænadjupet Slide area offshore Norway.* Marine Geology, 188: 35-60.

Lastras, G., Canals, M. and Urgeles, R., 2003. *Lessons from sea-floor and subsea-floor imagery of the BIG'95 debris flow scar and deposit.* In: Locat J, Mienert J (eds) Submarine Mass Movements and Their Consequences. Kluwer Academic Publishers, Dordrecht, pp 425-431.

Lindberg, B., Laberg, J. S. and Vorren, T. O., 2004. *The Nyk Slide - morphology, progression, and age of a partly buried submarine slide offshore northern Norway.* Marine Geology, 213: 277-289.

Strout, J. M. and Tjelta, T. I., 2005. *In situ pore pressures: What is their significance and how can they be reliably measured?* Marine and Petroleum Geology, 22(1-2): 275-285.

Sultan, N., Cochonat, P., Foucher, J. P. and Mienert, J., 2004. *Effect of gas hydrates melting on seafloor slope instability.* Marine Geology, 213: 379-401.

Vanneste, M., Mienert, J. and Bünz, S., 2006. *The Hinlopen Slide: A giant, submarine slope failure on the northern Svalbard margin, Arctic Ocean.* Earth and Planetary Science Letters, 245(1-2): 373-388.

Varnes, D. J., 1978. *Slope movement types and processes.* In: Schuster RL, Krisek RJ eds) Landslides, Analysis and Control. National Academy of Sciences, Transportation Research Board, Special Report 176, pp 11-33.

FLOOD-INDUCED TURBIDITES FROM NORTHERN HUDSON BAY AND WESTERN HUDSON STRAIT: A TWO-PULSE RECORD OF LAKE AGASSIZ FINAL OUTBURST FLOOD?

G. ST-ONGE

Institut des sciences de la mer de Rimouski (ISMER), Université du Québec à Rimouski, Rimouski, Québec, Canada & GEOTOP-UQAM-McGill, Montréal, Québec, Canada

P. LAJEUNESSE

Centre d'études nordiques & Département de géographie, Université Laval, Québec, Québec, Canada

Abstract

In Hudson Bay and Hudson Strait, the rapid collapse of the Laurentide Ice Sheet (LIS) culminated in the catastrophic drainage of proglacial Lake Agassiz into the North Atlantic around 8500 cal BP. It has been suggested that this catastrophic event may have triggered the 8200 cal BP cold event recorded in Greenland ice cores. Evidence for that outburst flood was the identification of a centimeter to decimeter-thick hematite-rich red layer that was observed in Hudson Strait sediments around 8000 yr BP. In this paper, we have identified a sequence of two flood-induced turbidites (i.e., hyperpycnites) in a reddish layer from two cores collected in northern Hudson Bay (core AMD0509-27bLEH) and western Hudson Strait (core AMD0509-28PC) in 2005 onboard the ice-breaker CCGS Amundsen. These two reddish layers can be correlated to a red bed previously identified as a regional isochron in Hudson Strait and associated with the final drainage of Lake Agassiz around 8500 cal BP. Regardless of the exact timing of the catastrophic drainage, the hyperpycnites described in this paper suggest that they were deposited following two pulses, which is in agreement with the one of the scenarios proposed by Clarke et al. (2003) [Science 301, 922-923] for the drainage of Lake Agassiz. Finally, this study demonstrates for the first time the turbiditic and the flood-induced nature of the Hudson Strait red bed isochron.

Keywords: hyperpycnites, turbidites, Lake Agassiz, outburst flood, 8.2 ka event, Hudson Bay, Hudson Strait

1. Introduction

Flood-induced turbidites are now recognized as an important part of the sedimentary records (see Mulder et al., 2003 for a review). They can be deposited following a river flood, but also after the breaching and rapid draining of a natural dam previously generated by mass wasting (St-Onge et al., 2004), a jökulhaup or the rapid draining of a subglacial lake. These events can suddenly release millions of cubic meters of sediment laden freshwaters. One characteristic criterion to identify such flood-induced turbidites is a sequence of reverse and normal grading. Kneller (1995) demonstrated that a basal coarsening-upward unit can be deposited by a depletive waxing flow, i.e. a flow decelerating with distance but accelerating with time, whereas Mulder and Syvitski (1995) showed that such flows can be generated by flooding rivers if the suspended

V. Lykousis, D. Sakellariou and J. Locat (eds.), Submarine Mass Movements and Their Consequences, 129–137.

matter concentration at the river mouth is above a threshold value. Beds consisting of coarsening and then fining upward units have been interpreted as hyperpycnal turbidity current deposits, or hyperpycnites, and associated with flooding both in lacustrine (Schneider et al., 2004; Chapron et al., 2006; Guyard et al., 2007, this book.; Chapron et al., subm.) and marine environments (Mulder et al., 2001a; 2001b; Mulder et al., 2002; St-Onge et al., 2004). In these deposits, the coarsening upward basal unit is deposited during the rising limb of the flood hydrograph and the top fining upward unit during the falling limb.

In Hudson Bay and Hudson Strait, the rapid collapse of the Laurentide Ice Sheet (LIS) culminated in the catastrophic drainage of proglacial Lake Agassiz into the North Atlantic around 8500 cal BP. It has been suggested that this sudden outburst of freshwater may have triggered the 8200 cal BP cold event recorded in Greenland ice cores (e.g., Alley et al. 1997; Barber et al., 1999; Alley and Ágústsdóttir, 2005). Evidence of this outburst flood was the identification of a centimeter to decimeter-thick hematite-rich red layer in Hudson Strait sediments that was dated around 8000 yr BP (radiocarbon years corrected by -450 yrs for marine reservoir effect; Andrews et al., 1995; Kerwin, 1996; Barber et al., 1999). The provenance study of Kerwin (1996) suggested that this red bed likely originated from the reworking of sediments originating from the predominantly red rocks of the Dubawnt Group. These rocks outcrop on land northwest of Hudson Bay and were transported in the western central sector of the bay by an ice stream of the LIS (Shilts, 1986; Fig. 1). In this paper, we report and describe two red-colored flood-induced turbidites (i.e., hyperpycnites) from northern Hudson Bay and western Hudson Strait that we associate with a two-pulse final outburst flood of Lake Agassiz as proposed by Clarke et al. (2003; 2004a).

2. Materials and methods

2.1 CORE LOCATION AND SAMPLING

Cores AMD0509-28PC and AMD0509-27bLEH (hereinafter referred to as cores 28PC and 27bLEH) were collected by piston and gravity coring onboard the ice-breaker CCGS Amundsen in 2005 (Fig. 1). The sites were carefully selected using the 3.5 kHz subbottom profiler (Knudsen 320M) and multibeam sonar (Kongsberg-Simrad EM-300) to avoid areas affected by mass wasting deposits or iceberg scouring. Core 28PC was collected in western Hudson Strait (63°02'49.8"N, 74°'18'62.9"W) at 430 m water depth in an acoustically weakly stratified mud unit located in a deep basin containing thick (>30 m) Quaternary deposits; core 27bLEH was collected in northern Hudson Bay (61°03'11.8"N, 86°'12'49.9"W) at 245 m water depth in a thin (< 4 m) veneer of mud deposits overlying till. Acoustic stratification is not perceptible in this unit due to its small thickness.

2.2 CORE PROCESSING AND CONTINUOUS LOGGING

In the laboratory, the cores were run for wet bulk density and low field volumetric whole core magnetic susceptibility using a GEOTEK Multi Sensor Core Logger (MSCL).

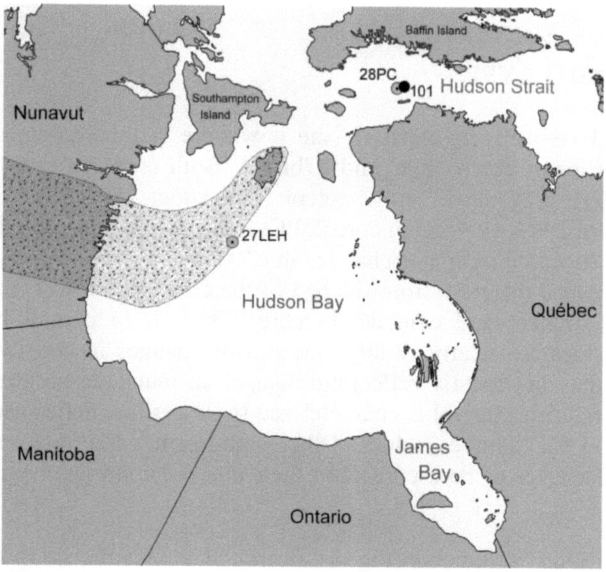

Figure 1. Location of core 27bLEH and 28PC sampling sites (grey circles). Also illustrated is the location of core 101 (black circle; Kerwin, 1996, see text for details). The location of the ferruginous till dispersion train associated with erosion and reworking of the red rocks of the Dubawnt Group is also illustrated (From Shilts, 1986).

The cores were also ran through a CAT-scan (computerized axial tomography) at INRS-ETE for the identification of sedimentary structures and extraction of CT-number profiles (see St-Onge et al., 2007 for details) and then split, photographed and described. CT number profiles primarily reflect changes in bulk density (St-Onge et al., 2007; see also Figs. 2-3). Finally, color reflectance measurements were performed using a hand-held X-rite DTP22 digital swatchbook spectrophotometer. Reflectance data are reported as a* from the widely used *Commission Internationale de l'Éclairage* (CIE) color space, whereas a* ranges from +60 (red) to -60 (green). Variations in a* values are often associated with changes in the concentration of red minerals, such as hematite (e.g., Helmke et al. 2002). These variations were previously used by Hall et al. (2001) to identify the red bed layer in Hudson Strait sediments.

2.3 GRAIN SIZE ANALYSES

Grain size analyses were carried out at ISMER with a sampling interval ranging from 10 to 1 cm depending on the facies characteristics. Prior to grain size analyses, the samples were added to a Calgon electrolytic solution (sodium hexametaphosphate) and rotated for about 3 hours using an in-house rotator. The samples were then sieved (2 mm) and disaggregated in an ultrasonic bath for 90 s prior to their analysis. Disaggregated samples were then analyzed with a Beckman-Coulter LS-13320 (0.04 to 2000 μm) laser sizer. The results of at least three runs were averaged. The average continuous disaggregated particle size distribution output was then processed using the Gradistat software for sediment parameters (Blott and Pye, 2001).

3. Results

3.1 THE REDDISH LAYER

Figures 2 and 3 respectively illustrate the downcore sedimentological, physical and magnetic properties of cores 28PC and 27bLEH. Both cores are mostly composed of olive grey to dark grey clayey silts, except from about 150 cm to the base of core 27bLEH and from 296 to 304 cm in core 28PC. In these two intervals, the sediments are red, as clearly illustrated by major changes in a* values. Grain size values within these two intervals are also different from the rest of the core, with lower mean grain size in core 28PC and higher sand contents in core 27bLEH. In core 27bLEH, magnetic susceptibility values are relatively high and mirror changes in density and grain size, indicating that they are mostly reflecting changes in magnetic grain size, whereas in core 28PC, the reddish interval is characterized by lower magnetic susceptibility, grain size, density and CT number values. Other significant variations are also observed above the reddish layer of both cores, but their interpretation lies beyond the scope of this paper.

A zoom in core 28PC sediments (Fig. 4) clearly illustrates the sharp upper and lower contacts as well as the presence of laminations in the reddish interval. The silt percent content in this interval also depicts two distinctive sequences of reverse and normal grading. Similarly, the sand percent, magnetic susceptibility, density and CT number profiles also highlight two clear sequences of reverse and normal grading in core 27bLEH. Parallel horizontal laminations are also observed in the reddish sediments of this core. While the base of the reddish layer was not recovered in core 27bLEH (not fully penetrated by coring), the upper contact is gradual.

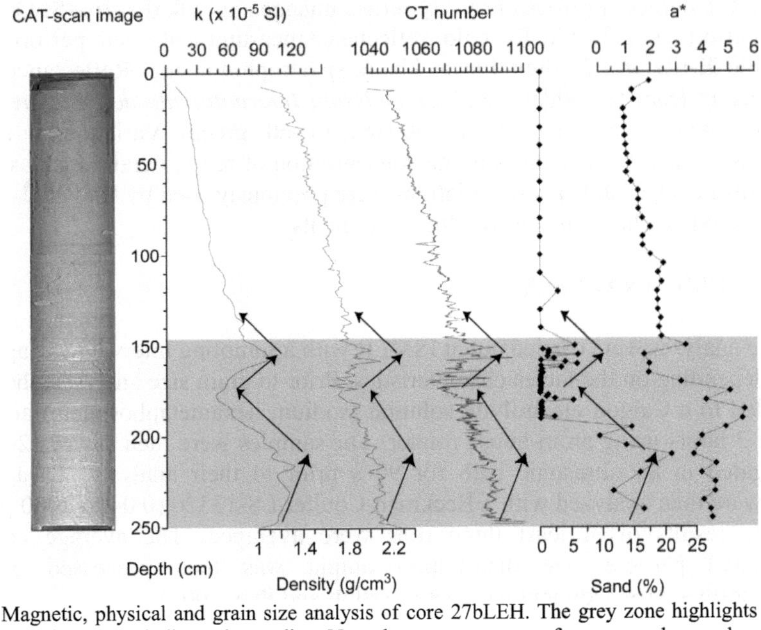

Figure 2. Magnetic, physical and grain size analysis of core 27bLEH. The grey zone highlights the reddish layer, whereas the arrows indicate the grading. Note the two sequences of reverse and normal grading in the reddish layer.

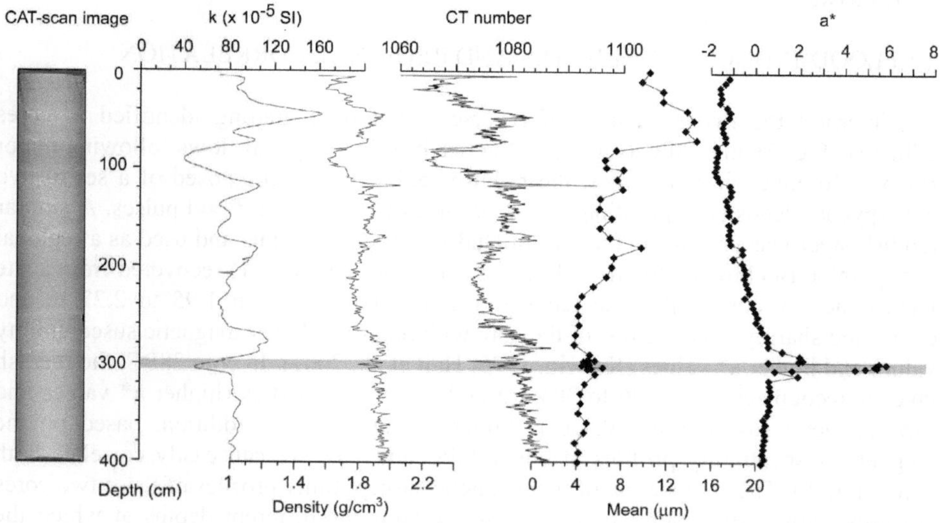

Figure 3. Magnetic, physical and grain size analysis of core 28PC. The grey zone highlights the reddish layer.

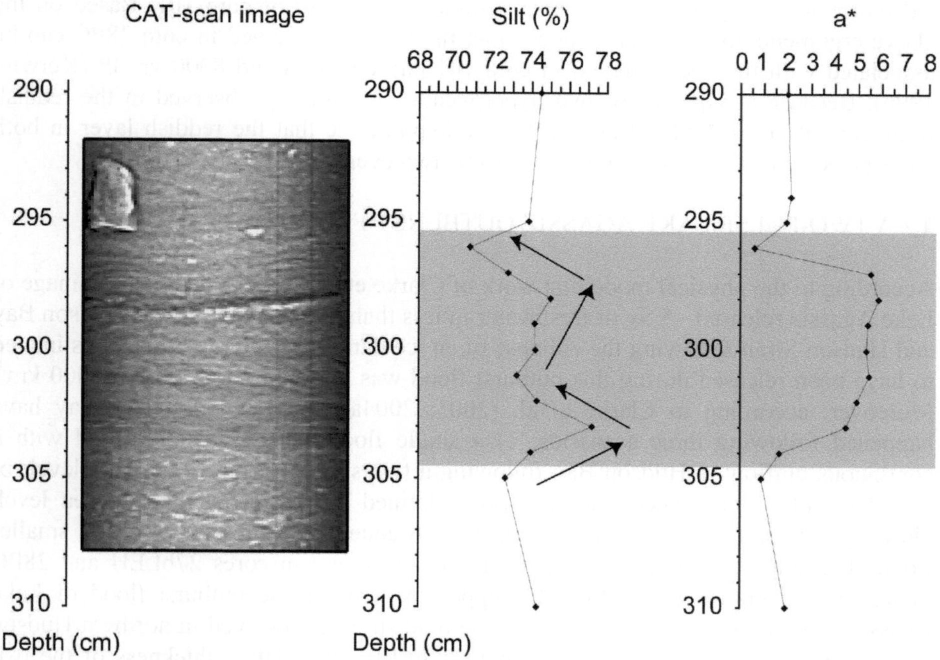

Figure 4. Zoom in the reddish layer of core 28PC. The grey zone indicates the reddish layer, whereas the arrows indicate the grading. Note the two sequences of reverse and normal grading in the reddish layer.

4. Discussion

4.1 FLOOD-INDUCED TURBIDITES AND REGIONAL CORRELATION

We interpret the two sequences of reverse and normal grading identified in cores 27bLEH (Fig. 2) and 28PC (Fig. 4) as the result of hyperpycnal flows following major floods. Moreover, in both cores, the reddish sediments are composed of a set of two hyperpycnal deposits, suggesting the occurrence of at least two flood pulses. A similar reddish layer was previously identify in Hudson Strait sediments and used as a regional stratigraphic isochron (Kerwin, 1996). In a core (core 101; Fig. 1) recovered from a site located nearby core 28PC, the red bed was observed between 1.95 to 2.23 m and contrasted sharply from the rest of the core with relatively lower magnetic susceptibility values and higher a* values (Kerwin, 1996; Hall et al., 2001). In core 28PC, the reddish layer is recognized from 296 to 304 cm and has characteristics (higher a* values and low magnetic susceptibility values) similar to core 101. In addition, based on the magnetic susceptibility profiles of cores 28PC and 101, we can easily correlate both cores (Fig. 5). The comparison of the magnetic susceptibility profiles of these two cores not only allows their correlation, but also explains the different depths at which the reddish layer is observed, as several tens of centimetres seem to be missing from the top of core 101 (Fig. 5). This point is also revealed by the relatively old date of 2205±45 yr BP (corrected ^{14}C years) in the upper sediments (2-5 cm) of core 101. Based on the above argumentation, we thus propose that the red bed observed in core 28PC can be associated with the one observed in core 101 and dated around 8000 yr BP (Kerwin, 1996). Because a sequence of two hyperpycnites is similarly observed in the reddish sediments of cores 27bLEH and 28PC, we hypothesize that the reddish layer in both cores are synchronous and related to the same two events.

4.2 A TWO-PULSE LAKE AGASSIZ OUTBURST FLOOD?

According to the physical modelling work of Clarke et al. (2004b), the final drainage of Lake Agassiz released ~5 Sv of freshwater in less than a year into northern Hudson Bay and Hudson Strait following the collapse of an ice dam. The volume of water estimated to have been released during this outburst flood was between 40 000 to 151 000 km^3. Moreover, according to Clarke et al. (2003; 2004a), this final drainage may have happened following three scenarios: 1) a single flood; 2) a two-pulse flood with a continuous outflow to Hudson Bay following a two-step lowering of the lake level; or 3) a two-pulse flood where the first flood drained the proglacial lake to sea level, allowing the reformation of an ice dam, its subsequent refilling and a second smaller flood. The sequence of two hyperpycnal deposits found in cores 27bLEH and 28PC indicates a two-pulse flood, strongly supporting a two-pulse outburst flood of Lake Agassiz and the flood-induced nature of the reddish layer observed in northern Hudson Bay and Hudson Strait. Moreover, in core 27bLEH, the relative thickness of the two hyperpycnites may hint at a larger first flood and a subsequent smaller second flood, as postulated in the third scenario of Clark et al. (2003; 2004a). Finally, the coarser grain size and the prominent thickness of core 27bLEH reddish layer likely indicate the more proximal location of the sediment source.

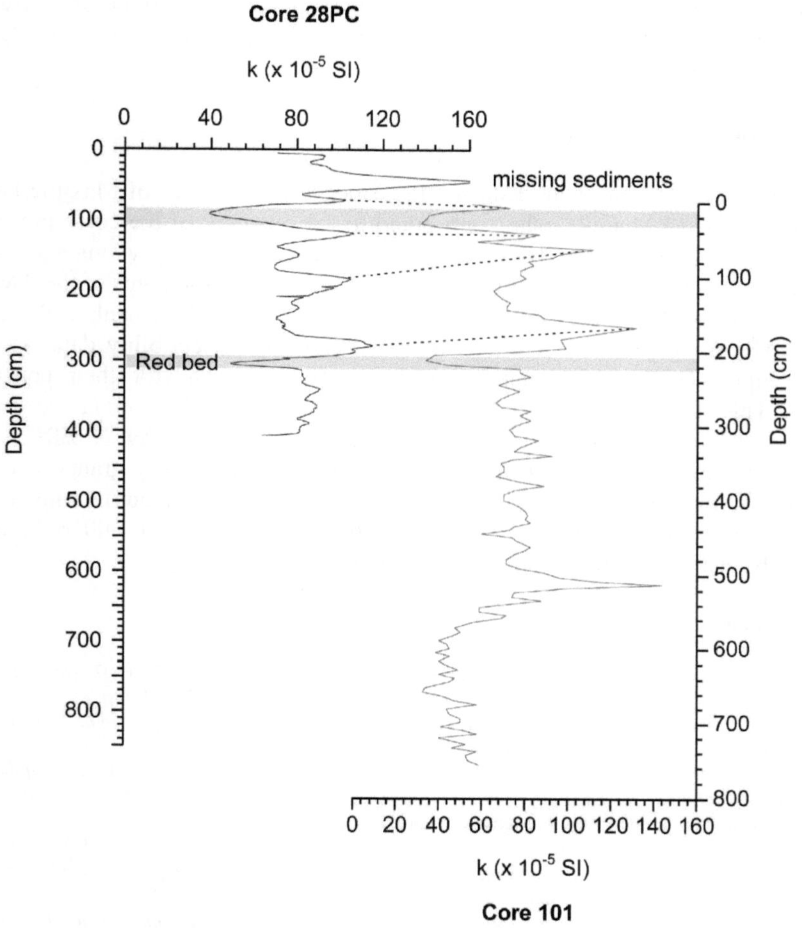

Figure 5. Correlation of core 28PC to core 101 (Kerwin, 1996) from Hudson Strait. Two low magnetic susceptibility horizons of similar values, including the reddish layer, are highlighted and used for correlation. Several other magnetic susceptibility highs can be correlated and are illustrated by dashed lines. Core 101 age model was constructed using 7 radiocarbon dates and the red bed was dated around 8000 yr BP (Kerwin, 1996).

5. Conclusions

In this paper, we identified a sequence of two flood-induced turbidites in reddish sediments of cores from northern Hudson Bay and western Hudson Strait. These two reddish layers can be correlated to a red bed previously recognized as a regional isochron in Hudson Strait and associated with the final drainage of Lake Agassiz. Regardless of the exact timing of the catastrophic drainage, the hyperpycnites described in this paper suggest that they were deposited following two pulses, which is in agreement with the two-pulse outburst flood scenarios of Clarke et al. (2003; 2004a). Finally, this study demonstrates for the first time the turbiditic and the flood-induced nature of the Hudson Strait red bed isochron. Paleomagnetic and radiocarbon analyses are currently undertaken to independently assess the age of these flood-induced turbidites. Further investigations on the morphology of the seafloor of Hudson Bay using multibeam and subbottom profiler data should bring information on the dynamics of this catastrophic drainage.

6. Acknowledgments

P. Lajeunesse and G. St-Onge dedicate this paper to the memory of Maxime Paiement (GEOTOP UQAM-McGill), who contributed to the recovery of the cores presented in this paper, but most importantly, a friend who passed away at a too young age. We wish to sincerely thank the captain, officers, crew and scientific participants of the AMD0509 cruise on board the ice-breaker CGCS Amundsen. We also thank J.T. Andrews (INSTAAR) for generously sharing core 101 magnetic susceptibility data, as well as D.J.W. Piper (GSC) and Emmanuel Chapron (ETH-Zurich) for their constructive reviews. This study was supported by a FQRNT (*Fonds québécois de la recherche sur la nature et les technologies*) new researcher grant to G.S., by NSERC (Natural Sciences and Engineering Research Council of Canada) discovery grants to G.S. and P.L. and by ArcticNet (Network of Centers of Excellence of Canada) grants and ship-time. This constitutes GEOTOP-UQAM-McGill contribution n° 2007-0016. Finally, the authors also acknowledge the help of V. Gagnon in the laboratory.

7. References

Alley, R.B., Mayewski, P.A., Sowers, T., Stuiver, M., Taylor, K.C., Clark, P.U., 1997. *Holocene climatic instability: a prominent, widespread event 8200 years ago.* Geology 25, 483-486.

Alley, R.B., Agustsdottir, A.M, 2005. *The 8k event: cause and consequences of a major Holocene abrupt climate change.* Quaternary Science Reviews 24, 1123-1149.

Andrews, J.T., Maclean, B., Kerwin, M., Manley, W., Jennings, A.E., Hall, F., 1995. *Final stages in the collapse of the Laurentide Ice Sheet, Hudson Strait, Canada, NWT: ^{14}C AMS dates, seismics stratigraphy, and magnetic susceptibility logs.* Quaternary Science Reviews 14, 983-1004.

Barber, D.C., Dyke, A., Hillaire-Marcel, C., Jennings, A.E., Andrews, J.T., Kerwin, M.W., Bilodeau, G., McNeely, R., Southon, J., Morehead, M.D., Gagnon, J.M., 1999. *Forcing of the cold event of 8,200 years ago by catastrophic drainage of Laurentide lakes.* Nature 400, 344-348.

Blott, S.J., Pye, K., 2001. *Gradistat: a grain size distribution and statistics package for the analysis of unconsolided sediments.* Earth Surface Processes and Landforms 26, 1237-1248.

Chapron, E., Ariztegui, D., Mulsow, S., Villarosa, G., Pino, M., Outes, V., Juvignié, E., Crivelli, E., 2006. *Impact of the 1960 major subduction earthquake in Northern Patagonia.* Quaternary International 158, 58-71.

Chapron, E., Juvigné, E., Mulsow, S., Ariztegui, D., Magand, O., Bertrand, S., Pino, M., Chapron, O., subm. *Clastic sedimentation processes in Lago Puyehue: Implications for natural hazards and paleoenvironment reconstructions in the Chilean Lake District, 40.5°S.* Sedimentary Geology.

Clarke, G., Leverington, D., Teller, J., Dyke, A., 2003. *Superlakes, megafloods and abrupt climate chance.* Science 301, 922-923.

Clarke, G.K., Seidov, D., Haupt, B.J., 2004a. *Freshwater ocean forcing scenarios for the megaflood from glacial Lake Agassiz.* Eos Trans. AGU, 85 (17), Jt. Assem. Suppl., Abstract GC12A-02.

Clarke, G.K.C., Leverington, D.W., Teller, J.T., Dyke, A.S., 2004b. *Paleohydraulics of the last outburst flood from glacial Lake Agassiz and the 8200 BP cold event.* Quaternary Science Reviews 23, 389-407.

Guyard, H., St-Onge, G., Chapron, E., Anselmetti, F., Francus, P., 2007. *The AD 1881 earthquake-triggered slump and late Hocoloce exceptional flood-induced turbidites from proglacial Lake Bramant, Western French Alps.* See this book.

Hall, F.R., Andrews, J.T., Kerwin, M., Smith, L.M., 2001. *Studies of sediment colour, whole-core magnetic susceptibility, and sediment magnetism of Hudson Strait-Labrador Shelf region: CSS Hudson cruises 90023 and 93034. In* B. MacLean (Ed.), Late Quaternary sediments, depositional environments, and Late Glacial/Deglacial history derived from marine and terrestrial studies. Geological Survey of Canada Bulletin 566, pp. 161-170.

Helmke, J.P., Schulz, M., Bauch, H.A., 2002. *Sediment-color record from the Northeast Atlantic reveals patterns of millennial-scale climate variability during the Past 500,000 Years.* Quaternary Research 57, 49-57.

Kerwin, M.W., 1996. *A regional stratigraphic isochron (ca. 8000 ^{14}C yr B.P.) from final deglaciation of Hudson Strait.* Quaternary Research 46, 89-98.

Kneller, B.C., 1995. *Beyond the turbidite paradigm: Physical models for deposition of turbidites and their implications for reservoir prediction. In* Hartley, A.J., Prosser, D.J. (Eds.), Characterization of deep marine clastic systems. Geological Society [London] Special Publication 94, pp. 31-49.

Mulder, T., Syvitski, J.P.M., 1995. *Turbidity currents generated at river mouths during exceptional discharges to the world oceans.* Journal of Geology 103, 285-299.

Mulder,T., Migeon, S., Savoye, B., Jouanneau, J.-M., 2001a. *Twentieth century floods recorded in deep Mediterranean sediments.* Geology 29, 1011-1014.

Mulder,T., Migeon, S., Savoye, B., Faugères, J.-C., 2001b. *Inversely graded turbidite sequences in the deep Mediterranean: a record of deposits from flood-generated turbidity currents?* Geo-marine Letters 21, 86-93.

Mulder,T., Migeon, S., Savoye, B., Faugères, J.-C., 2002. *Inversely graded turbidite sequences in the deep Mediterranean: a record of deposits from flood-generated turbidity currents? Reply.* Geo-marine Letters 22, 112-120.

Mulder, T., Syvitski, J.P.M., Migeon, S., Faugères, J-C., Savoye, B., 2003. *Marine hyperpycnal flows: initiation, behavior and related deposits. A review. Marine and Petroleum* Geology 20, 861-882.

Shilts, W.W., 1986. *Glaciation of the Hudson Bay region. In* I.P. Martini, (Ed.), Canadian Inland Seas. Elsevier Oceanography Series 44, Elsevier, New York, pp. 55-78.

Schneider, J.-L., Pollet, N., Chapron, E., Wessels, M., Wassmer, P., 2004. *Signature of Rhine Valley sturztrom dam failures in Holocene sediments of Lake Constance, Germany.* Sedimentary Geology 169, 75-91.

St-Onge, G., Mulder, T., Piper, D.J.W., Hillaire-Marcel, C., Stoner, J.S., 2004. *Earthquake and flood-induced turbidites in the Saguenay Fjord (Québec): a Holocene paleoseismicity record.* Quaternary Science Reviews 23, 283-294.

St-Onge, G., Mulder, T., Francus, P., Long., B. 2007. *Continuous physical properties of cored marine sediments.* In C. Hillaire-Marcel and A. de Vernal (Eds.), Proxies in Late Cenozoic Paleoceanography, Elsevier, pp. 65-101.

UNDERWATER ROCKFALL KINEMATICS: A PRELIMINARY ANALYSIS

D. TURMEL AND J. LOCAT
Dept. of Geology and Geological Engineering, Université Laval, Québec, Canada, G1K 7P4, dominique.turmel.1@ulaval.ca, locat@ggl.ulaval.ca

Abstract

The marine environment presents various settings in which talus slopes are formed via a rock fall process similar to what exists on land. This is the case along fjords and submarine canyons in particular. Although many studies have been carried out on land, surprisingly very little is known for the submarine environment. We propose here the first kinematics analysis of underwater rockfall. It is postulated that the block have a diameter of more than one meter. As it can be expected, the main addition to the subaerial case is the consideration of the role of water (e.g. density or current). For submarine rockfall analysis, the effect of the ambient fluid cannot be neglected. Hydrodynamic constraints are controlled by the speed, shape, and size of the moving mass. Wind does not have a significant role in subaerial rockfall analysis, but currents must be considered in the subaqueous environment. In addition, coefficients of restitution are not only controlled by the elastic properties of the material, but also by impact Stokes number. This paper provides a summary of underwater rockfall kinematics in order to formulate underwater rockfall governing equations.

1. Introduction

Kinematics of falling rocks have been studied analytically and experimentally for several decades (Ritchie, 1963; Pfeiffer and Bowen, 1989; Azzoni *et al.,* 1995). Almost all of these works have been done for subaerial environment, but rockfalls also happen in underwater environment (Edmunds *et al.*, 2006). In order to protect submarine installations such as pipelines or to be able to protect the environment, for example coral reef (Edmunds *et al.*, 2006), underwater rockfall analysis must be undertaken. To the authors' knowledge, the only work on submarine rockfalls is from Beranger *et al.* (1998): they have introduced two constants to account for water effects (drag and added mass forces). In its preliminary form, we propose here the first kinematics analysis of underwater rockfall. As a first step, geometry of the problem and forces will be defined. Rock shape will be taken as a disk for simplification reasons. Effect of hydrodynamic forces on every part of the movement will then be presented in order to develop governing equations. A discussion on how coupling these equations will then be done. Further work will be to develop a mathematical model and laboratory or in-situ validation.

2. Physics

It can be expected that for underwater rockfalls, general modes of motion will be the same as for subaerial rockfalls: rolling or sliding, bouncing and freefall (Figure 1). Since the values of water density and viscosity are higher than those of the air, hydrodynamic forces cannot be neglected. They will influence modes of motion. Among all hydrodynamic forces, lift, drag and added mass will be discussed.

V. Lykousis, D. Sakellariou and J. Locat (eds.), Submarine Mass Movements and Their Consequences, 139–147.

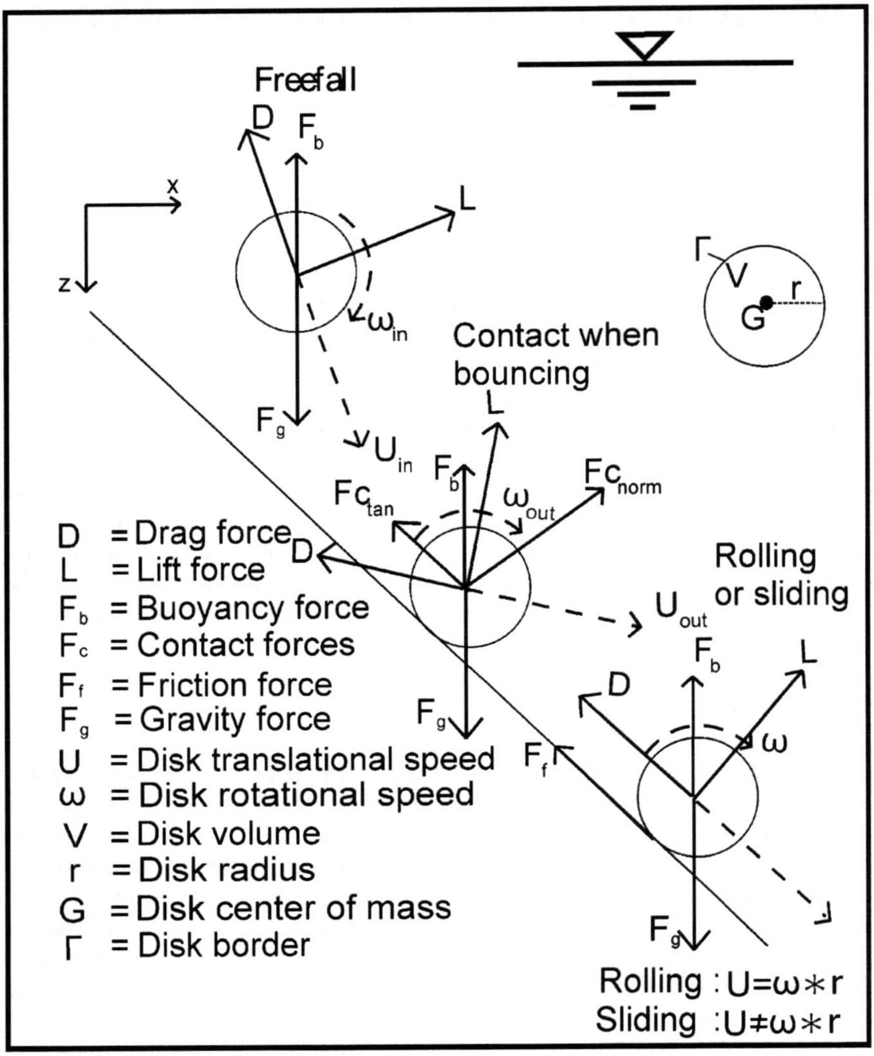

Figure 1 General modes of motion and corresponding forces.

These forces are shown in Figure 1. Basset history force, force due to the instationarity of the flow, can be neglected because of large particle diameter and density of the rock mass (Thomas, 1992).

2.1 HYDRODYNAMIC FORCES

2.1.1 Buoyancy forces

Buoyancy force is the force that obeys Archimedes principle which states that a body immersed in a fluid is buoyed up by a force equivalent to the mass of displaced fluid. This force is directed upward, and is defined by: $F_b = -\rho_w V g$, where ρ_w is water density, V is the body volume, and g is the gravitational acceleration directed downward.

2.1.2 *Lift and drag forces*

Lift and drag forces are related forces acting in perpendicular directions. When a body moves through a fluid, an interaction between the body and the fluid occurs. This interaction can be described in terms of shear stress due to viscous effects and normal stress due to pressure. (Munson *et al.*, 1998). Lift force (L) is oriented perpendicular to the direction of motion and drag force (D) is oriented in a direction opposite to that of motion.

It is not always possible to directly calculate these two forces, without complex numerical analysis. Two adimensional terms have been introduced to be able to evaluate lift and drag forces for different object shapes, lift coefficient (C_l) and drag coefficient (C_D):

$$C_l = \frac{L}{0.5\rho U^2 A} \qquad C_D = \frac{D}{0.5\rho U^2 A} \qquad \text{(1) and (2)}$$

Where U is the relative velocity of the object in the fluid and A is a characteristic area of the body. Numerical values of these coefficients have been found after several laboratory experiments (Allen, 1900, Wieselsberger, 1922, Liebster, 1927, Achenbach, 1965). Both coefficients are function of Reynolds number (Re) which represents the ratio between inertial and viscous forces: $Re = \frac{\rho U l}{\mu}$, roughness of the object and rotation of the object (Magnus effect). One can show that for a moving disk without rotation, lift force is zero so $C_l = 0$.

A standard drag curve (SDC) has been developed by Lapple and Shepherd (1940) (Figure 2). When it exceed $Re = 10^5$, the drag coefficient shows a sudden drop, called drag crisis, due to the transition from laminar to turbulent boundary layer. In our case, because of the large radius of the disk and its speed, Re will vary from around 10^5 to much higher Re numbers (Fig. 2). With such high Re numbers, C_D may vary from 0.08 to 0.7.

2.1.3 *Robins effect*

Robins effect is the lift force observed when a spinning sphere moves through a fluid. The effect of spin is to delay separation on the retreating side and enhance it on the advancing side (Mehta, 1985). Little work has been done on smooth spheres, and they have only been done for a very narrow range of Re numbers ($Re \leq 2000$).

Work of Kurose and Komori (1999) shows that both C_l and C_D varies with Re and a rotational parameter defined as $\Omega = \omega r / U$ where ω is the rotation speed of the sphere, r is the radius of the sphere, and U is the relative velocity of the sphere in the fluid. Most work has been done on Magnus effect for rotating cylinders or for sports balls such as golf or cricket ball. Up to now, no correlation seems to exist to link rotation rate, Reynolds number and lift or drag coefficient in viscous flow when Re number is high.

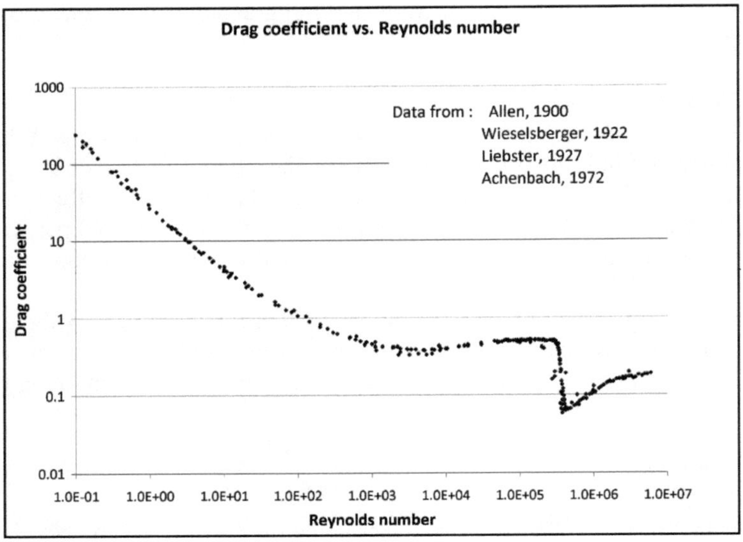

Figure 2. Standard drag curve.

2.1.4 Added mass force

When a body accelerates in a fluid, it must also accelerate part of the fluid. The kinetic energy associated with the moving fluid will be changed. Kinetic energy can be defined as (Brennen, 1982) (with Einstein notation):

$$T = \frac{\rho}{2} \int_V u_i u_i dV \tag{3}$$

Where u_i *(i=1,2,3)* represents components of fluid velocity. If the object speed is constant, this kinetic energy will be constant and dependant of the square of the object translation velocity (U^2). If the flow is a potential flow, it can be said that when U is altered, the velocity u_i at each point in the fluid varies proportionally with U (Brennen, 1982) then:

$$T = \rho \frac{I}{2} U^2 \quad \text{where } I = \int_V \frac{u_i}{U} \frac{u_i}{U} dV \tag{4}$$

If the body accelerates, then an additional work will have to be done to increase the kinetic energy of the fluid and can be expressed as *dT/dt*. This extra work result in an additional drag experienced by the object, *-FU* and is equal to *dT/dt*, so (Brennen, 1982):

$$F = -\frac{1}{U}\frac{dT}{dt} = -\rho I \frac{dU}{dt} \tag{5}$$

The resultant form of the added mass force is expressed as (modified from Gondret *et al.*, 2002):

$$F_{am} = -\frac{4}{3}\pi r^3 C_a \rho_f \frac{dU}{dt} \tag{6}$$

where C_a is the added mass force coefficient, varying from 0.5 to 1.05 dependant of the ratio ρ_s / ρ_f (Odar and Hamilton, 1964).

2.2 CONTACT FORCES

In most of the modeling software for rockfall analysis and because in any non-perfectly elastic collision some kinetic energy is lost (Pfeiffer and Bowen, 1989), contacts are modeled by use of two adimensional restitution coefficients: normal and tangential. Normal coefficient of restitution is defined as the ratio between rebound to impact velocity normal to the slope. Normal coefficient of restitution is function of material properties, angle of contact (α) and impact (Chau *et al.,* 2002). Tangential restitution coefficient is also function of slope inclination, impact angle, impact velocity, and material properties.

CRSP formulation (Pfeiffer and Bowen, 1989) takes most of these parameters into account and reproduces well experimental data (Heidenreich, 2004).

2.2.1 Effect of ambient fluid on restitution coefficients

Few studies have been done on particle-collision in fluids, but all of them show the same results. The viscous liquid dissipates the energy and may weaken the restitution process in collision (Yang, 2006). McLaughlin (1968), Joseph *et al.* (2001), and Gondret *et al.* (2002) investigated normal (*i.e.* without tangential velocity) particle-wall collision in liquid and found that normal restitution coefficient is also function of Stokes number. Stokes number is defined by the ratio of viscous to inertial forces of an object in a fluid:

$$St = (\rho_s / p_f) Re / 9 \tag{7}$$

For $St < St_{critical} = 10$, there is no rebound and for very high St (>1000), viscous forces are negligible and wet coefficient of restitution equals dry coefficient of restitution (Figure 3). When there is a tangential component, Joseph and Hunt (2004) showed that normal coefficient have the same behaviour as the one with normal collision and is not affected by tangential speed component. Stokes number must be modified to take into account normal component of impact velocity. For the tangential component, two cases can occur: (1) there is solid-solid contact or (2) there is no solid-solid contact. Joseph and Hunt (2004) showed that, when mean surface roughness is larger than elastohydrodynamic (EHD) lubrication minimum distance of approach, there will be solid-solid contact and the collision will exhibit the same behaviour as subaerial collision. Otherwise, when the surface roughness is smaller than EHD lubrication minimum distance of approach, there is a substantial decrease in the rotational impulse, when compared to collisions in air. This will also cause a reduction of friction coefficient by almost an order of magnitude (Joseph and Hunt, 2004).

2.3 WALL EFFECTS

When an object moves close to a wall, the classical lubrication theory predicts a force that increase as the inverse of the gap width (Yang, 2006). This force is a lift force and is directed normal to the wall.

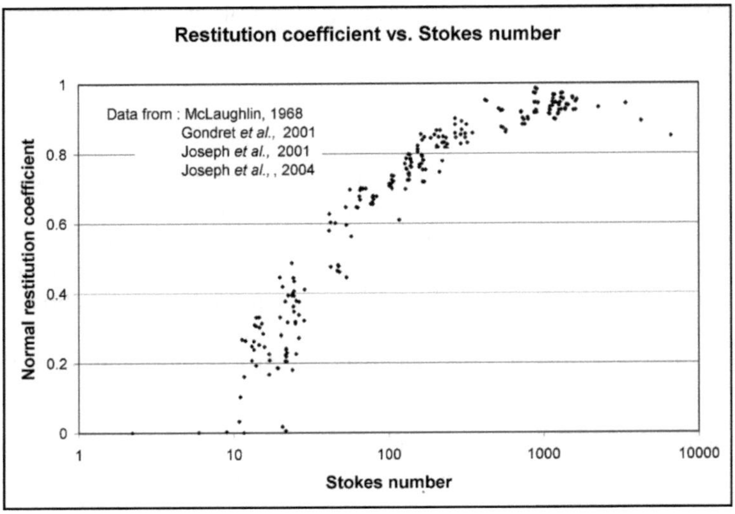

Figure 3. Effect of fluid on restitution coefficient.

It has been shown by Jan and Chens (1997) and Chhabra and Ferreira (1999) that for a sphere rolling down a smooth plane, drag coefficient in the supercritical regime (C_D) is about 0.74. For the SDC, C_D is about 0.45. Unfortunately, no data seems to exist for critical and transcritical regimes. Drag coefficient is much larger near a wall than in the freefall case.

Jan and Chens (1997) have also studied the effect of wall proximity for added mass. They have shown that for a sphere rolling down an incline added mass coefficient is larger than that in the freefall. $C_a = 2$ have better agreement with their experimental data.

3. Governing equations

Underwater rockfalls is a multi-physic problem involving a moving rigid body in a Newtonian fluid. Body motion is due to a constant force: gravity and hydrodynamic forces. Contact forces also have to be taken into account. We have shown that no hydrodynamic forces are constant throughout the same problem. They are function of moving body speed, rotation, proximity to a wall, etc.

3.1 FREEFALL MOVEMENT

Governing equation for the spherical body will be expressed as (forces balance and moment balance) (Glowinski *et al.*, 1999):

$$M\frac{dV}{dt} = Mg + F_h + F_c \qquad I\frac{d\omega}{dt} = T_h \qquad \frac{dG}{dt} = V \qquad \text{(8), (9) and (10)}$$

Where V is the rigid body speed, M the body mass, T the hydrodynamic moment, G the center of mass and ω the angular velocity. F_h is hydrodynamic forces (lift force, drag force and added mass force) and F_c is contact force.

For the surrounding fluid, governing equations are Navier-Stokes equations and continuity equation:

$$\rho_f\left(\frac{\partial u}{\partial t}+u\cdot\nabla u\right)=\rho_f g+\nabla\cdot\sigma \qquad \nabla\cdot u=0 \qquad \text{(11) and (12)}$$

Where σ is the stress tensor given by:

$$\sigma=-pI+\mu_f\left[\nabla u+(\nabla u)^T\right] \tag{13}$$

Where p is pressure, I is identity matrix and μ_f is fluid viscosity. To model the problem, we must set boundary conditions on the moving block and on the wall. This condition is called "no-slip condition". Fluid is moving at the same speed than the solid on each boundary.
Hydrodynamic forces and moment are defined as (Glowinski *et al.,* 1999):

$$F_h=\int_\Gamma \sigma n \mathrm{d}\gamma \qquad T_h=\int_\Gamma (x-G)\times(\sigma n)\mathrm{d}\gamma \qquad \text{(14) and (15)}$$

n is a unit vector normal to Γ. Presence of water current can be added to this model by imposing an initial water speed.
By coupling Navier-Stokes and motion equations, it is possible, at each instant, to numerically calculate hydrodynamic forces acting on the moving rigid body. It also allows using any kind of block geometry (while modifying *eq.* 9). Coupling these equations have already been done for simulating particle-fluid system such as fluid-induced erosive failures or debris flows, but have never been done for underwater rockfall analysis. Such an analysis is complicated by the fact that fluid motion neither block motion is known a priori, because block motion influence fluid motion and *vice-versa.*

3.2 BOUNCING AND ROLLING MOTION

Contact forces have to be seen as independent of the freefall case. When distance between the moving body and a wall approach block radius, new velocities will be computed to take into account contact forces. CRSP formulation will be modified to consider Stokes number. After collision, if normal speed after the collision is small compared to tangential speed, the block will be rolling (or sliding). Otherwise, a new freefall trajectory will be calculated.

When the block is rolling or sliding on or near a wall, another set of equations will have to be developed to take into consideration friction coefficient of the wall and EHD theory. If a rebound occurs because of a change in the shape of the wall, then a new freefall part of motion will be computed.

4. Discussion

In their work, Beranger *et al.* (1998) have assumed constant drag and mass coefficients: 0.4 and 0.5, respectively. This seems to be a good approximation for boulders in freefall subcritical regime. In our paper, it can be seen that for underwater rockfall, where expected *Re* number will be over 10^5, drag coefficient will be less than 0.5 and will increase with speed. If rotation is not neglected, Magnus effect should play an important role in the motion. When the block is near a wall two different forces will interact. Added mass force will be greater, so block acceleration or deceleration will decrease because of fluid momentum. In addition, drag coefficient will increase as well as lift coefficient.

Equations 8 to 13 can be joined via *eq.* 14 and 15. To solve this system, with specified boundary conditions, two main different methods can be used: Arbitrary Lagrange-Euler method or Distributed Lagrange multiplier. The last one seems to be the best one to use in our case, if we want to optimize computation time (Carlson *et al.*, 2004). Carlson *et al.* (2004) used finite differences to solve these equations, which is easier to implement than finite elements. Distributed Lagrange Multiplier method takes the solid as a rigid fluid (Patankar *et al.*, 2000; Patankar, 2001). In our case, a turbulent model for the fluid will have to be added to account for high Re number flow.

5. Concluding remarks

It has been shown that for underwater rockfalls effect of ambient fluid cannot be neglected. The shape of the block, speed, and presence of a wall control hydrodynamic constraints. Presence of the fluid also controls bouncing motion and friction coefficient. Lift force, drag force, and added mass force are all non-constant forces. It seems necessary, in order to model behaviour of underwater rockfall, to join block equations of motion (*eq.* 8 and *eq.* 9) with Navier-Stokes (*eq.* 11) and continuity equations (*eq.* 12), and to modify collision equations with Stokes number dependence. A turbulence model will have to be added to Navier-Stokes equations. It has not been done so far for submarine rockfall analysis, but it will be the next step of this work.

6. Acknowledgements

The authors would like to thank both reviewers, J. Hadjigeorgiou and M. Quirion for their constructive remarks. We would also like to thank the Natural Sciences and Engineering Research Council of Canada for their financial support.

7. References

Achenbach, E. (1972). *Experiments on the flow past spheres at very high Reynolds numbers*. Journal of Fluid Mechanics, pp. 565-575.

Allen, H. (1900). *The motion of a sphere in a viscous fluid.* Philos. Mag, pp. 323-338;519-534.

Azzoni, A., La Barbera, G., Zaninetti, A., (1995). *Analysis and Prediction of Rockfalls Using a Mathematical Model.* International Journal of Rock Mechanics and Mining Sciences and Geomechanical Abstracts 32(7) pp.709-724.

Beranger, M., Carpaneto, R., Boote, D., & Zuccarelli, F. (1998). *Probabilistic assessment of rockfall impact on subsea pipelines.* OMAE 1998: Proceedings of the 17th international conference on offshore mechanics and arctic engineering. Portugal.

Brennen, C.E. (1982). *A review of added mass and fluid intertial forces.* California: Naval Civil Engineering Laboratory, Port Hueneme.

Carlson, M., Mucha, P.J., Turk, G. (2004). *Rigid fluid: animating the interplay between rigid bodies and fluid,* ACM Transactions on Graphics (TOG), 23(3) pp. 377-384.

Caughey, D.A. (2006*). Incompressible potential flow,* M&AE 305, Cornell University

Chaabra, R.P, & Ferreira, J.M. (1999). *An analytical study of a sphere rolling down a smooth inclined plane in an incompressible Newtonian fluid,* Powder technology (104), pp. 130-138.

Chau, K.T., Wong, R.H.C., & Wu, J.J. (2002). *Coefficient of restitution and rotational motion of rockfall impacts,* International Journal of Rock Mechanics and Mining Sciences 39(1) pp. 69-77.

Edmunds, M., Bryant, C., Crozier, J. Gilmour, P., Pickett, P., Stewart, K., & Williams, J. (2006). *Port Phillip Bay Channel Deepening Project. Trial Dredging Experiment: Deep Reef Impact Assessment.* Report to Port of Melbourne Corporation. Australian Marine Ecology Report 334, Melbourne.

Gondret, P., Lance, M., & Petit, L. (2002, February). *Bouncing motion of spherical particles in fluids.* Physics of fluids, 14 (2), pp. 643-652.

Heidenreich, B. (2004). *Small- and half-scale experimental studies of rockfall impacts on sandy slopes.* Lausanne: PhD Thesis, EPFL.

Jan, C., & Chen, J. (1997). *Movements of a sphere rolling down an inclined plane.* Journal of Hydraulic Research, 35 (5), pp. 689-706.

Joseph, G., & Hunt, M. (2004). *Oblique particle-wall collisions in a liquid.* Journal of Fluid Mechanics, 510, pp. 71-93.

Joseph, G., Zenit, R., Hunt, M., & Rosenwinkel, A. (2001). *Particle-wall collisions in a viscous fluid.* Journal of Fluid Mechanics, 433, pp. 329-346.

Kurose, R., & Komori, S. (1999). *Drag and lift forces on a rotating sphere in a linear shear flow.* Journal of fluid mechanics, 384, pp. 183-206.

Lapple, C., & Shepherd, C. (1940). *Calculation of particle trajectories.* (A. C. Society, Éd.) Industrial and Engineering Chemistry, 32 (5), pp. 605-617.

Liebster, H. (1927). *Uber den Widerstand von Kugeln.* Annals of Physics , pp. 541-562.

McLaughlin, M. (1968). *An experimental study of particle-wall collision relating of flow of solid particles in a fluid.* Engineer's degree thesis, California Institutie of Technology, Pasadena, California.

Mehta, R.D. (1985). *Aerodynamics of Sports Balls,* Annual Review of Fluid Mechanics, 17 (1), pp.151-189.

Munson, B.R., Young, D.F., & Okiishi, T.H. (1998). *Fundamentals of Fluid Mechanics* (éd. 3rd edition). New-York: John Weiley & Sons, Inc.

Odar, F., & Hamilton, W.S. (1964). *Forces on a sphere accelerating in a viscous fluid.* Journal of fluid mechanics (18), pp. 302-314.

Patankar, N.A., Singh, P., Joseph, D.D., Glowinski, R., & Pan, T.-W. (2000). *A new formulation of the distributed Lagrange multiplier/fictious domain method for particulate flows.* International Journal of Multiphase Flow, 26(9), pp. 1509-1524.

Patankar, N.A., (2001). *A formulation for fast computations of rigid particulate flows.* Center for Turbulence Research Annual Research Briefs 2001, pp.185-196.

Pfeiffer, T., & Bowen, T. (1989). *Computer Simulation of Rockfalls.* Bulletin of the Association of Engineering Geologists , XXVI (1), pp. 135-146.

Ritchie, A.M. (1963). *Evaluation of rockfall and its control.* Highway Research Record (17), pp. 13-28.

Thomas, P.J. (1992). *On the influence of the Basset history force on the motion of a particle through a fluid. Brief communications,* Phys. fluids A, 4 (9), pp. 2090-2093.

Turton, R., & Levenspiel, O. (1986). *A short note on the drag correlation for spheres.* (Elsevier, Éd.) Powder technology, 47 (1), pp. 83-86.

Wieselsberger, C. (1922). *Weitere Feststellungen uber die Gesetze des Flussigkeits- und Luftwiderstandes.* Phys. Zeitschr, pp. 219-224.

Yang, F. (2006). *Interaction law for a collision between two solids particles in a viscous liquid.* California: PhD Thesis, California Institute of Technology, Pasadena.

ANTHROPOGENIC TURBIDITY CURRENT DEPOSITS IN A SEISMICALLY ACTIVE GRABEN, THE GULF OF CORINTH, GREECE: A USEFUL TOOL FOR STUDYING TURBIDITY CURRENT TRANSPORT PROCESSES

M. IATROU, G. FERENTINOS, G. PAPATHEODOROU,
Department of Geology, University of Patras, 26500, Rio, Patras, Greece
iatroum@upatras.gr, George.Papatheodorou@upatras.gr

D.J.W. PIPER, E. TRIPSANAS
Geological Survey of Canada (Atlantic), Bedford Institute of Oceanography, P.O. Box 1006, Dartmouth, Nova Scotia B2Y 4A2, Canada

Abstract

A detailed marine survey in Antikyra Bay in the northern margin of the Gulf of Corinth graben in Greece was carried out to examine the distribution and dispersion of bauxitic "red mud" tailings. The red mud tailings have been discharged at a rate of 500 000 to 640 000 tonnes/year via submerged outfalls on the shelf at a water depth of 100 m. The red mud tailings at the mouth of the outfalls have formed three oval shaped mounds. These mounds have a maximum thickness of about 27 m and thin out radially in the downslope direction along the longitudinal axis of the bay in a south-southwestward direction forming a common depositional lobe. The "red mud" tailings are transported by turbidity currents, via channels which incise the slope, to the basin floor at a water depth of 800 m and over a distance of 17 km, where they form small sheet-like deposits. High-resolution sedimentological analysis of 30 short cores using visual inspection, X-radiography, grain size measurements, X-ray diffraction and smear slides have shown the presence of seven (7) distinct turbidity flow events which were activated on the shelf and the slope and have occurred during the 24 years of tailing discharge. Five of these have their source at the "red-mud" mounds at the mouths of the outfalls and two of them have their source at the shelf break and the upper slope. The analysis revealed that the former consist of (a) graded structureless red mud deposits and/or bioturbated red mud deposits, or (b) laminated red mud deposits, whereas the latter are mostly structureless and poorly graded.

Keywords: anthropogenic turbidity currents, Gulf of Corinth, mud, tailings

1. Introduction

Submarine gravity flows are important mechanisms to transport sediment from shallow to deep environments and are the dominant depositional processes on many basin floors. The sedimentation processes in the tectonically active fjord-like basin of the Gulf of Corinth, in western Greece, are dominated by gravitational mass movements, such as slumps, debris and turbidity flows which are mostly triggered by earthquakes (Ferentinos et al. 1988; Papatheodorou 1991; Papatheodorou and Ferentinos 1997; Lykousis 1991; Perissoratis et al. 2000; Hasiotis et al. 2002, Papatheodorou et al. 2003).

V. Lykousis, D. Sakellariou and J. Locat (eds.), Submarine Mass Movements and Their Consequences, 149–157.

On the northern coast of the Gulf of Corinth, a Submarine Tailing Disposal system (STD) is operated by a bauxitic processing plant, the Aluminium de Grèce STD systems have been developed to discharge tailing slurry as a quasi-steady density current which flows by gravity to the final deposition area (Ellis et al. 1995). Bourcier & Zibrowius (1972) studied the discharges of "red mud" tailings from aluminum smelters in the Canyon de la Cassidaigne, near Marseille and reported deposits of the anthropogenic turbidity current at 1200 m water depth. A man-made turbidity current, produced at a constant rate from a single source and with constant chemistry, provides an unusual opportunity for deep sea research. It is a natural laboratory for studying turbidity current flows, stability of continental slope deposits and materials and energy transfer from shallow to deep environment.

The distribution of red-mud tailings in the Corinth Gulf has been studied previously by Papatheodorou (1991) and Papatheodorou et al. (2003). The results of 1991 and 2003 studies were based on the visual inspection of core samples collected in 1984 (that is 14 years after the beginning of the STD operation). The present work supplements the previous studies by presenting the evolution of "red-mud" distribution and dispersion 24 years after the start of tailings discharge.

2. Tailings discharge in Antikyra Bay

The metalliferous "red-mud" tailings are the by-product of bauxite processing for the production of aluminum. The aluminum plant has discharged tailings in Antikyra Bay with a mean annual production from 500 000 to 640 000 tonnes, since 1970 (Papathedorou 1991). The tailings were initially discharged on the shelf through two 2 km long submerged pipelines at a water depth of 85-100 m. In 1989 the two pipelines were replaced by a new one which releases tailings at 120 m depth until today. At the outfall site the "red-mud" slurry has a concentration of 500 g/l and a bulk density of 1.3 g/cm^3 and comes in contact with sea water of salinity ~38.4 % (Varnavas et al. 1986)

3. Materials and Methods

The data set presented in the present study was collected during a 1994 survey which took place in Antikyra Bay and the deep basin of the Gulf of Corinth. The investigated area was surveyed using a 30 kHz hydrographic echo sounder, a 3.5 kHz high resolution sub-bottom profiler and a side-scan sonar system. A large number of seafloor samples were collected from the shelf, slope and basin areas using a 3-m long Benthos corer and a Day-grab sampler. High resolution sedimentological examination was carried out using visual inspection, photography, X-ray imaging and sieve-pipette and laser granulometric analysis.

4. Results

4.1 BATHYMETRY AND MORPHOLOGY OF THE STUDY AREA

Bathymetric and high resolution seismic data showed that the northern flank of the central Gulf of Corinth consists of three physiographic provinces: the shelf, the slope and the abyssal plain or basin floor, as also reported by Brooks and Ferentinos (1984).

The shelf reaches a maximum width of 10 km in Antikyra Bay and dips gently southwards, with a gradient less than 1.2° to a depth of 300 m. In the western and eastern extremities of the study area, off Pangalos and Velanidia capes, the shelf narrows to less than 0.5 km width and extends to the 100 m isobath (Figure 1).

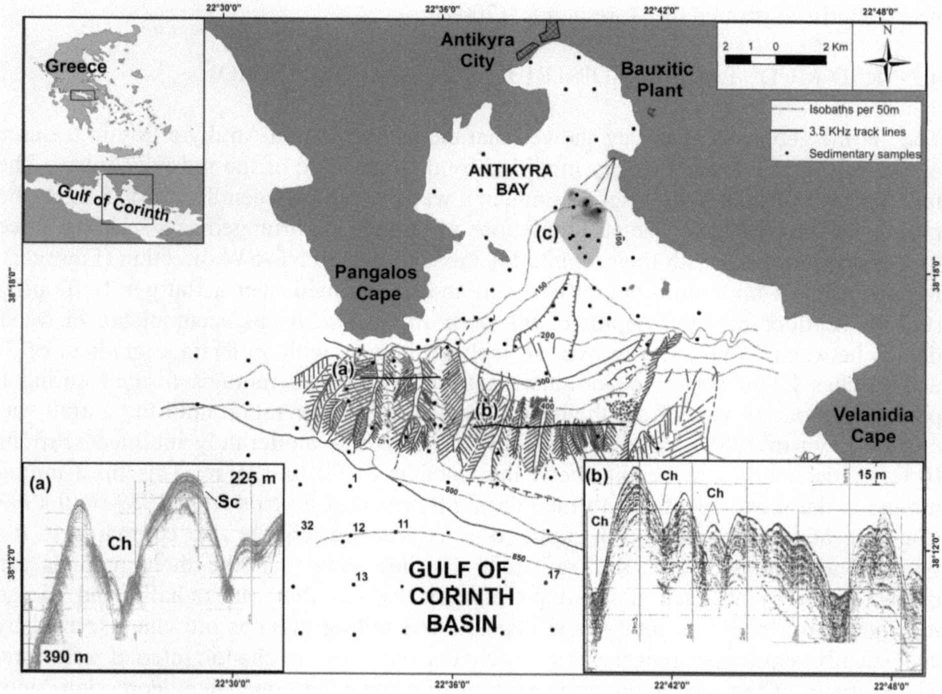

Figure 1. Bathymetric and morphologic map of the study area showing the location of the sampling stations. Numbered stations correspond to Figure 3 with representative core samples. (a), (b) seismic sections collected parallel to the slope trend, which show the channels incising the floor. (c) the lobe-shaped common depositional area of "red-mud" tailings formed at the mouth of the outfalls in Antikyra Bay. For details see Figure 2.

The central part of the Antikyra slope extends to a water depth of 700 m across a width of 4 km, with an average gradient of 5°-7.5°. As shown in reflection profiles collected parallel to the slope trend, the seafloor is dissected by numerous canyons and channels with NE-SW direction (Figure 1a, b). The V-shaped canyons start at the shelf break and have side walls 50-100 m high with gradients ranging from 5° to 10°. The U-shaped channels begin further down-slope at a water depth of 400-500 m and the height of their side walls ranges from 15 to 75 m with a maximum gradient of about 10°. The thalwegs of the channels are filled with debris flow deposits and coarse-grained sediments, as indicated by the absence of internal reflectors and the strong surface reflection in the seismic records. South of Pangalos and Velanidia capes, the slope narrows to less than 3 km and dips with gradients ranging from 10° to 16°. The shelf floor is also dissected by V-shaped deep canyons which evolve downslope into channels. The ridges in the inter-canyon areas are covered by stratified sediments.

The central basin of the Gulf of Corinth has a maximum depth of about 865 m and has a gradient of less than 0.1°. The flat basin floor is covered by a thick, well stratified turbiditic sequence characterized by closely spaced, strong seismic reflections overlain by a more acoustically transparent interval 15-20 m thick with few internal reflectors. The presence of lacustrine turbiditic deposits beneath the floor of the Gulf of Corinth was recently confirmed by Moretti et al. (2004).

4.2 "RED-MUD" TAILINGS DISTRIBUTION AND DISPERSION

The marine geophysical survey showed that the bathymetry in Antikyra Bay in the area of the outfalls has been strongly modified by the discharge of the red-mud slurry. The tailings have formed high-relief mounds in a water depth between 85 and 125 m, at the mouths of the pipelines. The accumulations of tailings are arranged as a series of three oval-shaped mounds with their longitudinal axis in N-S to NE-SW direction (Figure 2). *Mound A* has a maximum height of 14 m and has formed over a flat gently dipping (0-2.9°) seafloor in water depths of 90 to105 m. *Mound B* has accumulated in water depths between 75 and 120 m over an inclined bottom with an average gradient of 3° and reaches 27 m above the original seafloor. These two mounds formed during a period of about 19 years from the beginning of the STD operation until the outfall was closed down in 1989. *Mound C* has developed over a moderately inclined seafloor (0-6.5°) reaching 23 m height above the isobaths of 90 to 125 m. This third tailing mound is the most recent and formed during a period of 5 years from 1989 until 1994 (the date of the present study), supplied at a greater discharge rate compared to the two previous mounds. Seismic reflection profiles show that the three mounds are characterized by wedged down-slope margins, which thin out radially and merge together into a common lobe shape deposit. The tailing mounds are characterized by acoustically semitransparent facies with few discontinuous to chaotic internal reflectors. The surfaces of the mounds present patterns of scarps and crater-like depressions with arcuate to circular shapes about 45 m in diameter.

The red-mud tailings disperse from the mounds on the shelf floor in a down-slope direction and form an elongated main surface deposit with a minimum thickness of 1 m, an areal extent of about 5 km^2 and a volume of about 13×10^6 m^3. The distal tailings deposit, which is distributed beyond the main deposit as far as the shelf-break, covers 30 km^2 of the shelf floor. This layer thins radially downslope to 1 cm thick.

On the base of slope and the basin floor, the "red-mud" tailings have formed successive layers, which are intercalated with olive gray natural sediments. The surface and subsurface layers have a thickness from few mm to 5 cm and they cover an area of about 273 km^2. The volume of the "red-mud" deposits in the Corinth basin is estimated at about 3.5×10^6 m^3.

Considering the fact that the bulk volume of red-mud tailings deposits in the study area is estimated about 16×10^6 m^3, it can be suggested that the 81 % of the discharged red mud tailings are deposited in the outfall termination site and 19% are transported to the basin floor.

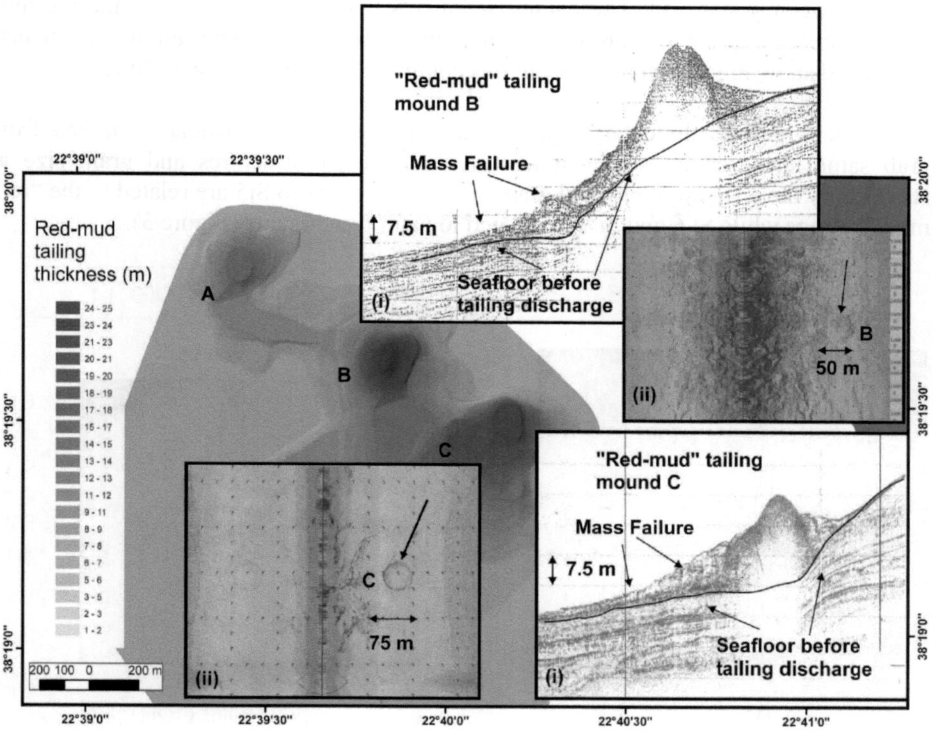

Figure 2. Map illustrating the thickness distribution of the three "red-mud" tailings mounds in Antikyra Bay. A, B, C are the three mounds. Insets (i): seismic records collected across the mounds showing the thickness of the tailings. The transparent acoustic character, the hummocky relief and the wedging out downslope reflectors indicate the formation of mass failures. Insets (ii): side scan sonar records collected across the mounds which show the crater-like depressions indicated by arrows.

4.3 THE METALLIFEROUS TAILINGS

The "red mud" submarine tailing deposits are highly enriched in Fe_2O_3, TiO_2, Cr_2O_3, Ni, Co, Pb and Cu compared to the surrounding natural sediments. In contrast, the natural sediments are characterized by high concentrations of Mn, Zn and $CaCO_3$, compared to the tailings (Ferentinos et al. 1995). The highest concentrations of Fe_2O_3, TiO_2, Cr_2O_3, Ni, Co, Pb and Cu are detected at the mouths of the outfall in Antikyra Bay. On the basin floor, the tailing deposits are characterized by much lower concentration in the above mentioned heavy metals compared to the outfall area. This indicates that the "red-mud" tailings during their transport from the shelf to the basin are mixed with natural sediments.

4.4 FINE-GRAINED SEDIMENTARY TYPES ON THE BASIN

The "red-mud" tailings on the shelf and the basin floor consist of medium- to fine-grained sediments with 40-60 % silt, 45-50 % clay and < 5% sand contents. Tailings samples generally show unimodal grain size populations with an average mean size of 7 to 8.8 Ø. However, several red-mud samples show bimodal populations with a coarse mode between 4 and 5 Ø. The natural sediments, which intercalate with the red-mud tailings are fine-grained muds with similar mean size range. The tailings are poorly sorted and show higher standard deviations compared with the natural sediments.

Seven fine-grained sedimentary types (St) are distinguished in surficial sediments from grab samples, based on sediment texture, sedimentary structures and grain size as revealed by the X-radiography and grain-size analysis. St1 to St5 are related to the "red-mud" tailings while St 6 and St7 are related to natural sediments (Figure 3).

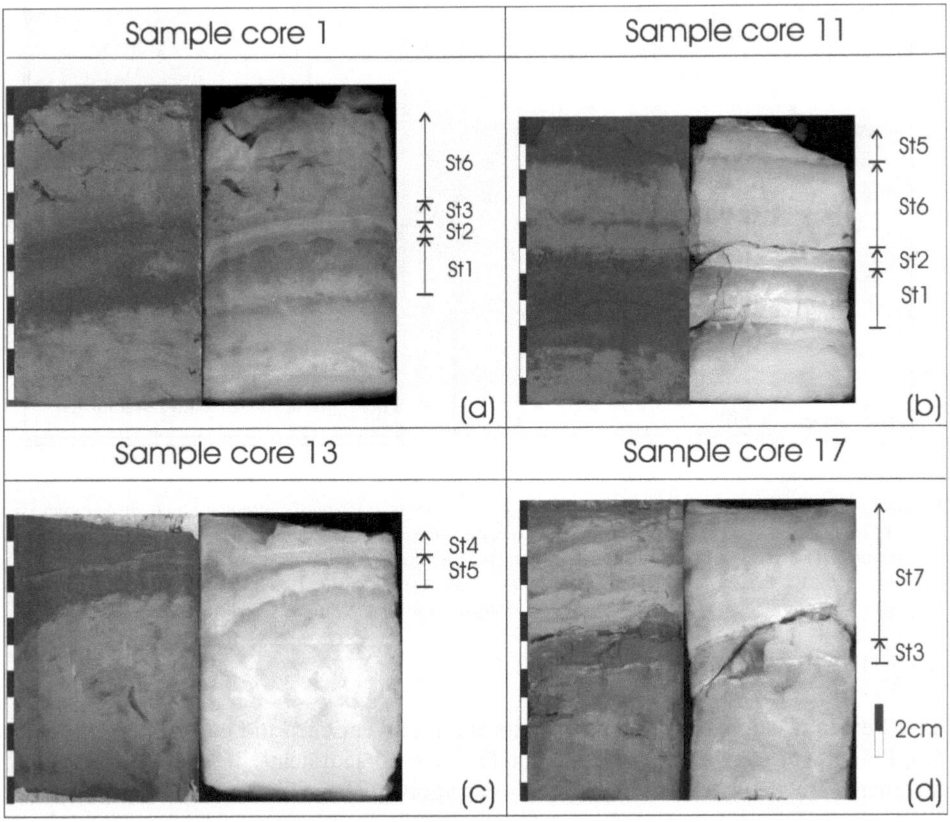

Figure 3. Grab sample cores representative of the seven sedimentary types (St). Arrows indicate the limits of each sedimentary type and the normal size grading. Position of the cores is given in Figure 1.

St1 consists of a normally graded basal silty-clay layer, 0.5-1 cm thick. The basal contact is moderately sharp with micro-relief and is disturbed by burrows. The basal layer passes upward to silty-clay mud layer 0.1-1 cm thick, with poor grading (Figure 3a, b).

St2 consists of a thin (1-3 mm) discontinuous lenticular silty basal lamina which occasionally displays convolute micro-lamination and grades upward into silt laminae that decrease in frequency and thickness. The basal contact is sharp with micro-relief (Figure 3a, b).

St3 consists of a basal sequence of parallel to sub-parallel fine silt laminae, a few mm thick. The basal sequence has a gradational bottom contact and passes upward into a structureless clayey silt layer 0.8-1.5 cm thick, with inverse size grading (Figure 3a, d).
St4 consists of a sharp based silty-clay layer 0.5 to 1.5 cm thick, which is characterized by thin parallel silt or clayey-silt laminae, which become indistinct upward (Figure 3c).
St5 consists of a basal silty-clay layer 0.5 – 1 cm thick, which grades into a thin (< 0.3 cm) massive mud layer. Type 5 deposits have sharp bottom contacts (Figure 3b, c).
St6 consists of a single thin (<1 mm) lenticular basal lamina, occasionally indistinct, which grades upward into uniform moderately bioturbated silty-clay mud. Bioturbation intensifies in the top 2 cm of the layer (Figure 3a, b).
St7 resembles type 6 but instead of a single silt lamina at the base, the basal layer consists of well to poorly defined alternating laminae of silt and mud, which decrease upward in frequency and thickness and grade upward into thick graded structureless mud. Bioturbation, where present, is limited to the top of the mud (Figure 3 d).

The overall examination of the collected grab samples shows that: (i) the sediment types correspond to distinct layers which form small scattered sheet like deposits, (ii) each layer is between 0.5 and 6 cm thick, (iii) the bottom and the top contacts of the layers are well defined and (iv) the surficial sedimentary cover in the upper 8 cm below the sea floor is built up by these layers.

The total thickness of the recent sedimentary layers in the basin of the Gulf of Corinth ranges from 2 to 8 cm and has been attained during the last 24 years of tailings discharge. The accumulation rates of the recent sedimentary cover thus range from 0.8 to 3.3 mm/yr. They are significantly higher than the rates of hemipelagic sedimentation during Holocene from 0.1-0.3 mm/yr in basins in the Aegean and Ionian Seas (Geraga et al. 2000). Lykousis et al. (2007) estimated that the recent accumulation rate for the late stratified turbiditic sequence in the Gulf of Corinth ranges from 1.3 to 1.4 mm/yr which is similar to the average sedimentation rate of about 2 mm/yr estimated from the present study. The high rates of accumulation indicate that the sedimentary layers were deposited by turbidity flows rather than by hemipelagic sedimentation. Furthermore, the presence of escape burrows indicative of rapid deposition and the restriction of bioturbation to the upper part of the layers which according to Stow and Piper (1984) distinguish the fine-grained turbiditic deposits, also attest to the fact that the sedimentary layers were deposited rapidly by turbidity flows.

In addition, the structures which are observed in the sedimentary types such as: sharp-based silt beds that grade up into mud (St1 and 5), silt laminated muds that fine and fade upwards (St3 and 4) or grade upwards into uniform mud (St7), and a single lamina which grades upward into mud (St2 and 6) are indicative of mud turbiditic deposits according to Stow and Piper (1984).

5. Discussion

Evidence from seismic data across the "red-mud" mounds indicates that the mounds are affected by mass failures (Figure 2i). The high rates of accumulation at the mouth of the pipelines in association with cyclic external loading by earthquakes are responsible for the instability of the tailings and the triggering of rotational slumps. The transformation of failed tailings slurry into mud-flows (Mulder and Cochonat 1996) is indicated by the seismic profile across the mounds of the "red-mud" tailings on the shelf, and subsequent conversion into turbidity currents is indicated by the deposits on the basin floor.

The formation of the recent sedimentary cover in the basin plain which consist of successive thin-bedded "red-mud" and natural sediment layers is attributed to turbidity flows. The five "red-mud" turbidity flows are related to discharge of metalliferous tailings and can be attributed to anthropogenic activities. The other two turbidity flows transport natural sediments and are not related to tailing discharge therefore they were not caused by human intervention. The natural turbidity flows intercalate with the "red mud" turbidity flows. The fine-grained turbidites were recognized by distinctive sedimentary structures and can be documented by either Piper's (1978) or Stow's (Stow and Shanmugam 1980) structural schemes. Each sedimentary type has been formed by a different turbiditic event, as indicated by the sharp contacts and the presence of bioturbation between the successive types. The full sequence of structures $T_{0\text{-}7}$ introduced by Stow and Shanmugam, which represents a single turbidity event, cannot be identified in a single layer due to the their small thickness of only 0.5 to 6 cm. Instead it was observed that St1 and St5 are characterized by top cut-out sequences while St2, St3, St4, St6 and St7 are characterized by more commonly base-cut-out sequences similar to those described as mud turbidites by Stow and Piper (1984). The latter authors suggested that such mud turbidites could result from muddy turbidity currents driven entirely by fine-grained sediments. This may be the case for the turbidity flows, which either originate from failure in the "red-mud" tailings mounds or generated at the shelf edge and upper slope.

6. Acknowledgments

The leading author would like to thank the National Scholarships Foundation (NSF) of Greece for the financial support.

7. References

Brooks M, Ferentinos G (1984) *Tectonics and sedimentation in the Gulf of Corinth and the Zakynthos and Kefallinia channels, western Greece.* Tectonophysics 101:25-54

Bourcier M, Zibrowius H (1972) *Les "Boues Rouges"deversées dans le Canyon de la Cassidaigne (region de Marseille). Observation en soucoupe plongeante SP 350 (juin 1971) et resultats de dragages.* Tethys 4:811-842

Ellis DV, Poling GW, and Baer RL (1995) *Submarine tailings disposal (STD) for mines: an introduction.* Marine Georesources and Geotechnology 13:3-18

Ferentinos G, Papatheodorou G, and Collins M (1988) *Sediment transport processes on an active submarine fault escarpment: Gulf of Corinth Greece.* Marine Geology 83:43-61

Ferentinos G, Papatheodorou G (1995) *Marine environmental study of bauxite tailings in the Gulf of Corinth, Greece.* Technical report submitted to Aluminium de Grèce, University of Patras

Geraga M, Tsaila-Monopolis S, Ioakim G, Papatheodorou G, and Ferentinos G (2000) *Evaluation of palaeoenvironment changes during tha last 18.000 years in Myrtoon basin SW Aegean Sea.* Palaeogeography, Palaeoclimatology, Palaeoecology 156:1-17

Hasiotis T, Papatheodorou G, Bouckovalas G, Corbau C, and Ferentinos G (2002) *Earthquake-induced coastal sediment instabilities in the western Gulf of Corinth, Greece.* Marine Geology 186:319-335

Lykousis V (1991) *Submarine slope instabilities in the Hellenic Arc Region, Northeastern Mediterranean Sea.* Marine Geotechnology 10:83-96

Lykousis V, Sakelariou D, Moretti I, and Kaberi H. *Late Quaternary evolution of the Gulf of Corinth: Sequence stratigraphy, sedimentation, fault-slip and subsidence rates.* Tectonophysics (2007), doi:10.1016/j.tecto.2006.11.007

Mulder T, Cochonat P (1996) *Classification of offshore mass movements.* Journal of Sedimentary Research 66:43-57

Moretti I, Lykousis V, Sakellariou D, Reynaud, J-Y, Benziane B, and Prinzhoffer, A (2004) *Sedimentation and subsidence rate in the Gulf of Corinth: what we learn from the Marion Dufresne's long piston coring.* C.R. Geoscience 336: 291-299.

Papatheodorou G (1991) *Sedimentation in the Gulf of Corinth.* PhD thesis, University of Patras

Papatheodorou G, Ferentinos G (1997). *Submarine and coastal sediment failure triggered by the 1995, Ms= 6.1R Aegion earthquake, Gulf of Corinth, Greece.* Marine Geology 137:287-304

Papatheodorou G, Stefatos A, Christodoulou D, and Ferentinos G (2003) *Small scale present day turbidity currents in a tectonically active submarine graben, the Gulf of Corinth (Greece): their significance in dispersing mine tailings and their relevance to basin filling.* In: Locat J, Mienert J (eds) Submarine mass movements and their consequences. Advances in Natural and Technological Hazards Research Series. Kluwer, pp 459-468

Perissoratis C, Piper DJW, and Lykousis V (2000) *Alternating marine and lacustrine sedimentation during late Quaternary in the Gulf of Corinth rift basin, central Greece.* Marine Geology 167:391-411

Piper DJW (1978) *Turbidites, muds and silts on deep-sea fans and abyssal plains.* In: Stanley DJ, Kelling G (eds) Sedimentation in Submarine Canyons, Fans and Trenches Dowden, Hutchinson and Ross, Stroudsburg PA, pp 163-176

Stow DAV, Piper DJW (1984) *Deep-water fine-grained sediments: facies models.* In: Stow DAV, Piper DJW (eds) Fine-Grained Sediments: Deep-Water Processes and Facies. Geol. Soc. London Spec. Publ., 15:611-645.

Stow DAV, Shanmugam G (1980) *Sequence of structures in fine –grained turbidites: comparison of recent deep-sea and ancient flysch.* Sedimentary Geology 25:23-42

Varnavas S, Ferentinos G, and Collins M (1986) *Dispersion of bauxitic red mud in the Gulf of Corinth, Greece.* Marine Geology 70:211-222

Section 3 - New techniques, approaches and challenges in submarine slope instability analysis

PROBABILITY STUDY ON SUBMARINE SLOPE STABILITY

SHAOLI YANG, FARROKH NADIM AND CARL FREDRIK FORSBERG
International centre for Geohazards, Norwegian Geotechnical Institute, Oslo, Norway

Abstract

Most of the parameters used in slope stability analyses, in particular the mechanical soil properties, are uncertain. Probability theory and reliability analyses can provide a rational framework for dealing with uncertainties. Different methods for doing reliability analysis for slopes are discussed in this study and applied to case studies. The results obtained from FOSM, PEM, and FORM via response surface method combined with the finite element method are compared, and the parameters which contribute most to the uncertainty in the factor of safety are identified.

Keywords: Submarine slope, probability analysis, reliability index, factor of safety.

1. Introduction

Submarine slope failures occur frequently both on active and passive margins (Mienert, 2004). The mechanical properties of the sediments, especially the soil shear strength, play an important role in the stability of a submarine slope. However, the mechanical properties of natural sediments are variables that depend on the way sediments are formed. Uncertainty in slope stability evaluation is due to inherent spatial and temporal variability of soil properties; measurement errors (random and/or systematic); statistical fluctuations; model uncertainty; uncertainty in load and load effects and omissions.

The stability situation for natural and man-made slopes is often expressed by the factor of safety. The factor of safety is defined as the ratio of the characteristic resisting force (commonly referred to as "resistance" or "capacity") to the characteristic load or driving force (commonly referred to as "load" or "demand"). The conventional approach does not address the uncertainty in load and resistance in a consistent manner. The ambiguous definition of "characteristic" values allows the engineer to implicitly account for uncertainties by choosing conservative values of load (high) and resistance parameters (low). The choice, however, is somewhat arbitrary. Slopes with nominally the same factor of safety could have significantly different safety margins because of the uncertainties and how they are dealt with. A low safety factor by deterministic analyses does not necessarily correspond to a high probability of failure and vice versa (Nadim and Lacasse, 1999; Christian, 2004). Probability theory and reliability analyses provide a rational framework for dealing with uncertainties and decision making under uncertainty. Depending on the level of sophistication, the analyses provide one or more of the following outputs: probability of failure; reliability index; the most probable combination of parameters leading to failure and sensitivity of result to any change in parameters. The reliability based approach has been extensively used in slope engineering (Chowdhury, 1984; Christian, et al., 1994 among others). The purpose of this study is to compare the results from different probabilistic methods for slope stability analyses.

V. Lykousis, D. Sakellariou and J. Locat (eds.), Submarine Mass Movements and Their Consequences, 161–170.

2. Probabilistic analysis method

As mentioned earlier, the conventional factor of safety does not address the uncertainties explicitly. Therefore a simple prescription of a factor of safety to be achieved in all instances is not realistic and may lead to over-design or unsafe situations. Well-established reliability methods, such as the first-order, second-moment approximation (FOSM), point estimate method (PEM), and first order reliability method (FORM), which are discussed below; and Monte Carlo simulation are useful techniques for determining the reliability of geotechnical designs for estimating the probability of failure. The reliability methods also reveal which parameters contribute most to the uncertainty and probability of failure.

The first step in estimation of failure probability using any probability method is to decide what constitutes unsatisfactory performance or failure. Mathematically, this is achieved by defining a performance function G(*X*), such that G(*X*) ≥ 0 means satisfactory performance and G(*X*) < 0 means unsatisfactory performance or "failure". *X* is a vector of basic random variables. For many geotechnical problems and related deterministic computer programs, the output is in the form of the factor of safety, and the capacity and demand on the system (i.e. resistance and load) are not explicitly separated.

The "reliability index", defined as

$$\beta = \frac{u_G}{\sigma_G} \tag{1}$$

in which μ_G and σ_G are respectively the mean and standard deviation of the performance function, is often used as an alternative performance measure to the factor of safety (Li & Lumb 1987, Christian *et al.* 1994, Duncan 2000). The reliability index provides more information about the reliability of a geotechnical design or a geotechnical structure than is obtained from the factor of safety alone. It is directly related to the probability of failure and the computational procedures used to evaluate the reliability index reveal which parameters contribute most to the uncertainty in the factor of safety. This is useful information that can guide the engineer in further investigations. Li & Lumb (1987)shows the reliability indices for different formats of the performance function using the FOSM method.

The reliability index may also be formulated in terms of the expected value of FS, (*E[FS]*) and standard deviation of FS, (σ_{FS}) . The reliability index is obtained using the following steps, assuming FS is log-normal distributed in the FOSM and PEM.

$$V_{FS} = \frac{\sigma_{FS}}{E[FS]} \tag{2}$$

$$\sigma_{\ln FS} = \sqrt{\ln(1 + V_{FS}^2)} \tag{3}$$

$$E[\ln FS] = \ln E[FS] - \frac{1}{2}\ln(1 + V_{FS}^2) \tag{4}$$

$$\beta = \frac{E[\ln FS]}{\sigma_{\ln FS}} = \frac{\ln(E[FS]/\sqrt{1 + V_{FS}^2})}{\sqrt{\ln(1 + V_{FS}^2)}} \tag{5}$$

where V_{FS} is the variance of the *FS*; σ_{lnFS}, $E[lnFS]$ are the standard deviation and expected value of the *lnFS*, and β is the reliability index.

2.1 FIRST-ORDER, SECOND-MOMENT APPROXIMATION (FOSM)

The first-order second moment (FOSM) approximation is based on the Taylor series expansion of the safety factor or the performance function about the mean value of the parameters, and neglecting the higher order terms (Ang and Tang, 1984). It provides analytical approximations for the mean and standard deviation of a parameter of interest as a function of the mean and standard deviations of the various input factors, and their correlations. This is a simple method and exact for linear performance functions, and one must assume the distribution function for the FS beforehand to estimate the failure probability (Christian (1996), Duncan (2000), Maia and Assis (2005)).

2.2 POINT ESTIMATE METHOD (PEM)

An alternative method to estimate moments of a performance function based on the moments of the random variables is the point estimate method (Rosenblueth, 1975). The PEM is a simple, direct and effective method of computing the low-order moments of functions of random variables. It can be used in any slope stability problem regardless of how complex the expression for the FS is. 2^n evaluations of the FS have to be made when there are n random variables (Hassan and Wolff, 2000; Christian and Baecher, 2002; Maia and Assis, 2005). The PEM is more accurate than the FOSM because it is based on higher order expansions.

2.3 FIRST ORDER RELIABILITY METHOD (FORM)

The performance function is explicit in the FORM (Hasofer and Lind, 1974). With a known joint probability density function of all random variables, the probability of failure is given by :

$$P_f = \int_L F_x(X)dX \tag{6}$$

Where L is the unsafe domain of X where the performance function $G(X) < 0$. $F_x(X)$ is the joint probability density function.

This method is exact when the limit state surface is planar and the parameters follow normal distributions. The reliability index and probability of failure can be obtained directly from the FORM. The contribution of uncertainty in different parameters in the total uncertainty in the safety factor can be given by the sensitivity factor by the FORM at the same time (Nadim et al. (2005); Nadim and Locat (2005)).

2.4 RELIABILITY ANALYSIS VIA RESPONSE SURFACE METHOD

In many geotechnical problems, an explicit functions for FS cannot be derived. In these situations, a polynomial function may be used to approximate the *true* performance function. Experiments or numerical analyses are then performed at various sampling points, x_i, to determine the unknown coefficients in the *approximate* polynomial function. The following two polynomial functions are commonly used:

$$\hat{g}(X) = b_0 + \sum_{i=1}^{k} b_i X_i + \sum_{i=1}^{k} b_{ii} X_i^2 \tag{7}$$

$$\hat{g}(X) = b_0 + \sum_{i=1}^{k} b_i X_i + \sum_{i=1}^{k} b_{ii} X_i^2 + \sum_{i=1}^{k-1} \sum_{j>i}^{k} b_{ij} X_i X \tag{8}$$

where X_i (i = 1, 2, . . . , k) = ith random variable; and b_0, b_i, b_{ii}, and b_{ij} = unknown coefficients to be determined either by solving a set of simultaneous equations or by using regression analysis. The number of unknown coefficients in (1) and (2) are $2k + 1$ and $(k + 1)(k + 2)/2$, respectively (Huh and Haldar, 2001).

The sampling points have to be designed in the response surface method (Bucher et al., 1989; Bucher and Bourgund, 1990; Rajashekhar and Ellingwood, 1993). Saturated design consists of only as many experimental sampling points as the total number of coefficients necessary to define a polynomial. Only saturated design was used in this study. For a polynomial without cross terms, the total number of required experimental sampling points is $2k$+ 1 (Fig. 1, 2a), where k is the number of random variables. For a polynomial with cross terms, the total number of required experimental sampling points is $(k + 1)(k+2)/2$ (Fig. 1, 2b).

Wong (1985) obtained the FS of a homogeneous slope by the Finite Element Method (FEM) with a gravity increasing approach and the failure probability was calculated by Monte Carlo simulation via the response surface method. Two random parameters (cohesion and friction angle) were considered. The response surface was approximated by a second order polynomial function in which the first order and the cross terms are used. Xu and Low (2006) used the FEM combined with a strength reduction technique to obtain the FS, while they calculated the reliability index by the FORM via response surface method. No iteration was used to obtain the response surface function. The response surface function was approximated by a second order polynomial function without cross terms. The sediment strength parameters, unit weight and the thicknesses of the layers involved in their case study were considered random variables.

The Monte-Carlo simulation method is used to simulate stochastic processes by random selection of input values to an analysis model in proportion to their joint probability density function. It is a powerful technique that is applicable to both linear and non-linear problems, but can require a large number of simulations to provide a reliable distribution of the response (El-Ramly et al., 2002; El-Ramly et al., 2003). The Random FEM (Griffiths and Fenton, 2004; Fenton and Griffiths, 2005) can also be used for the probability study of the slope stability. These two methods are not discussed in this study.

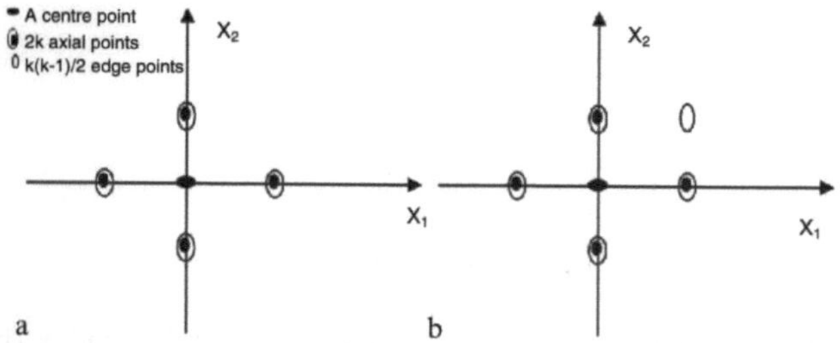

Fig. 1. Design points for k = 2.

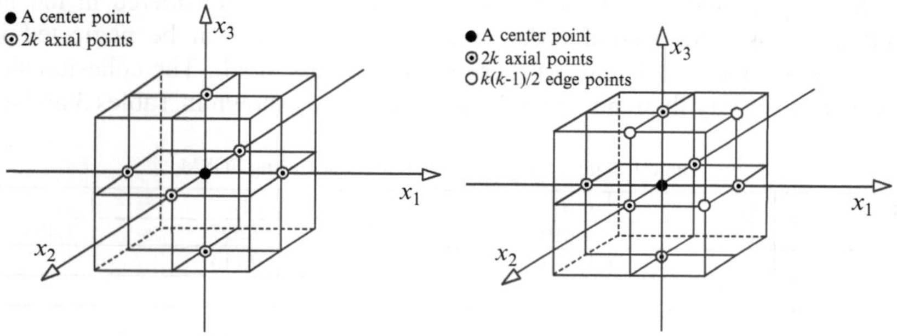

Fig. 2. Design points for k = 3.

3. Case studies

The FOSM, PEM, and FORM via response surface method, combined with the FEM (Plaxis, 2001) were used for slope stability analysis in this study. The strength reduction technique (Zienkiewicz et al. 1975) was used in Plaxis to obtain the safety factor.

The procedure used in this study was as follows:

- Determine the variables, their mean and standard deviation.
- Run Plaxis to get safety factors at selected values of variables.
- Determine the reliability index for the FOSM and PEM approximations.
- Determine the coefficients of the polynomial equations for the design methods.
- Run the FORM to obtain the reliability index.

The first slope has a simple shape (Fig. 3), and is simply characterized by its cohesion, c, and friction angle, ϕ, which are assumed to be normally distributed random variables. The mean value and standard deviation of c are respectively 15 kPa and 4 kPa, while the mean value and standard deviation of ϕ are respectilvey 20° and 2°. $c_{+,}$ means cohesion plus one standard deviation, c_{-} means cohesion minus one standard deviation, and the same applies for ϕ_{+} and ϕ_{-} in the following case study. Model uncertainty is due to errors introduced by mathematical approximations and simplifications. The model uncertainty ε was considered as a normally distributed random variable, with a mean of 1, and a standard deviation of 0.05.

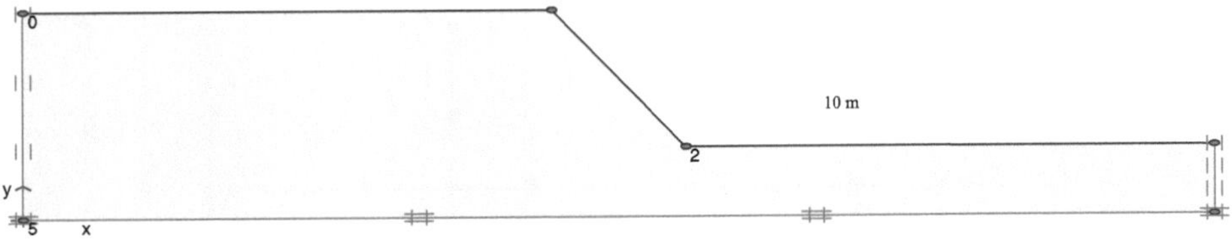

Fig. 3. Slope geometry for the first case study.

The outcome of the computations shows that a larger reliability index (β) is calculated using the FOSM and PEM compared to the FORM via response surface method (Tables 1-4). A lower β is obtained if the model uncertainty (ε) is considered in the FORM reliability analysis. The second order and the cross terms can be neglected in the polynomial, since the coefficients of these terms are very small. The cohesion shows a much higher sensitivity than the internal friction angle, as shown in Tables 3 and 4.

Table 1. Safety factor and reliability index using FOSM.

Variables	$c\phi$	$c_+\phi$	$c_-\phi$	$c\phi_+$	$c\phi_-$
FS	1.540	1.810	1.271	1.597	1.484
ΔFS		0.539		0.113	
β	2.348				

Table 2. Safety factor and reliability index using PEM.

Variables	$c_+\phi_+$	$c_+\phi_-$	$c_-\phi_+$	$c_-\phi_-$
FS	1.868	1.748	1.321	1.211
β	2.314			

Table 3. Safety factor and reliability index from saturated design and second order polynomial without cross terms.

Variables	$c\phi$	$c_+\phi$	$c_-\phi$	$c\phi_+$	$c\phi_-$
FS	1.540	1.810	1.271	1.597	1.484
Polynomial equation	$FS=0.021+0.066c+0.023\phi+3.12E\text{-}5c^2+0.000125\phi^2$				
β and P_f	1.937/2.64%, 1.904/2.85% with ε				
Sensitivity factors	c (0.98), ϕ (0.2) c(0.96), ϕ (0.2), ε (0.19)				

Table 4 Safety factor and reliability index from saturated design and full second order polynomial.

Variables	$c\phi$	$c_+\phi$	$c_-\phi$	$c\phi_+$	$c\phi_-$	$c_+\phi_+$
FS	1.540	1.810	1.271	1.597	1.484	1.868
Polynomial equation	$FS=0.059+0.064c+0.021\phi+3.12E\text{-}5c^2+0.000125\phi^2+0.000125c\phi$					
β and Pf	1.946/2.58%, 1.913/2.78% with ε					
Sensitivity factors	c (0.98), ϕ (0.2) c(0.96), ϕ (0.19), ε (0.19)					

The second example is a four-layer slope based on the geometry observed in the Storegga Slide, off mid-Norway (Yang, et al., 2007). The undrained shear strength for the clay was modelled using the SHANSEP approach (Ladd and Foott, 1974). Normalization of the undrained shear strength with respect to consolidation stress and overconsolidation Ratio (OCR) can be expressed with the following relationship:

$$\left(\frac{s_u}{\sigma_v^{'}}\right)_{OC} = \left(\frac{s_u}{\sigma_v^{'}}\right)_{NC} .OCR^m = \alpha OCR^m \tag{9}$$

where s_u is the undrained shear strength, σ_v' is the effective vertical consolidation stress, and m is a soil parameter. The indices OC and NC mean overconsolidated and normally consolidated respectively. Only α was regarded as a random variable for the case study. In the slope stability analysis within Plaxis, the shear strength anisotropy can be taken into account and calculated according to the following relationships (Kvalstad, et al., 2005):

$$s_{uDSS} / s_{uC} = 0.8 \tag{10}$$

$$s_{uE} / s_{uC} = 0.7 \tag{11}$$

where s_{uC} is the undrained shear strength from anisotropic undrained triaxial compression tests, s_{uDSS} from direct simple shear tests and s_{uE} from anisotropic undrained triaxial extension tests.

The example slope has a total of four layers (Fig. 4). However, only three layers (three variables) were considered in this study because the contribution to the safety factor of the fourth topmost layer, downslope from the main scar was very small from the preliminary study. The second layer is a marine layer with low shear strength, while the first (bottom) and third (top) layers are glacial till layers. The mean and standard deviation for the three variables were respectively as follows:
α_1 (0.25, 0.05); α_2 (0.2, 0.035); α_3 (0.25. 0.05); ε (1, 0.05)

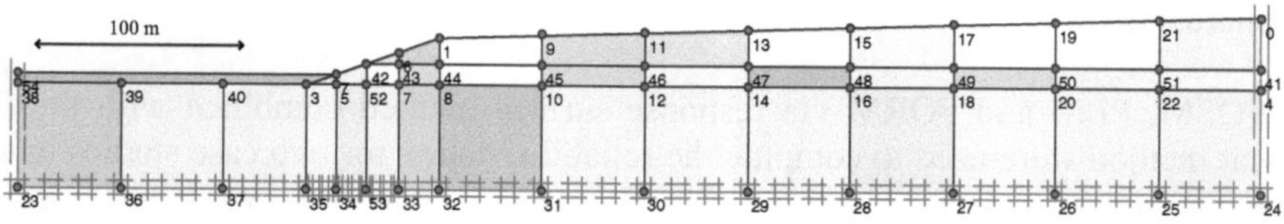

Fig. 4. Slope geometry for the second case study.

Table 5. Safety factor and reliability index using FOSM.

Variables	$\alpha_1\alpha_2\ \alpha_3$	$\alpha_{1+}\alpha_2\ \alpha_3$	$\alpha_{1-}\alpha_2\ \alpha_3$	$\alpha_1\alpha_{2+}\alpha_3$	$\alpha_1\alpha_{2-}\alpha_3$	$\alpha_1\alpha_2\alpha_{3+}$	$\alpha_1\alpha_2\alpha_{3-}$
FS	1.583	1.689	1.457	1.64	1.516	1.681	1.445
ΔFS		0.232		0.124		0.236	
β	4.071						

Table 6. Safety factor and reliability index using PEM

Variables	$\alpha_{1+}\alpha_{2+}\alpha_{3+}$	$\alpha_{1+}\alpha_{2-}\alpha_{3+}$	$\alpha_{1+}\alpha_{2+}\alpha_{3-}$	$\alpha_{1+}\alpha_{2-}\alpha_{3-}$	$\alpha_{1-}\alpha_{2+}\alpha_{3+}$	$\alpha_{1-}\alpha_{2-}\alpha_{3+}$	$\alpha_{1-}\alpha_{2+}\alpha_{3-}$	$\alpha_{1-}\alpha_{2-}\alpha_{3-}$
FS	1.880	1.695	1.583	1.483	1.625	1.447	1.388	1.286
β	3.803							

Table 7. Safety factor and reliability index for saturated design and second order polynomial without cross terms.

Variables	$\alpha_1\alpha_2\alpha_3$	$\alpha_{1+}\alpha_2\alpha_3$	$\alpha_{1-}\alpha_2\alpha_3$	$\alpha_1\alpha_{2+}\alpha_3$	$\alpha_1\alpha_{2-}\alpha_3$	$\alpha_1\alpha_2\alpha_{3+}$	$\alpha_1\alpha_2\alpha_{3-}$
FS	1.583	1.689	1.457	1.64	1.516	1.681	1.445
Polynomial equation	$FS=-0.855+4.32\alpha1+3.404\alpha2+6.36\alpha3-4\,\alpha_1^2-4.082\,\alpha_2^2-8\,\alpha_3^2$						
β and P_f	2.69/0.357%, 2.638/0.417% with ε						
Sensitivity factors	α_1 (0.57), α_2 (0.27), α_3 (0.78) α_1 (0.56), α_2 (0.26), α_3 (0.76), ε (0.2)						

This case study also showed that a larger β is obtained using the FOSM and PEM than with the FORM via response surface method (Tables 5-8). A lower β is calculated if the model uncertainty is considered, as it happens in the first case study. However, in this case study the second and cross terms could not be neglected. The stress ratio α_3 shows the highest sensitivity (Tables 7-9).

Table 8. Safety factor and reliability index for saturated design and full second order polynomial.

Variables	$\alpha_1\alpha_2\alpha_3$	$\alpha_{1+}\alpha_2\alpha_3$	$\alpha_1\alpha_2\alpha_3$	$\alpha_1\alpha_{2+}\alpha_3$	$\alpha_1\alpha_{2-}\alpha_3$	$\alpha_1\alpha_2\alpha_{3+}$	$\alpha_1\alpha_2\alpha_{3-}$	$\alpha_{1+}\alpha_{+}\alpha_3$	$\alpha_1\alpha_{2+}\alpha_{3+}$	$\alpha_{1+}\alpha_2\alpha_{3+}$
FS	1.583	1.689	1.457	1.64	1.516	1.681	1.445	1.745	1.756	1.809
Polynomial	$FS=0.181+2.234\alpha1+0.976\alpha2+2.103\alpha3-4\alpha_1^2-4.081\alpha_2^2-8\alpha_3^2-0.571\alpha1\alpha2+8.8\alpha1\alpha3+10.286\alpha2\alpha3$									
β and P_f	3.110/0.0936%, 3.026/0.124% with ε									
Sensitivity factors	α_1 (0.32), α_2 (0.05), α_3 (0.95) α_1 (0.35), α_2 (0.07), α_3 (0.90), ε (0.24)									

4. Conclusions

The FOSM, PEM and FORM via response surface method combined with the finite element method were used to compute the reliability index for two case studies in order to compare results.

- A lower probability of failure was obtained using simple methods such as the FOSM and PEM compared to the more sophisticated FORM via response surface method.
- A lower probability of failure is calculated using the FORM via response surface method, if the cross terms are considered in the polynomial equation.
- A higher probability of failure is obtained when the model uncertainty is considered in the FORM.
- In the first case study, the cohesion has a greater contribution to the total uncertainty in the calculated safety factor than the frictional angle and the model uncertainty. A linear polynomial can be used in such case.
- In the second case study, the stress ratio α, at the bottom layer has a greater contribution to the total uncertainty in the calculated safety factor than the stress ratio at the top and medium layers, and the model uncertainty.

FOSM and PEM can be used as a preliminary check for the failure probability of a slope stability. However, the more sophisticated method like FORM should be used in the analysis if the slope has a great importance.

5. Acknowledgements

This is publication number 146 of the International Centre for Geohazards (ICG), and was partly supported by Euromargin project (European Science Foundation 01-LEC-EMA14F).

6. References

Ang, A.H.S., and Tang, W.,1984. *Probability concepts in engineering planning and design*. Vol. 1, Basic principles. John Wiley and Sons, New York, USA.

Bucher, C.G., Chen, Y.M., and Schuller, G.I.,1989. "*Time variant reliability analysis utilizing response surface approach.*" Proc., 2nd International Federation for Information Processing, Working Group 7.5 Conf., Springer, Berlin.

Bucher, C.G., and Bourgund, U., 1990. "*A fast and efficient response surface approach for structural reliability problems.*" *Struct. Safety*, Amsterdam, 7, 57-66.

Chowdhury, R.N., 1984. *Recent development in landslide studies: Probability methods state of the art report.* The forth international symposium on landslides, Vol.1, 209-228.

Christian, J.T., Ladd, C.C., and Baecher, G.B. 1994. *"Reliability applied to slope stability analysis." J. Geotech. Eng.*, Vol.120, No.12, 2180-2207.

Christian J.T., 1996. *Reliability methods for atability of existing slopes. Proceedings of Uncertainty in the geologic environment: From theory to Practice.* Geotechnical special publication No.58, 409-418.

Christian, J.T., and Baecher, G.B., 2002. *The point-estimate method with large numbers of variables.* International Journal for Numerical and analytical methods in geomechanics , 26, 1515-1529.

Christian, J.T., 2004. *Geotechnical Engineering reliability: How well do we know what we are doing?* ASCE *Journal of Geotechnical Engineering*, 120(12):2180-2207.

Duncan, J.M., 2000. *Factors of safety and reliability in geotechnical engineering*. Journal of Geotechnical and Geoenvironmental Engineering, Vol. 126, No.4, 307-316.

El-Ramly H., Morgenstern N.R., and Cruden D.M., 2002. *Probabilistic slope stability analysis for practice*, Canadian Geotechnical Journal, 39, 665-683.

El-Ramly H., Morgenstern N.R., and Cruden D.M., 2003. *Probabilistic stability analysis of a tailings dyke on presheared clay–shale* Canadian Geotechnical Journal, 40 192-208.

Fenton, G.A., and Griffiths, D.V., 2005. *A slope stability reliability model*, Proceedings of the K.Y. Lo Symposium, on CD, London, Ontario, Jul 7-8.

Griffiths, D.V., and Fenton, G.A., 2004. *Probabilistic slope stability by finite elements*, ASCE Journal of Geotechnical and Geoenvironmental Engineering, 130(5), 507-518.

Hasofer, A.M., and Lind, N.C., 1974. *An exact and invariant first order reliability format.* ASCE Journal of Engineering Mechanics Division, 100, EM1, 111-121.

Hassan A.M., and Wolff, T.F., 2000. *Effect of deterministic and probabilistic models on slope reliability index.* Geotechnical special publication No.101, Slope stability, ASCE194-208.

Huh, J., and Haldar, A., 2001. *Stochastic finite element based seismic risk of nonlinear structures*. Journal of structural engineering, Vol. 127, No.3, 323-329.

Kvalstad T.J., Nadim F., Kaynia A.M., Mokkelbost K.H., Bryn P., 2005. *Soil conditions and slope stability in the Ormen Lange area*, Marine and Petroleum Geology, Vol. 22, 299-310

Ladd, C.C., Foott, R., 1974. *New design procedures for stability of soft clays*. Journal of the Geotechnical Engineering Division, ASCE 100 (GT7), 763-786.

Li, K.S., and Lumb, P., 1987. *Probabilistic design of slopes*. Canadian Geotechnical journal, 24: 520-535.

Maia, J.A.C., and Assis, A.P., 2005. *Reliability analysis of iron mine slopes*. Proceedings of the international conference on landslide risk management, Vancouver, Canada, 31 May-3 June, 623-627.

Mienert, J., 2004. *COSTA-continental slope stability : major aims and topics*. Marine Geology, Vol. 213, No 1-4, 1-8.

Nadim, F., and Lacasse, S., 1999. *Probabilistic slope stability evaluation*. Geotechnical risk management, Geotechnical division, Hongkong Institution of Enginners, 179-186.

Nadim, F., and Locat, J., 2005. *Risk assessment for submarine slides*. Proceedings of the international conference on landslide risk management, Vancouver, Canada, 31 May-3 June, 321-333.

Nadim, F., Kvalstad, T.J., and Guttormsen, T., 2005. *Quatification of risks associated with seabed instability at Ormen Lange.* Marine and Petroleum Geology, 22, 311-318.

Plaxis, 2001. http:/www.plaxis.nl.

Rajashekhar, M.R., and Ellingwood, B.R. (1993). "*A new look at the response surface approach for reliability analysis.*" Struct. Safety, Amsterdam, 12, 205-220.

Rosenblueth, E. 1975. "*Point Estimates for Probability Moments*," Proceedings of the National Academy of Science, USA, 72(10), 3812-3814.

Wong, F.S., 1985. *Slope reliability and response surface method.* Journal of Geotechnical Engineering, Vol. 111, No.1, 32-53.

Xu, B., and Low, B.K., 2006. *Probability stability analyses of embankments based on finite element method.* Journal of geotechnical and geoenviromental engineering, vol. 132, No.11, 1444-1454.

Yang, S.L., Kvalstad T.J., Solheim A., and Forsberg C.F., 2007. *Slope stability at Northern Flank of Storegga Slide.* Accepted by the international conference on offshoe and Polar engineering, July, Lisbon, Portugal.

Zienkiewicz, O.C., Humpheson, C., and Lewis, R.W., 1975. "*Associated and non-associated viscoplasticity and plasticity in soil mechanics.*" *Geotechnique*, 25:4, 6.

MARINE DEEP-WATER FREE-FALL CPT MEASUREMENTS FOR LANDSLIDE CHARACTERISATION OFF CRETE, GREECE (EASTERN MEDITERRANEAN SEA) PART 1: A NEW 4000M CONE PENETROMETER

S. STEGMANN AND A. KOPF
Research Centre Ocean Margins, Bremen University, Leobener Strasse, 28359 Bremen, Germany

Abstract

The *in situ* measurement of seafloor physical properties such as pore pressure, shear strength or compressibility poses a challenge to engineers, in particular in the marine realm. We present the design and first use of a marine, deep-water free-fall instrument for cone penetration testing (CPT). The probe can be operated in up to 4000 m water depth to measure cone resistance, sleeve friction, deceleration, temperature and tilt as well as pore pressure in u_1 and u_3 position. In this paper we discuss the advantages and disadvantages of the current prototype design, and dwell on the differences between quasi-static versus dynamic cone penetration testing.

Keywords: CPT, free-fall instrument, shear resistance, pore pressure

1. Introduction

Mechanical properties of seafloor sediments (e.g. undrained shear strength, compressibility and permeability) are influenced by a variety of characteristics such as grain-size distribution, bulk density, effective-stress history and *in situ* pore pressure. All of these properties are typically non-linear functions of the strain history that must be evaluated for each scenario and measurement technique. Likewise, there are differences in measurement techniques, if it comes to cone penetration tests for geotechnical soil characterisation. The two principal techniques are quasi-static (i.e. pushed) cone penetration tests and dynamic (i.e. free drop) tests (e.g. Stoll and Sun, 2005). While industry favours the first where the probe is jacked into the soil at a constant rate of 2 cm/s (see summary in Lunne et al., 1997), several workers have started to develop instruments for marine application which are lowered in free drop, or at high winch speeds into the seafloor (Christian et al., 1993; Osler et al., 2006). The results of penetrometer tests show that there can be a wide spread in the penetration resistance that is measured depending on the degree of sediment inhomogeneity and the rate of penetration (Lunne et al., 1997; Sultan et al., 2007; Stegmann et al., 2007). Moreover the dilative response of sediments appears to further complicate matters because of the sudden, large changes in shear strength and pore pressure that may occur. All the above findings suggest that further development of seagoing penetrometers for geotechnical characterisation of the uppermost seabed sediments is an emerging need.

One parameter, which is rather difficult to measure *in situ* is pore pressure (e.g, Fang et al., 1993, Strout and Tjelta, 2005), mainly because the stress field is disturbed once a probe is lowered to carry out the measurement. A tremendous engineering effort for

V. Lykousis, D. Sakellariou and J. Locat (eds.), Submarine Mass Movements and Their Consequences, 171–177.

several decades in both industry (see www.fugro.com) and academia (e.g. Davis et al, 1991; Christian et al., 1993; Meunier et al., 2004; Wright, 2004) was undertaken to develop seagoing tools operable in several hundred meters of water depth. Different designs have been constructed in order to meet regional or budget requirements. One common example is a rig, where a CPT probe is pushed hydraulically into the sediment at 2 cm/s, following ASTM Standard No. D3441. The deployment of these systems represents a huge technological and logistical challenge, and results in several tens of meters of penetration depth (e.g. Fugro, Penfeld [Meunier et al., 2004]). The disadvantage, however, is the destruction of the uppermost sediments by additional loading of the instrument. For that purpose, other groups developed lance-shaped free-fall instruments (FF-CPT) that penetrate several meters into the sediment by its own weight, providing a time and cost efficient tool for exploration (e.g. Christian et al., 1993; Osler et al., 2006).

This paper describes a new marine free-fall penetrometer for deep-water applications (4000 m maximum water depth), which was recently designed at the Research Centre Ocean Margins, Bremen, Germany. It complements an already established free-fall shallow-water instrument (Stegmann et al., 2006a, b), which is limited to 200 m water depth.

2. Instrument Design and Measurement Methodology

The 380 cm long deep-water (DW) free-fall instrument is equipped with a standard 15 cm^2 CPT piezocone (Geomil) with strain gauges inside the probe to measure cone resistance q_c (25 MPa range) and sleeve friction f_s (0.25 MPa range) (Fig. 1). The instrument has two pore pressure ports, which are located at the cone (u_1) and 80 cm above the cone (u_3, following the CPT nomenclature); both ports are connected to high-resolution (10 Pa) differential pressure transducers (Validyne DP215, ±82 kPa range) via stainless steel tubing to a sea bottom water reference port. An absolute pore pressure sensor measures hydrostatic pressure (i.e. water depth) down to 4 km water depth. To prevent the entrapment of gas inside the tubing, especially during the initial phase of deployment when the instrument is lowered through the water column, valves are used to bleed the gas from the tubing. The pressure sensors are protected by valves in case high excess pore pressures are encountered. Temperature sensors and 2 bi-axial acceleration sensors are also connected to the Tiger Basics microcontroller (>40 Hz operating frequency).

The DW-CPT may either be run self-contained on batteries and equipped with a flash card, or could be run with a Seabird Electronics (SBE36) telemetric system, which powers the microcontroller, sensors, and valves (Fig. 1). The telemetry provides real-time data acquisition as well as control of the instrument via an attached PC with custom-programmed LabView control software. The autonomous mode is utilised for long-term observation of the dissipation behaviour of pore pressure by disconnecting the deployed lance from the winch and recovering it after several hours. The raw data protocol was implemented for time, acceleration, absolute pore pressure (hydrostatic pressure), cone resistance, sleeve friction and differential pore pressures (u_1, u_3). Using the vertical component of acceleration, penetration velocity (1st integration) and depth (2nd integration) can be derived.

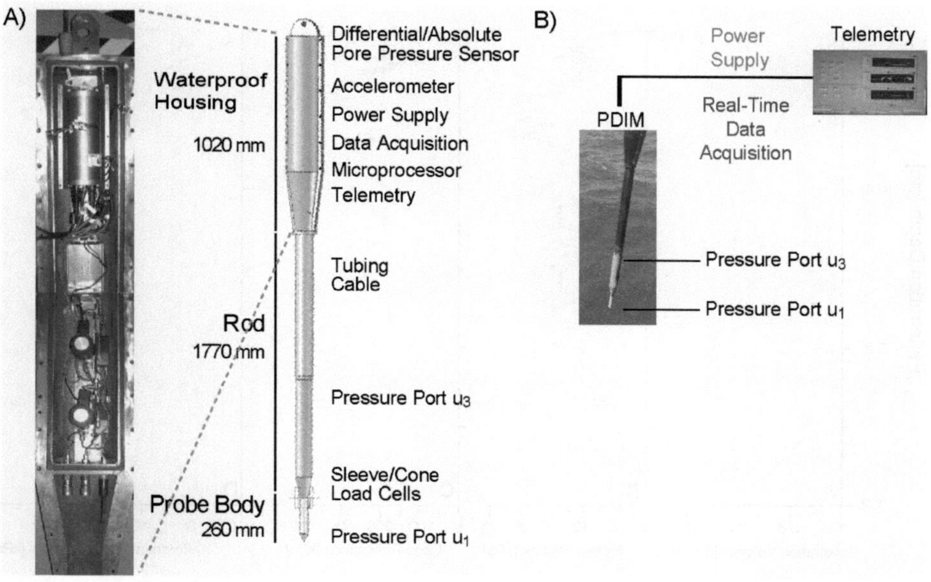

Figure 1. Photograph and schematic sketch of the deep-water lance (A) and its configuration (B).

During cruise P336 in the Cretan Sea, the instrument was lowered on a cable at an average rate of 1.5 m/s through the water column until it penetrates the seafloor and underlying sediment. The lance remained embedded in the sediment after insertion for 10 minutes to observe the dissipation of the pore pressure. This can be quantitatively expressed as T_{50} value (the time needed for a 50% decay of the maximum pore pressure), and serves as a first-order indicator of permeability (e.g. Bennett et al., 1985). Penetration velocity and depth was derived from the y-component of the acceleration sensor. After the lance was recovered through the water column, sediment stuck to the cone was measured and used as a control for the calculated penetration depth. Regarding water depth, most of the measurements were carried out in pogo-style. CPT data have been filtered (low-pass) using custom-made Matlab routines.

3. Results

We here summarise the results from a total of 40 CPT deployments we carried out during cruise P336 with RV *Poseidon*. 35 of the deployments were done with the prototype deep water instrument, while 5 were carried out with the shallow water device in the upper slope area. Since this paper is dedicated mostly to the new DW FF-CPT, we summarise only data from that instrument.

Figure 2 represents a typical data protocol, including penetration velocity, sleeve friction f_s, cone resistance q_c and pore pressure data obtained by ports u_1 and u_3 collected in undisturbed sediments at the northern Cretan margin. It can be seen that the values are largely synchronous, with q_c and f_s being followed with minor delay at port u_1, and a somewhat larger delay at port u_3 (Fig. 2d). The delay reflects some compliance of the pressure sensors, however, can be largely related to the insertion progress.

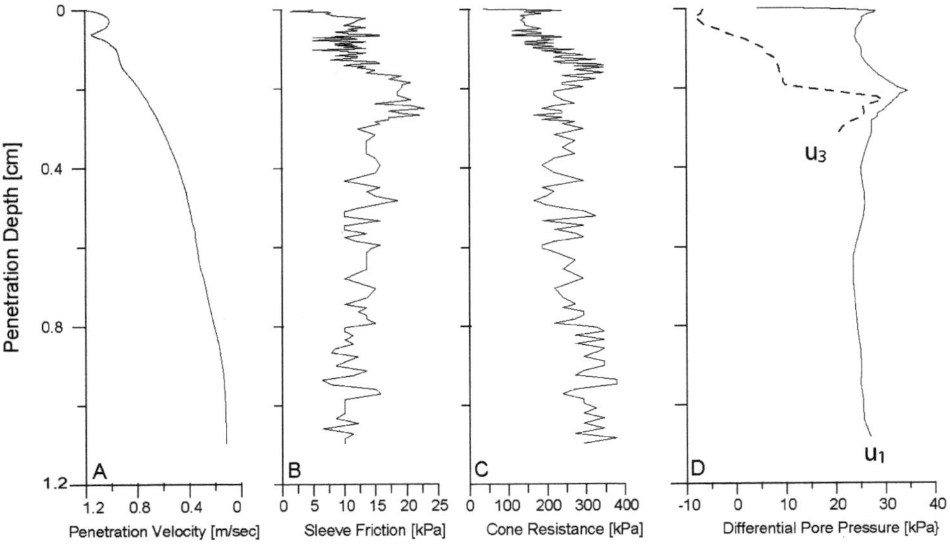

Figure 2. Typical penetration protocol showing (a) penetration velocity, (b) sleeve friction, (c) cone resistance, and (d) differential pore pressure at the port position u_1 and u_3 taken at the Cretan slope.

CPT measurements were carried out with an initial penetration velocity between 1.1 m/s and 1.8 m/s, which was controlled by winch speed (max. 2 m/s) and influenced by external conditions (waves, swell). Absolute penetration depth of the complete CPT data set (2nd integration of acceleration) ranged between 0.6 m and 1.6 m with 1 m in average. In nearly all measurements the tilt of the penetrated lance did not exceed ±9°. Unfortunately, the CPT cone failed during the campaign, so that the strength parameters (q_c, f_s) could not be measured in each location. Consequently, we focus mainly on differential pore pressure signal at u_1 and u_3 position (Fig. 1). Regarding the u_1-signal vs. time, pore pressure evolution reflects different physical properties (Fig. 3). During penetration the signal is generally characterised by an insertion maximum followed by a sudden drop (possibly, but not necessarily, to sub-hydrostatic values) (see the red part in Fig. 3). With the halt of the lance, pore pressure rises again and later asymptotically dissipates towards equilibrium (Fig. 3). If we regard all our test results, the difference between the maximum and the minimum of the drop ranged between 7 and 145 kPa. By comparing the pore pressure profile during penetration with cored sediments (see Kopf et al., this issue, section 4.3), the excess pore pressure peak, which varied between 13 to >82 kPa, coincides with an increase in sedimentary strength. Moreover, this higher strength is accompanied by decreasing sediment permeability (influencing the magnitude of the pore pressure drop) and larger T_{50} values during pore pressure decay. In order to overcome a potential bias resulting from the impact of the probe, we used the second excursion maximum in the pore pressure curve when calculating T_{50} (following Burns and Mayne, 1998).

There are two shortcomings in our DW CPT instrument. First, several measurements exceeded the upper limit of the differential pore pressure sensor during insertion, maybe as a result of the stiffness of the sediment. This means that we may have potential errors in case the second pore pressure maximum still exceeded 82 kPa. Second, the lower portion of the instrument (Fig. 1) seems too sturdy to allow deep penetration. As a consequence, we obtained only u_3 signals in 80% of the deployments. The insertion signal at u_3 was significant lower than at u_1 (i.e. 9.5-52 kPa), which maybe resulted from the different shear resistance along the sturdier portion of the lance during penetration. The lower values of u_3 represented the shear-induced pore pressure, whereas the high u_1 signal is generated by compression in the vicinity of the tip (see discussion in Song and Voyiadjis, 2005).

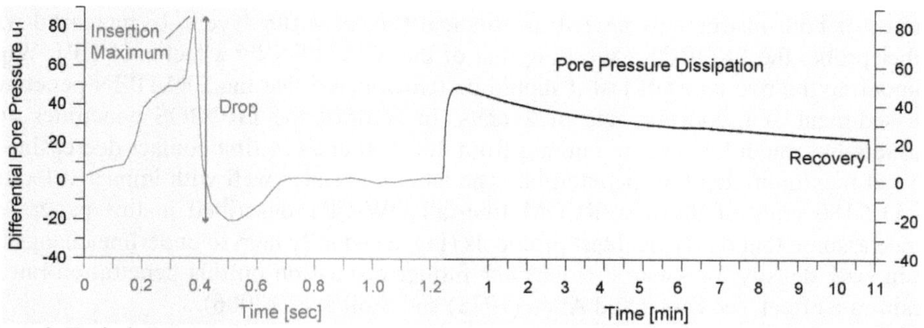

Figure 3. Typical pore pressure response measured in u_1 position during the insertion (red portion of graph) and within the 10 minutes of the lance remaining stuck in the sediment (black portion of graph). Please note the different time scales on the x-axis.

4. Discussion

Compared to the shallow-water CPT device (e.g. Stegmann et al., 2006a, b, 2007), there are still a number of shortcomings in the proto-type deep-water FFCPT design. First, the differential pressure sensors need to accommodate for higher excess pore pressure ranges, since they repeatedly maxed out when inundated surface sediments were hit (current range ±82 kPa differential pressure). However, this can easily be achieved by replacement of the diaphragm and recalibration of the Validyne DP215 transducers. Second, the overall layout of the lower part of the instrument is too sturdy and hampers deep penetration. We are currently building a new instrument where the lower portion is as narrow as the actual 15 cm^2 probe (Geomil). On the positive side, we observe consistent results with the shallow-water and deep-water FF-CPT instruments. With either probe, both q_c and pore pressure allow us to make a distinction between undisturbed vs. remobilised sediments off Crete, which is consistent with what is expected based on the seismic reflection profiles. Still, given the overall volume of mass wasting deposits at the Cretan slope (see Kopf et al., this volume), the most emerging need for improvement is a larger penetration depth, especially given that some amalgamated, remobilised deposits may be covered by background sediment again. This can be achieved by either designing a smaller diameter tool, increase total weight, or a combination of the two.

On the pro-side, we feel confident that the free-fall probe will provide us with interesting *in situ* soil physical properties. The instrument fell down on the seafloor only four times, while all other measurements led to stable positioning, even if penetration was as low as 0.6 m. We are particularly optimistic that the high impact velocities help to accentuate changes in physical parameters. This has been found earlier in controlled laboratory tests as well as seagoing deployments in the Gulf of Mexico. Here, Stoll et al. (2006) used a quasi-static instrument called STATPEN and a freely-falling dynamic instrument called PROBOS. Typically, the STATPEN results show a gradual, more or less uniform increase in cone resistance whereas the PROBOS record contains many large peaks of high penetration resistance suggesting an inhomogeneous sediment structure that is not obvious from the quasi-static test results (Stoll et al., 2006; their Fig. 6). No matter how different the absolute values of cone resistance, etc. were, the curves of both instruments agreed in principal (i.e. in stiffer layers, q_c increased with either probe, the PROBOS exceeding that of the STATPEN by a factor of 10!). When comparing the two types of test it should be remembered that the STATPEN penetrates the sediment at a constant rate of 2 cm/s. In contrast the PROBOS penetrates at a variable but much higher rate ranging from 400-600 cm/s at first contact decreasing to zero at maximum depth of penetration. The latter correlates well with impact velocities of 110-180 cm/s of the new RCOM free-fall DW-CPT described in this paper. We hence assume that our typical test protocols (Fig. 2) equally help to underline changes in grain size, density, or shear strength. For further discussion on this penetration-rate or strain-rate effect, see Dayal and Allen (1973) and Stoll et al. (2006).

In an effort to explain the differences between static vs. dynamic data acquisition, we postulate that the sediment must contain lenses or thin layers of coarser, more granular sediment that exhibits a dilative response when penetrated at a high rate. This can be seen when the pore pressure data are regarded, where subhydrostatic excursions are not uncommon. They are believed to results from displacement of pore fluid when the probe penetrates at high rate. Examples include the Gulf of Mexico (Stoll et al., 2006), the Baltic Sea (Seifert, unpublished data), or the Cretan Sea (Kopf et al., this volume). Each of these studies further demonstrate that the *in situ* strength measured with the FF-CPT agrees well with laboratory-derived values using the fall cone apparatus or vane shear device. Similar results were published earlier by Johnson et al. (1988), or in recent work concerning the stability of slope sediments in lakes by Stegmann et al. (2007) and Strasser et al. (2007).

5. Acknowledgements

We thank Master Michael Schneider and his crew for the help on deck and excellent manoeuvring during CPT deployments during cruise P336 with RV *Poseidon*. We also thank Matthias Lange at RCOM Bremen for technical support when building the DW-CPT. The manuscript benefited from constructive reviews and suggestions by Katrin Huhn and Nabil Sultan. Funding for this work was received by DFG through RCOM Bremen (project C8). This is RCOM publication #500.

6. References

Bennett, R.H., Li, H., Valent, P.J., Lipikin, J., Esrig, M.I., 1985. *In-Situ Undrained Shear Strengths and Permeabilities Derived from Piezometer Measurements, Strength Testing of Marine Seidments: Laboratory and In-Situ Measurements,* ASTM STP 883, R.C. Chany, K.R., Demars, ASTM Philadelphia, p. 83-100.

Burns, S.E., Mayne, P.W., 1998. *Penetrometers for Soil Permeability and Chemical Detection.* School of Civil and Environmental Engineering, Georgia Institute of Technology, Report No. GIT-CEEGEO-98-1.

Christian, H.A., Heffler, D.E., Davis, E.E., 1993. *Lancelot – an in situ piezometer for soft marine sediments.* Deep-Sea Research, 40/7, p.1509-1520.

Davis, E.E., Horel, G.C., MacDonald, R.D., 1991. *Pore Pressure and Permeabilities Measured in Marine Sediments With a Tethered Probe.* J. Geophys. Res., 96, B4, p.5975-5984.

Dayal, U., Allen, J. H., 1973. *Instrumented Impact Cone Penetrometer.* Can. Geotech. J., 10, p.397-409.

Fang, W.W., Langseth, M.G., Schultheiss, P.J., 1993. *Analysis and Application of in situ Pore Pressure Measurements in Marine Sediments.* J. Geophys. Res., 98, p.7921-7938.

Johnson, G.W., Hamilton, T.K., Ebelhar, R.J., Mueller, J.L., Pelletier, J.H., 1988. *Comparison of In-situ Vane, Cone Penetrometer, and Laboratory Test Results for Gulf of Mexico Deepwater Clays.* In: Richards, A.F. (ed.), ASTM Spec. Tech. Publ. 1014, Vane Shear Strength Testing in Soils, Field and Laboratory Studies, p.293-305.

Meunier, J., Sultan, N. Jegou, P., Harmegnies, F., 2004. *First Test of Penfeld: a New Seabed Penetrometer.* Proc. 14th Int. Offshore Polar Eng. Conference, Toulon, France, May 23-28, p.338-344.

Osler, J., Furlong, A., Christian, H., Lamplugh, M., 2006. *The integration of the Free Fall Cone Penetrometer (FFCPT) with the Moving Vessel Profiler (MVP) for the rapid assessment of seabed characteristics.* Canadian Hydrographic Conference, The Westin Nova Scotian, Halifax, N.S., p.11.

Song, C.R., Voyiadjis, G.Z., 2005. *Pore pressure response of saturated soils around a penetrating object.* Computers and Geotechnics, 32, p.37-46.

Stegmann, S., Villinger, H., Kopf, A., 2006a. *Design of a modular, marine free- fall cone penetrometer.* Sea Technology, 47/02, p.27-33.

Stegmann, S., Moerz, T., Kopf, A., 2006b. *Initial Results of a new Free Fall-Cone Penetrometer (FF-CPT) for geotechnical in situ characterisation of soft marine sediments.* Norwegian Journal of Geology, 86/3, p.199-208.

Stegmann, S., Strasser, M., Anselmetti, F.S., Kopf, A., 2007. *Geotechnical in situ characterization of subaquatic slopes: The role of pore pressure transients versus frictional strength in landslide initiation.* Geophys. Res. Lett., 34, L07607, doi:10.1029/2006GL029122.

Stoll, R.D., Sun, Y.-F., 2005. *Using Static and Dynamic Penetrometers to Measure Sea Bed Properties.* Office of Naval Research, Science & Technology, Ocean Battlespace Sensing (32), Coastal Geosciences Annual Reports FY05: 1-5, also available at http://www.onr.navy.mil/obs/321/docs/cg/04/cgstoll.pdf.

Stoll, R.D., Sun, Y.-F., Bitte, I., 2006. *Seafloor properties from penetrometer tests.* IEEE J. Ocean Eng., http://www.onr.navy.mil/obs/reports/docs/05/cgstoll.pdf.

Strasser, M., Stegmann, S., Bussmann, F., Anselmetti, F.S., Rick, B., Kopf, A., 2007. *Quantifying subaqueous slope stability during seismic shaking: Lake Lucerne as model for ocean margins*: Marine Geology, in press.

Strout, J.M., Tjelta, T.I., 2005. *In situ pore pressure: What is their significance and how can they be reliably measured?* Mar. Petrol. Geol., 22, p.275-285.

Sultan, N., Voisset, M., Maresset, B., Marsset, T., Cauquil, E., Colliat, J.-L., 2007. *Potential role of compressional structures in generating submarine slope failures in the Niger Delta.* Mar. Geol., 237/3-4, p.169-190.

Wright, I., 2004. *Geotechnical Investigations Using Mini-Cone Penetrometer Testing.* Sea Technology, 45/7, p.49-52.

Section 4 - Monitoring stress on submarine slopes and sediment physical properties

LINKING GEOTECHNICAL AND RHEOLOGICAL PROPERTIES FROM FAILURE TO POST-FAILURE: THE POINTE-DU-FORT SLIDE, SAGUENAY FJORD, QUÉBEC

J. LOCAT, P. LOCAT, A. LOCAT AND S. LEROUEIL
Université Laval, Québec, Qc, Canada

Abstract

The Pointe-du-Fort submarine mass movement likely took place at the time of the February 5th 1663 earthquake as a sidewall slope failure which generated a mudflow with a run out distance of 1070m and a final flow thickness of 10-15m resting on a slope of 1.4 degrees. The slide involved about 1.95 Mm^3 of clayey sediments from an original slope of 24 degrees. The slide took place in normally consolidated sediments composed of stratified low organic Laflamme Sea clay at the base overlain by progressively more organic rich recent sediments. *In situ* strength testing and sampling on the tidal flat, morphological analysis and remolded strength of the debris lobe can be related to rheological tests to model the mobility of the debris. For the first time, it has been possible to link the mobility of a submarine slide with the characteristics of the sediments at the time of failure with no need to consider water content increase to explain the observed mobility.

Keywords: Rheology, undrained strength, yield strength, modeling, liquidity index, submarine slide

1. Introduction

Many submarine mass movements do evolve into debris or mud flows (Locat and Lee 2002, 2005). As part of both understanding the phenomena and predicting the consequences there is a growing interest in modeling the flow (*e.g.* deBlasio *et al.* 2005, Elverhoi *et al.* 1997). As more interest develops to understand the consequences of these phenomena as tsunamigenic event, there is also a growing need to integrate flow modeling into the evaluations of tsunamis potential for a given event (Locat *et al.* 2004). When the flow has to be modeled as a Bingham, Herschel Bulkley or Bi-linear fluid (Imran *et al.* 2001), one needs to obtain the rheological parameters. Earlier studies by Locat *et al.* (2004) have shown that using morphological input could provide some ways at estimating the yield strength of the flow, *i.e.* in the final stages. Then, the viscosity, if required, can be estimated from the empirical relationships using the liquidity index (I_L, Locat 1997). The work on the Point-du-Fort slide, triggered by the 1663 earthquake (Locat *et al.* 2003b), provides a unique occasion to compare and relate the geotechnical properties in the starting area to those in the depositional area.

Details of the geomorphological and geotechnical characteristics of the Pointe-du-Fort slide can be found in Locat *et al.* (2003b). This paper benefits of later field and laboratory work which focus on the comparison between the properties of the sediments in the source area compared to those in the depositional area (Locat 2005). One interesting outcome of this additional work was to validate the use of empirical relationships linking rheological parameters and liquidity index used to predict the flow

V. Lykousis, D. Sakellariou and J. Locat (eds.), Submarine Mass Movements and Their Consequences, 181–189.

properties. In addition, for this specific case, the properties in the source area are compatible with the properties in the deposit area suggesting that no significant water entrainment took place at the time of the event although some mixing likely occured.

2. Morpho-stratigraphy

The slide debris is tongue shaped and is slightly curved following the topographic gradient of the fjord bottom (Fig. 1a). The displaced mass rests at an average water depth of 160m. The horizontal and vertical distances from the crown to the tip of the displaced material are respectively 1070m and 160m (see Cruden and Varnes 1996 for definition of landslide features). The fahrböschung angle is 7.5°. The surface of rupture is inclined at 24° and the sea floor in the zone of accumulation is 1.4° (Fig. 1b). The displaced mass is 730m long and 300m wide, an has average thickness of 13m for total estimated volume of 1.95 Mm^3.

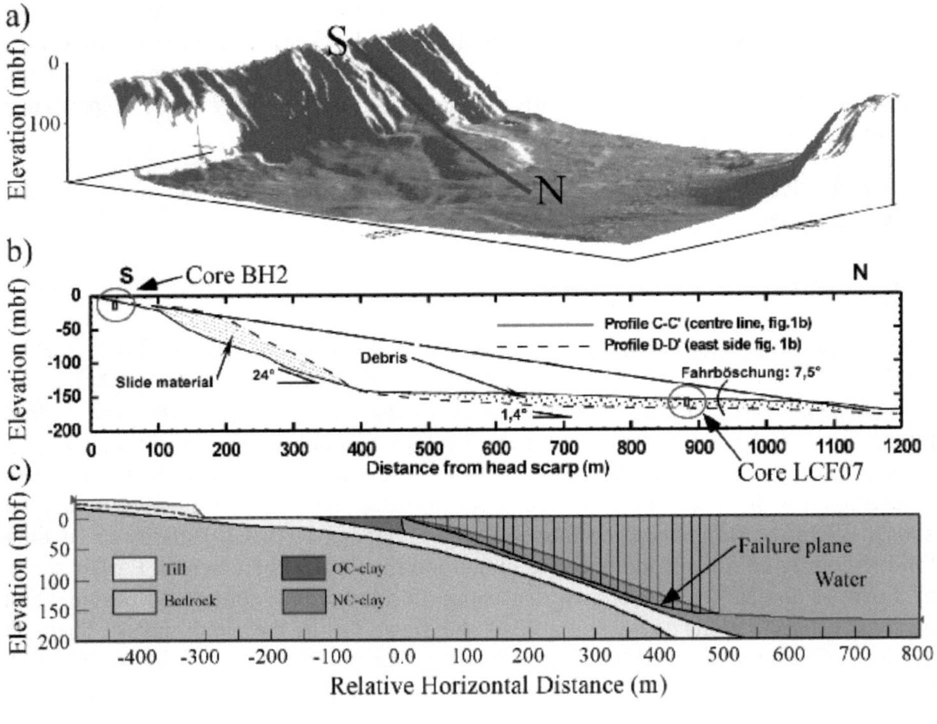

Figure 1. (a) 3D view of the Point-du-Forts slide; (b) geometric reconstruction of the slide area before failure, and (c) geological model used for slope stability analysis using *SlopeW*.

The stratigraphy of the slide has been reconstructed in Figure 1c on the basis of marine seismic survey and gravity coring of sediments along with *in situ* vane testing and coring on the tidal flat just above the scarp of the slide (*e.g.* Figs. 2 and 3; Locat *et al.* 2003b). From the morphology, it appears that the slide took place mostly underwater so that most of the sediments involved here were normally consolidated (see section 3). To validate the initial slide volume, a slope stability back analysis of the slide was carried out, under undrained conditions, using SlopeW and with the stratigraphic model shown in Figure 1c. The circular failure providing the lowest factor of safety (F = 1.8), without an earthquake, is shown in Figure 1c. This geometry gives a maximum thickness of 25m and length of 480m for the initial sliding mass. Using the width of the scarp, a volume of about 2 Mm^3 is obtained, which is comparable to the volume of the debris and supports the above observation.

3. Geotechnical properties

On the tidal flat above the slide scarp (Fig. 1c), Laflamme Sea clays are exposed (Fig. 2). They are easily recognized from the more recent clays by their low organic content (<1.0%), plasticity (<27%), by their grey color and the fine layers of interbedded sands. The more recent sediments (Fig. 3) have a higher organic content (1 to 3%) and plasticity index (30-46%). The undrained shear strength on the tidal flat, measured with the field vane (*Su*), increases from 30 kPa to 80 kPa at a depth of 17m (Fig. 2). According to this strength profile, and its relation to the $Cu/\sigma'_{v0} > 0.2$ line, about 10 to 15m of Laflamme Sea sediments has been eroded. From the closest sea level curve available (Locat *et al.* 2003a), it is possible to estimate that erosion of the tidal flat started between 5000 to 7000 yBP. So, this erosion took place at a time when fjord sedimentation was similar to actual conditions. Therefore, as a first approximation, continuous sedimentation over most of the failure zone since deglaciation supports considering the sediment column to being normally consolidated.

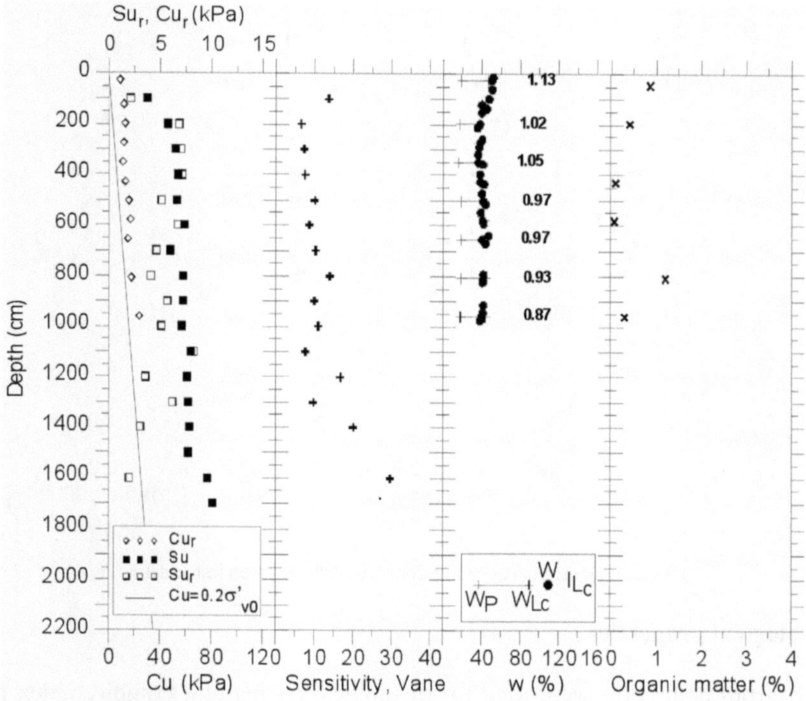

Figure 2. Geotechnical profile for core BH2 taken on the tidal flat.

On the tidal flat, the remolded strength (Cu_r) increases with depth from 1.2 to 3.1 kPa. In the debris area, there are two units. The lower unit, from the bottom of the section to a depth of 1.2m, consists of sediments involved in the sliding and contains many clasts of Laflamme Sea clay (grey) embedded either because of mixing during the slide or from erosion of the tidal flats. In that section, the strength (Cu) varies from about 15 kPa at the top to 60 kPa at the bottom with large variations caused by fall cone measurements on clasts. The remolded strength (Cu_r) in the lower unit varies between 1.3 and 5 kPa for an average of about 3.0 kPa. That section is clearly over-consolidated with a Cu/σ'_{v0} close to 0.8. The top layer consists of soft mud with intact strength increasing to about 2.5 kPa and represents sediment accumulation since 1663. In other parts of the debris, lower strength values have been reported in other parts of the debris (Locat *et al.* 2003b) which may be indicative that the debris involved in Figure 3 may represent a portion of the failed mass which has been less remolded. If we consider that the base of the slide was at a depth of about 25m, it is likely that the failure plane have cut through Laflamme Sea clay sediments.

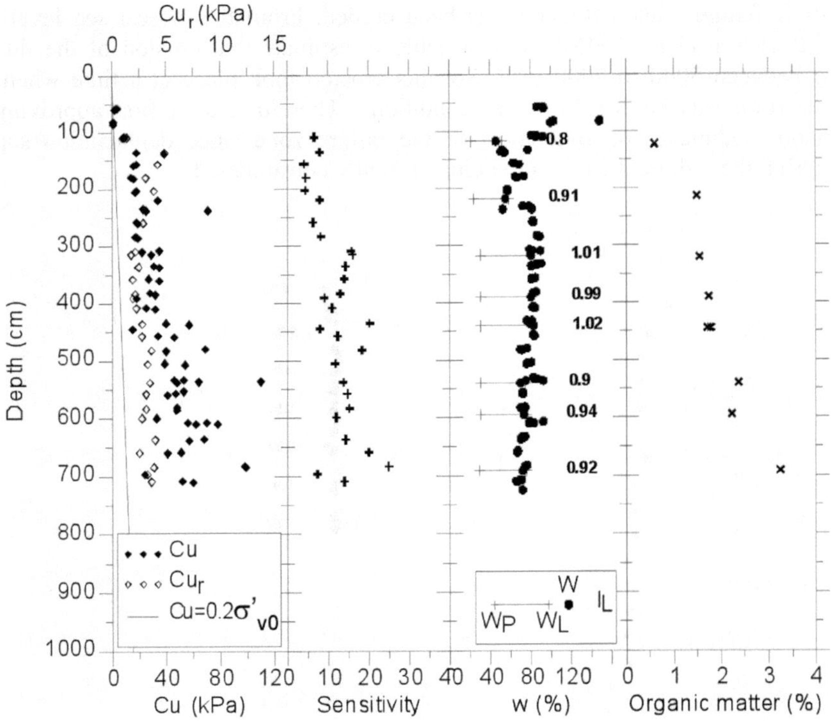

Figure 3. Geotechnical profile for core LCF_07 taken in the debris.

4. Rheological properties

Samples from both cores were used for rheological testing at a liquidity index ranging from 1.2 to 3.0, following the methodology proposed by Locat and Demers (1988). Values of yield strength (τ_c) and dynamic viscosity (μ) were computed considering a

Bingham type flow material. Results for various relationships, between the liquidity index, yield strength, viscosity and remolded undrained shear strength *(Cu_r)* are shown in Figure 4. For the source area they are:

$$\tau_c = \left[\frac{17.9}{I_L}\right]^{2.38} \text{(1)};\ \mu = \left[\frac{1.10}{I_L}\right]^{3.23} \text{(2)};\ \mu = 0.001\tau_c^{0.95}\text{; (3)},\ \tau_c = \left[\frac{Cu_r}{1.1}\right]^{0.94} \quad (4)$$

and for the debris area:

$$\tau_c = \left[\frac{9.1}{I_L}\right]^{2.94} \text{(5)};\ \mu = \left[\frac{0.63}{I_L}\right]^{2.70} \text{(6)};\ \mu = 0.001\tau_c^{0.83} \text{(7)};\ \tau_c = \left[\frac{Cu_r}{1.1}\right]^{1.11} \quad (8)$$

As we compare these relationships in Figure 4, we can see that the main difference in the rheological response of both samples is when we related the viscosity to the liquidity index (Eq. [2, 6]). Otherwise, the actual geotechnical properties and the rheological properties for the two cores are quite similar although their physico-chemical properties differ substantially (*e.g.* Atterberg's limits and organic matter content). According to Locat (2005), both samples also differ in their stress-strain behavior: tidal flat sediments behave more like a visco-plastic fluid, while debris sediments exhibit a Bingham behavior.

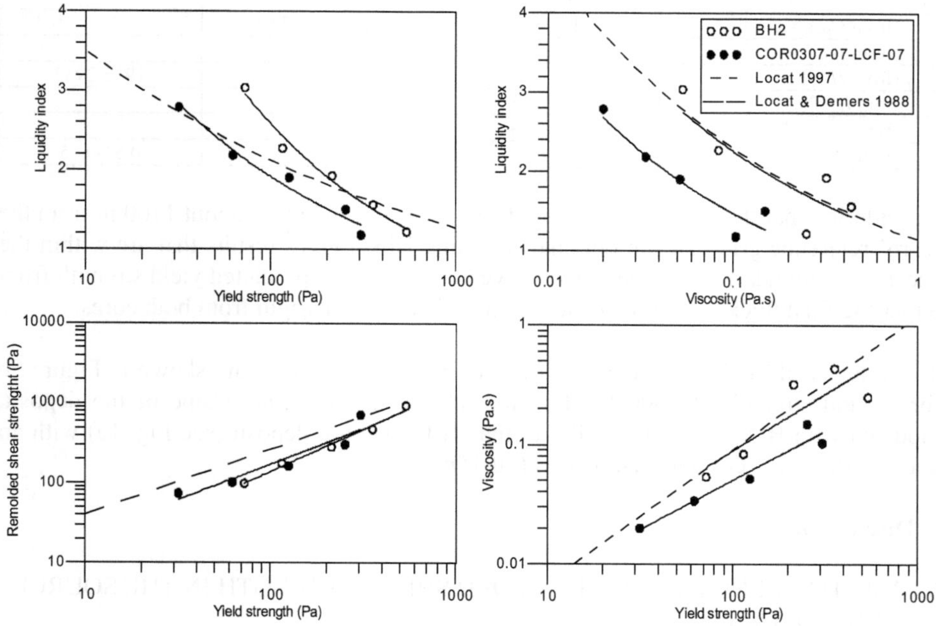

Figure 4. Rheological parameters obtained for mixtures of sample BH2 and LCF_07.

5. Post-failure analysis

For post-failure analysis, computation was carried out with BING (Imran *et al.* 2001) and by considering the flow as a Bingham fluid with an initial length 480m, and the following properties: a unit weight of 1800 kg/m^3 and a sea water density of 1000 kg/m^3. The mobilized yield strength can be estimated in two ways. One uses the remolded strength or liquidity index from the geotechnical profiles (Figs. 2 and 3) using Eqs [1, 4, 5, and 8]. The other one uses the average thickness in the depositional area to estimate the yield strength from Hampton (1972) equation relating the yield strength and the critical flow height H_c (in metres) such as:

$$\tau_c = H_c \gamma' \sin\beta \tag{9}$$

where γ is the buoyant unit weight (kN/m^3) and β the slope angle at which the flow stopped. A compilation of all the possibilities of estimating the mobilized yield strength and viscosity is given in Table 1. Results show that the values are within the same range with the yield strength being higher for the debris than the intact Laflamme Sea sediments while for the viscosity it is the opposite.

The range of potential rheological parameters (Table 1) can now be validated by using BING in order to see if they can predict the observed run out distance and flow thickness. (Figure 5).

Table 1. Summary of rheological properties determined with various methods.

Properties	BH2 (Tidal flat)	LCF07 (debris)	H_c Eq. [9] (10-15m)
Cu_r *(Pa)*	1200-3100	1300-5000	
I_L	0.87-1.13	0.80-1.02	
□$_c$ (from Cu_r), Pa	717-1750	2574-11479	
□$_c$ (from I_L), Pa	717-1336	623-1272	
□$_c$ (from H_c), Pa			1955-2932
□ (from I_L), Pa.s	0.92-2.13	0.27-0.52	
□ (from □$_c$-I_L), Pa.s	0.52-0.93	0.21-0.38	
□ (from H_c), Pa.s			0.53-0.75

Considering that the maximum run out distance for the slide is at about 1100 m from the 'zero' reference grid, the run out analysis using BING gives results that are within the acceptable boundaries provided by observed thickness and predicted yield strength from either the final thickness or the material remolded shear strength from both cores.

The velocity of the frontal elements and the final flow thickness are shown in Figure 5b for a yield strength of 3000 Pa. Results indicate that the final shape of the deposit, modeled with BING is quite similar to that of the natural deposit (see Fig. 1a) with an average thickness close to field values (10-15m).

6. Discussion

6.1 YIELD STRENGTH AND REMOLDED SHEAR STRENGTH IN THE SOURCE AND THE DEBRIS

Yield strength values from different areas are compatible with the observed and computed geometry of the deposits. For the first time it was possible to acquire some geotechnical information of sediments close to a failure zone.

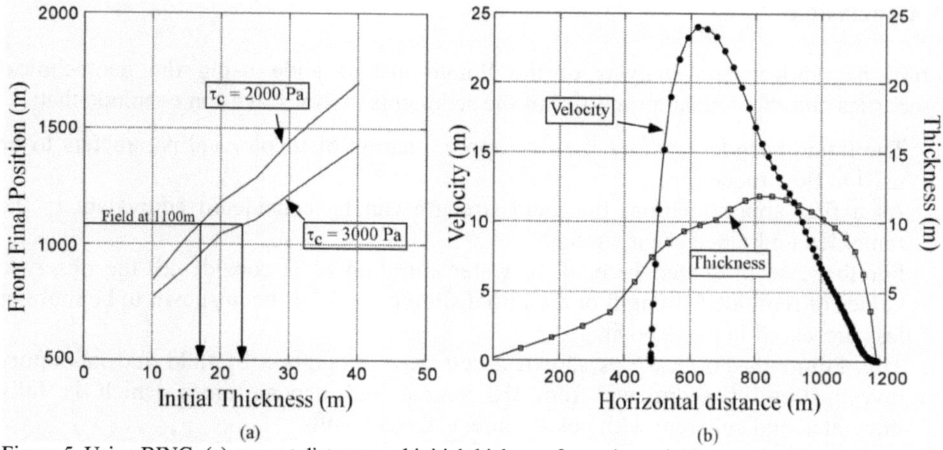

Figure 5. Using BING, (a) run out distance and initial thickness for various yield strength; (b) velocity profile for the front element and final shape of the debris considering a yield strength of 3000 Pa, a viscosity of 1 Pa.s and an initial flow thickness of 25m.

We were not able to core the debris section completely, but what has been obtained clearly indicate that the strength is lower in the debris area but still within the range of what has been measured on the tidal flat so that at both sites we do get apparent over-consolidation. The sediments in the debris area have a higher water and organic content mainly because they corresponds to the upper section of the failed mass involved in the slide. This suggest that the slide was not totally remolded even tough Laflamme Sea clay clasts are present.

It is also interesting to note here that various relationships between the liquidity index, the yield strength, and the viscosity have been successfully extrapolated to values outside of the range achieved during the rheological testing. It also indicates that in many circumstances, the yield strength can be considered equivalent to the undrained remolded shear strength and the value of the viscosity can be estimated based on know empirical relationships where the ratio varies from 10 to 1000 (Jeong *et al.* this book). Computation using BING clearly shows that if lower values of the yield strength were to be used, it would results in unrealistic run out values and a much lower thickness of the mud flow deposit (Fig. 5a).

6.2 POST-FAILURE BEHAVIOR

The peak velocity of the front modeled with BING was computed to be in the range of about 24m/s. Mohrig *et al.* (1988) indicate that the minimum velocity to initiate hydroplaning is about 6 m/s, and this has been exceeded here. However, in such a case, the run out distance would have been much longer, as speculated for the Storegga slide by deBlasio *et al.* (2005). Since the slide was likely triggered by an earthquake (Locat *et al.* 2003b), the initiation of the post-failure may have taken place on a weakened layer. In such a case, it is very likely that the most of the mass did behave more or less as a plug flow over a thin remolded layer.

7. Conclusion

From the above retro-analysis of the Pointe-du-Fort slide using the geotechnical properties and rheological properties of the sediments involved we can conclude that:

1. The liquidity index enables the first approximation of rheological parameters to be used in flow modeling.
2. As a first approximation, the yield strength can be considered equivalent to the remolded undrained shear strength.
3. For the case presented here, if no water entrainment is considered, the observed values of remolded strength of the initial sliding mass has been shown to be more or less preserved in the deposit.
4. This Pointe-du-Fort case as shown a complete integration of field and laboratory investigation of sediments from the source and deposit areas which is fully integrated and coherent with post-failure analysis results.

8. Acknowledgements

The Authors would like to thank their reviewers, J. Imran and H. Lee for their constructive suggestions. We also thank the National Science and Engineering Research Council of Canada (project COSTA) for their financial support.

9. References

Cruden, D.M. & Varnes D.J. 1996. *Landslide types and processes.* In: Landslides: investigation and mitigation, Special Report 247, Transportation Research Board, pp. 36-75.

De Blasio, F.V., Elverhøi, A., Issler, D., Harbitz, C.B., Bryn, P., & Lien, R., 2005. *On the dynamics of subaqueous clay rich gravity mass flows—the giant Storegga slide, Norway.* Marine and Petroleum Geology, 22: 179-186.

Elverhoi, A., Norem, H., Andersen, E.S., Dowdeswell, J.A., Fossen, I., Haflidason, H., Kenyon, N.H., Laberg, J.S., King, E.L., Sejrup, H.P., Solheim, A., & Vorren, T., 1997. *On the origin and flow behavior of subaqueous slides on deep-sea fans along the Norwegian-Barents Sea continental margin.* Geo-marine Letters, 17: 119-125.

Hampton, M.A., 1972. *The role of subaqueous debris flows in generating turbidity currents.* Journal of Sedimentary Petrology, 42: 775-793.

Imran, J., Parker, G., Locat, J., & Lee, H.J., 2001. *1-D numerical model of muddy subaqueous and subaerial debris flows.* ASCE Journals of Hydraulic Engineering, 127: 959-968.

Jeong, S.W., Locat, P., & Leroueil, S., 2007. *Rheological properties of fine-grained sediments in modeling submarine mass movements: the role of texture.* (This volume).

Locat, J., 1997. *Normalized rheological behavior of fine muds and their flow properties in a pseudoplastic regime.* In: Debris-Flow Hazards Mitigation: Mechanics, Prediction, and Assessment. Water Resources Engineering Division, ASCE, pp. 260-269.

Locat, A., 2005. *Étude comparative des matériaux de la source et des débris du glissement sous-marin de Pointe-du-Fort, Fjord du Saguenay, Québec, Canada.* Bachelor Thesis Project, Department of Geology and Geological Engineering, Laval University, Québec, p50.

Locat, J. & Demers, D., 1988. *Viscosity, remolded strength and liquidity index relationship for high water content soils.* Canadian Geotechnical Journal, 25: 799-806.

Locat, J., & Lee, H.J., 2002. *Submarine landslides: advances and challenges.* Canadian Geotechnical Journal, 39: 193-212.

Locat, J., & Lee, J., 2005. *Subaqueous Debris Flows.* Chapter 9, In: Debris-flows Hazards and Related Phenomena, Mathias and Hungr, ed., Springer, pp. 203-245.

Locat, J., Lee, H.J., Locat, P. & Imran, J. 2004. *Analysis of the Palos Verdes Slide, California, and its implication for the generation of tsunamis.* Marine Geology, 203: 269-280.

Locat, J., Tanaka, H., Tan, T.S., Dasari, G.R., & Lee, H., 2003a. *Natural soils: geotechnical behavior and geological knowledge*. Keynote In: Characterisation and Engineering Properties of Natural Soils, Tan et al. Ed., Balkema, vol. 1:3-28.

Locat, P., Locat, J., Leroueil, S., & Duchesne, M.J., 2003b. *Characterisation of a submarine flow-slide at Pointe-du-Fort, Saguenay Fjord, Québec, Canada*. In: Submarine mass Movements and Their Consequences, J. Locat & J. Mienert, Eds., Kluwer series on Natural and Technological Hazards, vol. 19, pp. 509-518.

Mohrig, D., Whipple, K.X., Hondzo, M., Ellis, C., & Parker, G., 1998. *Hydroplaning of subaqueous debris flows*. Geological Society of America Bulletin, 110: 387-394.

RHEOLOGICAL PROPERTIES OF FINE-GRAINED SEDIMENTS IN MODELING SUBMARINE MASS MOVEMENTS: THE ROLE OF TEXTURE

S.W. JEONG, J. LOCAT, S. LEROUEIL
Laval University, Québec, Canada, G1K 7P4

J.-P. MALET
School and Observatory of Earth Sciences, Institute of Global Physics, UMR 7516 CNRS-ULP, 5 rue René Descartes, F-67084 Strasbourg Cedex, France

Abstract

The rheological behavior of soils depends on many factors, including their mineralogy and grain size distribution. This work comprises an extensive search of data collected from the literature, an experimental work on about 17 samples. These results, along with a compilation of existing data, have been used to show that, as a first approximation, the yield strength/viscosity ratio is about 1000, 100 and less than 10 for clayey, silty and sandy fine-grained sediments mixtures, respectively. Our research results on the rheological properties of fine-grained sediments indicate that they are very sensitive to the variation in grain size, shear rate, and geometry of the system.

Keywords: Rheological properties, grain size, shear rate, Bingham yield stress, plastic viscosity

1. Introduction

Submarine landslides are commonly observed on the continental margins, particularly becoming recognized as a potential source of damaging tsunamis. These submarine mass movements are distinguished by several characteristics: sediment concentration, sediment-support mechanism, flow rate, and sediment rheology (Locat and Lee, 2002; Niedoroda et al., 2006). Practically, the Bingham yield stress and plastic viscosity are often used to describe the modern deep-sea features and to estimate future debris flow runout behavior with a trial-and-error process. These rheological parameters are also influenced by soil types, solid concentration, salinity and liquidity index, not mention the effect of the geometry of rheometer. In particular, the influence of coarse-grained soils in fine-laden debris flow is essential. Debris flow materials with large size particles ranging from clay-size to boulder-size might have various rheological behaviors. As a result, a number of more or less complex rheological models to describe flow behavior may be found in the literature. However, the determination of rheological properties still faces the problem of the presence of large size particles in highly concentrated sediment-water mixtures (Coussot et al., 1998). In this paper we summarize the general flow behaviour and characteristics of fine-grained sediments encountered in the various environments. It is a very important topic when we are dealing with high speed mass movements such as subaqueous debris flows, because submarine landslides are much mobile and tend to involve larger volumes than subaerial slides. The aim of this paper is to examine a possible range of rheological properties with respect to the grain sizes.

V. Lykousis, D. Sakellariou and J. Locat (eds.), Submarine Mass Movements and Their Consequences, 191–198.

2. Materials and methods

The rheological behavior of selected soils was evaluated and separated into the three groups. The first group is natural clays taken from the Jonquière, St-Alban, Saguenay Fjord including Baie des Ha!Ha!, Pointe-du-Fort, and Confluence in Upper Saguenay Fjord, Adriatic, Mediterranean, Beaufort Sea, Cambridge Fjord, and Hudson Apron samples. As for the materials in Group 1, the liquid limit (w_L) and plasticity index (I_P) are approximately 35 - 75 % and 15 - 50 %, respectively. Group 1 is described as soft clays, i.e., inorganic clays of medium to high plasticity. The second group is two montmorillonite-rich materials in natural clays, i.e., the Black Sea sediments and the natural sodium-rich montmorillonite clay supplied by Black Hills Bentonite, LLC (Wyoming, USA). In Group 2 that contains highly plastic soils, the liquid limit and plasticity index are 100 - 360 % and 80 - 300 %. The third group is silt-dominated mixtures analyzed or examined for unflocculated/flocculated iron tailings samples. In Group 3, the liquid limit and plastic limit of a soil are less than 23 % and less than 19 %, respectively. The plasticity indices (I_P) are 3.7 and 6.2% for unflocculated and flocculated samples, respectively. They are typically classified as inorganic clayey silts of low plasticity. Group 3 is only limited here to the test results of iron tailings, because of more complex soil characteristics of more coarser mixtures and unclear determination of Atterberg limits of soils. In short, Group 1 has low to medium activities (i.e., $0.5 < A_c < 1.0$), Group 2 is classified as highly active (i.e., $1.3 < A_c < 4$), and Group 3 is inactive soils (i.e., $0 < A_c < 0.3$), reflecting the low content in clay minerals.

The rheological analysis were carried out using either a coaxial cylinder viscometer (Rotovisco RV-12) or a parallel plate rheometry on the <400 μm fraction. The latter is used for comparison: the rheological characteristics depending on shear rate and testing apparatus, using the complex debris flow in a clay-shale basin (Alpes-de-Haute-Provence, France). A detailed analysis of the rheological properties and complete methodology are explained by Malet et al. (2003). After proper sample preparation, the liquidity index was slowly increased at a constant salinity of the pore water by adding water with the same salinity. For each step the required geotechnical parameters, e.g., water content and remoulded undrained shear strength using the Swedish fall cone, were measured according to ASTM and BNQ (Bureau de normalisation du Québec) standards. The detailed procedures followed for viscometric measurements are described in Locat and Demers (1988). These procedures may include three types of tests: (1) constant shear rate; (2) dynamic response; and (3) hysteresis.

3. Results

3.1 RHEOLOGICAL BEHAVIOR OF FINE-GRAINED SEDIMENTS

Many studies showed that non-swelling natural clays present classically shear thinning and thixotropic behavior with a yield stress (e.g., Coussot and Piau, 1994; Locat, 1997). Typical viscometric results can be presented in a semi-logarithm or log-log diagram of rheological and geotechnical parameters with regard to liquidity index. Empirical relationships were presented by Locat (1997) for: (1) the apparent yield stress and plastic viscosity; (2) liquidity index and plastic viscosity; and (3) liquidity index and apparent yield stress. These relationships are much better for a sensitive soil, even if the scatter is important. Results given here may be useful for a preliminary numerical analysis for subaerial and submarine mass movements (e.g., Kvalstad et al., 2005). Assuming that the flow behaves as a Bingham fluid, there is the relationship between plastic viscosity (η_h, in mPa.s) and yield stress (τ_c, in Pa) measured at various liquidity indices. More detailed flow behaviors and characteristics for fine-grained sediments will be investigated in comparison with those of the previous data. As for sensitive clays, within given range of liquidity indices, the plastic viscosity only represents about 1/1000 of the total shearing resistance of the mixture (Locat, 1997).

A comparison of test results with those of the previous data is shown in Figure 1 with the relationships given by Locat (1997) as a reference. The representative materials in each group generally exhibited a shear thinning flow behavior: i.e., viscosity decreases with increasing shear rate. Even if the scatter is important, results are close to the empirical relationships presented by Locat (1997). Those described herein it seems reasonable to temporarily use at least for non-swelling fine muds. The results in Figure 1 can be used hereafter to provide a first estimate of the relationships between liquidity index and rheological parameters. There is also a positive relationship between liquidity index and pseudo-Newtonian viscosity (Jeong, 2006).

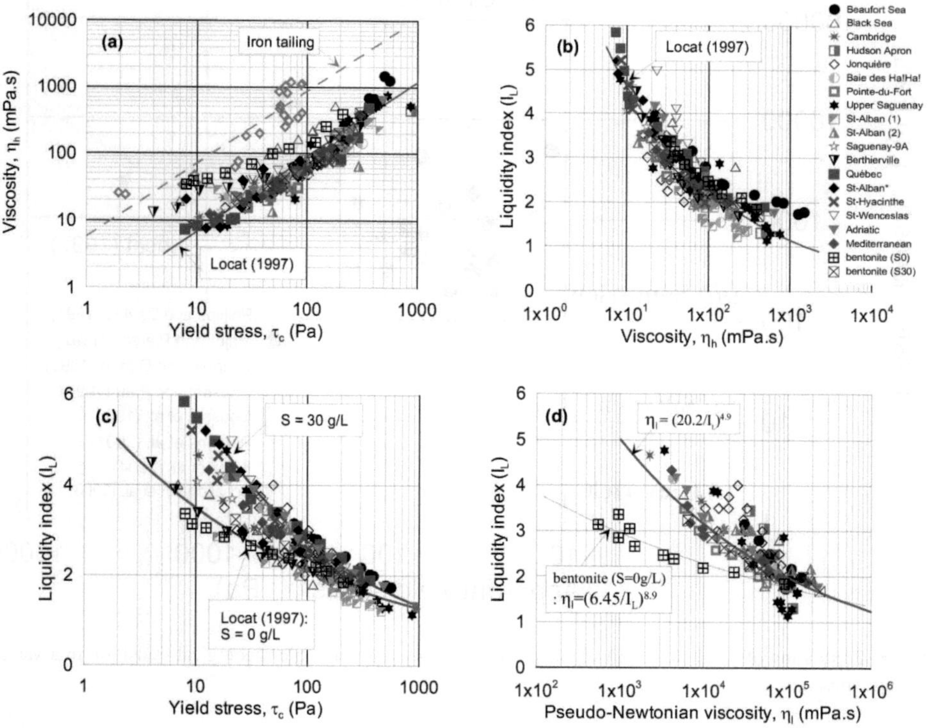

Figure 1. Rheological relationships among the yield stress, viscosity, and liquidity index. Lines correspond to the best power-law fit (equations presented by Locat, 1997; Jeong et al, 2004). Note that viscosity η_l and η_h are pseudo-Newtonian and Bingham viscosity, respectively.

3.2 POSSIBLE RANGE OF RHEOLOGICAL TRANSTION FROM FINE-GRAINED TO COARSE-GRAINED SOILS

It would be interesting to compare the rheological parameters of iron tailings with those obtained with a large-scale rheometer on coarser mixtures. The iron tailing has one of the largest grain sizes (i.e., silt-rich samples) in selected Groups of studied soils, as previously indicated. The plastic viscosity could represent about 1/100 of the total shearing resistance of the mixture (Jeong et al., 2004). Figure 2 shows the possible range of rheological transitions between grain sizes, often encountered in soil mixtures. It should be noted that the values of apparent yield stress and plastic viscosity presented here were approximately determined using ideal Bingham model. As far as the time-independent rheological behavior is concerned, a comparison of data obtained from conventional viscometer used in this study and the results presented in Figure 2 as a reference leads to similar results over a wide range of shear rates. Although there is data scattering with increasing grain size, on the basis of result of non-swelling natural clays, at the same yield stress, the viscosity increases stepwise with the same slope.

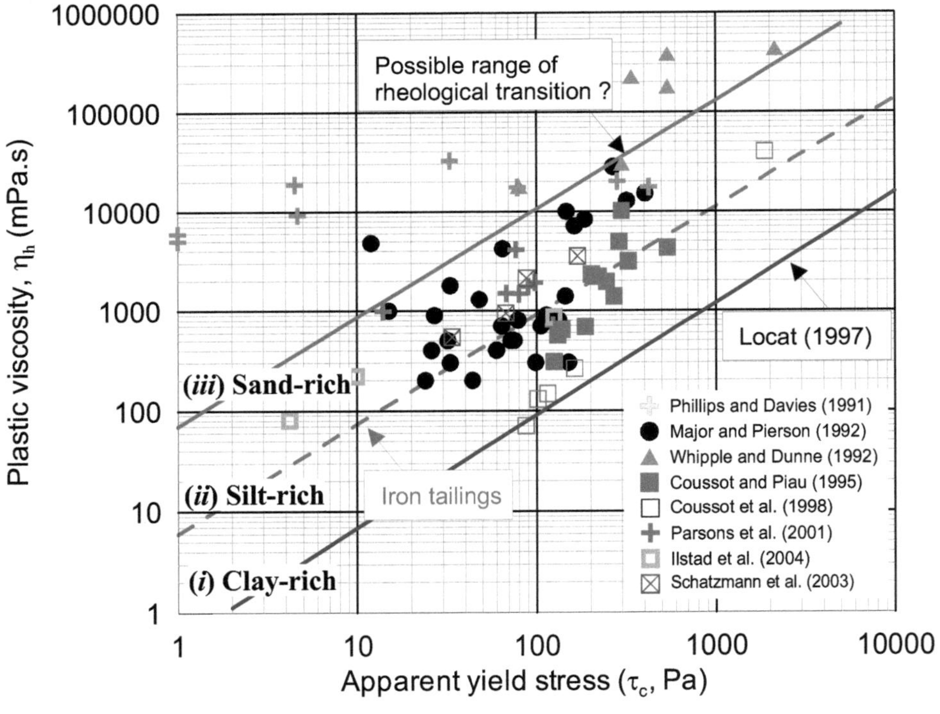

Figure 2. Relation of yield stress (τ_c) versus plastic viscosity (η_h): possible rheological transition in a variety of soil types.

Rheological properties presented by Schatzmann et al. (2003) and Ilstad et al. (2004) are similar to those obtained from test results of iron tailings. It is worth noting here that Ilstad et al. (2004) presented more coarser artificial mixtures, for examples, water (35%) - clay (28.7%) - sand (36.3%), with rheological properties appearing in the range of given values close to the limit of silt-rich materials. The rheological parameters obtained from Coussot and Piau (1995) are varied between the range of sand and clay sizes, but those obtained from Coussot et al. (1998) are getting closer to the limit of natural clays. The latter is evidently close to the behavior of fine-grained soils, but increasing large particles with solid concentration of 73% can result in the critical limit of silt-rich material. The rheological parameters determined in various flow types (e.g., Bingham, shear thinning and few dilatant) on artificial silt-and-sand rich materials presented by Major and Pierson (1992) are also shown with a similar range, even though the scatter is significant. Except for the parameters obtained from debris flows with large sand and gravel contents (e.g., Whipple and Dunne, 1992 and Parsons et al., 2001), the others fall well within the defined range between sand-rich and the fine-grained material. Consequently, the critical limits identified in Figure 2 can represent the possible range of rheological transition from clay to sand-size particle. It would be interesting to plot the ratio of plastic viscosity to yield stress as a function of the grain size. The ratio obtained from Locat (1997) and from the present study is 1 for clayey soils. The ratio for silt-rich material with low percentage of sands is 10. On the other hand, Phillips and Davies (1991) and Whipple and Dunne (1992) who tested gravelly soils found a ratio in the order of 200 - 2000 (see Figure 2). It results mostly from the fact that in cohesionless materials yield stress (τ_c, in Pa) is small compared to plastic viscosity (η_h, in mPa.s) and consequently η_h/τ_c ratio is large.

As shown in Figure 2, the influence of the grain size distribution on the flow curves is of paramount importance. The rheological characteristics of black marl slopes in the French Alps were examined in the same way. It is well known that these landslides, in most cases, show complex nature by changes in behavior in which the slides transform into various flow types (Malet et al., 2005). According to Malet et al. (2003, 2005), all natural and artificial mixtures have a higher silt and clay content and the textural classes vary from silty clay to silty sand. No swelling clays were detected. The mixtures were taken from deposits: earthflows (Super-Sauze, C1a, IND; La Valette, VAL; Poche, POC), muddy debris avalanches (Super-Sauze, COU), and debris flows torrents (Faucon, FAU; Riou-Bourdoux, RBX). In addition, several artificial mixtures of moraines, marls and sandstones were examined. The rheological characterization was conducted with two geometries: a coaxial viscometer (CO) on the < 0.075 mm and a parallel plate rheometer (PP) on the < 400 μm fraction. The rheological parameters were determined by the Bingham model. There are, of course, many experimental difficulties associated with the dynamic response with low liquidity index resulting in much scatter in data. Scatter may have partly resulted from the mineralogy of the clay fraction, the structure of the clay, uncertainty of steady state regime using a specific shear period, loss of material, and possible migration and sedimentation of the particles within the muds in a parallel plate rheometer.

For the mixtures tested in the coaxial geometry, results show a very good agreement between the ratios of yield stress to plastic viscosity obtained from black marls hillslopes and those obtained from the relationship presented by Locat (1997). This is mainly due to the samples are "clay-rich" mixtures, and they differ depending on the amplitude of the imposed shear rate. As shown in Figure 3, except for the limited number of cases that particularly for very low solid concentration the mixture behaves as a liquid, the difference is large between the ratios obtained from coaxial (CO) and parallel plate viscometer (PP). For the mixtures containing a low clay fraction, the imposed shear rate results in a higher viscosity for a given yield stress. However, at the yield stress of 100 Pa, CO/PP is about 1.2, even though the Bingham yield stress was determined in the similar ranges of shear rate (e.g., 500 - 1200 s^{-1}). The disagreement may be due to the difference in: (1) grain size, (2) shear rate (e.g., see Møller et al., 2006), and (3) geometry. Shear rate is considered as the main reason resulting in the inconsistency of results. Due to limitations in the experimental devices, the range of shear rates (e.g., γ = 0.01 - 20000 s^{-1}) is two or three times higher than that met with debris flows in the field (O'Brien and Julien, 1988). Thus, the smaller yield stresses are often estimated by rheometry due to a higher shear rate and smaller grain sizes in comparison with those based on field observations (Malet et al., 2005). More laboratory data and field observations needed to define the transition from viscous to granular debris flow-like behavior.

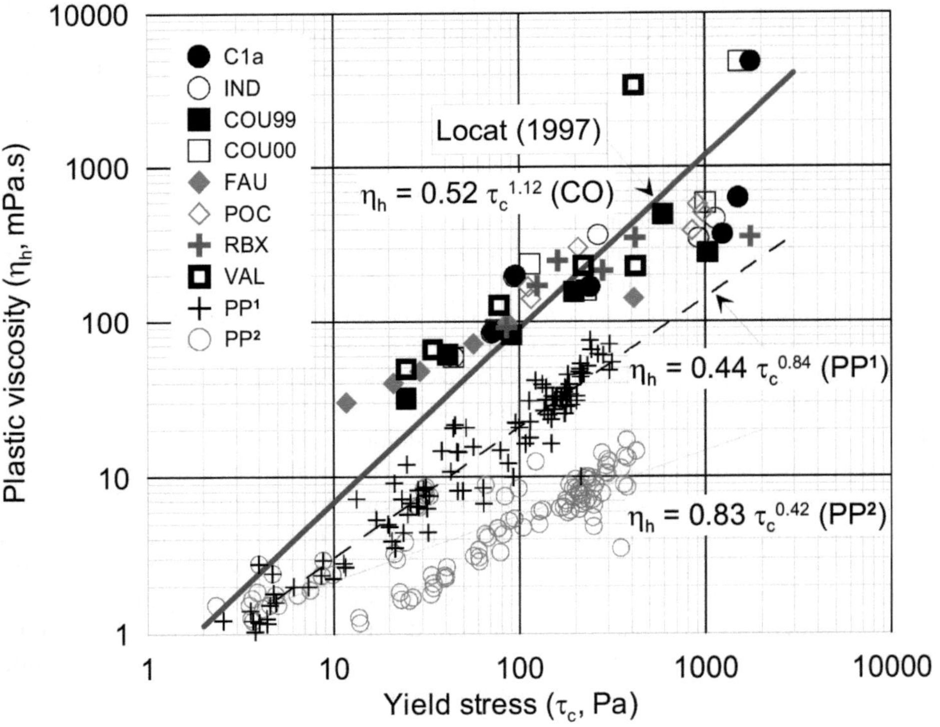

Figure 3. Relationship between yield stress and plastic viscosity depending on the geometry and imposed shear rate. The Bingham yield stress and plastic viscosity were determined in the (1) coaxial viscometric (CO) system having a maximum shear rate of 1200 (s^{-1}), (2) parallel plate (PP^1) rheometer having a maximum shear rate of about 20000 (s^{-1}), but the ratios were determined by the Bingham model in the similar range of coaxial viscometer; (3) parallel plate (PP^2) rheometer having a maximum shear rate of about 20000 (s^{-1}).

4. Conclusion

Rheological properties of fine-grained sediments and the possible rheological transition between viscous and granular flows were studied. The results are intended to provide guidance in selection of resistance parameters in those studies where numerical models are used to simulate the high speed deep-sea debris flows. Non-swelling and swelling materials have characteristics of a pseudoplastic (shear thinning) fluid, even in the case of the iron tailings. It is worth noting here that, for the non-swelling natural clays, the ratio of yield stress to viscosity is about 1 to 1000, whereas the ratio varies depending on the grain size distribution. Results suggest that it may be possible to estimate the rheological parameters, on the basis of result of non-swelling natural materials. The possible transitional ranges of rheological properties implemented by a linear relation of apparent yield stress (in Pa) and plastic viscosity (in mPa.s) are $\tau_c / \eta_h = 1000$ for clays (Group 1), 100 for silts (Group 3), and 10 for sands. Due to limitations in the experimental devices, the difference in ratio of yield stress to viscosity is related to the change in: (i) grain size, (ii) shear rate, and (iii) geometry. Shear rate is considered as the main reason resulting in the inconsistency of results. More laboratory data and field observations are needed to confirm the relationships proposed in this study.

5. Acknowledgements

The authors would like to thank the Natural Sciences and Engineering Research Council of Canada for their support into COSTA-Canada projects.

6. References

Coussot, P., and Piau, J.-M. 1994. *On the behavior of fine mud suspensions*. Rheologica Acta, **33**: 175-184.

Coussot, P., and Piau, J.-M. 1995. *A large-scale coaxial cylinder rheometer for the study of the rheology of natural coarse suspensions*. Journal of Rheology, **39**: 105-124.

Coussot, P., Laigle, D., Arattano, M., Deganutti, A., and Marchi, L. 1998. *Direct determination of rheological characteristics of debris flow*. Journal of Hydraulic Engineering, **124**: 865-868.

Ilstad, T., Elverhøi, A., Issler, D., and Marr, J.G. 2004. *Subaqueous debris flow behavior and its dependence on the sand/clay ratio: a laboratory study using particle tracking*. Marine Geology, **213**: 415-438.

Jeong, S.W., Locat, J., and Leroueil, S. 2004. *A preliminary analysis of the rheological transformation due to water infiltration as a mechanism for high mobility of submarine mass movements*. 57th Canadian geotechnical conference, Québec, Session 7G: 15-22.

Jeong, S.W. 2006. *Influence of physico-chemical characteristics of fine-grained sediments on their rheological behavior*. Ph.D thesis, Department of Civil Engineering, Laval University, Québec.

Kvalstad, T.J., Andresen, L., Forsberg, C.F., Berg, K., Bryn, P., and Wangen, M. 2005. *The Storegga slide: evaluation of triggering sources and slide mechanics*. Marine and Petroleum Geology, **22**: 245-256.

Locat, J. 1997. *Normalized rheological behaviour of fine muds and their flow properties in a pseudoplastic regime*. Proceedings 1st International conference on Debris-Flow Hazards Mitigation, San Francisco. ASCE, New York: 260-269.

Locat, J., and Demers, D. 1988. *Viscosity, yield stress, remoulded strength, and liquidity index relationships for sensitive clays*. Canadian Geotechnical Journal, **25**: 799-806.

Locat, J., and Lee, H.J. 2002. *Submarine landslides: advances and challenges*. Canadian Geotechnical Journal, **39**: 193-212.

Major, J.J., and Pierson, T.C. 1992. *Debris flow rheology: experimental analysis of fine-grained slurries*. Water Resources Research, **28**(3): 841-857.

Malet, J.P., Remaître, A., Maquaire, O., Ancey, C., and Locat, J. 2003. *Flow susceptibility of heterogeneous marly formations. Implications for torrent hazard control in the Barcelonnette basin (Alpes-de-Haute-Provence, France). In* Rickenmann, D. and Chen, C.L. (*Editors*). Proceedings of the 3rd International conference on Debris-Flow Hazards Mitigation, Millpress, Rotterdam: 351-362.

Malet, J.P., Laigle, D., Remaître, A., and Maquaire, O. 2005. *Triggering conditions and mobility of debris flows associated to complex earthflows.* Geomorphology, **66**: 215-235.

Møller, P.C.F., Mewis, J., and Bonn, D. 2006. *Yield stress and thixotropy: on the difficulty of measuring yield stresses in practice.* Soft Matter, **2**: 274-283.

Niedoroda, A.W., Reed, C., Das, H., Hatchett, L., and Perlet, A.B. *Controls of the behavior of marine debris flows.* Norwegian Journal of Geology, **86**: 265-274.

O'Brien, J.S., and Julien, P.Y. 1988. *Laboratory analysis of mud flow properties.* Journal of Hydraulic Engineering, ASCE, **114**: 877-887.

Parsons, J., Whipple, K., and Simoni, A. 2001. *Experimental study of the grain flow, fluid-mud transition in debris flow.* Journal of Geology, **109**(4): 427-447.

Phillips, C.J., and Davies, T.R.H. 1991. *Determining rheological parameters of debris flow material.* Geomorphology, **4**: 101-110.

Schatzmann, M., Fischer, P., and Bezzola, G.R. 2003. *Rheological behavior of fine and large particle suspensions.* Journal of hydraulic engineering, **129**(10): 796-803.

Whipple, K., and Dunne. T. 1992. *The influence of debris-flow rheology on fan morphology, Owens Valley, California.* Geological Society of America Bulletin, **104**: 887-900.

MARINE DEEP-WATER FREE-FALL CPT MEASUREMENTS FOR LANDSLIDE CHARACTERISATION OFF CRETE, GREECE (EASTERN MEDITERRANEAN SEA) PART 2: INITIAL DATA FROM THE WESTERN CRETAN SEA

A. KOPF, S. STEGMANN, S. KRASTEL, A. FÖRSTER
Research Centre Ocean Margins, Bremen University, Leobener Strasse, 28359 Bremen, Germany

M. STRASSER
Geological Institute, ETH Zurich, Universitätstrasse 16, CHN, 8092 Zurich, Switzerland

M. IRVING
Picker Engineering Program, Smith College, 51 College Lane, Northampton, MA 01063, U.S.A.

Abstract

Pore pressure and shear strength are major controlling parameters for slope stability, which can be measured *in situ* using CPT (cone penetration testing) instruments. This paper presents results from initial tests with two free-fall CPT probes deployed in the neotectonically active submarine slope of northern Crete, Greece. Research expedition P336 investigated landslide-prone areas in the Cretan Sea using multibeam swathmapping, seismic reflection profiling, *in situ* CPT measurements, and gravity coring. Several large landslide complexes at the NE Cretan Margin as well as a small, but steep landslide scarp structure further east were identified on the seismic profiles. CPT devices were deployed in undisturbed slope sediments, across the slide scar, and in the main body of the slide, and remained stuck in the sediment for ~10 minutes to monitor pore pressure dissipation upon insertion. Excess pore pressure after insertion is in a range around 60 kPa in background sediment, and exceeds 80 kPa in the slide deposits. Cone resistance of the slope sediment ranges between 300 and 500 kPa, corresponding to undrained shear strengths of up to 40 kPa. The slid sediments (specifically the headwall material with <10 kPa strength) show velocity-weakening behaviour during ring shear experiments, indicating that those sediments are unlikely to show stable creep and instead may fail catastrophically.

Keywords: CPT, landslide, shear strength, pore pressure

1. Introduction

Sediment stability at continental margins depends on given different soil mechanical conditions and a variety of trigger mechanisms (e.g. Hampton et al. 1996). This complexity demands a multi-disciplinary research approach, which has been achieved by several studies that combined geophysical, sedimentological and geotechnical methods (e.g. Storegga Landslide [Kvalstad et al. 2005]; Niger Delta [Sultan et al., 2007]). In these studies, the sediment physical properties were assigned a high priority, with shear strength and pore pressure playing a key-role in the assessment of sediment

V. Lykousis, D. Sakellariou and J. Locat (eds.), Submarine Mass Movements and Their Consequences, 199–208.

stability. Equally, there are several landslide occurrences north of Crete, an area that is regularly struck by neotectonic activity and earthquake tremors (Lykousis et al., 2002). During cruise P336 in April/May 2006 in the Cretan Sea, we studied landslide processes in two areas (here termed D and E). Bathymetric mapping and seismic profiling served to characterise the landslide-prone slopes. Subsequently, *in situ* CPT measurements were made in undisturbed slope sediments as well as in the mass wasting deposits and were complemented by geotechnical measurements on adjacent sediment cores.

2. Geological background of the Cretan Sea (Eastern Mediterranean)

The Cretan Sea represents the northernmost portion of the forearc region in the Hellenic subduction zone (HSZ) between Africa and Eurasia, which is well recorded over the past ca. 35 million years (Le Pichon and Angelier, 1979). It is sandwiched between the Aegean back-arc basin and island arc volcanoes (e.g. Santorini) in the north and the island of Crete, a prominent forearc-high, in the south (Fig. 1a). The island of Crete comprises a stack of nappes of variable lithologies (for details, see e.g. Fassoulas, 1999), parts of which got exhumed some 19 Ma ago and now are extending in both E-W- and N-S-direction. The main extensional phase of the Cretan Sea occurred between the Late Miocene and Pliocene however the Late Pleistocene experienced only minimal extension phenomena (Mascle and Martin, 1990). Tectonic movements still occur today, as indicated by recent seismicity and volcanic activity in the area (McKenzie, 1978). Landslides are one of the most immanent hazards in the Cretan Sea, being triggered by both the tectonic movements of the Cretan block in the south (e.g. Chronis et al., 2000a, b) and the flank collapse of volcanic islands in the north (e.g. Dominey-Howes et al., 2000). Although the inherent mechanisms and factors governing slope stability and submarine landslides are comprehensively studied, their temporal and spatial variability are poorly understood.

Geomorphologically, the Cretan Basin is an elongated depression, trending E–W; it is bounded to the north by the Cyclades Plateau, a relatively shallow (500 m) complex of islands, and has water depths no larger than 1000 m (with localized, ca. 2500 m deep sub-basins; see Kopf et al., 2006). Sediment accumulation processes at the southern margin of the Cretan Basin represent pro-delta deposition in the inner middle shelf, settling from bottom nepheloid layers in the shelf and upper slope, calcareous sediment formation due to settling from suspension, and redeposition from suspension due to gravity processes and bottom currents (Chronis et al., 2000b). Hemipelagic sediments of the entire Cretan Sea are characterised by four different lithologies, regarded from top to bottom (Giresse et al., 2003):

(i) yellowish brown mud with bioturbation structures
(ii) grey mud mottled and moderate bioturbated with
(iii) brownish or olive black mud with >2% C_{org}, which represents sapropel S1 (9600-6400 yrs. BP)
(iv) yellowish, grey clay-rich mud.

Sedimentation rates in the Cretan trough are estimated to be 4.3-15 cm/ka (Giresse et al., 2003).

3. Methodology

Cruise P336 with the RV *Poseidon* focused on slope instability and sedimentation at the northern Cretan Margin (Fig. 1a). A variety of geophysical, sedimentological, and geotechnical methods were applied, of which only a few are relevant for this paper. A detailed report of this cruise is given in Kopf et al. (2006).

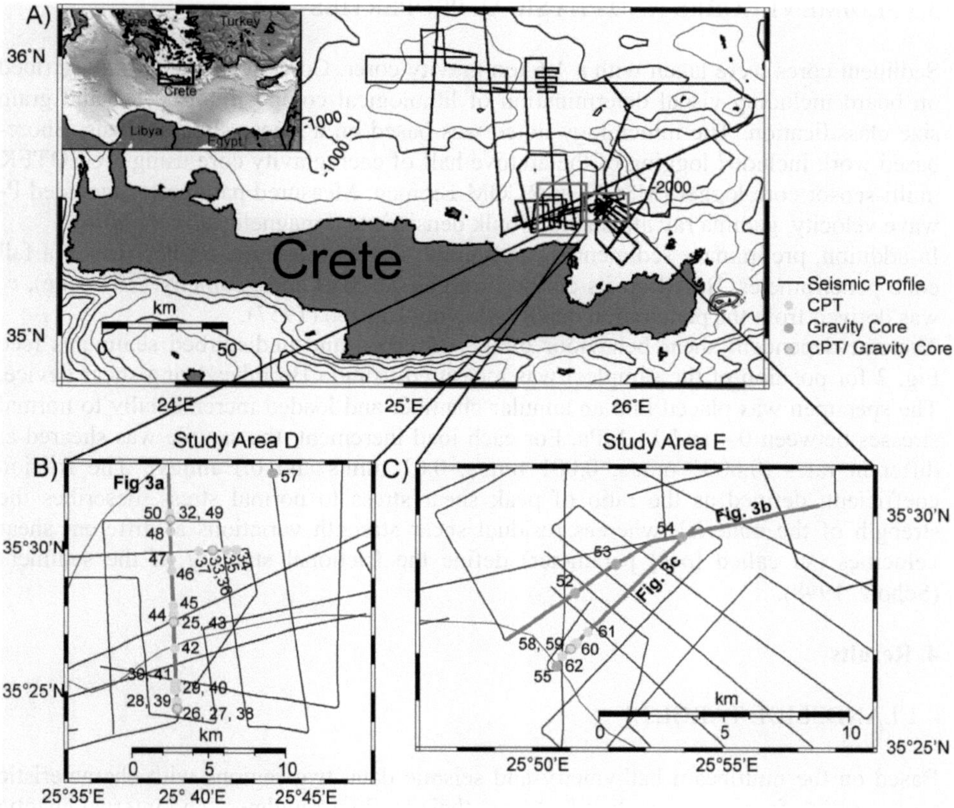

Figure 1. Map of the complete study area in the Eastern Mediterranean off Crete, Greece (A). Mass wasting deposits at the Cretan Margin are identified in study area D (B) and study area E (C, D). Numbers and lines represent the position of seismic profiles, gravity core stations and CPT locations.

3.1. GEOPHYSICAL CHARACTERISATION

Continuous seafloor mapping was routinely carried out with the multibeam echosounder ELAC SEABEAM 1050 in order to identify landslide scars. Seismic data were collected using a 3.5 kHz system and a high-resolution multi-channel system. The multi-channel seismic system consists of a Mini-Generator-Injector Airgun (frequency range 100-500Hz) and a 100-m-long 16 channel streamer. The presented seismic profiles (Fig. 1) are brute stacks generated by summing up the first three channels. The data were filtered with a wide bandpass (55/110-600/800 Hz). The combination of 3.5 kHz and seismic data were the basis for selecting coring and CPT-stations.

3.2 *IN SITU* CPT MEASUREMENTS

In situ measurements of sediment physical properties were carried out with two free-fall CPT devices. Their design and mode of deployment is described in the first part of this manuscript (see Stegmann and Kopf, this issue).

3.3 SEDIMENT CORES AND PHYSICAL PROPERTIES

Sediment cores were taken with a 1.6 ton gravity corer. Cores were split and described on board including visual determination of lithological composition, colour and grain size classification. The mineralogy noted was based on a smear slide analysis. Shore-based work included logging of the archive half of each gravity core using a GEOTEK multi-sensor core logger (MSCL) at RCOM Bremen. Measured parameters included P-wave velocity, gamma ray attenuation (bulk density), and magnetic susceptibility.
In addition, preliminary sediment shear strength c_u was measured on board with a fall cone penetrometer. Based on its defined weight (80.51g) and geometry (30° cone), c_u was derived from the penetration depth following Hansbo (1957).
The rate-dependent shear behaviour of the disturbed and undisturbed sediments (see Fig. 2 for position of the samples) was measured using a Bromhead ring shear device. The specimen was placed into an annular chamber and loaded incrementally to normal stresses between 0.4 and 16 MPa. For each load increment, the sample was sheared at different rates (0.0005 mm/s, 0.001 mm/s, 0.01 mm/s and 0.1 mm/s). The friction coefficient, defined as the ratio of peak shear stress to normal stress, describes the strength of the material, whereas residual shear strength variations at different shear velocities (so called [*a-b*] parameter) define the frictional stability of the sediment (Scholz, 1998).

4. Results

4.1 LANDSLIDE TARGETS

Based on the multibeam bathymetry and seismic data, two regions with characteristic mass wasting features were identified from their seafloor roughness and internal chaotic signatures (Fig. 2). Northeast of the island of Crete, area D shows a huge, roughly ~9 km long and 3-4 km wide lobe of displaced slope sediment consisting of two distinct slide units, with a relatively smooth surface (Fig. 2a). Some of the failed material seems to have slid as intact blocks while other portions appear to have been amalgamated. A headwall cannot be identified, but at the head of the slide body, the upper 20 m of sediments are missing and seem to be incorporated into the slide. Intact structures inside of the generally chaotic seismic facies of the slides suggest that the internal structure has not been totally destroyed and that the slide has not travelled very far.

Further east, in area E, a smaller slope failure event was identified based on its steep head scarp (Figs. 2b, c). Undisturbed, well-stratified sediments upslope are cut at the headwall, which has a height of ~50 m at this location. Directly adjacent to the headwall, a relatively thin (< 50 m) chaotic unit overlies well-stratified sediments. Approximately 4 km down-slope of the headwall the chaotic unit thickens to roughly 100 m, which most likely represents the main depositional area of the slide. However, as this unit does not appear as a continuous feature, an estimation of the depositional area is difficult. It could be possible that the bulk part of the slide material is transported much further down-slope and was deposited in the deep basin.

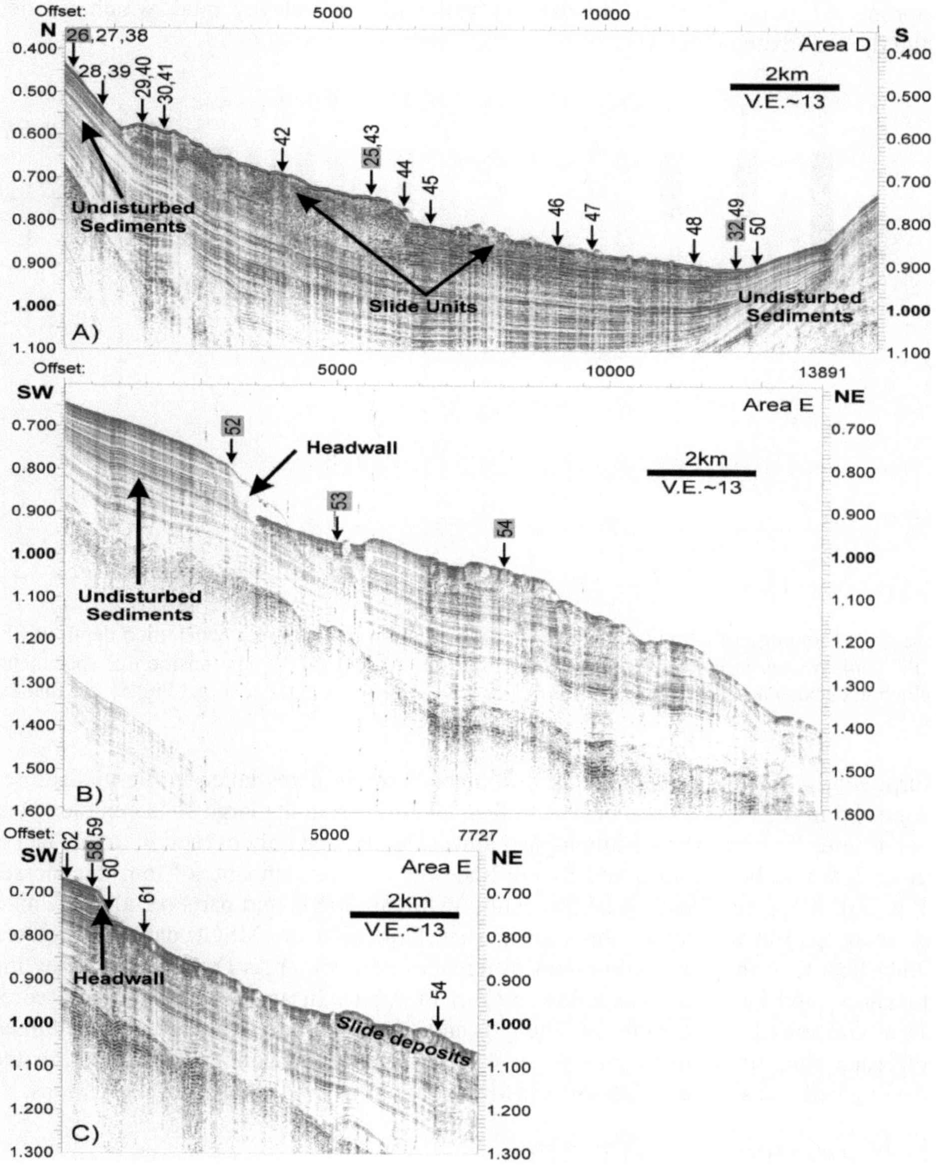

Figure 2. Airgun profile of mass wasting events at the Cretan Margin in area D (A) and area E (B, C) with the positions of the CPT and gravity cores (green marked signature).

In the two slides identified in areas D and E, a total of 11 gravity cores with lengths of 1-4.6 m were recovered (Figs. 2-3; Kopf et al., 2006). The sediment is generally comprised of four different lithologies: Yellowish brown bioturbated mud (Unit 1) is underlain by mottled and moderately bioturbated grey mud (Unit 2), sometimes containing a volcanic layer of the Thera eruption (3370 B.P.). Below that, greyish-brownish to olive grey mud with C_{org} >2% and no bioturbation has been identified as sapropel S1 (Unit 3). It is underlain by yellowish grey clayey mud, which is often slightly bioturbated (Unit 4).

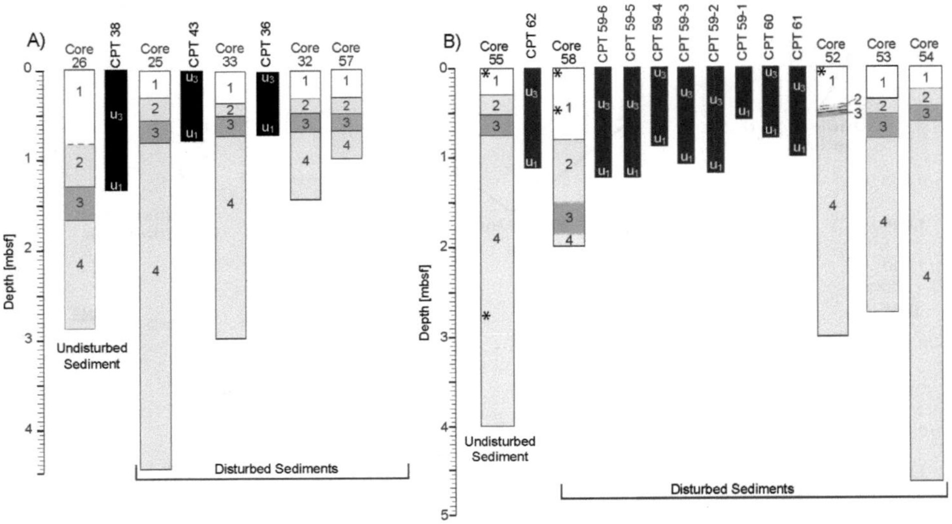

Figure 3. Lithology of cored sediments (see description in the text) and penetration depth of FF-CPT deployments in landslide sediments in areas D (A) and E (B). The position of specimens, which have been tested in the ring shear device (see description in the text and Fig. 5), are marked by *.

Surprisingly, there is no significant difference between cores taken in the undisturbed slope and in the wasted mass below. In fact, all cores from the large slide complex D as well as area E show the four lithological units (Fig. 3). The only exception to this fact is core 52, where both Unit 2 and S1 are traceable only as remnants of mm-thicknesses (Fig. 3b). It appears from visual inspection that Unit 3 (S1) and parts (or all) of Unit 2 are missing. We will revisit this aspect when looking at the MSCL data (see below). Other than that, there are only minor differences between areas D and E regarding the thickness (and hence sedimentation rate) of the units. In area D, the landslide cores show condensed successions of Units 1 and 2 when compared to the undisturbed reference site. In contrast, area E shows no systematic relationship, with both the hangingwall and slid mass deposits showing both normal and condensed successions.

4.2 *IN SITU* PHYSICAL PROPERTIES

During the cruise, 26 FF-CPT deployments were carried out in areas D and E (Figs. 2-3). Unfortunately, the CPT cone failed during some of the deployments so that the strength parameters (q_c, f_s) could not be measured in each location. Consequently, we focus mainly on the differential pore pressure data in regions D and E.

Initial penetration velocity of the complete CPT data set (derived from acceleration) ranged between 1.1 m/s and 1.8 m/s, which was mostly limited by winch speed (max. 2 m/s) and external conditions (waves, swell). Given the stiff nature of the sediments off Crete, total penetration depth was rather low. It appears as if the S1 layer, which has a higher strength than the surrounding sediments, is limiting the maximum penetration depth since it slows down the lance's momentum dramatically. In area D, penetration depth varied between 0.65 and 1.35 m (Fig. 3a), with the highest values in undisturbed sediments upslope of the scar, and between 0.6 and 1.25 m in area E (Fig. 3b). In nearly all measurements the tilt of the penetrated lance did not exceed ±9°.

In situ measured cone resistance is limited to undisturbed and failed sediments of area D because of problems with the CPT probe at the tip of the lance. However, based on the results collected, we can show that the undisturbed section shows higher strength than the remobilised portion. This is reflected by maximum q_c plotting around 400-500 kPa upslope and 300-380 kPa on the landslide body. These findings correspond to the working hypothesis that the remobilised sediment has higher water content and lower strength, which is indicated further by the MSCL data. These attest to lower p-wave velocities and bulk densities in the upper portion of cores in the landslide body (stations 25, 33, 32, 57), but higher p-wave velocities in Unit 4 in the lower part. Here, the undisturbed material ranges between 1575-1590 m/s while the landslide cores range from 1510 to 1560 m/s, possibly related to fluids trapped during remobilisation.

Measured pore pressure response generally shows an insertion peak followed by a sudden drop. At the time, the lance is still penetrating the sediment (Stegmann and Kopf, this issue, red portion in their Fig. 3). Pore pressure then rises again to a second maximum, which in turn is followed by an exponential decay (Stegmann and Kopf, this issue, black portion in their Fig. 3). Unfortunately, several of our measurements exceeded the upper limit of the differential pore pressure sensor immediately after the impact of the probe (again, maybe as a result of high excess fluid pressures in Unit 4; see previous paragraph). For all landslide measurements of u_1, maximum excess pore pressure values after insertion ranged between 24 kPa to more than 82 kPa (this latter value being the upper range of the sensor). Pore pressure signals in area D show maximum insertion pore pressure (u_1) between 40 and 60 kPa for the sapropel unit of area D. The sediment in area E is found less dense and with higher porosities compared to area D. Accordingly, T_{50} values range between 1.9 and 4.2 mins. in the sapropel unit (60-80% porosity) compared to T_{50}=6 mins. in the muddy sediment (40-60% porosity). The u_3 pore pressure signal was measured in only 76 % (area D) and 33% (area E) of the measurements, because penetration depth did not exceed 80 cm (see instrument design, Stegmann and Kopf, this issue). When recorded, the u_3 signal often resembles that in u_1 position (see Stegmann and Kopf, this issue, Fig. 3d). The insertion pressure values are higher in area D, varying between 27 kPa (station 40) and 52 kPa (station 39), than in area E with a range between 9.5 kPa (station 62) and 28.6 kPa (station 59-4).

4.3 LAB-BASED PHYSICAL PROPERTIES

Data from the MSCL do not allow a clear distinction between cores taken in the undisturbed slope cover (reference core) and that in the landslide material. In general, area D cores show low p-wave velocities (1500-1550 m/s) and smaller bulk densities of approximately 1.85 g/cm^3 than area E. Values increase gradually down section in the reference core (core 26) (ca. 1600 m/s, ca. 2 g/cm^3), but decrease in each of the landslide cores in Unit 4. Lab-determined physical properties such as undrained shear strength c_u (determined with the fall cone penetrometer) mirror this trend. In the upper portion (i.e. Units 1-3) c_u increases with depth from 10 to 20 kPa. In the deeper section (2-2.9 m) of the reference core, higher c_u (40 ±8 kPa) coincides with a significant jump to lower porosity (av. 40 %). Failed material of the Cretan Margin (stations 25, 32, 33, 57) can be different to very similar to the intact sediments located above the scarp. In contrast, the farthest removed deposits reveal a process of homogenisation as a result of displacement, expressed with a relatively high porosity of 60 % and a density in a range around 1.8 g/cm^3. c_u is more or less constant, which seems indicative of prograde consolidation history.

Although the scarp structure of the area E landslide is very recognisable in seismic data, landslide features are not very obvious to identify in the very homogeneous sediments with an average density of 1.8 g/cm^3. Upslope (station 55) and down-slope (station 54) materials are characterised by a linear increase of c_u from 20 to 40 kPa. Immediately near the scarp (station 52) and within the channel-like failure structure (stations 53, 58) a less pronounced linear trend of c_u is evinced.

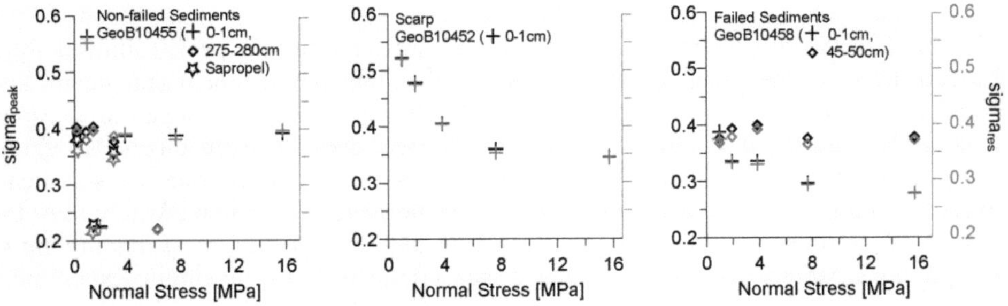

Figure 4. Frictional behaviour of undisturbed and disturbed superficial (crosses) and sediments from depth (lozenges) of the head scarp region in area E. The coefficient of friction is plotted vs. normal stress due to incremental loading during ring shear tests with a shear velocity of 0.01 mm/s. Black colours signify the coefficient of peak strength μ_{peak} while red colours show the coefficient of residual strength μ_{res}.

Ring shear data have only been collected in area E to characterise small-scale lateral variations across the headwall of the slide (Fig. 2b-c, 3b). The undisturbed sediments (station 55) indicate no significant difference between superficial (Unit 1) and deeper portion (Unit 4; 2.75-2.8 m) with an average μ_{peak} of 0.4 (Fig. 4a). Unit 3 (sapropel S1) shows a μ_{peak} range of 0.24-0.4, possibly reflecting a breakdown of the cohesive organic aggregates and fabric alignment. In contrast, μ_{peak} of the surface sediment from the headwall and landslide (stations 52, 58) are significantly lower, ranging around 0.35 (52; Fig. 4b) and 0.28 (58; Fig. 4c). Even the deeper portion of the landslide core 58 shows μ_{peak} ca. 0.36-0.4, which is slightly below that of the undisturbed core (Fig. 4a). This suggests to us that indeed some material is missing in the upper part of core 58 (see discussion below).

5. Discussion

Looking into the sedimentary and geotechnical results in more detail, we first revisit the seismic data. Despite the evidence for landslide features in area D with rough surface and internal features in the seismic images (Fig. 3), the gravity core description alone cannot unambiguously distinguish between the undisturbed vs. slid sediment. Based on sedimentological information, it can be speculated that:

(i) the lower part of the succession corresponds to amalgamated mud of the landslide body that would have occurred relatively shortly before the onset of sapropel deposition ~10 ka B.P., or that
(ii) the sedimentary succession represents primary sedimentary deposits and therefore, the landslide is either older and was not reached with coring, or all cores were recovered from an internally coherent slide, or out runner blocks. In case of the latter, the landslide can also be younger than S1 and the Thera volcanic deposit (3370 B.P.).

When further consulting the results from the MSCL and *in situ* measurements, we observe some variations that identify the landslide material. These include lower cone resistance in the remobilised section of landslide D, lower p-wave velocity in the deeper portion of cores from area D, and low frictional strength from ring shear tests at the head scarp materials and shallow landslide deposits in area E (cores 52 and 58; Fig. 4b, c). If we assume that these interpretations of the superficial measurements are correct, then the landslide should be relatively young. This is supported by the seismic reflection data where, despite lack of m-scale resolution, no seafloor-parallel, post-landslide reflections are found (Fig. 2). In study area E, the ~50 m high scarp is clearly visible in seismic reflection data (Fig. 2b, c). Apart from the low intrinsic friction (Fig. 4), mass wasting near the head scarp is confirmed by abundant clasts and carbonate concretions in core 58 immediately above sapropel S1 (see Kopf et al. [2006] for details). Also, sedimentation rates in that interval are roughly twice as high as in the other cores. However, given the overall lithostratigraphic similarities with S1 and other markers present, no final conclusion can be drawn on the timing and mechanism of scarp formation.

Since we were unable to perform long-term pore pressure dissipation experiments, we cannot safely propose the physical trigger mechanism of the two landslides. Neotectonic activity and regional seismicity make earthquake tremor a likely candidate. Earthquake magnitudes have been reported to be as high as M7.4 (e.g. in 1956; see Perissoratis and Papadopoulos, 1999), causing significant subsidence and sediment remobilisation. Excess pore pressures exceed 82 kPa, however, an unquantifiable portion of that number relates to the impact of the CPT instrument and is found to decrease rapidly (i.e. T_{50} values of several minutes only). In any case, pore pressures are believed to be lightly supra-hydrostatic because of the moderately high sedimentation rates. Hence, significant extra pore pressure from (pre-)seismic stress release is needed to cause landslide initiation. Sliding, however, is facilitated by friction coefficients of μ_{peak} ~0.3, or lower, as measured with the ring shear apparatus. Also, unstable sliding is likely given that these materials show velocity weakening behaviour when sheared at different rates, so that catastrophic mass wasting may occur.

6. Acknowledgements

We thank Master Michael Schneider and his crew for the friendly and efficient operations during cruise P336 with RV *Poseidon*. The manuscript benefited from constructive reviews and suggestions by Katrin Huhn and Nabil Sultan. Funding for this work was received by DFG through RCOM Bremen (project C8). This is RCOM publication #505.

7. References

Chronis, G., Lykousis, V., Georgopoulos, D., Poulos, M., Zervakis, V., Stavrakakis, S., 2000a. *Suspended particulate matter and nepheloid layers in the southern margin of the Cretan Sea (NE Mediterranean): seasonal distribution and dynamics.* Progress in Oceanography, 46, p.163-185.

Chronis, G., Lykousis, V., Anagnostou, C., Karageorgis, A., Stavrakakis, S., Poulos, S., 2000b. *Sedimentological processes in the southern margin of the Cretan Sea (NE Mediterranean).* Progress in Oceanography, 46, p.143-162.

Dominey-Howes, D., Cundy, A., Croudace, I., 2000. *High energy marine flood deposits on Astypalaea Island, Greece: possible evidence for the AD 1956 southern Aegean tsunami.* Mar. Geol., 163, p.303-315.

Fassoulas, C., 1999. *The structural evolution of central Crete: insight into the tectonic evolution of south Aegean (Greece).* Geodynamics, 27, p.23-43.

Giresse, P., Buscail, R., Charriere, B., 2003. *Late Holocene multisource material input into the Aegean Sea: depositional and post-depositional processes.* Oceanologica Acta, 26, p.675-673.

Hampton, M.A., Lee, H.J., Locat, J., 1996. *Submarine Landslides.* Reviews of Geophysics, 34/1, p.35-59.

Hansbo, S., 1957. *A new approach to the determination of the shear strength of clay by the fall-cone test.* Geotechnical Institute Proceedings, 14, p.50.

Kopf, A., Alves, T., Heesemann, B., Kaul., N.E., Kock, I., Krastel, S., Reichelt, M., Schäfer, R., Stegmann, S., Strasser, M., Thölen, M., 2006. *Report and preliminary results of POSEIDON cruise P336: CRESTS - Cretan Sea Tectonics and Sedimentation.* Berichte FB Geowiss., Univ. Bremen, No. 253: 140pp.

Kvalstad, T.J., Farrokh, N., Kaynia, A.M., Mokkelbost, K.H., Bryn, P., 2005. *Soil conditions and slope stability in the Ormen Lange area.* Marine and Petroleum Geology, 22, p.299-310.

Le Pichon, X., Angelier, J., 1979. *The Hellenic Arc and Trench system: a key to the neotectonic evolution of the Eastern Mediterranean area.* Technophysics, 6, p.1-42.

Lykousis, V., Roussakis, G., Alexandri, M., Pavlakis, P., Papoulia, I., 2002. *Sliding and regional slope stability in active margins: North Aegean Trough (Mediterranean).* Mar. Geol., 186, p.281-298.

Mascle, J., Martin, L., 1990. *Shallow structure and recent evolution of the Aegean Sea: a synthesis based on continuous reflection profiles.* Mar. Geol., 94, p.271-299.

McKenzie, D.P., 1978. *Active tectonics of the Alpine–Himalayan belt: the Aegean Sea and surrounding regions.* Geophysical Journal of the Royal Astrological Society, 55, p.217-254.

Perissoratis, C., Papadopoulos, G., 1999. *Sediment instability and slumping in the southern Aegean Sea and the case history of the 1956 tsunami.* Marine Geology, 161, p.287-305.

Scholz, C.H., 1998. *Earthquakes and friction laws.* Nature, 391, p.37-42.

Sultan, N., Voisset, M., Maresset, B., Marsset, T., Cauquil, E., Colliat, J.-L., 2007. *Potential role of compressional structures in generating submarine slope failures in the Niger Delta.* Mar. Geol., 237/3-4, p.169-190.

RECURSIVE FAILURE OF THE GULF OF MEXICO CONTINENTAL SLOPE: TIMING AND CAUSES

R. URGELES
Dept. of Stratigraphy, Paleontology and Marine Geosciences, University of Barcelona, 08028 Barcelona, Catalonia, Spain, urgeles@ub.edu

J. LOCAT
Dept. of Geology and Geological Engineering, Laval University, Québec G1K 7P4, Qc, Canada, locat@ggl.ulaval.ca

B. DUGAN
Dept. of Earth Sciences, Rice University, 6100 Main Street, MS-126, Houston TX 77005, USA, dugan@rice.edu

Abstract

Seismic and multibeam data have shown the occurrence of Plesitocene large-scale Mass Transport Deposits (MTDs) and Holocene failure events in Ursa Basin at ~1000 m depth in the eastern levee of the Mississippi Canyon, northern Gulf of Mexico. During IODP Expedition 308 Sites U1322 and U1324 were drilled adjacent to the Recent failures and through several MTDs of Holocene and Pleistocene age. A complete suite of logging, sedimentological and geotechnical data were acquired to reveal the factors controlling initiation of past sediment failures and to characterize the hazard potential of future slope instabilities. Fluid overpressure estimated from a variety of direct and indirect methods indicates that the vertical effective stress is 50 to 70 % lower than if hydrostatic conditions existed. Overpressure in Ursa Basin most likely results from deposition of fine-grained sediments with rates at least 1 m ky^{-1} with peaks up to 25 m ky^{-1}. The thickest and most widespread MTDs occur in periods of highest sedimentation rate. Considering the very low seismicity experienced by the margin it is most probable that the changes in sedimentation rate might be the primary driving force of slope failure. At Sites U1322 and U1324 a total of 14 MTDs were identified, which leads to a frequency of emplacement of 1 MTD/4.5 ka. Considering only the thickest MTDs the recurrence interval reduces to 1 MTD/10 ka.

Keywords: Mass Transport Deposits, Overpressure, Slope Stability, Gulf of Mexico

1. Introduction

Pleistocene sedimentation in the Gulf of Mexico from the Mississippi River is characterized by rapid sedimentation upon a mobile salt substrate (Worrall and Snelson 1989). Offshore Texas and Western Louisiana, individual slope minibasins are surrounded by elevated salt highs (Pratson and Ryan 1994) producing a remarkable hummocky topography. This morphology is obscured in front of the Mississippi delta, where sedimentation has been very rapid, exceeding 25 m ky^{-1} (Expedition 308 Scientists 2005).

Ursa Basin (~150 km due south of New Orleans, Louisiana, USA) lies in ~1000 m of water (Figure 1). The Pleistocene to Holocene sedimentary sequence in Ursa Basin consists of the lower Mississippi Canyon Blue Unit, a late Pleistocene, sand-dominated,

V. Lykousis, D. Sakellariou and J. Locat (eds.), Submarine Mass Movements and Their Consequences, 209–219.

"ponded fan" (Winker and Booth 2000, Sawyer et al. 2007). The Blue Unit is overlain by a mud dominated channel-levee assemblage that has dramatic along-strike variations in thickness (Figure 2). This mudstone package, belonging to the eastern margin of the larger channel-levee system of the Mississippi Canyon, has Mass Transport Deposits (MTDs) that record failure of the margin during the Pleistocene (Flemings et al. 2006, Sawyer et al. 2007).

Holocene failures have also been mapped all along the Gulf of Mexico, with the largest ones originating from Ursa Basin (Figure 3) (see also McAdoo et al. 2000). Ursa Basin is of economic interest because of its prolific oilfields, which are currently being exploited in the nearby tension leg platforms of Mars and Ursa. In consequence, geohazard characterization is essential for safe offshore activities. For the purpose of this paper, we use data from two boreholes (Figures 1-3) acquired during IODP expedition 308 (Expedition 308 Scientists 2005), including Logging While Drilling (LWD), measurements of physical properties and geotechnical analysis performed on whole round samples.

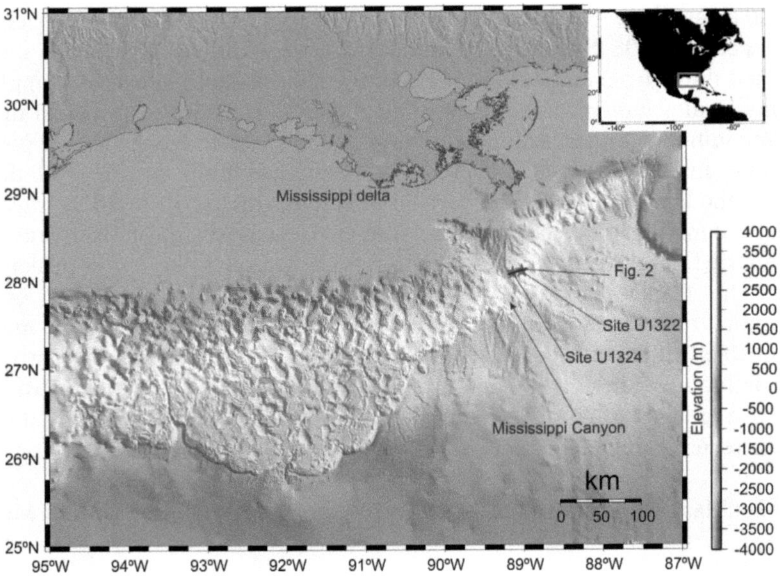

Figure 1. Swath bathymetry shaded relief image of the Gulf of Mexico showing location of Ursa drill sites U1322, U1324 and seismic line shown in Figure 2. Inset shows location.

The aim of this paper is to geotechnically characterize the sediments of this area, to determine the factors that most likely played a role in submarine landslide initiation, and to constrain the recurrence of these events.

2. Methods

The data used in this study are based on samples acquired on board the D/V *JOIDES Resolution* during IODP expedition 308. Visual core description during Expedition 308 was carried out on board shortly after core recovery. At sea, a complete set of LWD and wireline log data was acquired including resistivity, gamma radiation and density measurements that allow a high resolution characterization of sediment physical properties. Discrete sediment samples were taken on board to measure the natural water content, density, porosity (oven drying and pycnometer methods) and peak and remolded undrained shear strength (handheld penetrometer and motorized vane apparatus).

Shorebased experiments consisted of grain size analysis using a Coulter laser diffractometer and measurement of Atterberg limits using a British fall cone (Feng 2001). The consolidation analysis was carried out using an uniaxial incremental loading oedometer and the sediment strength parameters were determined on isotropically consolidated triaxial compression tests.

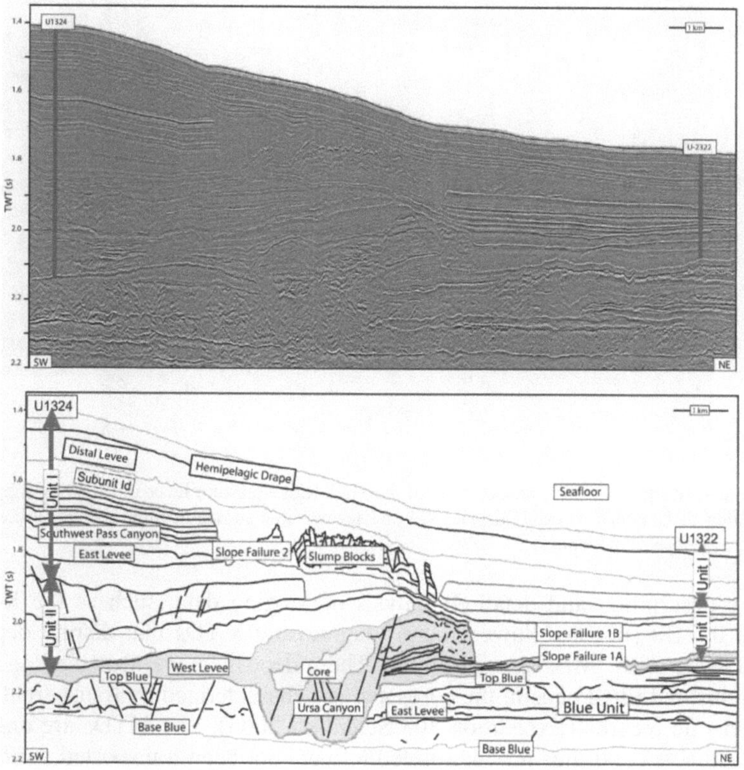

Figure 2. Top. Seismic cross-section (for location see Figures 1 and 3). Bottom. Interpreted cross-section. The sand-prone Blue Unit has been incised by a channel-levee complex and then overlain by a thick and heavily slumped hemipelagic mudstone wedge that thickens to the west (left). The Blue Unit sands are correlated to a distinct seismic facies. The thickness of the hemipelagic mudstone above the Blue Unit does not change significantly in the north-south direction. Major lithostratigraphic units and Subunit 1d are labelled for correlation with Figure 4. Seismic reproduced with permission of Shell Exploration and Production Company (after Flemings et al. 2006).

3. Results

Boreholes at Sites U1324 and U1322 penetrated a total 608 and 234 m respectively of the Pleistocene and Holocene mud-wedge above the Blue Unit. Drilling at both sites was halted shortly before reaching the top of the Blue Unit, dated at ~89 ka. The predominant lithology that was cored at Ursa Basin is mud and clay and significant amounts of silt and sand occur only at Site U1324 below 360 mbsf. Grain size analysis shows a mean grain size of about 4μm at both sites with the sediment being almost equally composed of silt and clay size particles. Sand content did not exceed 6% (Figure 4). At Sites U1324 and U1322 a slight increase in grain size is observed from 135 and 115 mbsf respectively, coinciding with a period of increase in sedimentation rate from ~24 ka BP.

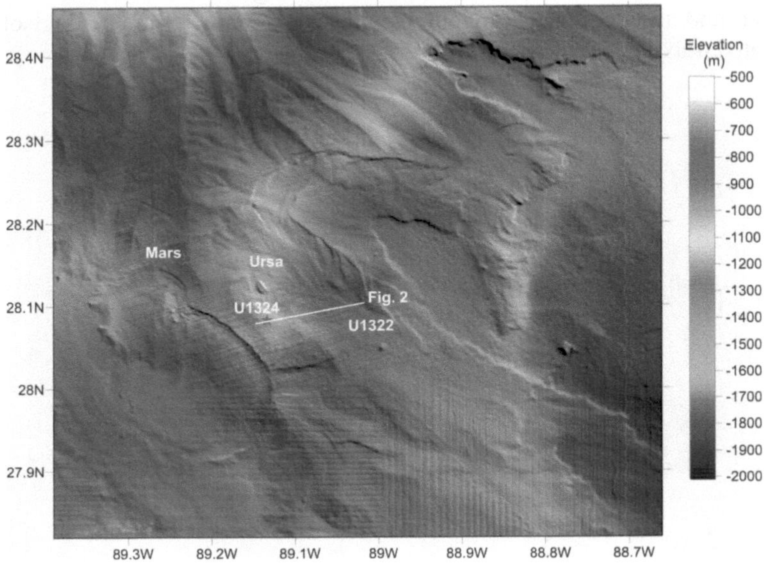

Figure 3. Detailed swath bathymetry shaded reliefof the Mars Ridge (eastern levee of the Mississippi Canyon) showing location of Ursa drill sites U1322, U1324 and seismic line shown in Figure 2. Profuse evidence of mass-wasting processes is evident.

Visual core description and detailed analysis of logging data, such as the Resistivity Imaging at the Bit (RAB), allowed identifying several MTDs that disrupt deposits on the eastern levee of the Mississippi Canyon. These deposits were identified on the basis of folds, faults and tilted beds in the cores and are characterized by transparent acoustic facies in seismic records (Expedition 308 Scientists 2005). The MTDs are only mildly deformed and tilted and thus are interpreted to have not been transported very far from their original position (Expedition 308 Scientists 2005). At Site U1324 (Figure 4) 5 MTDs were drilled in a sedimentary sequence that expands no longer than 65 ka, although some of the MTD may correspond to more than one episode (Dugan et al. 2007). At Site U1322 (Fig. 4) 9 MTDs were identified over the same time interval. Correlation of MTDs between both sites it is only possible for the two uppermost events. The base of the largest and thickest MTD (Subunit 1d; Figures 2 and 4) is at 125 mbsf (Site U1322) and 150 mbsf (site U1324) and can be dated at about 20 ka. The sediments involved in the landslide mass are of ages comprised between ~24 and 20 ka. Below this event, we will assume that all MTDs in Sites U1324 and U1324 are distinct events since reflectors are not laterally continuous from one site to another (Figure 2).

The trends in porosity and water content in the sedimentary section at Ursa Basin show a profile that more or less follows a power law trend, with water content values decreasing rapidly from 160% near the surface to 50% at about 50 mbsf and then decreasing in a more subdued way to the bottom of the respective holes down to values of about 35% (Figure 4). The porosity follows a similar trend with values near the seafloor of 80% which decrease to ~40% at the bottom of both holes.

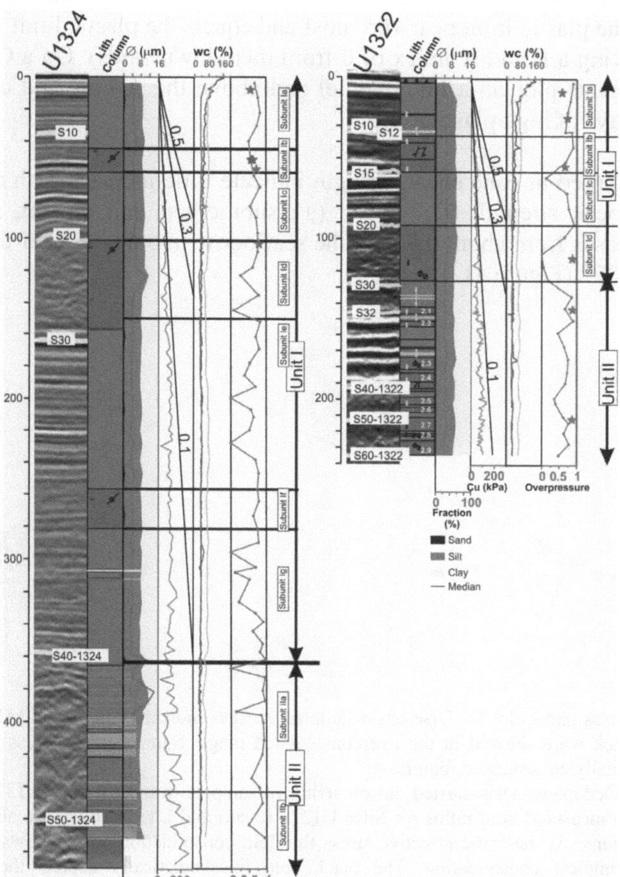

Figure 4. Stripe of seismic profile with major reflectors labeled and lithologic log (MTDs in red, non-failed sediments in green) for drill sites U1324 and U1322 (see also Expedition 308 Scientists 2005). Logs of physical properties correspond from left to right to: 1) Median grain size (red line) and grain size fraction abundance; 2) Undrained shear strength as determined from shipboard motorized vane tests and trends for $C_u{\cdot}(\gamma' z)^{-1}$ ratios of 0.1, 0.3 and 0.5; 3) Plastic limit (red), liquid limit (cyan) and shipboard measured water content (blue) and 4) Overpressure determined from Skempton's (1957) eq. and preconsolidation pressures measured in oedometer tests.

Major excursions from the trends in water content, porosity and bulk density occur at Site U1324, especially below 360 mbsf where most sandy and silt intervals are found (Figure 4). Some of these excursions might result from poor borehole conditions in sand-rich intervals. Consistent shifts in those properties are also found near the base of

MTDs, especially at Site U1322 where the overburden above the Blue Unit is thin. A 5% reduction in water content and porosity is observed near the base of the thickest MTDs compared to the non-deformed sediments immediately below, but reductions in porosity and water content are also observed near the base of other MTDs.

The sediment plastic limit is at about 35%, while the liquid limit, slightly more variable, generally oscillates between 55% and 70% (Figure 4). At both Sites the water content converges on the plastic limit near 105 mbsf and equals the plastic limit to the bottom of the hole (implying a liquidity index of 0 from then downwards). On a Casagrande plot, the Ursa sediments plot on a line parallel and above the A-line, and can be therefore classified as clays of high plasticity.

The trends observed in vane shear strength indicate little increase with depth. The ratio of undrained peak strength (C_u) to $\gamma'z$ (γ': submerged unit weight, z: depth below seafloor) decreases from about 0.3 near the seafloor to about 0.05 near the base of holes U1324 and U1322 (Figure 4).

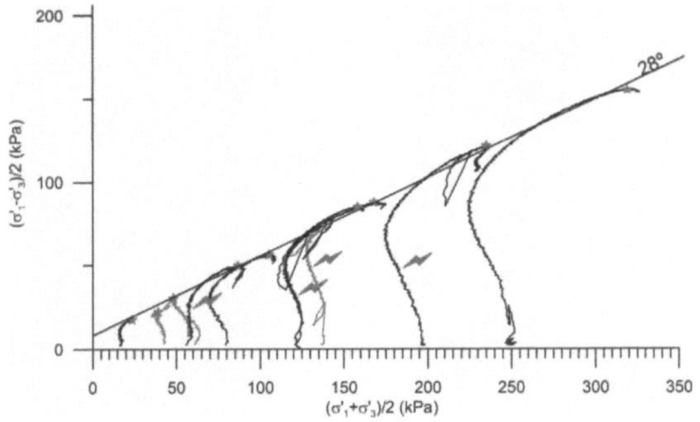

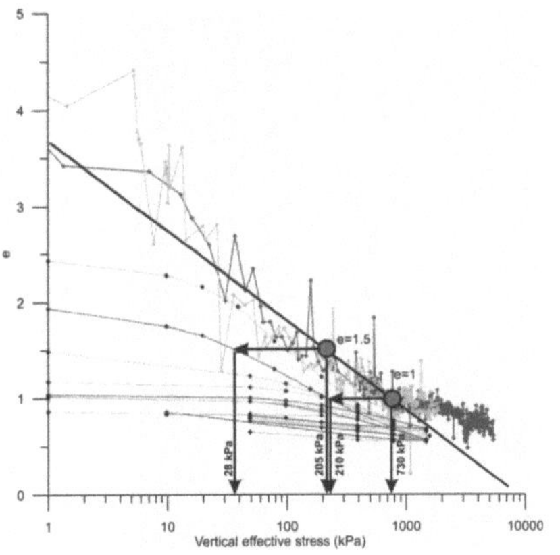

Figure 5. (left): Stress paths plot for Ursa basin samples. Arrows indicate samples on MTDs. Samples with stress paths in black were sheared in the overconsolidated range. Samples with stress paths in red where sheared in the normally consolidated regime.

Figure 6. (right): Oedometer tests carried out on sediment samples from drill sites U1324 (red) and U1322 (orange). The core measured void ratios for Sites U1322 (cyan) and U1324 (blue) are also shown assuming hydrostatic conditions. At the same effective stress the field consolidation curve shows higher void ratios, indicative of incomplete consolidation. The black bold lineshematically depicts the theoretical virgin consolidation assuming full hydrostatic conditions. It can be seen that for a certain void ratio the effective stress is much larger assuming these hydrostatic conditions, than that resulting from the odometer tests.

It is not clear however how well the observed trends represent the *in situ* values of undrained shear strength, since elastic recovery and gas expansion may have affected the measured values (see for instance Sultan et al. 2007). Nevertheless, major increases/deviations from the main trend occur near the base of MTDs, which show higher undrained shear strength with respect to surrounding sediments, coinciding with reductions in porosity.

Triaxial compression tests show similar results both in hemipelagic "normally" deposited sediments and MTDs: a sediment effective friction angle of 28° and cohesion of 7 kPa (Figure 5). The friction angle is consistent with the fracture angles observed in the samples when brittle behavior was observed. Stress-strain relationships show elasto-plastic behavior. MTDs show some strain softening behavior. Stress paths on the triaxial compression tests show that the sediments have mostly a contractive behavior if they are consolidated into the normally consolidated range previous to shearing, and a dilatant behavior if they are sheared in the overconsolidated range (Figure 5).

Oedometer tests indicate that the sediments in Ursa Basin have compression indices (C_c) in the range from 0.2 to 0.9 with the shallowest samples showing the highest values and the deeper samples showing the lowest ones, which probably reflects some degree of disturbance for the deepest samples. As illustrated in Figure 6, the effective pre-consolidation pressure is very low, 2 to 6 times lower than the respective effective stresses assuming fully hydrostatic conditions. Void ratios derived from measurements made along cores and those measured on the odometer at large stresses converge onto a void ratio of 0.6 (a porosity of 37.5%) probably corresponding to the void ratio at the plastic limit.

4. Discussion

4.1. PORE PRESSURE

Pore pressure is one of the major controls on slope instability. In absence of external triggers such as the cyclic loads imposed by earthquakes (the Gulf of Mexico is rather aseismic, Frohlich 1982) and storm waves in shallow waters, pore pressure and variations in slope angle resulting from depositional processes and salt tectonics are considered to be the main controlling factors in slope failure initiation in Ursa Basin. Pore pressures in excess of hydrostatic had been previously documented in relation to shallow water flow sands near the seafloor in the deepwater Gulf of Mexico (Eaton 1999, Pelletier et al. 1999, Ostermeier et al. 2000, 2001) and in the delta sediments (McClelland 1967). At Ursa Basin, pore pressures were estimated from a variety of methods (Figure 4) suggesting that fluid pore pressure in Ursa Basin is in excess of hydrostatic. Oedometer tests (Figures 4 and 6) suggest that the preconsolidation stress is 15-50% of the effective stress at hydrostatic fluid pressure. This equates to a normalized overpressure ($\lambda^*=(U-P_h)\cdot(\sigma_v-P_h)^{-1}$, where U is the pore pressure, P_h is hydrostatic pressure and σ_v is the lithostatic stress) of 0.5-0.85 (Figure 4).

Pore pressure estimates from trends in porosity and using Skempton's (1954) equation ($\sigma'_v=C_u\cdot(0.11+0.0037I_p)^{-1}$where σ'_v is vertical effective stress, C_u is undrained shear strength and I_p is the plasticity index; Figure 4) confirm a high overpressure at both sites. Preliminary interpretation of in situ measurements of the pore pressure with piezometers also point out the existence of pore pressures high above hydrostatic (Flemings et al. 2006). On the other hand, MTDs show significant reductions in porosity and water content (up to 5%; Figure 4) indicating that some consolidation and pore pressure release occurred during the failure process, especially at Site U1322.

Clearly, Sites U1324 and U1322 show similar overpressure (Figure 4) despite the difference in overburden and sedimentation rates during some periods of their sedimentation history. The similar overpressures might suggest that the pore pressure is laterally transferred through more permeable beds such as the Blue Unit from areas of high overburden (Site U1324, reduction in pore pressure) to areas of lower overburden (Site U1322, increase in pore pressure).

4.2. SEDIMENTATION HISTORY VS. SLOPE INSTABILITY

Preliminary biostratigraphic and magnetostratigraphic data suggest that the mud-dominated channel-levee assemblage above the Blue Unit in Ursa Basin deposited very rapidly as suggested by the very high sedimentation rates (Expedition 308 Scientists 2005). However, these sedimentation rates are far from being constant both in space and time (Figure 7). Spatial differences in sedimentation rate have resulted in large differences in mud thickness above the Blue Unit (~2 times difference in Site U1324 with respect to U1322). Sedimentation rates have varied largely through time too, from 1.8 to more than 30 m ky^{-1}. Most MTDs occur in periods of rapid accumulation, especially the thickest MTDs are embedded in periods of rapid accumulation. An anomaly is probably the large amount of MTDs occurring at Site U1322 in the lower part of the cored interval. The high amount of MTDs there suggests that, despite the relatively low sedimentation rates compared to Site U1324, pore pressures were high

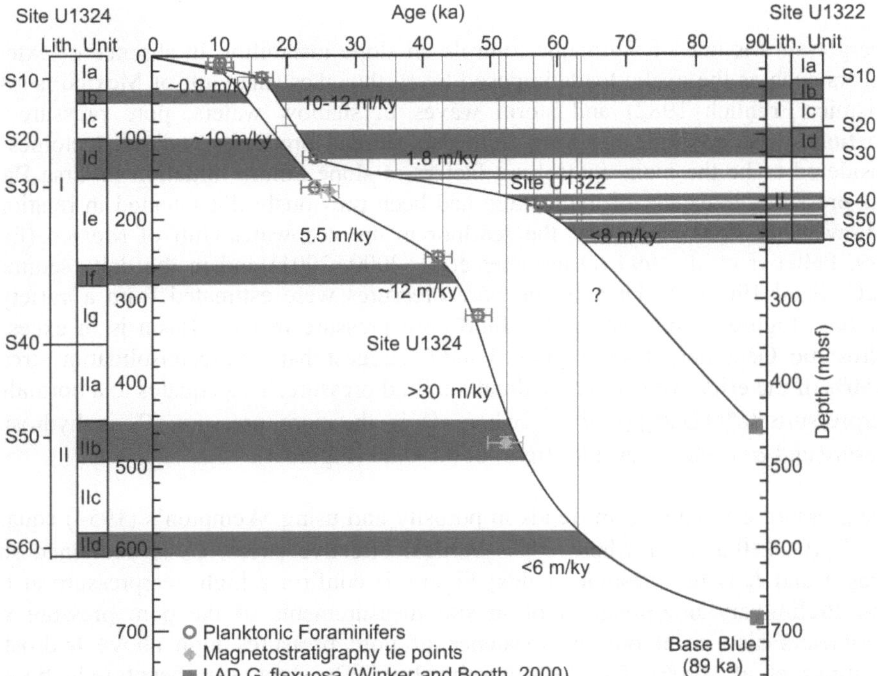

Figure 7. Summary of age constraints of MTDs (red shading) from microfossils and magnetostratigraphy from Sites U1322 and U1324. Age curve from cores is additionally projected to the base of the Blue Unit. The Blue Unit was not penetrated during Expedition 308, but its age is estimated from industry penetration in the area (Winker and Booth, 2000). LAD = last-appearance datum.

enough to trigger instability. This would confirm that lateral transfer of pore fluid pressures could occur from regions of high overburden to regions of lower overburden over a permeable bed.

At Site U1324 five MTDs have been cored, which yields a frequency of 1 MTD every 12 ky, while at Site U1322 eleven MTD have been cored producing a frequency of 1 MTD every 6 ka. A total of 14 MTD have been identified in Ursa basin yielding an average recurrence of 1 MTD per 4.5 ka. Considering only major events those which deposit exceed 10 m in thickness, the Ursa Basin records 6 major failures during the cored interval, which represents a recurrence interval of 1 major MTD every 10 kyrs.

4.3. SLOPE STABILITY

As pointed out earlier major controls in slope stability in Ursa Basin are considered to be the pore pressure and variations in slope angle resulting from depositional processes and salt tectonics. Regional slope gradients in this area of the Gulf of Mexico rarely exceed 2° but local slope may attain values of 15° or even higher (Figure 3). Normalized overpressure (λ^*) is of the order of 0.7 but we will consider that this value might differ from the time the failures occurred. We will consider drained and undrained failure through the cohesion and friction angle, and the ratio of $C_u{\cdot}(\gamma' z)^{-1}$.

For the MTDs observed in Ursa Basin the following observations can be made: the slope is relatively uniform, the failure planes are subparallel to the seafloor and the length of the failure surfaces is large compared to the failure thickness. Therefore, it is considered that the infinite slope approach provides a first approximation to the slope stability. In the worst case scenario, if undrained conditions are considered, very low $C_u{\cdot}(\gamma' z)^{-1}$ ratios of less than 0.05 are needed to trigger slope failure (Figure 9). The range

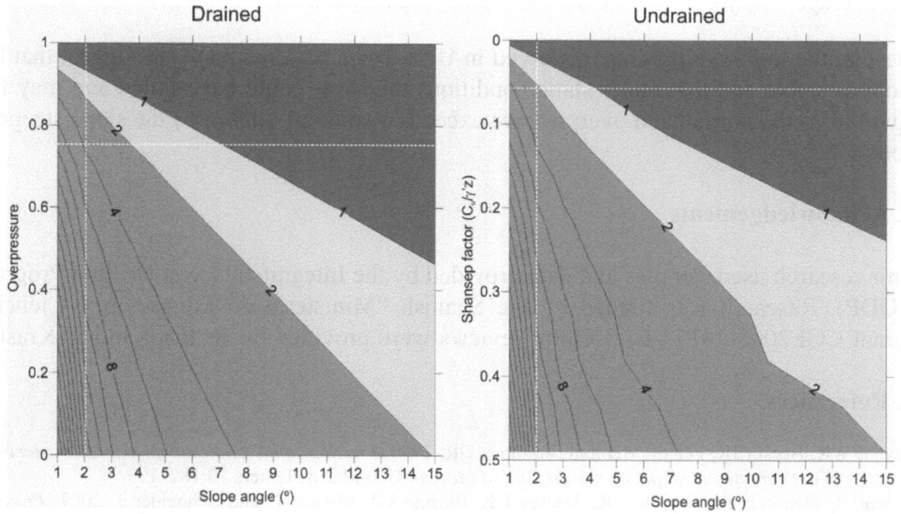

Figure 8. Infinite slope stability analysis using drained (left) and undrained (right) parameters for Ursa Basin. Contours inside plots represent the Factor of Safety. Dashed white lines represent Present day mean values in Ursa Basin.

of observed values in Ursa Basin is 0.3-0.05, but the lower values probably result from core disturbance. Considering drained failure it is observed that on a regional basis the slope is stable under present day conditions. A relatively high overpressure of ~0.9 is needed to trigger past and present day failure of the slope. This might indicate that higher overpressures might have existed during periods of rapid accumulation, and that these overpressures have partially dissipated at Present. Reduced sedimentation rates in the Holocene have resulted in lower pressure generation and this facilitates dissipation of previously generated pressure. It must be noted that local slopes that exceed 7° are not stable following the approach used in this study (Figure 8). However for locally steep slopes 2-D or 3-D limit equilibrium analysis, depending on slope geometry, is a minimum requirement.

5. Conclusions

Sediments in Ursa Basin are largely overpressured, implying effective stresses that are 25% of the effective stress that would be present if hydrostatic conditions existed.

MTDs appear to have lower porosity compared to non-failed sediments immediately above and below, and therefore appear to have experienced significant consolidation during the failure process.

The similar overpressures at sites U1324 and U1322, despite their differences in sedimentation rate suggest that pore pressure might be transferred through the Blue Unit from places of high overburden to places of lower overburden.

MTDs embedded in periods of high sedimentation rate suggest that high accumulation rates are the origin of the high pore pressures and instability of the margin.

On average 1 major MTD occurs every 10 kyrs in Ursa Basin.

Despite the high overpressure observed in Ursa, basic regional analysis suggest that the slope is stable and that under static conditions the slope could have failed and may fail again if a) the normalized overpressure exceeds a value of ~0.9 or b) the slope steepens above 7°.

6. Acknowledgements

This research used samples and data provided by the Integrated Ocean Drilling Program (IODP). Research was funded by the Spanish "Ministerio de Educación y Ciencia" (grant CGL2005-24154-E). Helpful reviews were provided by A. Kopf and S. Krastel.

7. References

Bryant WR, Bryant JR, Feeley MH and Simmons GR, 1990. *Physiographic and bathymetric characteristics of the continental slope, northwest Gulf of Mexico.* Geo-Marine Letters 10:182–199.

Dugan B, Flemings PB, Urgeles, R, Sawyer DE, Iturrino GJ, Moore JC and Schneider J, 2007. *Physical Properties of Mass Transport Complexes in the Ursa Region, Northern Gulf of Mexico, IODP Expedition 308, Determined From Log, Core, and Seismic Data.* Proceedings of the 2007 Offshore Technology Conference, paper OTC 18704, 8pp.

Eaton LF, 1999. *Drilling through deepwater shallow water flow zones at Ursa. Proc*—SPE/IADC Middle East Drill Conf, 153–164.

Expedition 308 Scientists, 2005. *Gulf of Mexico hydrogeology—overpressure and fluid flow processes in the deepwater Gulf of Mexico: slope stability, seeps, and shallow-water flow.* IODP Prel Rept, 308. doi:10.2204/iodp.pr.308.2005.

Feng T-W, 2001. *A linear log d–log w model for the determination of consistency limits of soils.* Canandian Geotechcal Journal 38:1335–1342.

Flemings PB, Behrmann JH, John CM and Expedition 308 Scientists, 2006. Proc IODP, 308: College Station TX (IODP Management International, Inc). doi:10.2204/iodp.proc.308.2006.

Frohlich C, 1982. *Seismicity of the central Gulf of Mexico.* Geology 10:103–106.

Mahaffie MJ, 1994. *Reservoir classification for turbidite intervals at the Mars discovery, Mississippi Canyon Block 807, Gulf of Mexico.* In Bouma AH, Perkins BG (Eds), Submarine Fans and Turbidite Systems: Proc GCSSEPM 15th Annu Res Conf, 233–244.

McAdoo BG, Pratson LF and Orange DL, 2000. *Submarine landslide geomorphology, US continental slope.* Marine Geology 169:103–136.

McClelland, B, 1967. *Progress of consolidaiotn in delta and pro-delta clays of the Mississippi River.* In: Marine Geotechnique, A.F. Richards, Ed., U.I.P., pp. 22-40.

Ostermeier RM, Pelletier JH, Winker CD, Nicholson JW, Rambow FH and Cowan KM, 2000. *Dealing with shallow-water flow in the deepwater Gulf of Mexico.* Proc Offshore Tech Conf, 32(1):75–86.

Ostermeier RM, Pelletier JH, Winker CD and Nicholson JW, 2001. *Trends in shallow sediment pore pressures.* Proc—SPE/IADC Drill Conf, 1–11.

Pelletier JH, Ostermeier RM, Winker CD, Nicholson JW and Rambow FH, 1999. *Shallow water flow sands in the deepwater Gulf of Mexico: some recent Shell experience* (International Forum on Shallow Water Flows Conference, League City, TX, 6–9 Oct, 1999).

Pratson LF and Ryan WBF, 1994. *Pliocene to Recent infilling and subsidence of intraslope basins offshore Louisiana.* AAPG Bulletin 78:1483–1506.

Pulham AJ, 1993. *Variations in slope deposition, Pliocene–Pleistocene, offshore Louisiana, northeast Gulf of Mexico.* In Posamentier H, Weimer P (Eds.), Siliclastic Sequence Stratigraphy: Recent Developments and Applications. AAPG Mem, 58:199–233.

Sawyer DE, Flemings PB, Shipp C and Winker C, 2007. *Seismic Geomorphology, Lithology, and Evolution of the Late Pleistocene Mars-Ursa Turbidite Region, Mississippi Canyon Area, Northern Gulf of Mexico.* AAPG Bulletin, 91:215-234.

Skempton AW. 1957. *Discussion: the planning and design of the new Hong Kong airport.* Proceedings Institute of Civil Engineers, vol. 7, pp 305–307.

Sultan N, Gaudin M, Berne S, Canals M, Urgeles R and Lafuerza S, 2007. *Analysis of slope failures in submarine canyon heads: An example from the Gulf of Lions,* Journal of Geophysical Research, 112: in press, doi:10.1029/2005JF000408.

Winker CD and Booth JR, 2000. *Sedimentary dynamics of the salt-dominated continental slope, Gulf of Mexico: integration of observations from the seafloor, near-surface, and deep subsurface.* Deep-Water Reservoirs of the World: Proc GCSSEPM 20th Annu Res Conf, 1059–1086.

Worrall DM and Snelson S, 1989. *Evolution of the northern Gulf of Mexico, with emphasis on Cenozoic growth faulting and the role of salt.* In Bally AW, Paler AR (Eds), The Geology of North America—An Overview (Vol. A): Boulder, CO (Geol Soc Am), 97–137.

GEOTECHNICAL CONSIDERATIONS OF SUBMARINE CANYON FORMATION: THE CASE OF CAP DE CREUS CANYON

M. SANSOUCY[1], J. LOCAT[1] AND H. LEE[2]
[1]*Laboratoire d'études sur les risques naturels, Department of Geology and Geological Engineering, Laval University, Québec, Qc, Canada, G1Y 2P1*
[2]*United States Geological Survey, Menlo Park, California, U.S.A.*

Abstract

A portion of the Cap de Creus canyon, situated in the Gulf of Lion, has been selected for a detailed analysis of slope instability. This sector has been chosen because of morphological evidence for slide. Three piston cores have been taken at different water depths along an axis perpendicular to the thalweg and a box core has also been taken in the thalweg. At the top of the flank, the geotechnical signature suggests that clay sedimentation has been continuous. In contrast, overconsolidated silty clay has been observed in the core taken on the flank, about 60 m below a headwall escarpment, on a failure plane. The geotechnical profile of the core taken at the toe of the flank suggests that rapid depositional events, such as debris flows and turbidity currents, occur frequently. A series of triaxial tests have been performed, and provided input parameters for analyzing the initial stability of the flank with *Slope/W*. The impacts of several natural processes on failure development have been tested. Drained failures initiated by axial incisions seem to correspond to the main active process of the canyon, for small shallow failures. Large failure can occur under undrained conditions during earthquakes.

1. Introduction

The Gulf of Lions margin is incised by numerous submarine canyons, which are known to be important preferential paths for downslope sediment transport, mostly through gravity flow mechanisms (Got and Stanley 1974; Berné and Loubrieu 1999; Locat and Lee 2002). Two major hypotheses are generally accepted for submarine canyon formation: (1) erosive processes, provided by hyperpycnal or other density currents at low sea-level, when rivers mouth are situated near the shelf break (Mulder et al. 2003), or (2) retrogressive slides, initiated on the slope (Pratson and Coakley 1996). These two processes have played a role in the development of the Gulf of Lions (Bellaiche et al. 1989; Berné et al. 1999).

The Rhône deep sea fan is a depositional system situated in the deep northwestern Mediterranean Sea, fed in part by the Cap de Creus canyon (Reis et al. 2004), which is the westernmost canyon of the Gulf of Lions (Figure 1). The axial incision observed in this canyon has been interpreted as the imprint of erosive currents during glacial maxima, and suggests recurrent activity (Rabineau et al. 2005). This small erosive path within the canyon's major valley has a major influence on canyon evolution as it may trigger mass wasting on flanks and within the head (Baztan et al. 2005). The Cap de Creus canyon's orientation and position reflects the Catalane fault (Figure 1) (Alonso et al. 1991; Reis et al. 2004), which is linked to Pyrenean formation and is presently tectonically active (Got and Stanley 1974).

V. Lykousis, D. Sakellariou and J. Locat (eds.), Submarine Mass Movements and Their Consequences, 221–230.

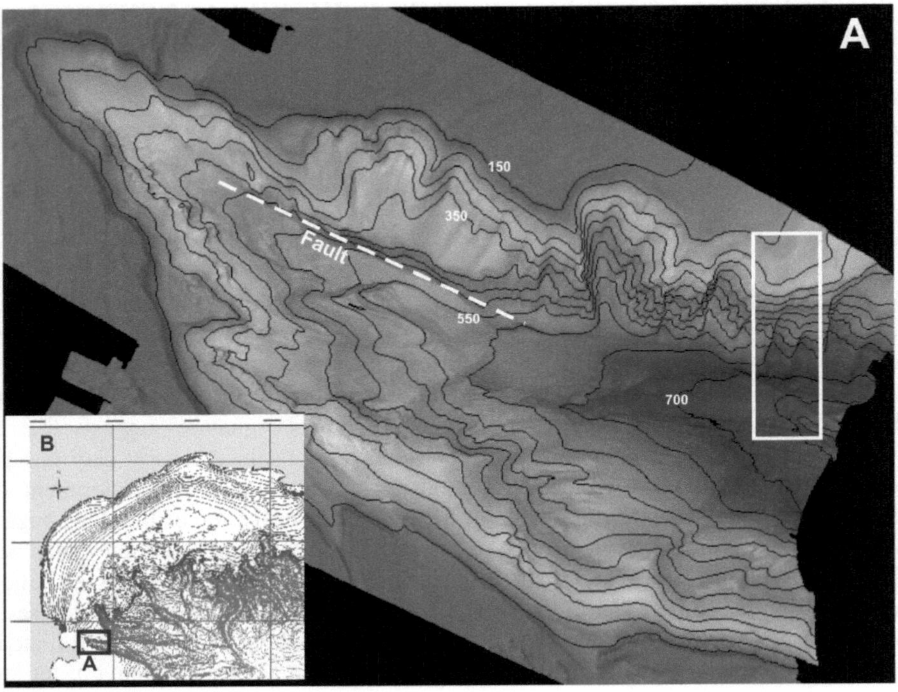

Figure 1. a) Enlargement of the upper Cap de Creus bathymetry and location of the study area (white box). b) Bathymetric map of the Gulf of Lions and location of the Cap de Creus canyon.

In order to evaluate active processes that control submarine canyons evolution, a portion of the Cap de Creus canyon has been selected for a detailed analysis of slope stability. The selected area is located on the north side of the canyon in water depths ranging from 150 m to 750 m, as shown by the white box in Figure 1b. This sector was chosen because of evidence for a slide, revealed by the presence of a headwall escarpment 30 m high and 450 m wide, with a slope of 27° and a typical bowl shape of 500m width, similar to what has been observed in the Bourcart canyon (Sultan et al. 2007). The mean slope of this part of the canyon is 20°. Three piston and box cores have been taken at different water depths along an axis perpendicular to the thalweg, providing samples from above the headwall escarpment, from the failure plane and from the debris accumulation zone (Figure 2). A box core (CTL-680) has also been taken in the thalweg. These cores have been analyzed in order to determine the nature and microstructure of the sediments along with their geotechnical and rheological properties (Sansoucy 2007). Geotechnical properties have been used for simulations of flank failures under various conditions.

2. Methodology

In October 2004, a cruise aboard the RV Oceanus surveyed the Cap de Creus canyon and collected several box cores and piston cores, whose length ranged from 3m to 5m. Multi-Sensor-Core-Logger (MSCL) profiles have been obtained on board. A few months later, the cores have been scanned by computed axial tomography (Cat-scan) at

INRS scanography laboratory in Québec City. These profiles provided a non-destructive visualization of internal structural features of the samples.

Multiple laboratory testing has been done since May 2005 on the four sampling sites located in the sector of interest (Figure 2). Since this paper focuses on slope stability, only the most relevant tests procedures are reported below. The cores, except for the last 50cm of piston cores, were first extruded vertically to measure water content and intact undrained shear strength (Cu) at every 2,5cm, with the Swedish Fall cone. Remolded undrained shear strength (Cu_r) measurements have also been carried out for each 5cm in the sediment column. A series of consolidation and triaxial tests have been carried out on intact sediments from the bottom (last 50cm) of each piston core.

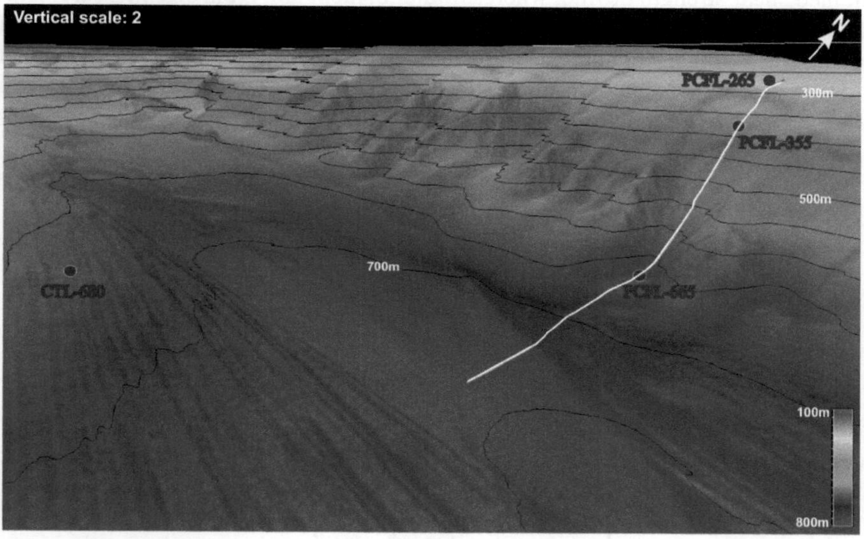

Figure 2. Bathymetric image of the study area showing the position of the cores. Numbers associated with core positions give sampling water depth. Isobaths intervals are 50m. The sector shown covers a width of 5km.

3. Sediment strength properties

Water content and undrained shear strength have been compiled in Figure 3 in order to visualize strength evolution with depth. From consolidation tests, the ratio of undrained shear strength to in situ effective overburding stress is estimated to 0.3 for normally consolidated sediments. The dash line shown on strength profiles represents this theoretical ratio ($c_u/\sigma'_{vo} = 0.3$). Results from consolidation tests are presented in Table 1.

Table 1 Consolidation tests results: Preconsolidation pressure (σ'_p), Overconsolidation ratio (OCR), Compression index (C_c), Recompression index (C_r), Theoretical height of erosion (H).

Core	Depth (cm)	Unit Weight (kN/m³)	σ'_p (kPa)	OCR	C_c	C_r	H (m)
PCFL-265	458	18.5	44	1,1	0,273	0,037	
PCFL-355	298	19	145	5,5	0,241	0,036	15-20
PCFL-665	471	17,23	38	1,1	0,455	0,048	
CTL-680	19	19,92	94	50,4	0,239	0,035	10-15

Piston core PCFL-265 has been taken near the top of the flank, above the escarpment (Figure 2). Many calcareous tubes and intact Pecten shells have been found throughout the length of this core. Apart from the first 70cm, grey clay has been observed. In contrast, the upper sediment layer consists of clayey sand with some sandy lenses. Strength and water content profiles of this core are shown in Figure 3a. Except for the first 150cm, intact shear strength approximately follows the relation for normally consolidated sediments, thereby suggesting that this sector experienced continuous sedimentation. This is confirmed by an OCR value of 1.1 for the sample taken at a depth of 458cm (Table 1).

Profiles of piston core PCFL-355, sampled about 60m below the escarpment, on the failure plane (Figure 2), are shown in Figure 3b. Two important peaks can be seen in the undrained shear strength profile, whereas water content and grain size measurements are quite constant. It may be due to some compaction at the top of core sections during transport and storage. Though, shear strength profile shows that below 20cm, the entire column of silty clay is overconsolidated. This is supported by an OCR of 5.5 measured on core sampled at a depth of 298cm (Table 1). These values suggest that a thickness of between 15m and 20m of sediments would have needed to be removed to provide this strength. However, the sediments from the first 20cm are quasi-normally consolidated, suggesting that this layer resulted from recent deposition since the failure event.

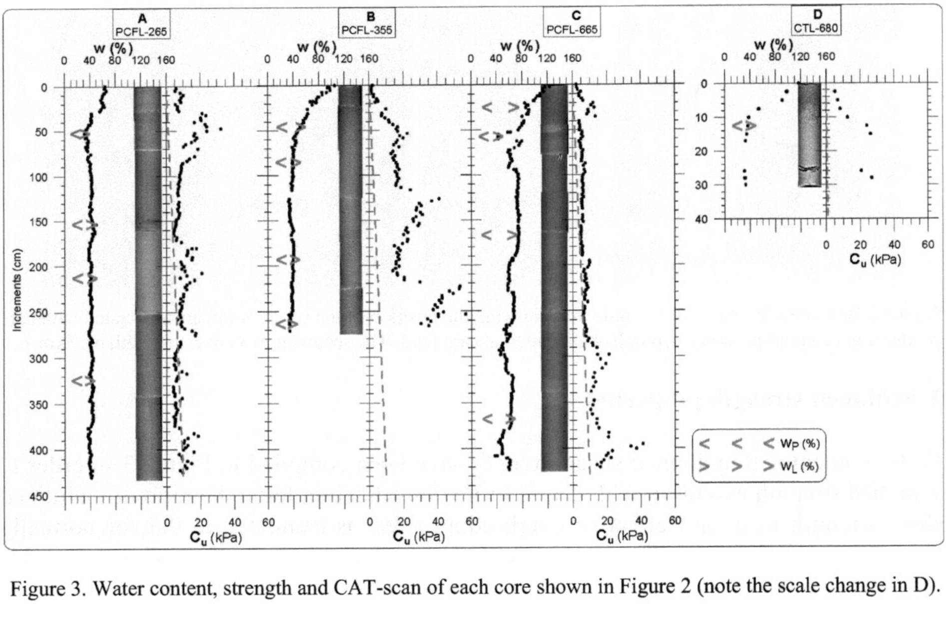

Figure 3. Water content, strength and CAT-scan of each core shown in Figure 2 (note the scale change in D).

Piston core PCFL-665 has been taken at the toe of the flank, on a slope of about 4°. The principal facies observed in this core consist of grey silty clay observed throughout the core, which contains a strong proportion of shells fragments. A second facies, consisting of bluish clay clasts, has been observed suspended within the grey matrix with an inverse grading, in the level from 300cm to 50cm depths. At levels in the sample where the two facies have been observed, water content values on the matrix are always higher than the one measured on clasts. At depths below 300cm in the core, the only facies observed corresponds to grey silty clay, having a small degree of overconsolidation. An OCR of 1.1 has been measured from a consolidation test on sediments from a depth of 471cm in this core and confirms the nearly normally consolidated state of the sediment, suggesting that this site corresponds to a zone of remolded debris accumulation.

In the thalweg, a box core of 30cm has been taken (Figure 3d). At a depth of 10cm in this core, an abrupt break has been observed, caused by a highly overconsolidated clay, overlain by very soft clay. The density contrast is also visible on the Cat-scan image. A consolidation test on sediments from a depth of 19cm yielded an OCR value of 50, which is really high, but consistent with the undrained shear resistance measured with the Swedish Fall cone. This indicates that a minimum of 10m of sediments have been eroded in this sector. However, the strength and density contrast could be interpreted as the result of a debris bloc, but it implies that it diameter would be more than 20cm.

Two triaxial tests have been carried out on sediment from each piston core, from which stress paths provided the failure envelope of Cap de Creus sediments. The critical state line can be described by an effective friction angle of 29.5° and a small amount of cohesion (less than 8kPa).

4. Failure Simulations

Geotechnical properties have been used for modeling of failures along the canyon flank with *Slope/W* software, by mean of Generalized Limit Equilibrium (GLE) method. A morpho-stratigraphic model, based on the profile that intersects the 3 piston cores (Figure 2), has been created in order to simulate initial failures and evaluate the influence of various possible scenarios on instability development. Initial failures, under undrained and drained conditions have first been analyzed. The position of the failure surfaces has been varied so that they exit at a specific point at the base of the slope and they enter through a wide range of entry points, in order to delimit the most critical slide plane, following an arbitrary shape. Hydrostatic conditions have been considered for each simulation.

4.1 INITIAL STABILITY

Because sediments from the sector of interest are characterized by a mean clay fraction around 40%, hydraulic conductivity must be low enough to allow for stability analysis under undrained conditions. In this model, undrained shear strength increases linearly from the surface, by 2.5kPa/m, which represents normal consolidation. However, surface undrained shear strength has been fixed to 30kPa for sediments on the flank, according to the overconsolidation measured in that area. The results for the actual morphology have shown that undrained failures, which give a minimal factor of safety of around 1.31, are unlikely given the geomorphic evidence. In fact, modeling the undrained condition produces deep failure planes, involving sediment layer to a depth of 200m (Figure 4a) and these are not observed in the failure on the actual canyon flank.

A Mohr-Coulomb strength model, with an effective friction angle of 29.5°, as derived from triaxial tests, has been used to represent drained conditions. The effect on stability of an apparent cohesion between 0 and 10 kPa has been tested, but since its influence was found to be negligible, this parameter has been fixed to 0 for simulations. Modeling these drained conditions leads to the formation of a superficial slide with a minimal factor of safety of 1.63, (Figure 4b). This geometry seems to better match canyon morphology. Moreover, the thickness of the sediments affected by this instability is approximately equal to 20m in the sector where core PCFL-355 has been sampled. This removal thickness is in agreement with what has been estimated from the preconsolidation pressure measured on these sediments (Table 1). However, initial condition analysis has shown that some destabilizing processes are necessary in order to trigger this slide.

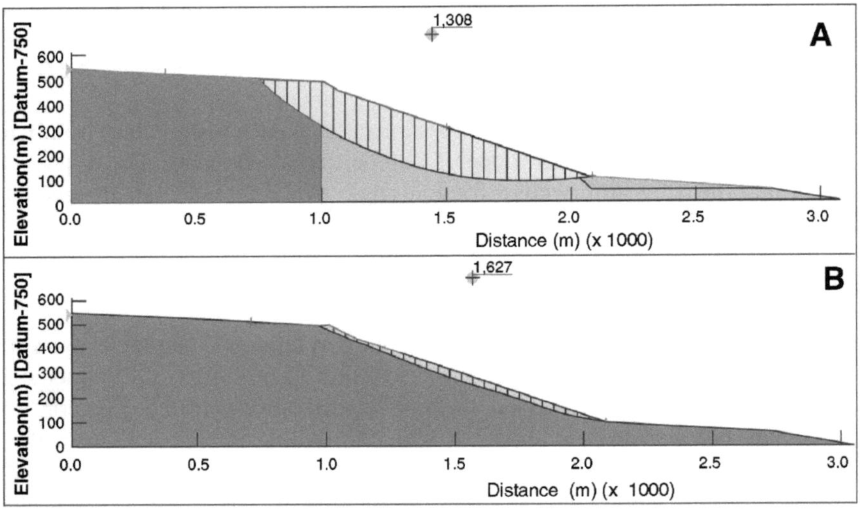

Figure 4. Modeled initial failure surface under a) Undrained conditions b) Drained conditions.

4.2 EROSION SIMULATIONS

Results of various scenarios of failure development have also been tested. The first natural process simulated corresponds to axial incision, as already proposed for the destabilization process (Baztan et al. 2005). Gradual modifications of the flank geometry have been modeled to visualize the effect of axial incision on stability. Erosion has been simulated with different cutback distances at the bottom of the slope. After testing several angles for the eroded escarpment, an angle of 35° has been found to represent the most probable case, because of the larger failed sediments volume that it can involve. The slope of the debris accumulation zone has been kept constant at 4°. Because the erosion process requires a long period to be significant, the analyses have been done assuming drained conditions. For each axial incision level simulated, the failure surface corresponding to the stability limit has been characterized by it length (L) and height (H). Figure 5a shows an example of the geometry for the case of an erosion length equal to 250m. The compilation of resulting slip surface geometries in relation to the erosion length is shown in Figure 5b, and illustrates the strong linear effect that flank toe erosion has on the volume implicated in the failure. Thus, axial incision can have a significant control on geomorphology development.

4.3 SEDIMENTS LOADING SIMULATIONS

The effect of sediment overloading on stability has been evaluated, by simulating sand accumulation at the top of the flank, as this process is suggested by the sandy layer observed for the first 70cm in the core sampled at the top of the flank. The thickness of the sandy layer has been increased progressively and its effect on stability has been analyzed quantitatively, by looking at the reduction ratio of the factor of safety. The long term duration of the sedimentation process has been simulated by assuming drained conditions. Results have shown that for an accumulation of 200m thick layer of sand, the factor of safety is only reduced by 14%. In fact, the loading due to sedimentation only affects a small portion of the suggested failure surface. Thus, accumulation of sand has a negligible influence on stability and cannot explain the observed canyon geomorphology.

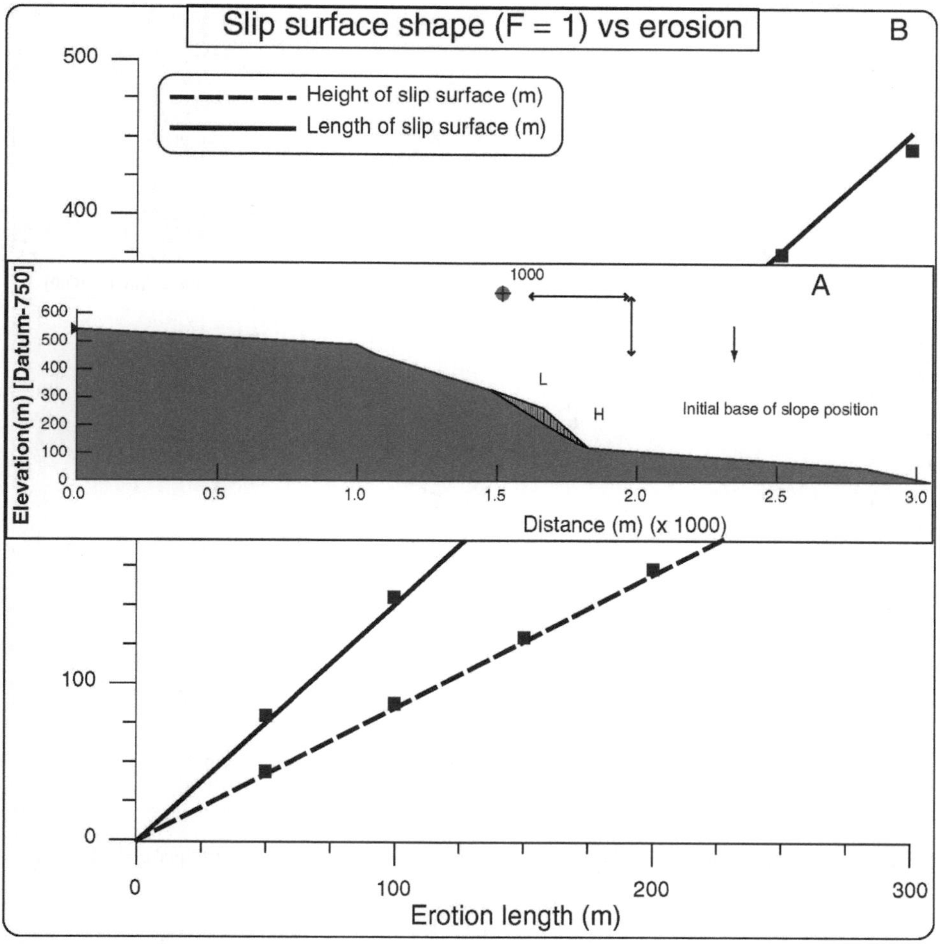

Figure 5: a) Example of erosion simulation, for the case of an erosion length equal 250m. b) Influence of erosion intensity on failure surfaces shape.

4.4 SEISMIC SHAKING SIMULATIONS

Instabilities under seismic loadings have been simulated, using the "pseudo-static" method. Various horizontal accelerations, provided by earthquakes, have been applied to the sediment as an external load, expressed in terms of a portion of the gravitational acceleration (g). These simulations have been done assuming undrained conditions, recognizing the rapid, dynamic loading. Results are presented in Figure 6, which shows that an earthquake with an acceleration corresponding to 0.05g is sufficient to destabilize an important volume of sediment. However, during the last 100 years, the maximum horizontal acceleration measured for the Gulf of Lions corresponds to only 0.02g (Sultan et al. 2007). Thus, for the last 100 years, earthquakes are insufficient for triggering instabilities. The presence of a fault along the canyon, (Figure 1) (Alonso et al. 1991; Reis et al. 2004) would support the argument that over a longer time scale, larger earthquakes may have occurred. Results suggest that these events, when they occur, may trigger large slides.

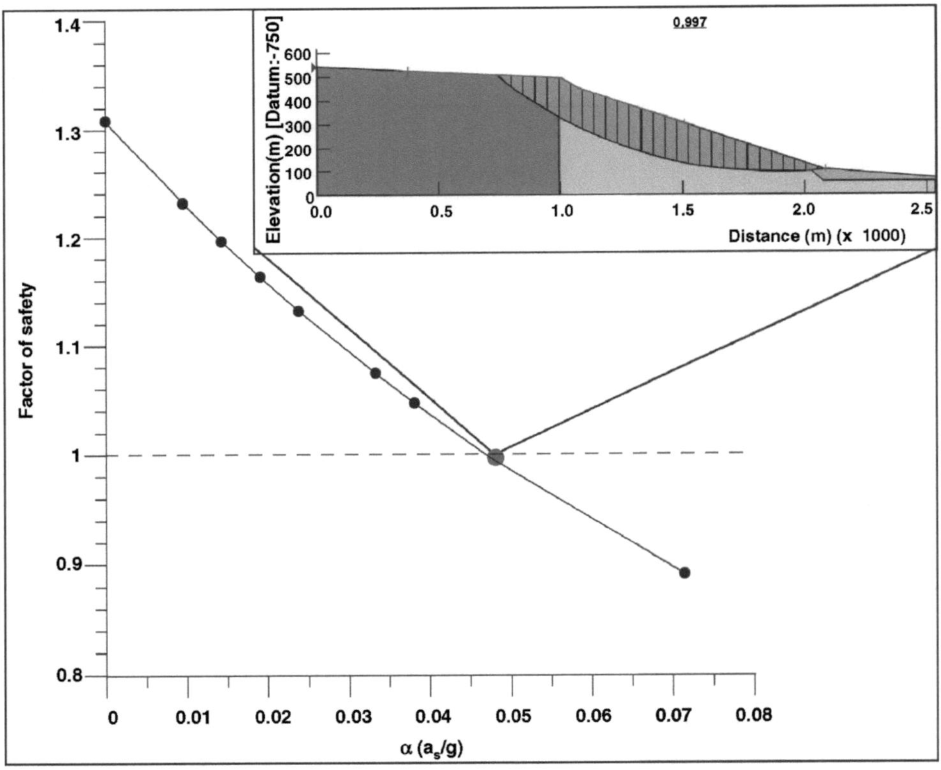

Figure 6: Influence of seismic loading on stability, where a_s is the pseudo-static earthquake acceleration.

5. Discussion

5.1 FAILURE MECHANISMS

Overconsolidation pressures measured for sediments from the thalweg (CTL-680) and the failure plane on the flank (PCFL-355) indicate that a thickness of only 10m and 20m respectively have been removed by erosive process.

A single drained failure event, initiated by erosion at the base of the flank, is sufficient to destabilize an equivalent sediment thickness. In fact, at the PCFL-355 sampling site, slides seem more likely to involve sediments through a depth of 20m. Also, an erosion length of 150m can lead to the removal of a layer of 10m thick sediments in the thalweg. Thus, in the Cap de Creus canyon, recent instabilities seem to affect only post-glacial sediments layers. Infrequent earthquakes could have provided the necessary ground acceleration to lead to voluminous slides (thickness up to 200m). However, earthquakes do not seem to be the most important recent active process, because of their long return period. Also sediment overloading cannot significantly influence flank stability.

5.2 CANYON FORMATION

Analysis of Cap de Creus canyon evolution is limited by the relatively short core length (less than 5m) available. Thus, only recent processes can be evaluated. Geotechnical data suggest that since the Quaternary, silty clay has been deposited as a draped layer on an inherited structure. Stability simulations showed that some modern shallow slides, possibly triggered by axial incision, have affected some portion of the canyon flank, but involved recently deposited sediments only. These small scale failures cannot have a strong influence on canyon enlargement, compared to the 7km width observed for Cap de Creus canyon opposite to the study sector. Therefore, since sea-level rose to its present level, Cap de Creus canyon morphology has probably remained relatively constant. Formation of the initial canyon entrenchment was likely initiated during the Messinian (6.7Ma to 5.2Ma) salinity crisis (Berné and Loubrieu 1998; Lofi et al. 2005) in order to provide the resulting morphology. However, since the Quaternary, significant enlargement of the canyon likely only results from rare larger failures, related to earthquakes.

6. Conclusion

Geotechnical properties of sediments from the study area of the Cap de Creus canyon suggest that at the top of the flank, clay follows a continuous sedimentation. At the base of the slope, debris accumulated at a normally consolidation state, which is interpreted as the result of remolding during failure events. However, the sector of the failure plane situated on the flank consists of overconsolidated clay, probably due to the removal of 15m to 20m layer of sediments since its post-glacial deposition. This erosion intensity seems low with respect to the canyon scale, thus confirming the long time period required to generate a depression such as the Cap of Creus canyon. These morphologic and laboratory observations have been used in order to simulate initial flank stability and to evaluate the influence of different natural processes on instabilities. Drained failures initiated by axial incision provide superficial slides, which seem to correspond to the main recent active process in the canyon evolution. However, larger slides could have been triggered occasionally by strong seismic shaking.

7. Acknowledgements

We thank scientists and crew members aboard the R.V. Oceanus; G. Cauchon-Voyer, C. Juneau, R. Mompin, J. Boisvert, B. Carkin and K. Orzech for their help during laboratory work; and the U. S. Office of Naval Research for financial support.

8. References

Baztan J. et al. (2005) *Axial incision: The key to understand submarine canyon evolution (in the western Gulf of Lions).* Marine and Petroleum Geology 22: 805-826.

Bellaiche G. (1993) *Sedimentary mechanisms and underlying tectonic structures of the northwestern Mediterranean margin, as revealed by comprehensive bathymetric and seismic surveys.* Marine Geology 112: 89-108

Bellaiche G. et al. (1989) *Detailed morphology, sedimentary structure and evolution of the continental margin of the western provençal basin since the late Miocene.* Marine Geology 89: 259-268.

Berné S. and Loubrieu B. (1999) *Canyons et processus sédimentaires récents sur la marge occidentale du golfe du Lion.* Comptes rendus de l'Académie de Sciences 328: 471-477.

Gorini C. et al. (1993) *Contribution to the structural and sedimentary history of the Gulf of Lions (Western Mediterranean), from the ecors profiles, industrial seismic profiles.* Bulletin de la société géologique de France 164(3): 353-363.

Got H. and Stanley D.J. (1974) *Sedimentation in two Catalonian canyons, northwestern Mediterranean.* Marine Geology 16: 91-100.

Locat J. and Lee H. (2002) *Submarine landslides: advances and challenges.* Canadian Geotechnical Journal 39: 193-212.

Locat J.et al. (2002) *Natural soils:geotechnical behavior and geological knowledge.* International Workshop on Characterization and Engineering, Properties of Naturals Soils, Singapore.

Lofi J. et al (2005) *Erosionnal processes and paleo-environmental changes in the Western Gulf of Lions during the Messinian Salinity Crisis.* Marine Geology 217: 1-30.

Mulder T. et al. (2003) *Marine hyperpycnal flows: initiation, behavior and related deposits.* A review. Marine and Petroleum Geology 20: 861-882.

Pratson L.F. and Coakley B.J. (1996) *A model for the headward erosion of submarine canyons induced by downslope-eroding sediment flows.* GSA Bulletin 108: 225-234.

Rabineau M. et al. (2005) *Sedimentary sequences in the Gulf of Lion: A record of 100 000 years climatic cycles.* Marine and Petroleum Geology 22: 775-804.

Reis A.T. et al. (2004) *Stratigraphic architecture of the Pyreneo-Languedocian submarine fan, Gulf of Lions, western Mediterranean Sea.* C. R. Geoscience 336:125-133.

Sansoucy M. (2007) *Geotechnical analysis of submarine canyon evolution: the case of Cap de Creus canyon.* in preparation

Sultan N. et al. (2007) *Analysis of slope failures in submarine canyon heads: An example from the Gulf of Lions.* Journal of geophysical research 112: doi :10.1029/2005JF000408.

Section 5 - Submarine slides in coastal areas, semienclosed seas (fjords, estuaries, gulfs) and lakes

SUBMARINE MASS MOVEMENTS IN THE BETSIAMITES AREA, LOWER ST. LAWRENCE ESTUARY, QUÉBEC, CANADA

GENEVIEVE CAUCHON-VOYER[1], JACQUES LOCAT[1] AND GUILLAUME ST-ONGE[2]
[1]*Department of Geology and Geological Engineering, Université Laval, Québec, Québec, Canada, G1K 7P4*
[2]*ISMER, Université du Québec à Rimouski, Rimouski, Qc, Canada, G5L 3A1*

Abstract

A complex submarine geomorphology was revealed from multibeam bathymetry and seismic reflection surveys conducted between 2001 and 2006 in the Lower St. Lawrence Estuary offshore Betsiamites River, Québec, Canada. In this paper, we describe the submarine morpho-sedimentology of an area of ~500 km^2 with focus on the consequences of three mass movement events. A chronology suggesting the ages for the failures is established. A major landslide scar is characterized by two large channels on the shelf and a sediments fan in the Laurentian Channel. This landslide is dated around 7.25 kyr cal BP. Morphological observations and sediment core analyses allow us to identify a least two different recent (*i.e.,* less than 1 kyr BP) debris flow accumulations on the shelf and in the Laurentian Channel. Two different ^{210}Pb-dated debris flow deposits were identified and associated to two recent earthquake episodes: (1) the AD 1663 (M~7) earthquake and (2) AD 1860 (M~6) or AD 1870 (M~6.5) earthquakes. The 1663 debris flow deposit is associated with a subaerial landslide observed on shore.

1. Introduction

Investigating submarine mass movements in order to evaluate slope stability for a region is required when carrying out risk assessment related to natural hazards. With the development of coastal and offshore activities there is an essential need to improve our understanding of the factors maintaining slope stability and those triggering mass movements. Submarine mass movements are widespread geomorphological processes found in many different oceanographic settings (Canals *et al.*, 2004; Locat and Lee, 2002). In Québec, comprehensive analyses of submarine mass movements have mostly been carried out in the Saguenay Fjord (Levesque *et al.*, 2006; St-Onge *et al.*, 2004; Locat *et al.*, 2003; Urgeles *et al.*, 2002). In the St. Lawrence Estuary (Figure 1), no exhaustive study has been undertaken with the primary goal of understanding submarine mass movements. As part of the COSTA-Canada project (COntinental slope STAbility) (Locat and Mienert, 2003), intensive field work was carried out in the St. Lawrence Estuary between the Betsiamites and Manicouagan deltaic systems (Cauchon-Voyer, 2007; Locat *et al.*, 2004; Duchesne *et al.*, 2003) and led to the recognition of significant evidence of submarine mass movements west of the Betsiamites River mouth (Figure 1). Considering the amount of historical earthquakes known to have disturbed the landscape across Eastern Canada since the last deglaciation (Levesque *et al.*, 2006; St-Onge *et al.*, 2004; Aylsworth *et al.*, 2000; Shilts et Clague, 1992; Smith, 1962) and the extent of the regional disturbance observed in the Betsiamites – Rimouski area (Figure 1), it would be expected to identify more than one failures in the area. Therefore, morphological, sedimentological, and seismostratigraphic analyses of the seafloor

V. Lykousis, D. Sakellariou and J. Locat (eds.), Submarine Mass Movements and Their Consequences, 233–241.

integrated with results from St-Onge *et al.* (2003) on the Holocene magnetic and sedimentological sequences of the St. Lawrence Estuary will provide basis for a chronology for the failure events.

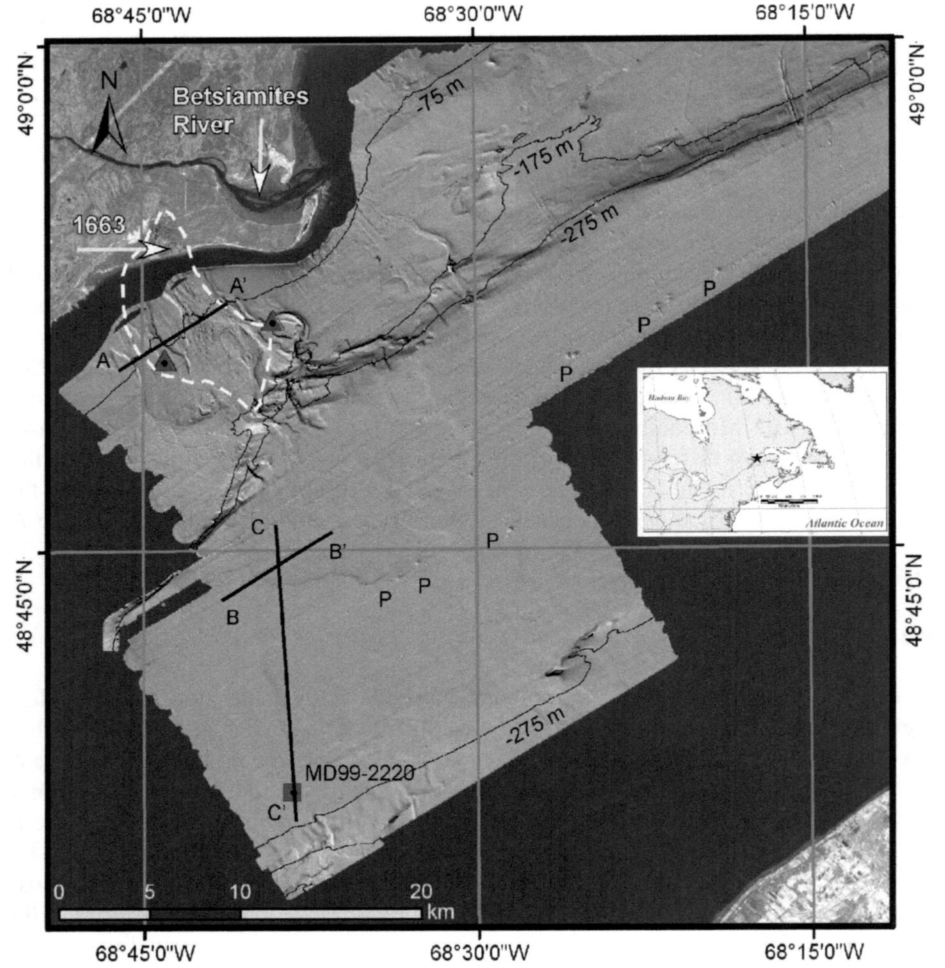

Figure 1. Sun illuminated map of the study area. The white dashed line indicates the location of the subaerial landslide scar and submarine debris associated to the 1663 earthquake. Gray triangles indicate position of two ^{210}Pb dated box cores presented in Figure 5. Notice the location of the Betsiamites River and the MD99-2220 coring station, indicated by the gray square. The black lines correspond to the position of the three high resolution profiles presented in this paper (A-A' Figure 3; B-B' and C-C' Figure 4). The letters P indicate the locations of large pockmarks.

2. Data and Methods

Bathymetric data was acquired using a SIMRAD EM1000 multibeam echosounder (Figure 1). The seismic reflection profiles presented in this study were obtained with an EG&G chirp system (2-12 kHz). 34 cores (box, Lehigh and piston) were recovered from 15 sampling stations between years 2003 and 2006. Low field volumetric magnetic susceptibility (k) and wet bulk density and porosity were measured on board using a GEOTEK multi-sensor core logger (MSCL). Digital X-ray images of all cores were obtained with computerized co-axial tomography (CAT-Scan) with a pixel resolution of 1 mm. Sedimentation rates were derived from ^{210}Pb measurements within sediments of two box cores (Figure 5) following routine procedures at the GEOTOP-UQAM-McGill research center (*e.g.*, Zhang, 2000).

3. Regional morphology

The study area is located along the North Shore of the Lower St. Lawrence Estuary, 400 km northeast of Québec City (Figure 1). On land, a subaerial landslide scar with an area of 6.5 km^2 can be observed, west of the Betsiamites River (Figure 1). With an area of 6.5 km^2 and a volume of more than 300 millions m^3 it is one of the largest historical subaerial landslides identified in Québec. It has been suggested by Bernatchez (2003) that this landslide was triggered by the earthquake (M~7) that shook the province of Québec in AD 1663 (Smith, 1962). The water depth in the study area ranges from the shoreline down to 375 m in the Laurentian Channel, a long U-shaped glacial valley (Loring and Nota, 1973). The shelf is a sub-horizontal surface and has an average width of 10 km and a maximum slope of 2°. shelf break occurs between 150 and 200 m water depth. A wide variety of landforms are revealed from seafloor investigation (Figure 2). We define four (4) main types of geomorphological features: mass movement morphologies, paleochannels, pockmarks, and undisturbed seafloor. A paleo-submarine channel and its meander are identified on the shelf, east of the landslide scar (Figure 2). This submarine channel is not currently related to the modern Betsiamites River discharge (Figure 1). This paleochannel is probably a vestige of the progradation of the deltaic system during the last sea level regressive phase (Hart and Long, 1996). Pockmarks were identified on the shelf and in the Laurentian Channel regions. On the shelf, they are concentrated at water depths ~ 140 m (Figure 2). Their diameters range between 50 and 75 m and depths from 2 to 4 m below surrounding seafloor. In the Laurentian Channel, the pockmarks are significantly larger, some having diameters of up to 400 meters and forming depressions reaching 12 m (Figure 1).

4. Mass movements morphologies

4.1 7.25 KYR CAL BP EVENT

In the study area, mass movements have modified the shelf, slope, and Laurentian Channel. On the shelf, an area of 70 km^2 was modified by mass movements (Figure 2). It stretches for 10 km from the shoreline to the shelf break. The landslide scar on the shelf is characterized by three main distinctive features: (a) a landslide scar with two large channel, (b) buttes of remnant deposits and (c) recent and shallow landslide debris (Figure 2).

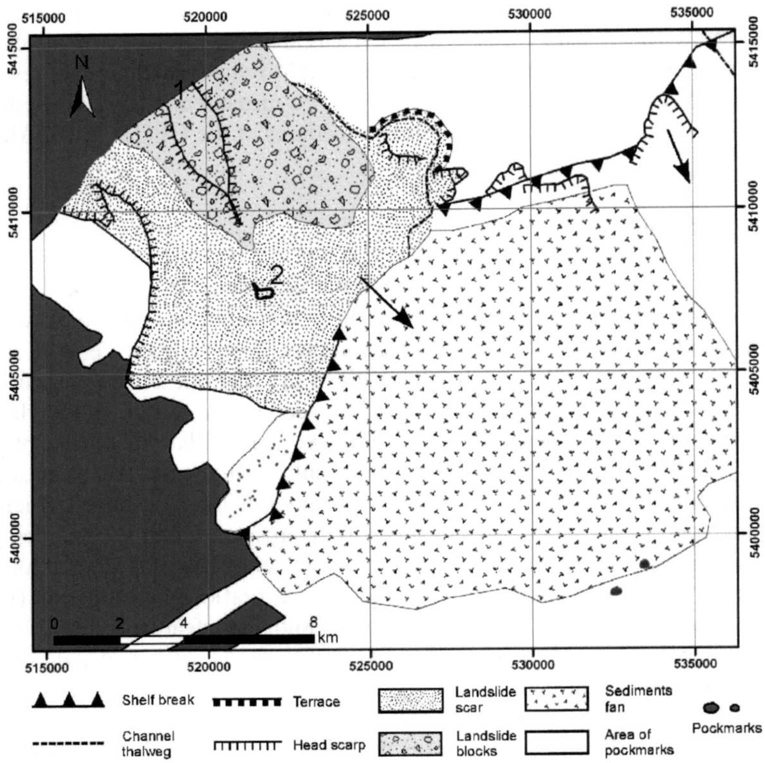

Figure 2. Geomorphological interpretations of the area, with focus on the mass movement morphologies. Arrows indicate landslide flow direction. 1 and 2 refer to the buttes, see text for explanation.

At least four different events modified this area of the shelf (Figure 3) and three of them were dated. An event created the two large channels separated by a butte of remnant stratified deposits (Figure 3). Sequential analysis of the seismic configuration of the reflectors at the base of the channels allows us to interpret them as being formed synchronously as a sliding mass. The West landslide channel has a width ranging from 2 to 3 km and a length of 5 km. The slope of the West landslide channel floor is 1°. The western flank heights of the West landslide channel ranges from 12 to 18 m, with average slope of 12°. For the eastern flank, heights range between 10 to 20 m with average slopes of 5°. The East landslide channel width varies from 2 to 4 km, has a length of 5 km, and a floor slope of 1°. The morphology of the eastern flank of the East landslide channel is irregular due to mass wasting processes.

Two buttes with steep flanks and flat tops are observed within the landslide scar (1 and 2 on Figure 2). They are remnant deposits of stratified sediments (above R2 on Figure 3). A continuous stratified sequence is interpreted on the seismic profiles (Figure 3) of butte 1, implying that it was kept mostly intact when the landslide channels were formed. Butte 1 extends over 5 km^2 with a maximal length and width of 4.5 km and 1.6 km, respectively (Figure 3). The average slope of the top of butte 1 is 1°. Butte 2 was kept intact within the lower section of the landslide scar (Figure 2).

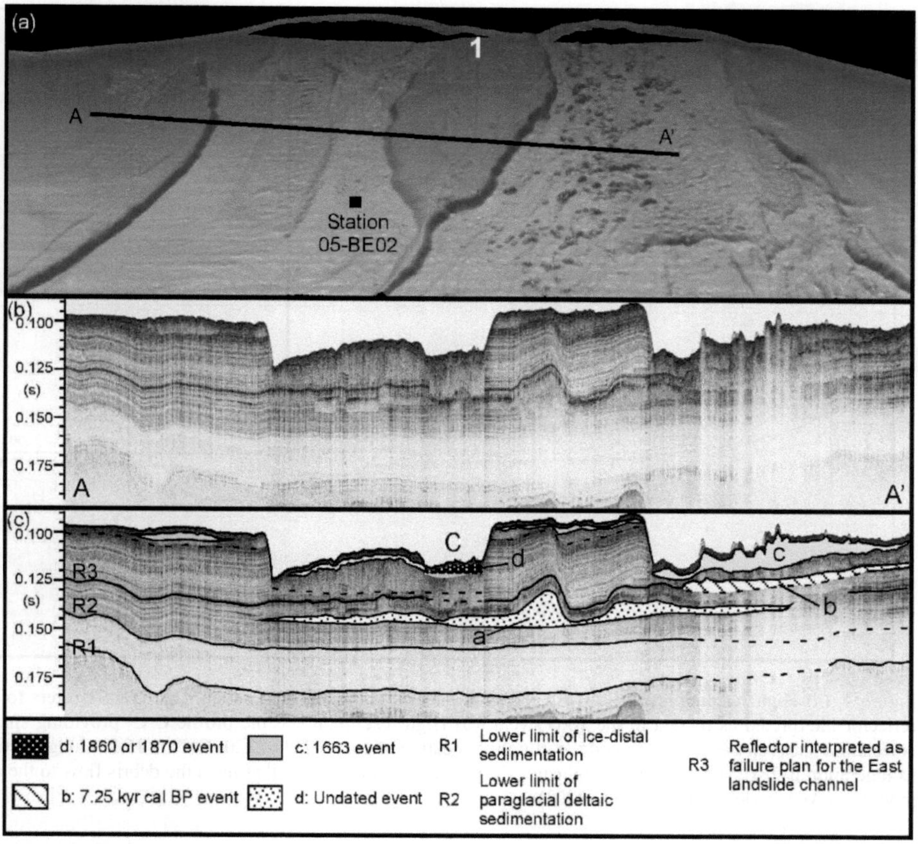

Figure 3. (a) 3D shaded bathymetry relief of the landslide channels and butte 1. View is looking north-west. Vertical exaggeration is 5x. Position of the 7.4 km long seismic profile A-A' is shown. (b) High resolution seismic A-A'. Scale is in second (two-way travel time). (c) Interpretations. Letter C indicates the position of a shallow landslide channel (event of AD 1860 or AD 1870) in the 1663 debris. R1 is interpreted as the lower limit of ice-distal sedimentation. R2 is interpreted as the lower limit of paraglacial deltaic sedimentation. Notice that both the undated and landslide channels events occurred within this unit of highly stratified sediments.

In the Laurentian Channel, traces of this event are interpreted in the seismostratigraphic sequence. In fact, a large sediment fan is observed at a water depth of 350 m (Figure 2). The fan has an area of 115 km^2 and a maximum diameter of 15 km. Seismic reflection profiles allow us to interpret this fan as the result of accumulation of debris (Figure 4). The last debris flow is buried under an average of 15 m of postglacial hemipelagic sediments, which implies that the fan is currently inactive. With an average thickness of 9 m, the debris flow has an estimated volume of 1 km^3. A high amplitude seismic reflector is interpreted as the upper boundary of the debris flow deposit, indicated by S1 in Figure 4 and extends out of the limit of the debris fan. A correlation to date this event can be done with the work of St-Onge *et al.* (2003) who established from 17 AMS ^{14}C dates an age model for core MD99-2220 sampled in the Laurentian Channel, ~ 15 km from the study area (Figure 1).

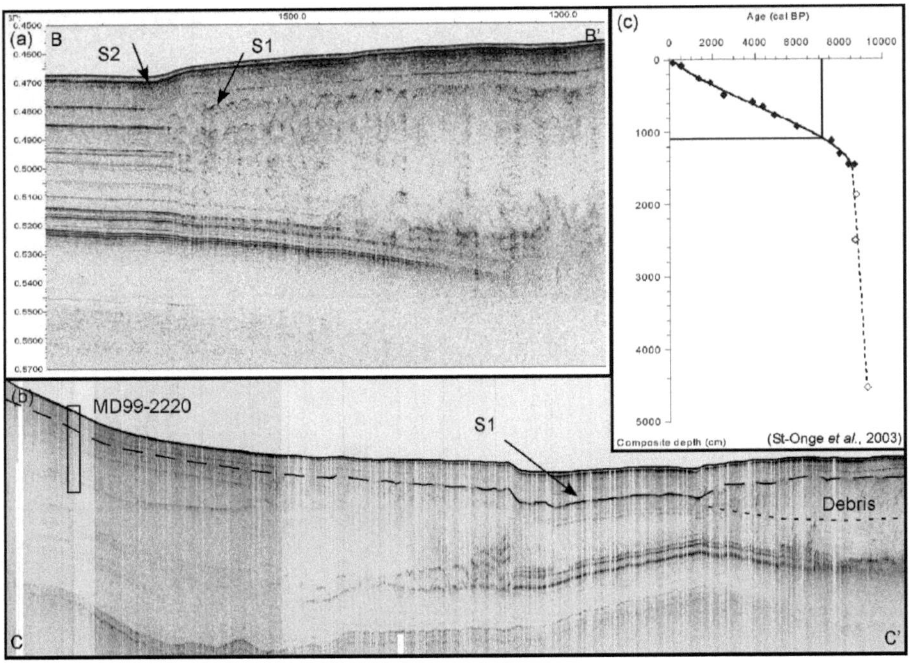

Figure 4. (a) High resolution profile B-B' across the debris fan in the Laurentian Channel. S2 refers to the reflector interpreted as related to the 1663 event. (b) High resolution seismic profile C-C' providing spatial correlation between the debris fan and the sampling location of core MD99-2220. C-C' is 15 km long, vertical exaggeration is 28x. The dashed line indicates the position of reflector S1 that links the debris flow to the age model. (c) Age model (St-Onge *et al.*, 2003).

S1 is observed at a depth of 1100 cm, leading to an age estimate of about 7250 cal BP (Figure 4). This age estimate is consistent with many observations linking glacio-isostatic rebound and earthquake-triggered landslides close to the study area. For example, St-Onge *et al.* (2004) suggested that at least 4 rapidly deposited layers possibly caused by earthquakes occurred in the Saguenay Fjord between 6800 and 7200 cal BP. Similarly, Aylsworth *et al.* (2000) associated observations of very disturbed terrain in a flat erosional plain in the Ottawa Valley to earthquake deformations and liquefaction of sensitive clays that occurred ca. 7060 yr BP.

4.2. RECENT EVENTS

Morphological observations and high resolution seismic interpretations led us to establish that more than one recent landslide (*i.e.,* less than 1000 years old) have occurred in the area. The floor of landslide channels East and West adjacent to the shoreline is covered with a chaotic layer having a rough surface and transparent seismic attributes (Figure 3). It is interpreted as recent debris flow deposits. The East landslide channel debris are a continuity of the subaerial landslide (Figure 1). It is covered by large rafted blocks impeding seismic penetration (Figure 3). The blocks extend downslope 8 km from the shoreline. Small wood branches, bark, and peat were found in the sediments recovered from the submarine debris, which attests of their subaerial provenance.

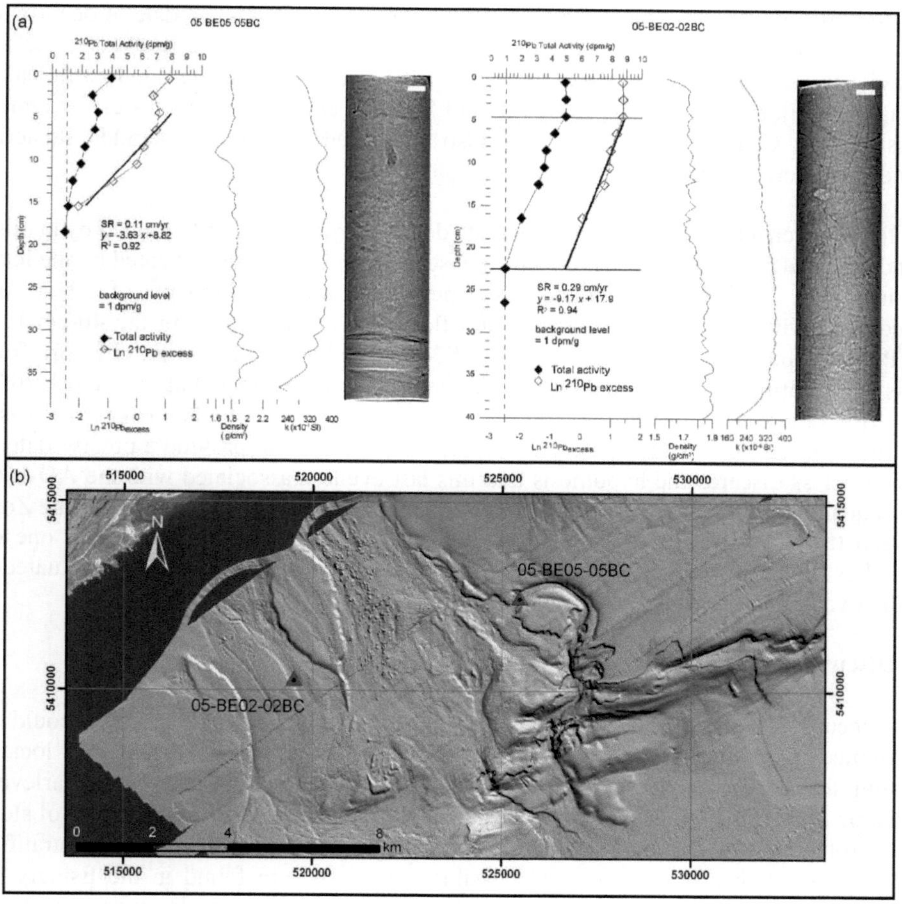

Figure 5. (a) ^{210}Pb measurements in box cores 05-BE05-05BC and 05-BE02-02BC. Regional ^{210}Pb supported value is from Zhang (2000). Bulk density, magnetic susceptibility profiles and CAT-Scan images are presented (white bar represents 2 cm). Active mixing occurs in the upper 4.5 cm of both cores. For box 05BC, SR = 0.11 cm yr^{-1} (R^2= 0.92). Notice the sand bed at 30 cm, associated with the 1663 event. For box 02BC, SR = 0.26 cm yr^{-1} (R^2= 0.94). (b) Sun illuminated indicating position of sampling stations.

Analysis of the sediments physical properties recovered from a box core (05-BE05-05BC) sampled in the meander area (Figure 5) led to the identification of recent homogenous hemipelagic sedimentation and of a sand bed associated to a debris flow (Figure 5). The debris flow deposit was identified at 30 cm bsf and 5 cm of compaction was recorded when the core was sampled, implying that the debris flow at this location is buried under a minimum of 35 cm of hemipelagic sediments. Using the sedimentation rate of 0.11 cm yr–^{1} calculated from the slope of the ln ($^{210}Pb_{excess}$) (Figure 5), the debris flow buried under 35 cm is dated at about AD 1685, suggesting it could have been triggered by the 1663 earthquake. This recent event is also observed within the sediments of the Laurentian Channel. In fact, a high amplitude reflector was interpreted at an average depth of 1 m in the sediments (S1 on Figure 4). The layer was sampled with gravity coring and interpreted as a rapidly deposited layer (RDL) following a landslide event. Using the ^{210}Pb derived sedimentation rate of 0.28 cm yr^{-1} determined by St-Onge *et al.* (2003) on box core AH00-2220, we can estimate a date of occurrence at AD 1646, which can reasonably be linked to the AD 1663 earthquake. As described previously, many landslides elsewhere in Québec are related to the 1663 earthquakes (Saguenay Fjord, Levesque *et al.*, 2006 and St-Onge *et al.*, 2004; Saint-Jean Vianney, Lasalle and Chagnon, 1968) as it is also suggested for the Betsiamites subaerial landslide (Bernatchez, 2003).

In the West channel, the AD 1663 landslide debris were subsequently eroded by another landslide that produced the shallow channel seen at the surface, as indicated by the letter C in Figure 3. ^{210}Pb measurements were performed on the sediments of a box core recovered within this channel. The debris flow was not identified in the 40-cm long ^{210}Pb-dated box but identified at a depth of 30 cm within a different gravity core from this same coring station. Based on the sedimentation rate estimated at this station (0.29 cm yr^{-1}; Figure 5), a depth of 40 cm would lead to AD 1872 for the event whereas a depth of 45 cm to AD 1855. Despite the fact that we can not provide a precise date, it nevertheless discards the hypothesis that this last event is associated with the AD 1663 earthquake. Two significant earthquakes were recorded in the Charlevoix Seismic Zone (CSZ) (Smith, 1962): a first one on October 17, 1860 (M~6) and a second one on October 20, 1870 (M~6.5). The epicenters for the 1860 and 1870 events are evaluated at 180 km and 200 km from the Betsiamites area.

5. Discussion and conclusion

The recurrence of slope instability within the Betsiamites – Rimouski area could be attributed to seismic activity due to glacio-isostatic rebound. As the area is located within the Lower St. Lawrence Seismic Zone (LSZ) and close to the Charlevoix Seismic Zone (CSZ), it is expected that seismicity will influence the occurrence of slope instability. However, other factors reducing slope stability such as highly stratified deposits, groundwater seepage, and possible gas escape are found in the Estuary. In combination with earthquakes they cumulate to create a less stable environment. As seen in Figure 3, the failure plan of the landslide channel is interpreted in a unit of dense stratified seismic reflectors. These highly variable seismic attributes could indicate different sediments composition, such as the alternation of silty and sandy layers. The layer identified by R3 in Figure 3 could have developed in a weak layer that would undergo liquefaction when subject to an earthquake and thus result in slope failure. Such liquefaction has likely occurred for the slide in the East landslide channel (Figure 3), as little sediments remained on the failure plane.

In our chronology, we were able to link 3 main events (7250 cal BP, AD 1663, and AD 1860 or 1870) to slope instability (*i.e.*, not to catastrophic river discharge), and to differentiate them from one another based on a sequential stratigraphy point of view and on ^{210}Pb or radiometric dates. Our analysis has raised many other questions such as the tsunamigenic potential of these events and the treat of future similar events elsewhere in the Estuary. There is thus a need to pursue our research to clearly define the mechanisms responsible for slope failures and to describe post-failure behavior in order to assess the risk of slope instability in the St. Lawrence Estuary.

6. Acknowledgements

The authors would like to thank P. Lajeunesse and R. Urgeles for their constructive reviews. We also thank the Natural Sciences and Engineering Research Council of Canada for their financial support for the Costa-Canada project and for ship time allocation. Finally, many thanks to all the crew members and scientific participants aboard the F.G. Creed and Coriolis II vessels.

7. References

Aylsworth, J.M., Lawrence, D.E. and Guertin, J., 2000. Did two massive earthquakes in the Holocene induce widespread landsliding and near-surface deformation in part of the Ottawa Valley, Canada? Geology, 28(10): 903-906.

Bernatchez, P., 2003. Évolution littorale holocène et actuelle des complexes deltaiques de Betsiamites et de Manicouagan-Outardes : synthèse, processus, causes et perspectives, Université Laval, Sainte-Foy, 460 pp.

Canals, M., et al., 2004. Slope failure dynamics and impacts from seafloor and shallow sub-seafloor geophysical data: case studies from the COSTA project. Marine Geology, 213 (1-4): 9-72

Cauchon-Voyer, G., 2007. Morpho-sédimentologie et mouvements de masse au large de la Rivière Betsiamites, estuaire du Saint-Laurent, Québec, M.Sc. Thesis, Université Laval, 196 pp.

Duchesne, M.J., Long, B.F., Urgeles, R. and Locat, J., 2003: New evidence of slope instability in the Outardes Bay delta area, Québec, Canada. Geo-Marine Letters 22, 233-242.

Hart, B.S. and Long, B.F., 1996. Forced regressions and lowstand deltas: Holocene Canadian examples. Journal of Sedimentary Research, 66(4): 820-829.

Lasalle, P., and Chagnon, J.-Y., 1968. An ancient landslide along the Saguenay River, Québec. Canadian Journal of Earth Sciences, 5(1):548 -549.

Levesque, C.L., Locat, J., and Leroueil S., 2006. Dating submarine mass movements triggered by earthquakes in the Upper Saguenay Fjord, Québec, Canada. Norwegian Journal of Geology, 86: 231-242.

Locat, J. and Lee, H.J., 2002. Submarine landslides: advances and challenges. Canadian Geotechnical Journal, 39(1): 193-212.

Locat J., Levesque C., Locat P., Cauchon-Voyer G., Leroueil S., Godin A., Sanfaçon, R., 2004. Aperçu des mouvements de masse sous-marins au large de la rivière Betsiamites, estuaire du Saint-Laurent, Québec, Canada, Conférence Canadienne de Géotechnique, Québec.

Locat, J. and Mienert, J. (Editors), 2003. Submarine Mass Movements and their Consequences 1st International Symposium. Kluwer Academic Publishers, Dordrecth, The Netherlands, 540 pp.

Loring, D.H. and Nota, D.J.G., 1973. Morphology and Sediments of the Gulf of St. Lawrence, Bulletin of the Fisheries Research Board of Canada 182, Ottawa, 147 pp.

Shilts, W.W. and Clague, J.J., 1992. Documentation of Earthquake-Induced Disturbance of Lake-Sediments Using Subbottom Acoustic Profiling. Canadian Journal of Earth Sciences, 29(5): 1018-1042.

Smith, W., 1962. Earthquakes of eastern Canada and adjacent areas, 1534-1927, 26. Publications of the Dominion Observatory, Ottawa, pp. 271-301.

St-Onge, G., Mulder, T., Piper, D.J.W., Hillaire-Marcel, C. and Stoner, J.S., 2004. Earthquake and flood-induced turbidites in the Saguenay Fjord (Québec): a Holocene paleoseismicity record. Quaternary Science Reviews, 23(3-4): 283-294.

St-Onge, G., Stoner, J.S. and Hillaire-Marcel, C., 2003. Holocene paleomagnetic records from the St. Lawrence Estuary, eastern Canada: centennial- to millennial-scale geomagnetic modulation of cosmogenic isotopes. Earth and Planetary Science Letters, 209(1-2): 113-130.

Urgeles, R., Locat, J., Lee, H.J. and Martin, F., 2002. The Saguenay Fjord, Québec, Canada: integrating marine geotechnical and geophysical data for spatial seismic slope stability and hazard assessment. Marine Geology, 185(3-4): 319-340.

Zhang, D., 2000. Fluxes of Short-lived Radioisotopes in the Marginal Marine Basins of Eastern Canada. Ph.D Thesis, Université du Québec à Montréal, Montréal, 193 pp.

SUBMERGED LANDSLIDE MORPHOLOGIES IN THE ALBANO LAKE (ROME, ITALY)

P. MAZZANTI, F. BOZZANO AND C. ESPOSITO
Università di Roma "Sapienza", Dipartimento di Scienze della Terra, P.le Aldo Moro 5, 00185, Rome, Italy.

Abstract

The geomorphological interpretation of the high resolution bathymetry of the Albano lake (central Italy), together with conventional geological and geomorphological investigations for the subaerial slope, allowed us to identify several subaerial and submerged morphologies due to slope failures of different size and presumably age. Two main landslide categories will be decribed in this paper: totally submerged, combined subaerial-submerged landslides. Furthermore a detailed description of two past large slope failures (volume of $10^6 m^3$) and the 1997 subaerial and submerged debris flow are presented. The wave induced by the 1997 debris flow testifies also the tsunamigenic potential of these phenomena which is still more serious if the presence of coastal settlements is taken into account.

Keywords: Albano lake, bathymetry, submerged landslide, submerged slope failure, debris-flow, tsunami

1. Introduction

Ongoing research activities were devoted to evaluate the landslide hazard of the slopes surrounding the Albano lake, located about 25 km southeast of the city of Rome in the Colli Albani volcanic district, and the related tsunami hazard in the shore villages. An high resolution multibeam swath bathymetric survey of the lake, performed using ultra high resolution instruments (Anzidei et al. 2006), allowed us to extend the geomorphological analysis of the area also to the submerged slopes of the Albano lake. In fact, both the impact of subaerial landslides on the water and the occurrence of submerged slope failures have to be taken into account in order to deal with tsunami hazard. This paper mainly focuses on the results of the morphological interpretation of the Albano lake submerged slopes, which is part of the landslide mapping undertaken for both subaerial and submerged slopes.

The Albano lake partially occupies a volcanic depression recently originated as a multiple maar. The overall morphology of this multiple maar is featured by a low aspect ratio edifice characterized by gently dipping outer slopes and steep inner slopes that correspond to the crater walls (fig.1). The latter form an elliptical crater rim, which has an axis maximum of about 4300 m and an axis minimum of about 2800 m; the Albano lake has a maximum water thickness of about 165 m.

As regards the geological setting, in the inner slopes of the Albano maar the hydromagmatic deposits (alternation of scoria lapilli beds and ash-rich layers, generally cemented for the zeolitisation and massive and chaotic, ash-matrix supported, up to 30 m thick ignimbrite deposits) related to the Albano maar activities locally overlay thick banks of lava and scoria deposits ascribed to previous volcanic phases.

V. Lykousis, D. Sakellariou and J. Locat (eds.), Submarine Mass Movements and Their Consequences, 243–250.

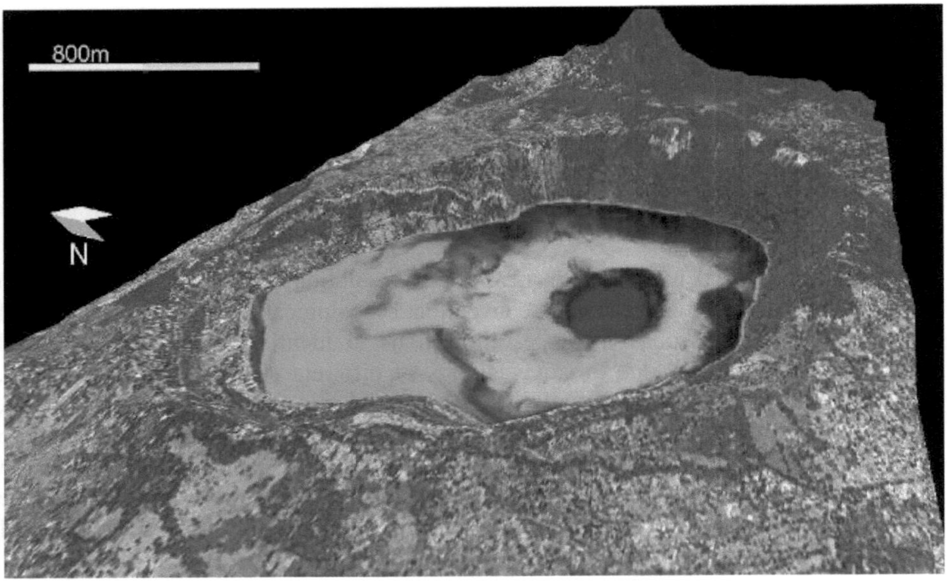

Fig. 1. 3D recontruction of the subaerial and subaqueous Albano lake morphologies.

2. Gravity-induced Landforms

Several gravity-induced landforms related to both past and ongoing landslide processes along the subaerial and submerged slopes of the lake have been found. Geological and geomorphological surveys of the subaerial landforms have been performed by coupling field activities and aerial-photograph interpretations in order to complete the frame of the whole inner crater slope. The geomorphological survey of the subaqueous slopes has been carried out by means of an analysis of the very detailed (1m x 1m square grid) LDEMs (Lacustrine Digital Elevation Models) and the LDEM-derived thematic maps (slope, aspect, curvature) together with specific software which permits a "virtual flight" above the bathymetry 3D model. These kind of analyses have been supported by morphometric computations of both surfaces and volumes of the main scars and accumulation areas. Such an integrated analysis of the whole subaqueous-subaerial slope system highlighted the presence of three main conditions: completely subaerial landslides, totally submerged landslides, combined subaerial-submerged landslide. The latter are represented by a subaerial detachment area with a transportation route and a depositonal area extending in the submerged slope. Only the totally or partially submerged landslides will be discussed, the small and medium sized ones in the paragraphs 2.1 to 3, and the large ones in the paragraph 4.

2.1 TOTALLY SUBMERGED LANDSLIDES

According to the classifications by Cruden and Varnes 1996, Mulder and Choconat 1996 and Hungr et al 2001, the completely submerged landslides mainly consist of:

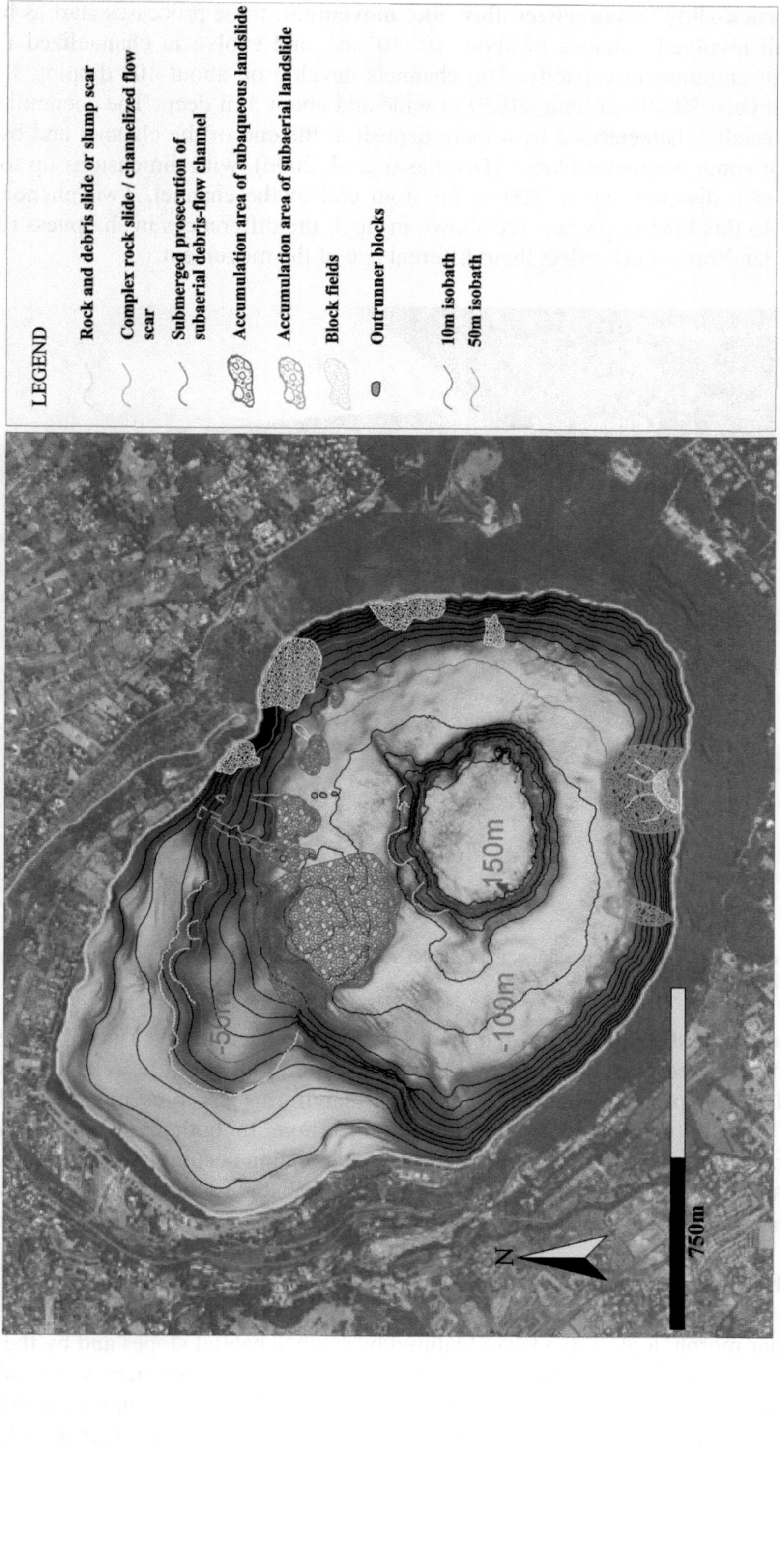

Fig. 2: Inventory map of the main subqueous gravity-induced landforms discussed in the text, plotted on the Albano lake bathymetric map produced in the frame of the INGV-DPC 2005-2007 project V3_1 by M. Anzidei (UR8, Responsible: F. Riguzzi).

Complex rock-slide / channelized flow-like movement: these processes start as block slides, with involved volumes of about 10^3-10^4 m^3, and evolve in channelized flows with a high entrainment capacity. The channels develop on about 10° dipping slopes and are between 50-200 m long, 20-50 m wide and about 5 m deep. The accumulation areas are usually characterized by a main deposit at the end of the channel and by the presence of some outrunner blocks (De Blasio et al. 2006), with dimensions up to 10^3 m^3, that reach distances up to 300 m far from end of the channel. Two phenomena ascribable to this kind of process are shown in fig.3; the differences in sharpness of the respective landforms may reflect their different age of the movement.

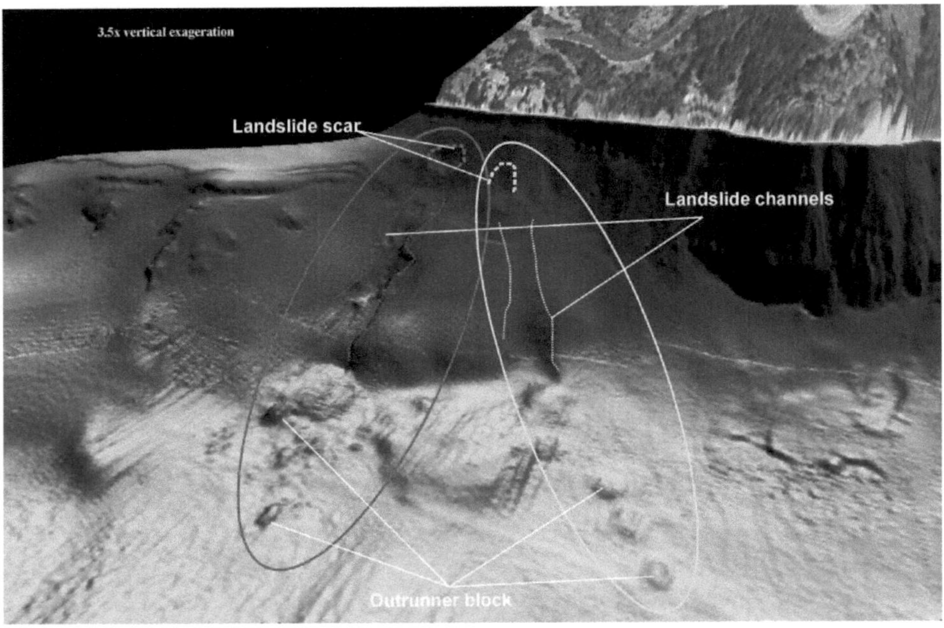

Fig. 3: Two examples of completely submerged complex rock-slide / channnelized flow in the northern part of the Albano lake floor.

Debris and rock slide or slump: the main landforms related to this kind of processes mainly affect the edge of the most internal maar crater, and could be related to local, syn-eruptive collapses (Németh 2000). Similar landforms are present upon a huge landslide deposit (par.4) and testify its remobilization. In both cases the detached volumes range between 10^4-$10^5 m^3$, while the corresponding accumulation areas are not clearly visible. Finally, also a large slope failure is ascribable to this kind of process and will be described in detail in par.4.

2.2 COMBINED SUBAERIAL-SUBMERGED LANDSLIDES

The peculiar morphology of the basin featured by steep subaerial slopes and by the lack of a well developed coastal platform or shore (except for the northern sector of the lake), frequently allows the subaerial landslides to extend their runout in the submerged slopes. In these cases the main depositional, submerged landforms are represented by:

Rockfall / topple deposits: areas with a diffuse presence of single blocks deriving from subaerial failures and that can be defined as **"block fields"**.
Rock-slide deposits: are featured by well defined accumulation areas located downslope the rock-slide scars. The most significant one is related to a large rock-slide that will be discussed in par. 4.
Complex rock-slide / debris flow deposits: are featured by accumulation areas located at the toe of the main transportation channels, that can whether completely develop in the subaerial slopes or partially continue in the submerged slopes. Most of these landslides affect the eastern sector of the lake: based on some geomorphological evidence, the most recent ones seem to be the partial reactivation of larger, past phenomena. From this point of view the 1997 debris flow occurred along the eastern slope is particularly significant and will be described in detail in the next paragraph.

3. The 1997 Debris-Flow

After an intense rainfall event (128 mm in 24 hours), a debris flow occurred in the eastern slope of the Albano lake on the 7th of November 1997.

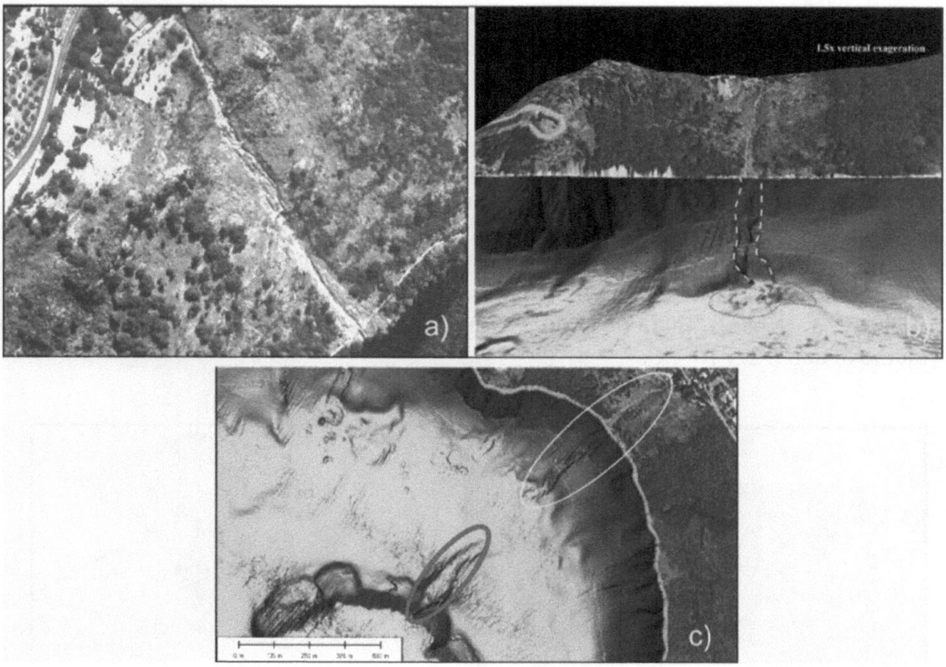

Fig. 4: a) Aerial view of the subaerial channel generated by the 1997 debris-flow; b) subaqueous continuation of the channel and the deposit; c) in red circle the continuation of the subaqueous channel.

According to local witnesses, a little tsunami wave generated after the impact of the mass onto the water and reached and flooded the ground floor of some houses along the coastline. This landslide began as a soil slide, which involved about 300 m^3 of eluvial material that slid along the contact with the bedrock. The so mobilized mass was channeled within a steeply dipping impluvium (about 40°) and thus evolved as a debris flow which entrained a large amount of debris material along the bottom of the channel and reached an estimated volume of some thousands of cubic meters during the subaerial path. The bathymetry data point out that the channel continues along the submerged slope until a depth of about 100 m below the lake level; in addition, it is wider than in the subaerial part and is about 3 m deep. Part of the landslide mass was deposited on the coastline immediately after the impact on the water, while the largest part of the deposit should have deposited in the submerged part. In addition, the above described submerged channel forms a lineation with a wider one that reaches a depth of about 130 m below the lake level (fig.4) and is up to 100 m wide, 10 m deep and has a slope angle less than 8°.

4. Evidences of great past landslides

Geomorphic evidence of past large landslide events with volumes of 10^6 m^3 are present in the Albano lake beyond the above described landslides, whose volumes do not exceed 10^5 m^3.

In the southern part of the lake inner slope, a wide, markedly concave slope sector, whose shape is ascribable to the scar area of a past subaerial massive rock slope failure is present (fig.5). In addition, in the submerged part just downslope this "negative" landform, a huge, convex positive landform is present and can be regarded as the debris accumulation corresponding to the above mentioned scar area. Furthermore, geological reconstructions point out that the involved material is constituted by massive and chaotic, ignimbrite deposits with a dip slope attitude, overlaying thick banks of plane parallel to low-angle scoria lapilli beds and lava lenses. Based on the morphologic evidence and on the geological-structural setting, this landslide can be classified, according to Hungr and Evans 2004, as a simple translational rockslide – structurally controlled. A reconstruction of the hypothesized pre-landslide topography allowed us to estimate a detached volume of about $3 \cdot 10^6 m^3$, that fits well with the submerged debris volume of about $2.5 \cdot 10^6$ m^3.

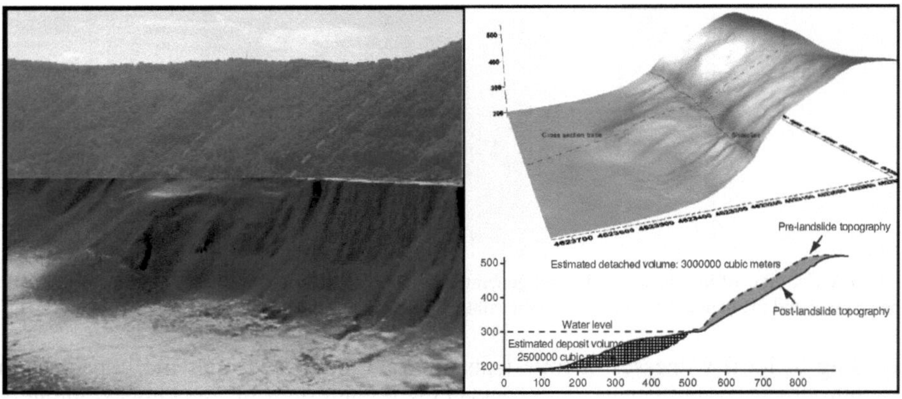

Fig. 5: On the left the subaerial scar and subaqueous deposit of the large Albano Lake rock-slide; on the right is the 3D subaqueous/subaerial model and the profile with the pre-failure topography reconstruction.

As regards the completely submerged slopes the largest landform explainable as a slope failure is present in the northern sector (fig.6). The total scar area is 365000 m^2 wide corresponding to an estimated total detached volume of about $7*10^6$ m^3. At the toe of the scar area a slight morphologic bulge is present and could represent the debris accumulation. This deposit shows a peculiar shape because of the presence of an inner depressed zone, similar to the ones surveyed in other kinds of landslides in closed basins such as fjords (Blikra et al. 2006) and lakes (Bacon et al. 2002). In our case, this shape could be related whether to the presence of a preexisting volcanic depression or to some peculiar depositional mechanisms.

As regards the dating some indirect constraints are given by the age of the maar activity responsible for the formation of the scarp involved in the slope failure (about 70 ky BP), and the first lacustrine deposits that mantle the scar area that is of about 30 ky BP according to Chondrogianni et al.1996.

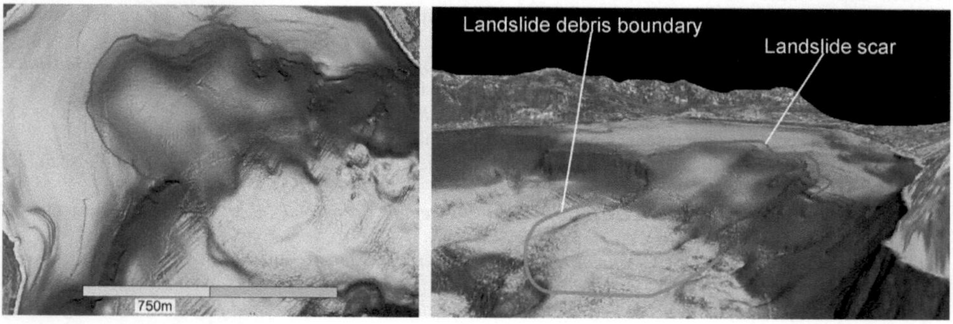

Fig. 6: Two different side view of the completely subaqueous big slope failure in the northern part of the lake Albano floor.

5. Conclusions

The relevance and importance of the landlside recognition and analysis in the Albano lake area is related to the presence of human activities around the coastline.

Several case studies highlight the tsunamigenic potential of subaquaeous and subaerial landslides that impact on the water (Papadopoulos and Kortekaas 2003), especially in closed basins related to both large-sized (Schnellmann et al. 2002; Panizzo et al. 2005) and smaller (10^5 m^3) events (Jorstad 1968; Wagner et al. 2003). The Albano lake slopes show an intense small- and medium-sized landslide activity. In addition evidence of past large slope failures is present. The tsunamigenic potential related to possible other failures is also enhanced by the considerable relief energy and the high velocity of such events as debris flows. The landslide mapping and preliminary interpretations presented in this paper should be thus considered as the basis on which further and more detailed studies will be carried out in order to achieve a final landslide and tsunami-related hazard assessment.

6. Acknowledgements

This research is financially supported by INGV-DPC 2005-2007 project V3_1- UR12 (Responsible: F. Bozzano). The bathymetry of the Albano lake is produced by UR8

(Responsible: F. Riguzzi) in the frame of the same research project. The authors thank M. Anzidei for the helpful discussion about the subaqueous topography.

7. References

Anzidei M, Esposito A, De Giosa F (2006) *The dark side of the Albano crater lake.* Annals of Geophysics, accepted July 2006.

Bacon CR, Gardner JV, Mayer LA, Buktenica MW, Dartnell P, Ramsey D, Robinson JE (2002) *Morphology, volcanism, and mass wasting in Crater Lake, Oregon.* GSA Bulletin vol 114, 6, pp 675-692.

Blikra LH, Longva O, Braathen A, Anda E, Dehls JF, Stalsberg K (2006) *Rock slope failures in norwegia fjord areas: examples, spatial distribution and temporal pattern.* In: Evans SG et al. (eds.), Landslides from massive Rock Slope Failure, pp 475-496.

Chondrogianni C, Ariztegui D, Niessen F, Ohlendorf C, Lister GS (1996) *Late Pleistocene and Holocene sedimentation in Lake Albano and Lake Nemi (central Italy).* In Guilizzoni P and Oldfield F (eds.), Palaeoenvironmental Analysis of Italian Crater Lake and Adriatic Sediments, Mem. Ist. ital. Idrobiol., vol 55, pp 23-38.

Cruden DM, Varnes DJ (1996) *Landslide types and processes.* In: Turner AK and Shuster RL (eds). Landslides Investigation and Mitigation; Tranportation Research Board. National Research Council, Special Report 247. Washington, DC., pp 36-75.

De Blasio FV, Engvik LE, Elverhøi A (2006) *Sliding of outrunner blocks from submarine landslides.* Geophysical Research Letters, vol 33, L06614, doi:10.1029/2005GL025165.

Hungr O, Evans SG, Bovis MJ, Hutchinson JN (2001) *A Review of the Classification of Landslides of the Flow Type.* Environmental & Engineering Geoscience, vol 7, 3, pp 221-238.

Hungr O and Evans SG (2004) *The occurrence and classification of massive rock slope failure.* Felsbau, 22: 16-23.

Jorstad F (1968) *Waves generated by landslides in Norwegian fjords and lakes.* Norwegian Geotechnical Institute Publication, n 69, pp 13-32.

Mulder T, Cochonat P (1996) *Classification of offshore mass movements.* Journal of Sedimentary research, vol 66, n 1, pp 43-57.

Németh K (2000) *Collapse structures of an eroded maar/diatreme volcanic field from Central Otago, New Zealand: The Crater as an example.* Terra Nostra, vol 6, pp 365-374.

Panizzo A, De Girolamo P, Di Risio M, Maistri A, Petaccia A (2005) *Great landslide events in Italian artificial reservoirs.* Natural Hazards and Earth System Sciences, vol 5, pp 733–740.

Papadopoulos GA and Kortekaas S (2003) *Characteristics of landslide generated tsunamis from observational data.* In Locat J. and Mienert J. (eds) Submarine Mass Movements And Their Consequences, 1st International Symposium. Kluwer Academic Publishers, pp 367-374.

Schnellmann M, Anselmetti FS, Giardini D, McKenzie JA, Ward SN (2002) *Prehistoric earthquake history revealed by lacustrine slump deposits.* Geology, vol 30, n 12, pp 1131–1134.

Wagner TP, McKee CO, Kuduon J, Kombua R (2003) *Landslide-induced wave in a small volcanic lake: Kasu Tephra Cone, Papua New Guinea.* Int. J. Earth Sci., vol 92, pp 405–406.

DYNAMICS OF THE DELTAIC CANYON AREA OF THE RV. CHOROKHI, GEORGIA

KAKHABER BILASHVILI
Institute of Geography, Iv.Javakhishvili Tbilisi State University, 0128 Tbilisi, Georgia
Grigol Ruso, Nana Megreli, Zurab Savaneli,
Scientific-Industrial Centre of Research and Governance of Coastal Processes "SAKNAPIRDATSVA", 150 D.Agmashenebeli Ave., 0112 Tbilisi, Georgia

Abstract

The Cape of Batumi belongs to accumulative morphological forms and mainly consists of terrigenous sediments of Rv.Chorokhi. It's location in the central part of the City of Batumi stipulates an importance of intensive research of phenomena. The natural growth of the cape towards the sea was stopped in 30-40-ties of last century, when it's frontal part closely approached the head of the Rv.Chorokhi submarine canyon. Investigation of the stability conditions of the bottom sedimentary layers deposited inside the canyon has been undertaken on the base of field observations, that repeatedly was held in 1960-2003. Sediment material from Rv.Chorokhi is a main component of the littoral cell of the Adjarian coastline of the Black Sea. The Chorokhi Canyion head locates at depth of 7-8 m at a distance of 70-140 m from the river mouth and invades the area of sediment laden currents. Its frontal area spreads along the shore line for as much as 1,5 km. More, than 90% of the river's load discharge were transported through the Chorokhi Canyon. The stability conditions of submarine sediments are disturbed by changes in the tangential reaction between soil particles or blocks, increased pore pressures, storms, instability of the underlying rocks, structural motions, tsunamis, earthquakes (e.g. stability conditions has been violated by strong earthquake occurred in Turkey on 14-th of January,1999).Conducted research deals with the analytical considerations of dynamical processes occurring in the given area.

Keywords: Cape of Batumi, sedimentary layers, submarine canyon, earthquake, stability

1. Introduction

General length of the global sea and ocean coastal line is 469 000 km. The length of genetically accumulated coastal line is around 133 000 km or 28% (Leontiev *et al.* 1977; Safianov 1978). At present its 70% is washed out or deviated on the average 10 cm during the year and 1m annually (Kaplin 1990; Meladze 2003).

It's obvious that sustainability of coastal line is in direct connection with the eustatic rise of the ocean level, however at present the increase of the washed out area and growth of intensity is observed also in comparatively stabile eustatic conditions.
It is also remarkable that intensification of washing out processes is noticed in the sections that are in the state of tectonic rise.

Taking into consideration the above mentioned we think, that the main reason of destabilization of sea coasts must be looked in disturbance of natural balance from

V. Lykousis, D. Sakellariou and J. Locat (eds.), Submarine Mass Movements and Their Consequences, 251–258.

human activities in coastal zone or in anthropogenic factors. They are similar for different regions of the world. In particular, extraction of inert materials from river's basin and sea coastal zone, artificial change or regulation of river mouth, building dams on the river or hydro-technical facilities (among them ports, stations and protecting constructions etc.) in active coastal zones, without taking into account the real conditions of natural development in concrete sections of coast.

Similar situation took place in Georgia. The length of the Black sea coastal line is approximately 315 km. Its biggest part (approximately 300 km) is typically accumulative and is built mainly with alluvial, terrigenous material, brought by rivers to the sea.

The mean long-term volume of rivers load discharge in Western Georgia is around 14-15.0mln m^3, from which 2.0mln m^3 deposits in artificial reservoirs, and 1 – 1,50mln m^3 is annually taken away from river beds' sand-pits. Accordingly, around 11.0mln m^3 reaches sea coastal area every year. From this volume about 2.0mln m^3 is lost in the submarine canyons in coastal zone and around 6,80mln m^3 of small fractions is carried out in the open sea and participates in contemporary processes of sedimentation.
So, for the formation of Georgian coastal zone nearly 2,20mln m^3/year of alluvium is needed (Jaoshvili 2003), that, obviously, cannot be distributed equally in different sections of the costal area.

2. Georgian littoral zone

Within the borders of Georgia the Black Sea littoral is divided into eight independent dynamic coastal systems (or littoral cells) identifiable as separate alongshore flows (Figure 1).

Submarine canyons are features associated with rivers and abundant discharge from them. They occur frequently in the eastern part of the Black Sea. There is a concentration of them, either as groups or as a series of isolated canyons, between the deltas of the Rv. Mzymta and the Rv. Chorokhi (e.g. investigation of the stability conditions of the bottom sedimentary layers deposited inside the submarine gorge has been undertaken on the base of field observations, that repeatedly was held in 1960-2003 years in the pre-deltaic submarine canyon of Rv.Chorokhi). The crests of most of these canyons lie at depths of 15–25 m. Several of them cut into the coastal zone, where they begin to slope steeply at 6–10 m. The crests of all the canyons lie in a Holocene unit and in contemporary silt, sand and pebble deposits. While slope angles generally vary between 6–20°, sheer escarpments also occur, the gradient of lateral declivities can be 45°, and there are individual occurrences of vertical walls. Offshore, the canyons run out to depths of more than 1 000 m. While it is difficult to be specific about the genesis of these submarine canyons, their origin and contemporary dynamics are undoubtedly dependent on a multiplicity of fluvial processes

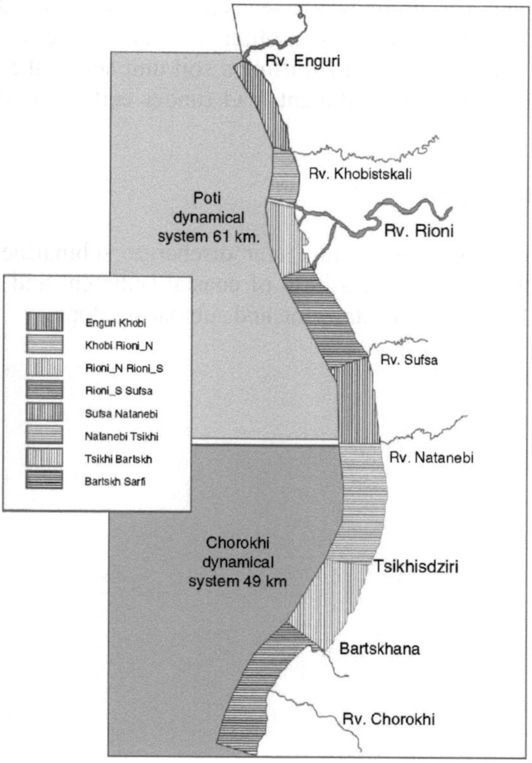

Figure 1. Scheme of littoral cells of Georgian Black Sea Coast.

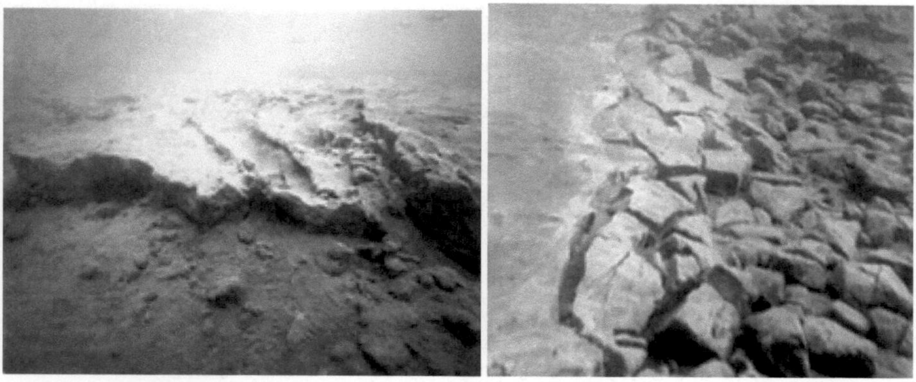

Figure 2. Images of submarine slides in the area of Rv.Chorokhi Canyon.

The forms of motion are frequently combined, e.g., land-slides, a viscous flows and turbidities form a gradational spectrum of allochtonous sediments (Figure 2). The stability conditions of submarine sediments are disturbed by changes in the tangential reaction between soil particles or blocks, increased pore pressures, storms, instability of the underlying rocks, structural motions, tsunamis, earthquakes, gas charging, in addition to erosion, and also groundwater seepage (Locat and Lee 2000; Lee, Schwab, and Booth 1993). The most common motion stimulators are waves which induce different changes in the stability, from shifting a soil unit across the accumulative form surface to suspending the whole sediment load (under certain conditions) (Bilashvili, Savaneli 2003).

3. Canyons of Adjara region

Irrevocably swallowing up part of the river discharge, submarine canyons exert an effective braking action on the deposition of coastal sediment and thus becoming the determining factor shaping the coastal zone and submarine slope.

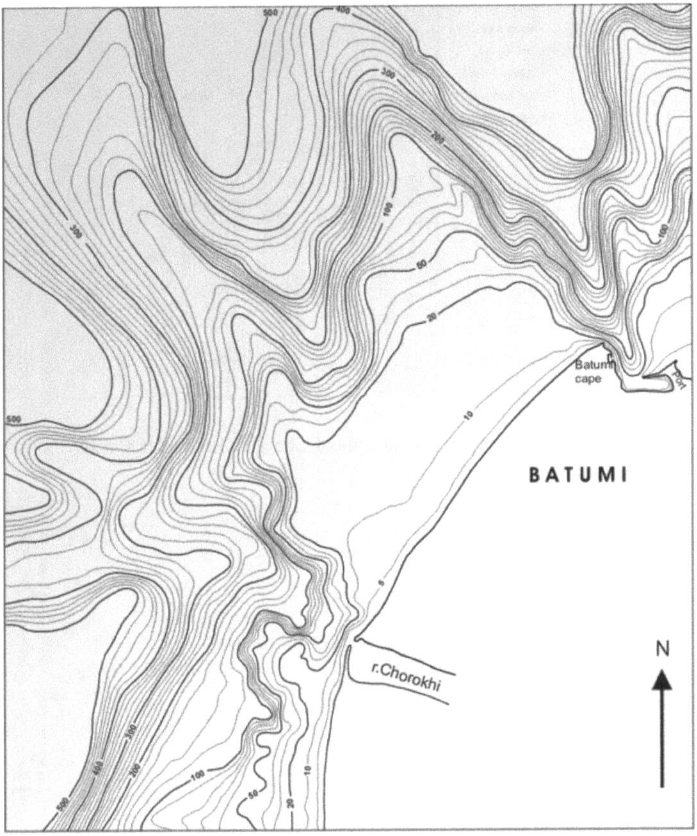

Figure 3. Topography of the adjacent to deltaic area submarine slope segments.

The length of Adjara sea coastal line is 50 km., from which around 45 km is built with alluvium brought by Rv.Chorokhi to the sea and the rest are the rocky capes.

Degradation of the Rv.Chorokhi litho-dynamic system began in the second part of the 19th century, when the mouth of the river was fixed in the area of submarine canyon, where the most part of alluvium was loosing (Russo, Safianov, Khorava 1987).

Further, as a result of active influence of construction of Batumi port in 1878 and other anthropogenic factors, common system of sea coastal zone divided into several autonomic sub-systems.

Sedimentary dynamics of the upper submarine slope regions mainly depend on the amount of terrigenous material, configuration of the shore line with respect to the local wave directions and the topography of the adjacent submarine slope segments, particularly with respect to the presence or absence of submarine canyons (Figure 3).

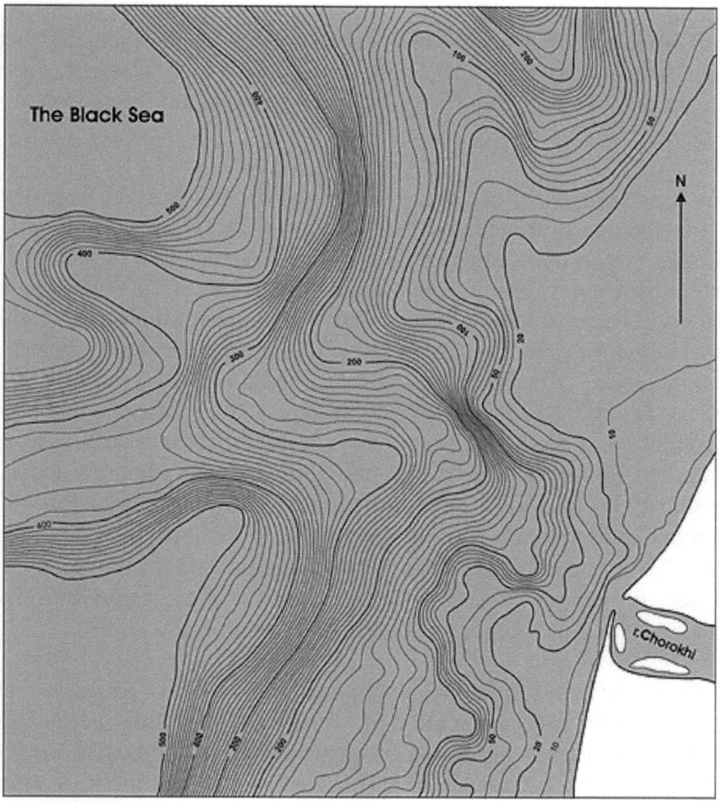

Figure 4. Bathymetrical map of the frontal part of the cape close to the head of the submarine canyon.

These factors in various combinations are known to facilitate the formation of sedimentary mass deposition on the submarine slopes and on the canyon bottom, which under some circumstances (long storm waves and tsunamis earthquakes, changes in tangential reaction between soil particles, increased pore pressure etc.) become unstable and can begin to move towards greater depths both as saturated plastic masses or as avalanches (debris flow) (Lykousis, Sakellariou, Roussakis 2003). These form deep-water depositional fan system observed downslope (in our case at depth 1600-1700m).

Investigation of the stability conditions of the bottom sedimentary layers deposited inside the submarine gorge (channel) has been undertaken on the base of field observations, that repeatedly was held in 1960-2003 years in the pre-deltaic submarine canyon of Rv.Chorokhi (Figures 5,6). Sediment material from Rv.Chorokhi is a main component of the littoral cell of the Adjarian coastline of the Black Sea.

The Chorokhi Canyon head locates at depth of 7-8 m at a distance of 70-140m from the river mouth and invades the area of sediment laden currents. Its frontal area spreads along the shore line for as much as 1,5 km. More, than 90% of the river's load discharge were transported through the Chorokhi Canyon (Russo, Bilashvili 2004).

The field research includes routine and periodic observations, including bathymetrical measurements (maps), hydrological, gravimetric and visual research. In addition an analytical approach has been used for quantitative evaluation of stability conditions of mass deposits on the bottom of the canyon.

The Batumi cape belongs to accumulative formation and is built from Rv. Chorokhi alluvium. It is situated in the central part of city of Batumi and south part of the port water area, that conditioned urgency of its study. At the end of 19th century in the coastal zone of the cape was built a spur aiming protection of the port water area as well as entering channel against silting by abundant drift and extreme waves. Construction of the spur accelerated the pace of a natural growth of the cape and already at the 30-40-ies of 20th century frontal part of the cape approached the river mouth of the submarine canyon (Figure 4). After this the growth of the cape actually stopped and beach-forming material, brought by alongshore stream from the south, started wasting in depths. This process is studied well by means of traditional instrumental methods and visually - by means of scuba divers (up to 40 m depth) and bathyal-diving (up to 350-400 m depth). At the different depths were observed fragments of the Batumi spur- blocks and sections of the accumulative masses of beach forming material. This is reliable marker to prove transfer of the terrigenous masses or constructions of any shape and size from the surface of the coastal zone into the sea depth. (Bilashvili, Russo 2003).

The nature of the waste of submarine canyon material varies for different sections of the cape. During the storm from the south part of the cape to the submarine slope with big inclination (up to 30°) takes place a permanent transfer of the beach-forming material. The material is gradually accumulating in the front part of the cape till it reaches critical mass. After that, even an insignificant impulse is sufficient to disturb gravitational balance and cause immediate move of the accumulated material into the depths. This process is analogous to submarine landslide or avalanche in the mountains. Above described process took place on the 14th of January 1999, when resonance of strong earthquake in Turkey in the conditions of quiet sea provoked immediate fall of a slide of a big volume from the Batumi cape into canyon. As moving to the depth accumulative mass took away part of the beach approximately with the length of 200m and width of 50-60 m.

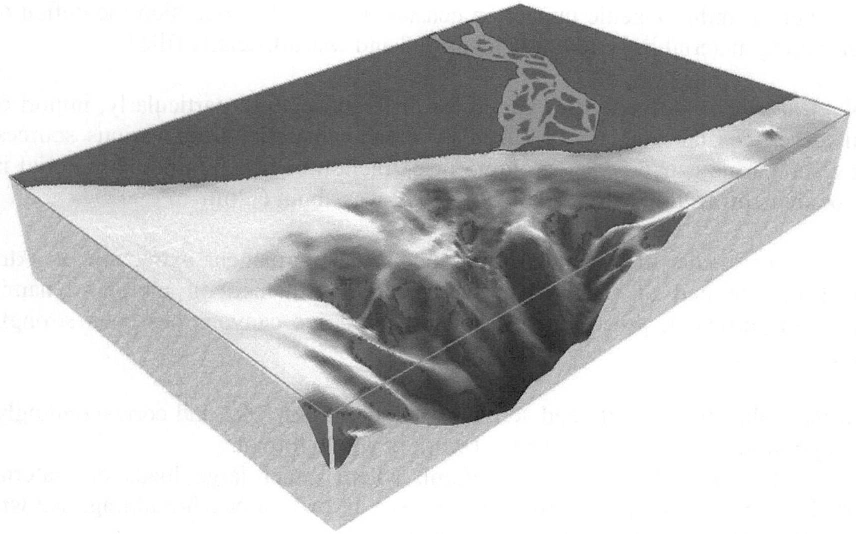

Figure 5. Sample of 3D block with overall morphology of Rv.Chorokhi Canyon.

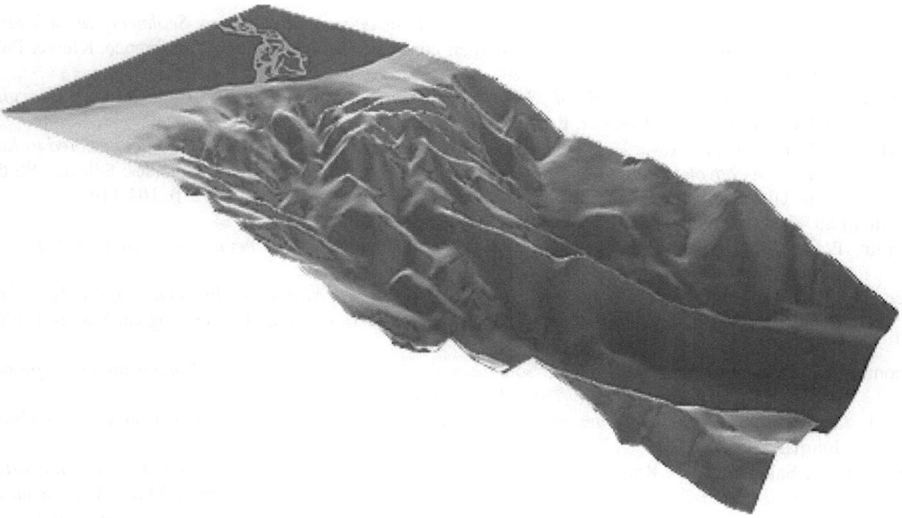

Figure 6. Sample of 3D block with overall morphology of Rv.Chorokhi Canyon.

4. Conclusions

From 1982 in Georgia a new method of reanimation of processes of the formation of the coast and its further management was elaborated. It took into consideration not only protection and fortification in different sections of the coast, but rehabilitation of existing natural processes in the frames of common dynamic system and its further regulation before anthropogenic impact on coastal zone. To this occasion the deficit of coast formatting material in system was identified and was artificially filled.

Technological schemes of activities varied for different regions, particularly, import of material to sea coastal zone was realized by passing material from various sources, including submarine canyons. The volume of extraction of beach-building material in Adjara for shore protection activities in 1982-1992 was about 6,2mln m^3.

The analyzes of results of the monitoring shows that permanent extraction of extra material from the bed of the canyon has positive influence on morpho-dynamic conditions. In particular, process of erosion in submarine canyons has been strongly decreased.

Unfortunately, the above-mentioned activities reduced from 1992 and correspondingly, it was not possible to carry out the planned work in whole format.
Though, to the north of the port of Batumi, where rather large loads of material transported by vessels took place, still exists reserve of load for beach-building, that will be enough to retain sustainability for 3-5 years more.

5. References

Bilashvili K.A., Savaneli Z. (2003). *"Stability conditions of accumulative Forms Sediments on submarine Slopes"*. Submarine mass movement and their consequences International Conference, Kluwer Publ., Nice, France.

Bilashvili K., Russo G. (2003) "*Slide of Mass Deposits on Submarine Slope of Cape Batumi, Georgia"*. OMARC Conference, Elsevier, Paris.

Carlson, P.R., Karl, H.A., Edwards, B.D., Gardner, J.V., and Hall, R., 1993. *Mass movement related to large submarine canyons along the Beringian Margin, Alaska.* In: Submarine Landslides: Selected Studies in the U.S. Exclusive Economic Zone. U.S Geological Survey Bulletin 2002, p. 104-116.

Jaoshvili Sh. (2003) "The rivers of the Black Sea". Tbilisi,Georgia.

Kaplin P. (1990) Предисловие редактора в книге Э.Барда. *"Изменение береговой линии"*. Л.: «Гидрометеоиздат».

Lee, H.J., Schwab, W.C., and Booth, J.S., 1993. *Submarine landslides: an introduction.* In: Submarine Landslides:Selected Studies in the U.S. Exclusive Economic Zone. U.S Geological Survey Bulletin 2002, p. 1-13.

Leontiev O., Lukianova S., Nikiforov L., Soloviova G., Kholodilin N. (1977) "*Карта типов берегов и побережья Мирового океана //Рельеф и ландшафты/"* М.:МГУ.

Locat, J.and Lee, H.J., 2000. *Submarine landslides: advances and challenges.* Canadian Geotechnical Journal, 39: 193-212.

Lykousis V., Sakellariou D., Roussakis G. 2003 *Prodelta slope stability and associated coastal hazards in tectonically active margins:gulf of corinth (NE Mediterranean),* Submarine Mass Movements and Their Consequences, Jacque Locat and Jurgen Mienert (Editors), Kluwer Academic Publishers.

Meladze F. (2003) *"Инженерные решения защиты морских берегов"*. Тб.

Russo G, Safianov G., Khorava A. (1987) *"Влияние искусственных изъятии пляжеобразующих наносов на динамику приустьевого взморья р.Чорохи"*. Moscow.

Russo G, Bilashvili K. (2004) *"Terrigenous Mass Dynamics in the Deltaic Canyon of the Rv. Chorokhi, Georgia"*. Poster Presentation. 32 Intern.Geological Congress, FLORENCE, Italy.

Safianov G (1978) "*Береговая зона океана в XX веке"*. М.:изд. «Мысль.

THE 1990 SUBMARINE SLIDE OUTSIDE THE NIDELV RIVER MOUTH, TRONDHEIM, NORWAY

J.-S. L'HEUREUX[1], O. LONGVA[2], L. HANSEN[2] AND G. VINGERHAGEN[1]
[1] *Norwegian University of Science and Technology (NTNU), 7491 Trondheim, Norway.*
[2] *Geological Survey of Norway (NGU), 7491 Trondheim, Norway.*

Abstract

The Trondheim harbor has been the locus for many large flow slides during the last century. The most recent of these occurred in 1990 just outside the mouth of the Nidelv River and mobilized ca. 5×10^6 m^3 of sediments. The mass movement took place as a liquefaction-induced flow slide outside the river outlet and developed into a lateral spread. The sediment mass slid along a weak layer of loose silty sand recognized by a distinct seismic reflection interpreted from high resolution seismic data. A combination of static and dynamic loading probably triggered the slide.

1. Introduction

Liquefaction flow slides are recurrent phenomena in coastal and deltaic soil deposits. They can be triggered by dynamic effects such as earthquakes, detonations and surface waves, and by static effects such as tidal variations and sedimentation. However, these flow slides mostly happen without much warning and the circumstances regarding the initiation of these slides are of great interest but often difficult to assess.

The Trondheim harbor, in Mid Norway (Fig. 1), has been the scene of many large flow slides during the last century. In this paper, we use a multidisciplinary approach including high resolution seismic and core data to describe the latest flow slide (1990). The aim is to explain the geotechnical character, the failure mechanisms and the possible triggers of the slide.

2. The 1990 slide at the mouth of the Nidelv River

In the history of the Trondheim harbor, three major slides are known to have occurred (Fig. 1). The slides from 1888 and 1950 are thoroughly described by Skaven-Haug (1955) and Bjerrum (1971). The latest of the three flow slides occurred at the mouth of the Nidelv River (Fig. 1) on April 25th of 1990 (Sand, 1990; Emdal *et al.*, 1996). At 16h21 of that day, the electricity and telephone cable and the water pipe joining the island of Munkholmen to the city was broken (Fig. 1). At exactly the same time, the electricity and phone cables to an emergency testing platform were also broken ("P"-Fig. 1). These cables were located in the road embankment southwest of the platform which, at that time, slumped into the fjord. Twenty minutes later, the beach east of the platform also slumped into the fjord. Parts of the shoreline to the east of the platform failed later that night.

V. Lykousis, D. Sakellariou and J. Locat (eds.), Submarine Mass Movements and Their Consequences, 259–267.

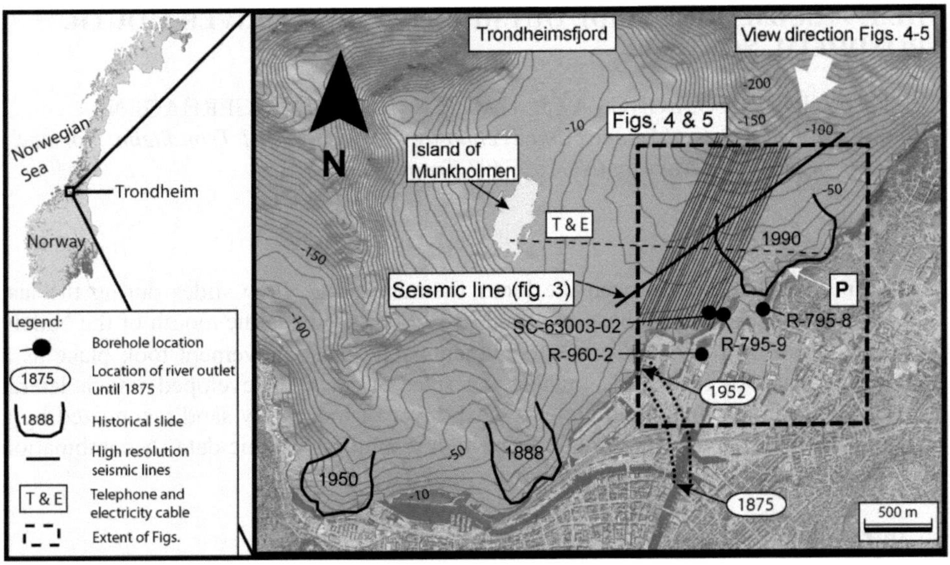

Figure 1. Site location, extent of figures, location of seismic line, boreholes locations and the position of the river outlet through time.

The slide affected an estimated area of 1 km^2 and the calculated volume of the slide is approximately 5 x 10^6 m^3. Many factors may have contributed to the sliding: 1) relocation of the river outlet and increase in sedimentation (Fig. 1), 2) initiation of a small fill along the shore some days before sliding and 3) vibration related to detonations along the shore 3 hours and 20 minutes before the event. Low tide was registered at 18h49.

3. Material and methods

High resolution bathymetric mapping of the Nidelv delta was carried out in 2003 by means of a 250 MHz GeoSwath interferometric side-scan sonar. More recently, in the fall of 2006, a grid of high resolution seismic lines at 20 m line spacing was collected in the slide area (Fig. 1). The survey was carried out using the Topas parametric system. Sediment coring has been carried out in the area over several years (Sand, 1990; Sand, 1996; and ScandiaConsult, 2003). The cores were taken in the shallower area of the harbor and are believed to be representative of the failed material (Fig. 1).

4. Physical settings

The sediments in the harbor consist of layered sequences of silt and sand resting partly on bedrock, moraine and marine clays. The deltaic material originates from the Nidelv River which starts at Lake Selbu (160 *m.a.s.l*) and runs for a distance of approximately 31 km before it ends in the Trondheimsfjord. Until 1875, the outlet of the river was further south in the city (Fig. 1). At that time a harbor development started, including a prolongation of the river bed more than 1 km in the northern direction, changing the river current and sedimentation conditions along the harbor (Emdal *et al.*, 1996). Again in 1952 the Nidelv River was redirected to a more north-eastern flow direction giving rise to the present day position of the river mouth in the north-eastern corner of the harbor (Fig. 1). The present Nidelv River is regulated and the course of the channel has been locked in place by the authorities in order to limit its tendency to migrate. From geophysical measurements, the sediment thickness at the present river outlet is approximately of 100-125 *m.*

5. Geotechnical characterization of the source area

Sediment core SC-63003-02 shows a stratified coarsening upward sequence, typical of deltaic deposits (Fig. 2A). The material is characterized by two units: A- 4.5 meters of medium to coarse gravelly sand indicating sedimentation proximal to the river outlet; and B- 7.5 meters of uniform medium to fine silty sand, with some laminated silt/clay, scattered shells and plant fragments. Facies changes are reflecting discharge events combined with relocation of the river outlet. The sole of a shoe was found at a depth of 15.8 m in a fine silty sand material corresponding to unit B in core R-960-2 (Sand, 1996) (Fig. 1). This indicates that the sediments are fairly recent with high sedimentation rates. The water content in core SC-63003-02 shows a tendency to increase with depth while the density of the soil decreases with depth (Fig. 2A). Geotechnical profiles corresponding to gravity cores R-795-8 and R-795-9 (see Fig. 1) are indicating very similar material properties (Emdal *et al.* 1996). The range of particle size corresponding to unit B and encountered in the different cores is presented in Fig. 2B. The uniform grain size distributions are typical for flowslide materials (Emdal *et al.*, 1996; and Kramer, 1988). Porosity values ranges from 35 to 38% for soil corresponding to unit A and values of 48-55% for soil matching unit B according to Emdal *et al.* 1996. Undrained shear strength values (*Su*) measured from the fall cone test are in the range 10-20 kPa for unit B (Fig. 2A). These values lead to ratio of *Su* to effective overburden stress in the order of 0.1-0.25.

A band of high amplitude reflectors can be identified at 4-5 m subbottom depth (Fig. 3). This may correspond to the coarser material of unit A. It is also possible to recognize the failure plane of the 1990 slide. It coincides with a transparent layer beneath a bright reflector that can be followed over most of the area. This potentially weak layer likely corresponds to a loosely compacted silty sand layer. Such sand layers have been recognized in the cores nearby (Fig. 2A). It is not possible, however, to make an exact correlation between cores and seismic data due to low acoustic penetration in the near shore areas.

6. Slide morphology and failure mechanisms

The combination of bathymetric soundings prior to the slide event (Fig. 4A) and swath bathymetry imagery collected after the slide took place (Fig. 5A) gives an exceptional opportunity to describe the geomorphological features of the slide and to reconstruct its development (Fig. 5B & 5C).

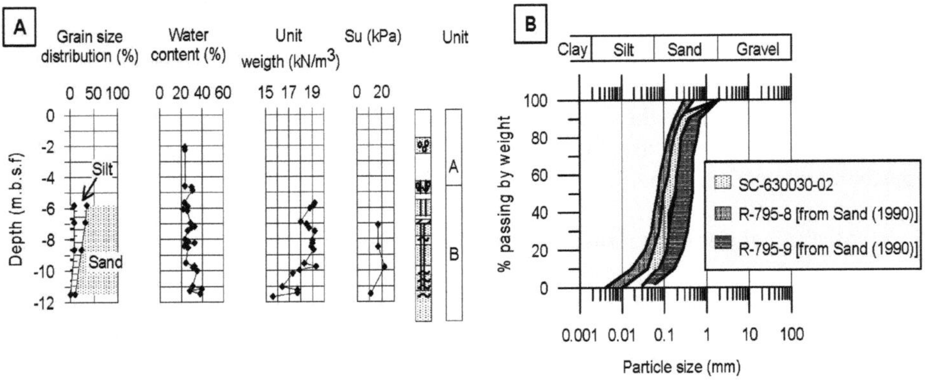

Figure 2. A) Geotechnical profile for gravity core SC-63003-02 (see Fig. 1 for location), and B) Range of grain size distribution corresponding to unit B for the different holes shown on the location map of Fig. 1.

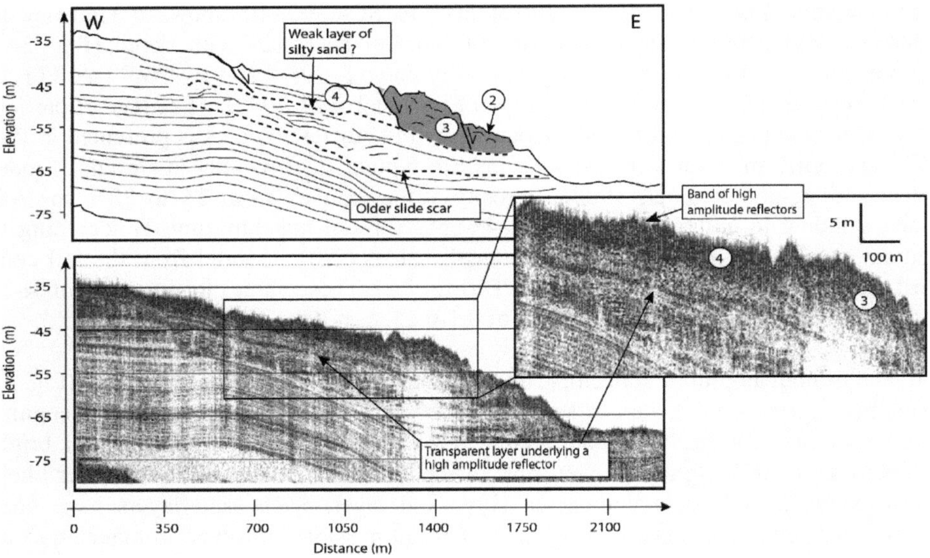

Figure 3. Seismic profile (location Fig. 1) in a west-east direction across the slide and the debris flow deposits. Notice slide blocks 2, 3 and 4 (see also Fig. 5), the weak layer of silty sand seen below the high amplitude reflector and the older slide scar. The main sliding movement is in a direction oblique to the figure.

6.1 STAGE 1 – CONDITIONS PRIOR TO FAILURE

The average slope angle, in the source area prior to failure, was in the order of 13 degrees near the coast and 2-3 degrees some 300 m off the coast (Figs. 4A & 4B).

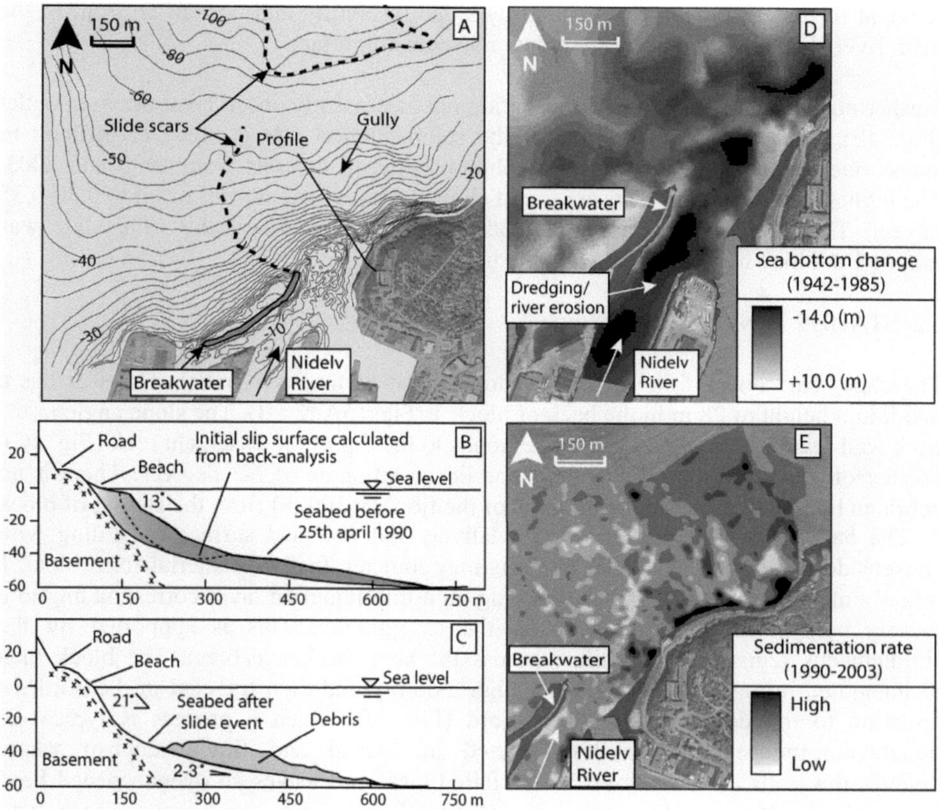

Figure 4. A) Situation map outside the mouth of the Nidelv River in 1985 [data from Sand (1990)], B) bathymetric profile before the slide, C) bathymetric profile after the slide D) estimated change in sea bottom in the period 1942-1985 and E) general sedimentation pattern in the period 1990-2003.

The bathymetry contours of Fig. 4A seem to indicate that already before 1985, the sediments outside the river outlet had been subjected to instability. At ca. 800 m north outside the river outlet and at a water depth of more than 80 m there appears to be a scar indicating sediment failure (Fig. 4A). This scar is well defined in the swath bathymetry imaging from 2003 (Fig. 5A).

Immediately outside the river mouth, and starting from the edge of the breakwater, a 750 m long and 1 m high scar indicates that sediments have settled vertically (Fig. 4A). These sediments correspond to block 4 in the bathymetry imaging of Fig. 5C. The volume of this block is roughly estimated to 1 x 10^6 m^3. The seismic data shows that detachment of block 4 occurred along a well-defined plane (Fig. 3), previously identified as a potential weak plane over the whole study area (Section 5). An important feature to observe on the bathymetry contour of 1985 is a gully feature in the sediment slope right outside the river mouth. This erosive element is possibly associated to a slide scar on the delta edge and testifies to early slope instability. Furthermore, bathymetry data shows important sea bottom changes in the period 1942-1985 (Fig. 4D). The removal of material may be explained by slope instability and current erosion. In the main river channel, dredging activity has also been important.

Furthermore, a relatively high sedimentation rate is presently observed at the river outlet (Fig. 4E). The absolute magnitude of the sedimentation rate is however difficult to assess due the different instrument resolution between datasets from 1990 and 2003. The highest sedimentation rates are found along the coast and are estimated to 0.30-0.35 m/year. This high sedimentation rate occurs in the area where the 1990 slide was initiated and may have contributed to the instability.

6.2 STAGE 2 – INITIATION

The slide scar can be followed 800 m to the east of the river outlet. The scar has a maximum height of 28 m in the back of block 1 (Figs. 5A & 5B). The slope angle in the back wall reaches 21 degrees and corresponds to the dip of the basement rock (Fig. 4C). Further off the coast (ca. 300m) the slope lies at an angle of 2-3 degrees. There is no debris to be found in the shallower parts of the fjord (< 100 m) from the sliding of block 1. The bathymetry imaging shows a relatively flat polished surface of sliding with slickenside features (Figs. 5A & 6). This may indicate that the material from block 1 initially moved as a coherent flake moving on a thin liquefied layer corresponding to a loosely compacted or labile layer in the stratigraphy. This is supported by the simultaneous registration of failure across the area (broken cables). The block then disintegrated retrogressively, gained a high mobility and then traveled in the northern direction to the deeper parts of the fjord (Fig. 5B). Such a process is typical of liquefaction-induced flow slides reported in coastal and fluvial deposit where liquefaction leads to the removal of the failed mass and leaves an oversteepened head scar, which itself can lead to another liquefaction failure. The retrogressive process may continue until a stronger material is reached. The steep head wall left by block 1 lends support to this. Moreover, the bathymetry imagery shows that a small slab of sediment to the east of block 1 was displaced and slipped until the slope flattened out (Fig. 5B). It is however, difficult to assess whether this smaller slide occurred simultaneously to the larger one.

6.3 STAGE 3 – CONTINUED MOVEMENT

Deposits from blocks 2 and 3 cover a large portion of the sea floor and are characterized by a lobe shape with hummocky surface and many compression ridges parallel to the slide scar (Figs. 5A & 5C). The surface of these blocks corresponds to the original sea floor which was deformed by compression and shearing during the down slope movement. On the seismic profile the displaced masses are recognized by an uneven topography and chaotic internal reflectors (Fig. 3). Seismic data suggests that the deposits are lying on a weak layer defined previously (Section 5) and on which also the detachment of block 4 occurred.

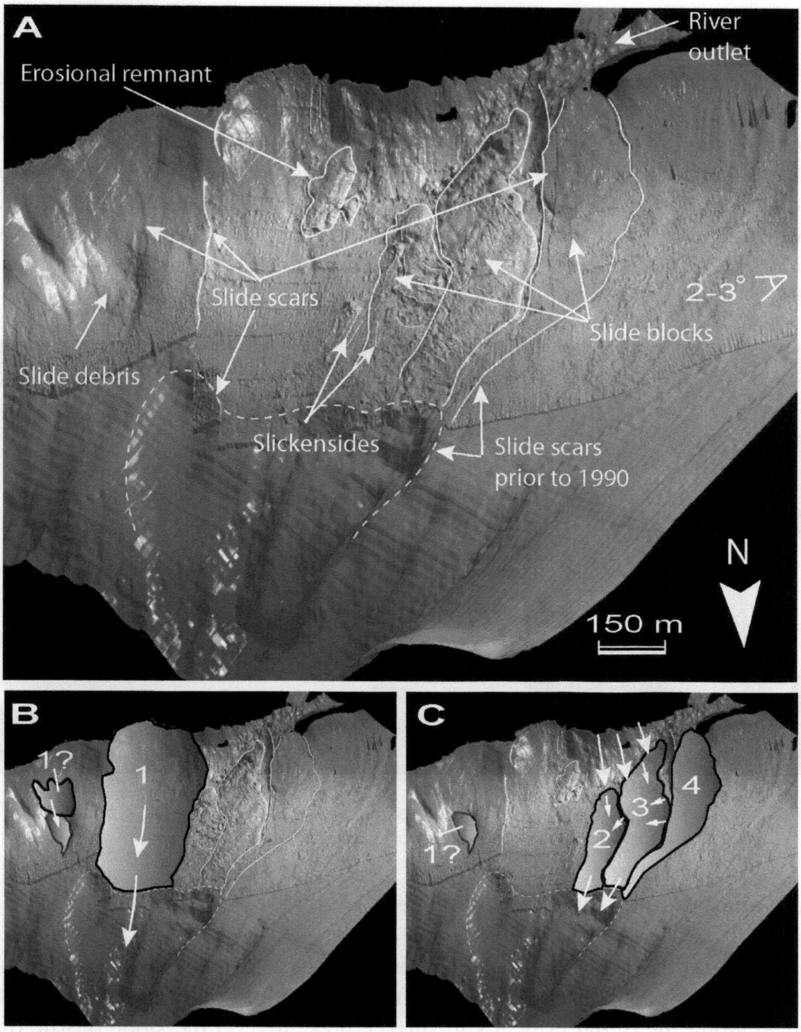

Figure 5. Shaded relief swath bathymetry imagery from 2003 outside the Nidelv River mouth showing: A) slide area morphology, B) Initial failure stage and C) final slide deposits.

It is difficult to assess whether blocks 1, 2 and 3 failed simultaneously. However the fact that the cable to the island of Munkholmen was broken precisely at the same time as the road slumped into the sea seems to confirm they failed within a very small time interval. The movement of blocks 2, 3, and likely also block 1 was predominantly translational along a weak plane and accompanied by retrogression (Fig. 6). The sediment package above the weak layer responded in a brittle manner to the underlying deformation to form compression ridges and blocks. The stratified block split internally along discrete layers as shown by retrogressive disintegration of block 2 and steps along the erosional remnant (Fig. 6).

6.4 STAGE 4 – ADJUSTMENT

Sand (1990) mentioned that witnesses saw a section of the shoreline, situated to the east of the testing platform, slumping into the sea 20 minutes after the initial slide. Parts of the shoreline also failed later that night. The striated surface southeast of the erosional remnant reflects evacuation of debris in the late stage of the slide (Fig. 6). Moreover, block 4 is likely to have experienced additional settlement in the latest stage of the slide. The swath imagery shows a 1 m high scar to the back of this block (Fig. 5A). It is however difficult to estimate on the amount of settlement due to the adjustment phase since this block had also settled prior to 1990 (Section 6.1).

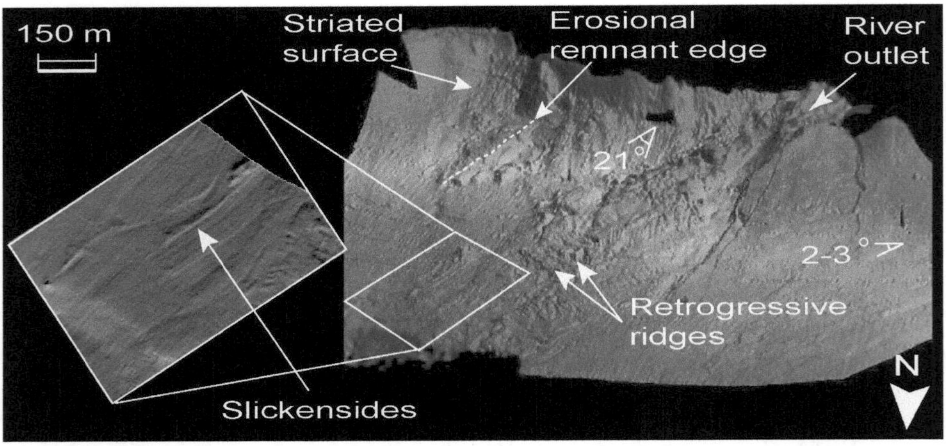

Figure 6. Bathymetry imaging showing slickensides on the sliding surface of block 1, retrogressive ridges in blocks 2 & 3 (see Fig. 5 for block location) and the erosional remnant.

7. Back-analysis and trigger mechanisms

Using limit equilibrium theory, a slope stability back-analysis shows that the mobilized friction angle (φ'_{mob}) is of 12 degrees at the initiation of failure. Calculation was performed along the initial critical slip surface presented on the slope profile of Figure 4B, and assuming hydrostatic conditions in the sediments. The back-calculated φ'_{mob} is much lower than effective friction angle values at critical state which usually ranges from 30 to 35 degrees in such fine sand deposits. This is typical of liquefaction flow slides and as been observed from earthworks in the Beaufort Sea (Sladen *et al.*, 1985b), instrumented model test (Eckersley, 1990) and during construction of the river training works at the Jamuna Bridge in Bangladesh (Hight *et al.* 1999). Moreover, the slope was subjected to dynamic loading from an explosion and, to static loading from filling along the shore on the day of the event. These conditions possibly caused the pore pressures to increase faster than they could dissipate, for a short moment, and resulted in instability of a weak layer. Once the instability was triggered, the weak layer liquefied and resulted in sliding, spreading and partly disintegration of a flake of denser material in proximity of the outlet. Other mechanisms which may have added to the instability are the renewed sedimentation-consolidation processes associated to the change in river outlet in 1952 and residual pore pressures in the sediments due to tidal drawdown. The effect of these mechanism are, however uncertain, and difficult to assess at present.

8. Conclusions

Already before 1985 slope instability could be observe outside the Nidelv river mouth. Newly acquired high resolution seismic and bathymetry imaging seem to show that a large block to the west of the river mouth had rotated along a well defined seismic reflector interpreted as a weak layer of loose silty sand. On April the 25th 1990, a large mass movement initiated as a liquefaction-induced flow slide ca. 600 m outside the river outlet. Subsequently, the sediment mass close to the river outlet slid along a distinct weak layer which presumably liquefied during or after the initial failure. The firmer and coarser sediments overlying this soft layer responded in a brittle manner to the deformation and the failure developed into a lateral spread in this latest stage. The slide seems to be similar to other described flow slides in the literature in such a way that it possibly initiated at a mobilized friction angle smaller than that at critical state. Mechanical testing of the sand from the source area would, however, provide a better appreciation of this collapse mechanism. A combination of dynamic loading from an explosion and static loading from filling along the shore probably triggered the slide.

9. Acknowledgements

This research project is financed by a consortium including the Norwegian road authorities, the Trondheim Harbor and the Norwegian Geological survey via the International Centre for Geohazards. Dr. Anders Solheim and Dr. Martyn Stoker are acknowledged for constructive reviews of the manuscript. Eiliv Danielsen and Oddbjørn Totland are thanked for collection of the seismic data. Kåre Sand is also thanked for the help in the compilation of geotechnical data.

10. References

Bjerrum, L., 1971. *Subaqueous slope failures in Norwegian fjords*. Norwegian Geotechnical Institute, Publication No. 88.

Eckersley, J.D., 1990. *Instrumented laboratory flow slides*. Geotechnique, Vol. 40, p. 489-502.

Emdal, A., Janbu, N. and Sand, K., 1996. *The shoreline slide at Lade*. Proceedings of the 7th International symposium on landslides, Trondheim, Norway, vol. 1.

Hight, D.W., Georgiannou, V.N., Martin, P.L., and Mundegar, A.K., 1999. *Flow slides in micaceous sands*. Proc. of the Int. Symp. on Problematic Soils, IS Tohoku, Sendai, Japan. Balkema, Rotterdam, Vol. 2: 945-958.

Kramer, S. L., 1988. *Triggering of liquefaction flow slides in coastal soil deposits*. Eng. Geol., 26:17-31.

Sand, K., 1990. *Raset ved Ladehammeren 25.04.90.* Report R-795-2. Trondheim Municipality.

Sand, K., 1996. *Pir II – Lade molo*. Report R-960. Trondheim Municipality.

Scandiaconsult, 2003. *Trondheim Havn - Lade Molo*. Ground investigation report No. 630030A.

Skaven-Haug, S., 1955. *Underwater slides in the Trondheim harbor area*. Norwegian Geotechnical Institute. Publication No. 7.

Sladen, J.A., D'Hollander, R. D. and Krahn, J., 1985 (b). *Back analysis of the Nerlerk berm liquefaction slides*. Canadian Geotechnical Journal, 22 (4): 579-588.

SUBMARINE SLOPE FAILURES NEAR SEWARD, ALASKA, DURING THE M9.2 1964 EARTHQUAKE

P.J. HAEUSSLER[1], H.J. LEE[2], H.F. RYAN[2], K. LABAY[3], R.E. KAYEN[4], M.A. HAMPTON[4] AND E. SULEIMANI[5]

[1] *U.S. Geological Survey, Anchorage, AK, USA, email: pheuslr@usgs.gov*
[2] *U.S. Geological Survey, Menlo Park, CA, USA*
[3] *Science Applications International Corporation (SAIC), Contractor to U.S. Geological Survey, Anchorage, AK, USA*
[4] *U.S. Geological Survey, Menlo Park, CA, USA*
[5] *Geophysical Institute, Univ. of Alaska, Fairbanks, AK, USA*

Abstract

Following the 1964 M9.2 megathrust earthquake in southern Alaska, Seward was the only town hit by tsunamis generated from both submarine landslides and tectonic sources. Within 45 seconds of the start of the earthquake, a 1.2-km-long section of waterfront began sliding seaward, and soon after, ~6-8-m high waves inundated the town. Studies soon after the earthquake concluded that submarine landslides along the Seward waterfront generated the tsunamis that occurred immediately after the earthquake. We analyze pre- and post-earthquake bathymetry data to assess the location and extent of submarine mass failures and sediment transport. New NOAA multibeam bathymetry shows the morphology of the entire fjord at 15 m resolution. We also assembled all older soundings from smooth sheets for comparison to the multibeam dataset. We gridded the sounding data, applied corrections for coseismic subsidence, post-seismic rebound, unrecovered co-seismic subsidence, sea-level rise (vertical datum shift), and measurement errors. The difference grids show changes resulting from the 1964 earthquake. We estimate the total volume of slide material to be about 211 million m^3. Most of this material was transported to a deep, flat area, which we refer to as "the bathtub", about 6 to 13 km south of Seward. Sub-bottom profiling of the bathtub shows an acoustically transparent unit, which we interpret as a sediment flow deposit resulting from the submarine landslides. The scale of the submarine landslides and the distance over which sediment was transported is much larger than previously appreciated.

Keywords: submarine landslides, 1964 earthquake, Alaska, multibeam bathymetry, tsunamis

1. Introduction

The 1964 M9.2 great Alaska earthquake is the second or third largest ever recorded, and it was the first well-documented megathrust earthquake (Plafker 1965). Shaking during the earthquake lasted about 4.5 minutes and the rupture area was about 800 km long by 250 km wide (Christensen and Beck 1994). The earthquake produced both tectonic and submarine-landslide-generated tsunamis. In Alaska, 106 of 115 deaths were tsunami related, but 83 of these 106 deaths were related to submarine landslide-generated tsunamis. The fjords and glacial landscape of coastal Alaska are an ideal geologic environment for producing submarine landslides, because they deposit sediment on the

V. Lykousis, D. Sakellariou and J. Locat (eds.), Submarine Mass Movements and Their Consequences, 269–278.

steep walls of fjords, which can subsequently fail. In order to understand better the local tsunamis produced during the 1964 earthquake we are studying two fjords where they occurred and in which recent high-resolution multibeam bathymetry was collected. This paper focuses on Resurrection Bay, near the town of Seward; another study by Lee et al. (this volume) focuses on Port Valdez, near the town of Valdez (Fig. 1).

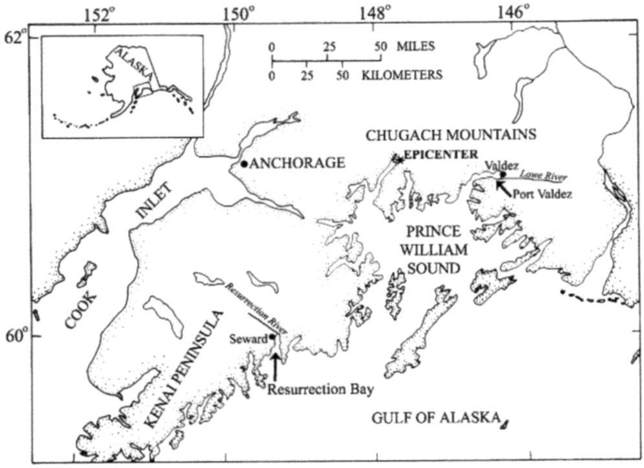

Figure 1. Location map showing Seward, Alaska, and the epicenter of the 1964 Mw = 9.2 earthquake.

There is a rich historical record of the 1964 earthquake and the tsunamis that struck Seward, Alaska. Seward was the only community hit by both submarine landslide-generated and tectonically generated tsunamis (Lemke 1967; Wilson and Torum 1972; Shannon and Hilts 1973). Within 30-45 seconds of the start of the earthquake, a section of the waterfront started sliding into the ocean accompanied by an 8-10 m drawdown of water. One or two large boils of water 3-5 m high were observed in the middle of the bay within 30 seconds of the start of earthquake. The boils produced large waves that traveled toward shore. Along with the initial drawdown of water along the Seward waterfront, fuel storage tanks fractured and oil ignited, resulting in flaming waves coming toward shore. After 1.5-2 minutes, a 6-8 m-high wave swept northward along shoreline. There was another high wave about 10 minutes after the earthquake. The tectonic tsunami arrived about 30 minutes after the earthquake, and at Seward had about the same maximum height as the initial landslide-generated waves.

Studies after the earthquake showed that an area along the Seward waterfront, 15 to 120 m wide by about 1200 m long, slid into the bay and contributed to the local tsunamis (Lemke 1967; Torum and Wilson 1972; Shannon and Hilts 1973). These studies described depth changes near the waterfront, examined the soil conditions in Seward, and inferred that a submarine landslide caused the initial tsunami. Plafker et al. (1969) catalogued tsunami run-up data.

The National Oceanographic and Atmospheric Administration (NOAA) recently completed a high-resolution multibeam bathymetry survey of all of Resurrection Bay, including the Seward waterfront (Fig. 2). The survey was completed as part of a program to update bathymetry at high-use ports. The survey shows features apparently related to submarine landsliding, and prompted the U.S. Geological Survey (USGS) to initiate research to better understand the submarine-landslide-tsunami events of 1964. In 2005 we obtained a grid of 'chirp' sub-bottom profiles, several gravity cores for dating, and vertical profiles of shear-wave velocity around the margins of the bay for geotechnical characterization. Our preliminary interpretations of the multibeam maps and sub-bottom profiles were published by Lee et al. (2006). In this paper, we reexamine the pre- and post-1964 bathymetry, and focus on assessing the changes before and after the 1964 earthquake.

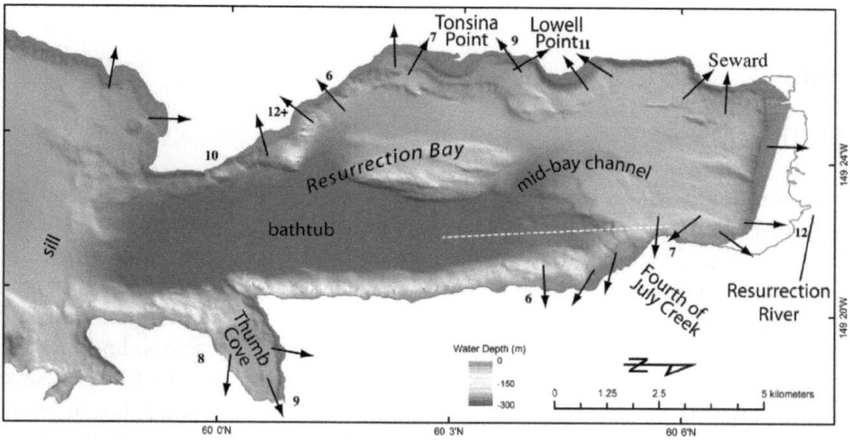

Figure 2. Multibeam imagery of Resurrection Bay showing features discussed in the text. Numbers and arrows around the margin of the fjord show the measured 1964 tsunami wave run-up height, in meters, and the run up direction (Plafker et al. 1969). Dashed white line is the location of seismic profile in Figure 6.

2. Methods

2.1 MULTIBEAM MAPPING

Multibeam echosounder systems are the preferred method for mapping seafloor bathymetry. Multibeam systems can provide complete bathymetric coverage of a swath of seafloor, with spatial errors of less than 1 m and vertical errors of less than 0.5% of the water depth. Complete multibeam coverage of Resurrection Bay was obtained in 2001 using high-resolution systems aboard the NOAA ship *Rainier* with differential GPS for navigation (Fig. 2). The sounding data were gridded at a 15-m cell size.

2.2 BATHYMETRIC DATA, 1905-1965

We digitized and combined 6905 soundings from eight NOAA smooth sheets from years: 1905, 1915, 1927, 1928, 1930, 1932, 1940, 1961. There was also a1965 survey, which had 11,186 soundings of the northern part of the fjord. All the older sounding locations were determined by triangulation from local features. Location errors are likely the largest single source of error. Depth measurements were recorded to the nearest fathom (6 feet, or 1.83 m) and rounded to the shallower fathom. We converted the depths to meters and combined the pre-1964 soundings into a single grid with a 15-m cell size. We maintained two significant figures of precision, although the error is likely around two meters (one fathom) on an individual measurement.

2.3 SUBBOTTOM PROFILES

We collected 157 km of 'chirp' high-resolution seismic reflection data in the fjord. Data were acquired using an Edgetech 512i Chirp System and recorded using Delph Seismic software. In 1992, we (MH) also acquired GeoPulse sub-bottom profiles for part of the fjord. The GeoPulse navigation data had issues that made it difficult to precisely locate some of the profiles.

3. Results

3.1 BATHYMETRY

The geomorphology of Resurrection Bay shows several alluvial fans spilling into the bay, a fjord-head delta of the Resurrection River, and a deep, flat trough, which we term 'the bathtub' (Fig. 2). The seafloor in the bathtub is now between 283-297 m below sea level. A sill, a maximum of 195 m below sea level, extends across the south end of the bathtub and effectively prevents sediment from leaving this depression. This fjord is like many others in coastal Alaska in that it was carved by glaciers, and after retreat, rivers entering the fjord formed fans and deltas at their margins. The town of Seward is built on the alluvial fan of Lowell Creek. Lowell Point, about 4 km south of Seward, is another alluvial fan, as is Tonsina Point, 2.5 km farther south. A significant feature of the 2001 bathymetry that is suggestive of submarine landslides is an area of blocks just east of Seward (Lee et al. 2006). The blocks have 5 to 10-m of relief, a few have up to 15 m relief, and they lie in the region of submarine landslides as identified by the post-1964 earthquake studies.

3.2 BATHYMETRY DATA ANALYSIS

We compared the 1965 sounding data with the 2001 multibeam data to evaluate the differences in the resolution of the datasets and to evaluate corrections applied to the data. We found that the 1965 survey is consistently shallower than the 2001 survey. For example, in the bathtub area, where the seafloor is relatively flat, and thus errors in the sounding location would be less significant, the 2001 survey is consistently deeper. We consider it unlikely that there has been much sedimentation in this area since the 1964 earthquake as it is far from the Resurrection River mouth. The median difference between the 1965 and 2001 surveys, at the 1965 sounding locations, is 2.17 m. This value incorporates post-seismic rebound of 0.32 m and takes into account 0.07 m sea level rise (Larsen et al. 2003). Given that the total difference is about 2 m, or about one fathom, we suggest that rounding the 1965 survey depths to the shallower fathom could account for some of the discrepancy. We infer that the same errors apply to the pre-1964 dataset, as much of it was collected with similar methods. The unrecovered part of the coseismic subsidence also needs to be applied to the pre-1964 data. Co-seismic subsidence was 1.07 m, but with 0.32 m post-seismic rebound, that leaves 0.75 m of unrecovered subsidence. Subtracting this value from the pre-1964 soundings should bring them to the same level as the 1965 and 2001 surveys. There has also been about 0.07 m of sea-level rise since 1964, so the vertical datum for the pre-1964 surveys has changed. Incorporating all of these corrections to the pre-1964 data yields a total adjustment of 2.99 m. After applying these corrections, we produced two maps that show the bathymetric differences between the 1965 adjusted grid and 2001 grid (the '1965-2001 difference map' (not shown)) and the pre-1964 adjusted grid and the 2001 grid (the 'pre-/post-1964 difference map' (Fig. 3)).

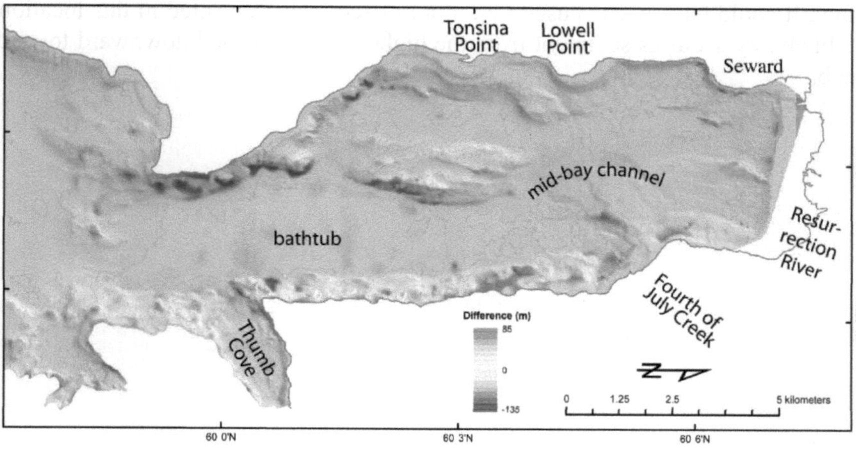

Figure 3. Difference between adjusted pre-1964 bathymetric grid and 2001 multibeam survey. The differences may be interpreted as sediment gain or loss during the 1964 earthquake. The zero change contour is shown. Cold colors indicate depth increase, and warm colors a depth decrease.

3.3 BATHYMETRY DIFFERENCE MAPS

The 1965 to 2001 difference map also highlights errors caused by the limited number of soundings in the older surveys. We see two common types, which we refer to as peak errors and base-of-slope errors (Fig. 4). Peak errors occur when the older survey had soundings on either side of a peak. The difference map, then shows positive values where the peak was missed by an old survey, which in an uncritical light could be interpreted as material added to the older surface. Base-of-slope errors are found where soundings from the older survey are absent from the base-of-slope. Thus, on the difference map, there are areas adjacent to the fjord walls, where negative values, suggest landsliding occurred, but this is just related to an insufficient number of soundings in the older survey.

Acknowledging these errors, the pre-/post 1964 difference map reveals at least seven areas where water depths increased substantially after the 1964 earthquake (blue colors, Fig. 3). These are: (1) Lowell Point, (2) Seward waterfront, (3) Resurrection River delta front, (4) old Fourth of July Creek delta front, (5) mid-bay channel, (6) Thumb Cove, and (7) Tonsina Point. Because of the significant deepening of the seafloor after the 1964 earthquake, we infer that most of locations with significant increases in depth are the source regions of the submarine landslides caused by the earthquake. The slide off of the Seward waterfront was previously known. GA Rusnak (unpublished, but cited in

Lemke 1967, p. E29) also concluded there were slides off the Resurrection River delta, Fourth of July Creek, and Lowell Point based on unpublished data. The other slides were not previously known. It is not surprising that there were submarine landslides off the Seward waterfront, Lowell Point, the Resurrection River delta, Fourth of July Creek, and the Tonsina Point fans. All locations have steep distal slopes and are actively aggrading fan-delta systems. A wave run up height of 9 m was reported at the head of Thumb Cove (Plafker et al. 1969), consistent with the presence of submarine failures in or near the Cove. Perhaps the most unexpected area of deepening is the mid-bay channel. It could have been caused by failure of sediment deposited in this location, or, more likely, by scour as sediment from the higher slides traveled downward toward the bathtub.

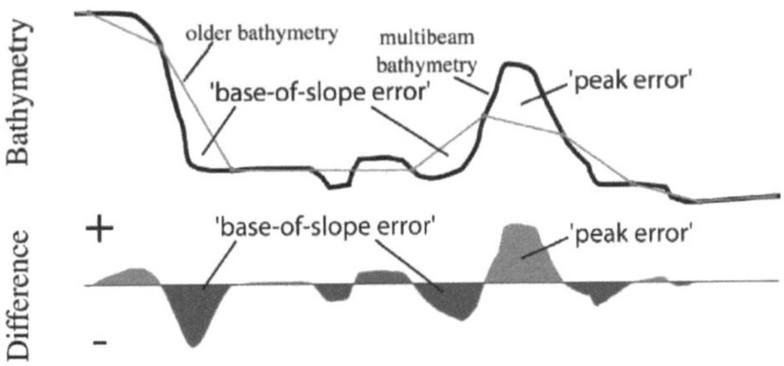

Figure 4. The origin of peak and base-of-slope errors. The upper, thick black line, shows a profile of high-resolution bathymetry. The thinner grey line shows a low resolution profile, which is more jagged, longer wavelength, and can miss peaks or the base-of-slope. Thus, when looking at the difference between surfaces constructed from both data sources, there can be apparent, and false, loss or gain of material where there are base-of-slope or peak errors.

Some slides are better documented than others. In Figure 5, we outline areas where the depth change contour was greater than 5 m and that contained three or more soundings (see also Table 1). We modified some contours such that they did not extend either onshore or into the bathtub. We also deleted some areas along the fjord walls, where the old data were variable, and where we infer that location errors were too large for us to be confident in the inference of sediment loss. Some slides, such as those off the old Fourth of July Creek delta (Fig. 3), have numerous pre-1964 soundings in the shallow areas, but few in the deeper areas. Base-of-slope errors cause the thickness and volume of the slide to be overestimated. Although the extent and volume of the submarine landslides cannot be more accurately determined with this map, we believe the general morphology and location of the 1964 submarine slides is correct. Some slides have well documented maximum depth increases, which are: Seward – 46 m; Lowell Point – 32 m; Fourth of July Creek – 48 m; mid-bay – 25 m; Thumb Cove – 38 m.

One large slide, located off the southern part of Fourth of July Creek (Fig. 2), appears to pre-date the 1964 earthquake. The morphology of it is clear in the 2001 multibeam survey, but the depths within it are similar in the pre-1964 data. The slide has a steep headscarp and appears related to an earlier failure of the Fourth of July Creek delta.

The main area where the depth decreased after 1964 is in the bathtub, which became an average of 3.5 m, and a maximum of 13 m, shallower (Fig. 3). This depth decrease indicates that much of the material from the submarine slides in the shallow areas of the fjord was transported up to 13 km and deposited in the bathtub.

3.4 EVIDENCE FOR LANDSLIDING ON CHIRP PROFILES

The chirp data show a number of features related to sedimentation and submarine landslides up to ~50 m below the sea bottom. The deepest imaged sedimentary units are typically finely layered, which we interpret as normal fjord sedimentation that occurred prior to the 1964 earthquake (see Fig. 4 in Lee et al. 2006).

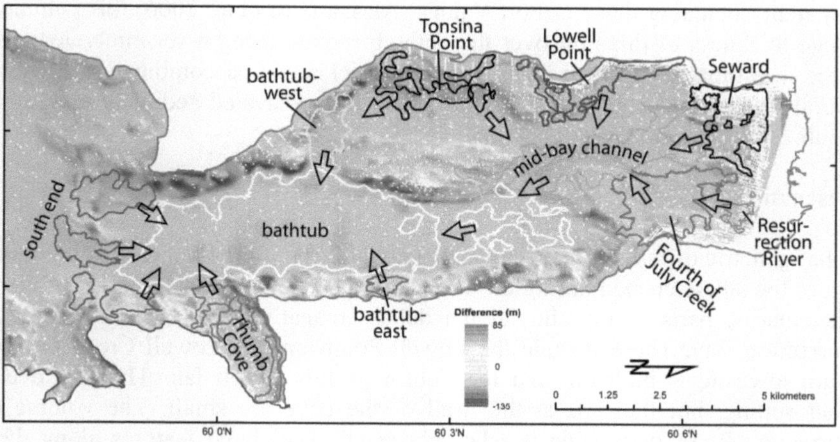

Figure 5. Pre-/post-1964 difference map as shown in Figure 3, but with inferred landslide areas, outlined with various colored lines. Slide names are adjacent to slide areas. Pre-1964 soundings are shown with white dots. Criteria for landslide areas area discussed in text. White lines outline bathtub areas where depth decreased in 1964. Arrows show interpreted sediment flow transport directions.

Table 1: Landslide and Bathtub Volumes, Areas, and Depth Change.

Landslide Volumes	**Volume ($\times 10^6$ m^3)**	**2D area ($\times 10^6$ m^2)**	**reliable maximum depth change (m)**
South End	33.3	2.93	---
Thumb Cove	16.5	1.08	---
Bathtub East	4.5	0.27	---
Bathtub West	15.3	1.06	---
Tonsina Point	16.8	1.45	-40
Lowell Point	18.1	1.49	-32
Seward	27.5	2.04	-46
Res. River	2.9	0.28	-47
Mid-Bay	40.7	4.18	-25
Fourth of July Creek	35.0	2.96	-48
All slides	210.6	17.74	---
Bathtub Volumes			
From bathymetric difference maps	53.4	15.16	13
From seismic data	188.0	---	20

East of Seward, the profiles show that the blocks, prominent on the multibeam bathymetry, have convolute layering at an angle to what we interpret as a slide plane beneath them (see Fig. 4 in Lee et al. 2006). Thus, because these blocks are located in the region that slid in 1964, they are a lag deposit left behind as the main slide mass moved downslope. The bathtub region has three units (Fig. 6). The deepest is a layered sequence, shallower is an acoustically transparent unit, and this is overlain by a thin unit of 1-2 narrow, laterally continuous, reflectors. The acoustically transparent unit has a few weak reflections and fills undulations in the deeper layered unit. As the depth of the bathtub decreased an average of 3.5 m after the 1964 earthquake, we infer that the acoustically transparent unit is a sediment flow deposit. Schnellmann et al. (2002) interpreted a similar facies for earthquake-induced landslides in Swiss lakes. Other sediment flow deposits imaged by high-resolution seismic data are also acoustically transparent, including those in Port Valdez, Alaska (Lee et al. 2006; this volume). The average thickness of this unit over the bathtub region, using a seismic velocity of 1.6 km/sec is estimated at 12 m. This sediment thickness is a combination of reworked material originating from landslides up slope and excavated sediment already in the bathtub.

4. Discussion

The bathymetric difference map shows that about 20% of the floor of Resurrection Bay, north of the sill, deepened during the 1964 earthquake. The most significant failures are the submarine parts of the alluvial fan deltas emanating from streams draining into Resurrection Bay. These include the Lowell Point fan, the Lowell Creek fan that the Seward townsite is built on, and the Fourth of July Creek fan. Thumb Cove has a similar setting, but the fans at the head of the cove are small. The volume of the Resurrection River delta slide is relatively small. The fluted features along the delta seen in the 2001 multibeam data suggest that small failures of the delta front occur relatively frequently. Deepening in the mid-bay channel is volumetrically similar to the other large failures, but occurred on remarkably shallow slopes of less than 1.5 degrees. This suggests that the channel is a result of being eroded by the flow of slide debris as it traveled toward the bathtub.

An integration of the volume of sediment in the bathtub determined by sub-bottom profiling (188 million m^3) is about 3.5 times the volume of sediment as determined from the bathymetric difference map (53 million m^3). We calculate the volume of the source regions of all slides as 211 million m^3. The cause of this discrepancy is not clear. We do not find evidence for a vertical datum issue as our pre-1964/2001 difference map produces nearly identical results as the lower resolution pre-1964/1965 difference map. Some sediment may have been deposited on the slopes between the slides and the bathtub, but neither the difference map nor the sub-bottom profiling indicate significant volumes of such sediment accumulations. Another explanation is that 70% of the sediment that slid became suspended and was transported down the fjord and beyond the bathtub. We consider such a high percentage as unreasonable. This leads us to consider that there may be still unresolved, systematic problems in the bathymetric surveys.

5. Conclusions

With careful analysis of the pre- and post-1964 bathymetry and chirp data we believe we have a good understanding of the location of the earthquake-induced submarine slides at Seward. The submarine failures initiated along the fjord walls in less than 50 m water depth and sediment was transported 6 to 13 km to the bathtub, likely as a series of coalescing sediment flows. These sediment flows probably scoured sediment from the mid-bay channel and upper bathtub and deposited it within the bathtub. The profiling data from the bathtub show a 12-m thick acoustically transparent unit, which we interpret as a sediment flow deposit. This deposit consists of both submarine landslide material and locally excavated material.

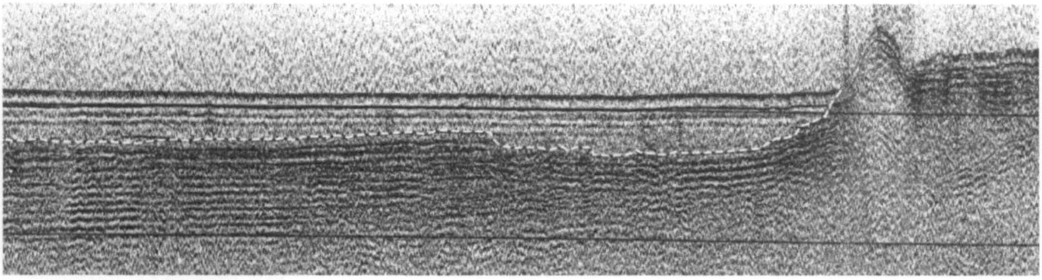

Figure 6. GeoPulse sub-bottom profile of bathtub region. Location shown by dashed white line on Figure 2. North is to the right. Horizontal black lines are 50 ms apart. Dashed white line is at the base of the acoustically transparent unit, which we interpret as a sediment flow deposited in the 1964 earthquake. Note the truncation of the layered reflectors beneath the acoustically transparent unit. We interpret this as a result of scour during deposition of the sediment flow.

These studies further underscore that fjords are an ideal environment for producing submarine landslides – especially when located above a megathrust. The nature and characteristics of the slides in Resurrection Bay, as well as the tsunami run up, are known well enough to provide useful test cases for models of submarine slide dynamics and tsunami generation. Lastly, the area affected by the submarine slides, the number of sediment slides, and the distance sediment was transported are much larger than previously appreciated.

6. Acknowledgements

We acknowledge help from the crew of the *R/V Montague* and support from the U.S. Geological Survey, the National Oceanic and Atmospheric Administration and the Geophysical Institute of the University of Alaska. We thank Chris Waythomas and Brian Bornhold for constructive reviews of the manuscript.

7. References

Christensen DH, Beck SL (1994) *The rupture process and tectonic implications of the Great 1964 Prince William Sound Earthquake*. Pageoph.: 142:29-53

Larsen CF, Motyka R, Freymueller J, Echelmeyer K (2003) *Tide gauge records of uplift along the northern Pacific-North American plate boundary, 1937 to 2001*. Journal of Geophysical Research: 108(B4), doi:10.1029/2001JB001685

Lee HJ, Ryan H, Kayen RE, Haeussler PJ, Dartnell P, Hampton MA (2006) *Varieties of submarine failure morphologies of seismically-induced landslides in Alaskan Fjords*. Norwegian Journal of Geology 86: 221-230

Lee HJ, Ryan H, Kayen RE, Hampton MA, Haeussler PJ, Locat J, Suleimani E, Alexander CR (2007) *Scale and style of seismically induced, tsunamigenic submarine slope failures in Port Valdez, Alaska, USA.* this volume

Lemke RW (1967) *Effects of the earthquake of March 27, 1964, at Seward, Alaska.* US Geological Survey Professional Paper 542-E. 43 pp

Plafker G (1965) *Tectonic deformation associated with the 1964 Alaska earthquake.* Science 148:1675-1687

Plafker G, Kachadoorian R, Eckel EB, Mayo LR (1969) *Effects of the earthquake of March 27, 1964 on various communities.* US Geological Survey Professional Paper 542-G. 50 pp. 2 plates

Schnellmann M, Anselmetti FS, Giardini D, McKenzie JA and Ward SN (2002) *Prehistoric earthquake history revealed by lacustrine slump deposits.* Geology 30:1131-1134

Shannon WL, Hilts DE (1973) *Submarine landslide at Seward.* In: The Great Alaska earthquake of 1964, Oceanography and Coastal Engineering. National Academy of Sciences, Washington, pp 144-156

Wilson BW, Torum A (1972) *Effects of the tsunamis: an engineering study.* In: The Great Alaska earthquake of 1964, Oceanography and Coastal Engineering. National Academy of Sciences, Washington, pp 361-523

THE AD 1881 EARTHQUAKE-TRIGGERED SLUMP AND LATE HOLOCENE FLOOD-INDUCED TURBIDITES FROM PROGLACIAL LAKE BRAMANT, WESTERN FRENCH ALPS

H. GUYARD [a,b], G. ST-ONGE [a,b], E. CHAPRON [c], F. S. ANSELMETTI [c], P. FRANCUS [d,b]

[a] *Institut des Sciences de la Mer (ISMER) à Rimouski, Université du Québec à Rimouski, 310 allée des Ursulines, Rimouski, Québec, Canada, G5L3AI*

[b] *GEOTOP-UQAM-McGill, C.P. 8888, Succursale Centre-ville, Montréal, Québec, Canada, H3C 3P8*

[c] *Geological Institute ETH Zürich, Universitätstrasse 16, CH-8092 Zürich, Switzerland*

[d] *Institut National de la Recherche Scientifique- Eau, Terre, et Environnement (INRS-ETE), 490 de la Couronne, Québec, Québec Canada, G1K 9A9*

Abstract

High-resolution seismic analyses on the sedimentary subsurface of the deep basin of proglacial Lake Bramant (Grandes Rousses Massif, Western French Alps) allowed the detection of a large lens-shaped body with chaotic internal reflections corresponding to a mass wasting deposit (MWD) triggered by the nearby AD 1881 Allemond earthquake (MSK intensity VII). This MWD was only retrieved at the base of a short gravity core and the top of a piston core. Sediments associated with this MWD are remoulded and laminated. Locally, blocks of sediment have preserved the original stratification. This earthquake-induced mass movement is an example of a slide that evolved into a slump. In addition, several Late Holocene turbidite and hyperpycnal deposits related to exceptional flood events were identified using high-resolution sedimentological, physical and geochemical analyses. However, the identification of hyperpycnites is sometimes complicated as erosion of the basal sequence can occur during the rising limb of the flood. While the precise dating of the oldest flood event is still ongoing, two flood events are coeval with the St. Sorlin glacier retreat following the end of the "Little Ice Age", suggesting outbursts of temporary ice contact lakes or subglacial lakes during warmer periods.

Keywords: slump, earthquake, hyperpycnites, turbidites, floods, proglacial lake, Western French Alps

1. Introduction

In lacustrine systems, mass wasting deposits such as slumps or turbidites can be generated by processes such as lake level change or slope overloading, but are more often triggered by the regional seismo-tectonic activity (Sims, 1975; Doig, 1986), especially in the Alpine region (Chapron et al., 1999; Schnellmann et al., 2002; Monecke et al., 2004; Nomade et al., 2005; Strasser et al., 2006). In addition, several major historic floods were previously recorded in Alpine lake sediments and were associated with hyperpycnal flows (Arnaud et al., 2002; Schneider et al., 2004). Hyperpycnal deposits (hyperpycnites) are characterized by the development of a coarsening-up basal unit during the increasing discharge period (up to the peak of the flood) and by a fining upward top unit during the decreasing discharge period (from the peak of the flood) (Mulder et al., 2001a, 2003; St-Onge et al., 2004; Chapron et al., 2006; St-Onge and Lajeunesse, this book). In proglacial environments, they can be deposited after the

V. Lykousis, D. Sakellariou and J. Locat (eds.), Submarine Mass Movements and Their Consequences, 279–286.

catastrophic drainage (i.e outburst) of a subglacial lake. Traditionally, the identification of catastrophic events resulting in mass movements such as earthquakes or exceptional floods is based on a range of sedimentological and textural criteria from visual description of sedimentary structures, on grain size measurements and on the establishment of a precise chronology. In this paper, we use a high-resolution multiproxy approach to identify, detail and determine the trigger mechanism of several rapidly deposited layers recorded in the deep basin of proglacial Lake Bramant in the Grandes Rousses Massif, Western French Alps (Fig. 1).

2. Setting

The Grandes Rousses Massif is affected by large tectonic features such as basement thrusts and strike-slip faults. The study area has been historically subjected to several moderate magnitude earthquakes (Chapron et al., 1999; Nomade et al., 2005) (Fig. 1a). Among them, the AD 1881 Allemond earthquake (intensity MSK VII) had an epicenter located less than 12 km away from Lake Bramant.

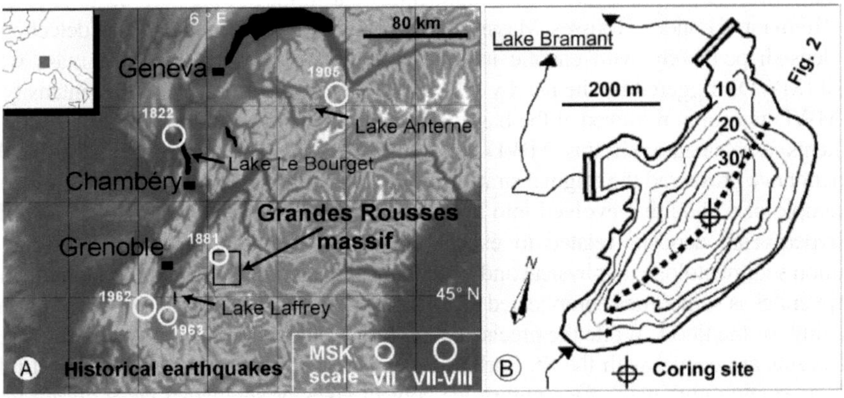

Figure 1: (A) General location of the Grandes Rousses Massif in the Western French Alps. Also shown are the MSK intensities at the epicentres of regional historical earthquakes previously recorded in lake sediments. (B) Bathymetry of Lake Bramant (thick isobaths: 5 m), situated on the northern part of the Massif. Hydroelectric dams were built at the lake outlets in 1918. Location of the coring site is based on seismic reflection profiling. The location of the seismic profile shown in Figure 2 is indicated by a thick dotted line.

This lake is situated at 2448 m a.s.l, is 600 m long, 400 m wide and has a maximum depth of ~39 m (Fig. 1b). It is the lowermost lake of a chain of three small proglacial lakes of the St-Sorlin glacier which is situated on the northern part of the Grandes Rousses Massif (see Guyard et al., subm. for details). In addition, temporary ice-marginal lakes are frequently observed at the margin of the St. Sorlin glacier (M. Vallon, pers. comm.) and subglacial lakes may have also formed in the past below the St. Sorlin glacier.

3. Methods

3.1 CORING SITE

The sedimentary infill of Lake Bramant was imaged with the ETH Zurich 3.5 kHz pinger system. Conventional GPS navigation allowed the acquisition of a dense grid of high-resolution seismic profiles with a mean line spacing ranging between 50 and 100 m (Fig. 2).

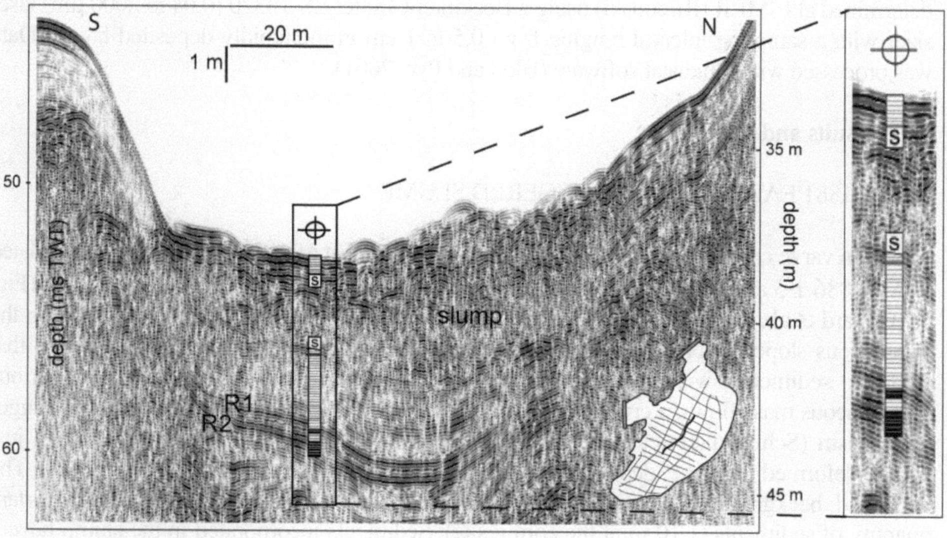

Figure 2: Seismic profile (3.5 kHz) across Lake Bramant deep basin and enlarged section illustrating the stratigraphy at the coring site. Undisturbed sediments were retrieved above and below a large mass wasting deposit (MWD) highly deformed as discussed in the text. This slump (S) and two high-amplitude reflections (R1 and R2) are clearly visible in the deepest and thickest part of the basin (see Guyard et al., subm. for details).

The system imaged an up to 20 ms two-way travel time (TWT) thick sedimentary succession and allowed the selection of a coring site in the depocenter of the basin. Following the seismic survey, a short gravity core (BRA03-1, 80 cm-long) and a long piston core (BRA03, cored interval 190-510 cm below lake floor) were taken in order to retrieve the best-possible undisturbed sediments above and below a large lens-shaped body with low-amplitude chaotic internal reflections that covers most of the deep basin and is the thinnest at the coring site (Fig. 2). This typical acoustic facies indicating mass wasting deposits (MWD) was only retrieved at the base of the short gravity core and the top of the piston core (Fig. 2 and 3), confirming the seismic interpretation that the non-cored interval (80-190 cm) only consists of MWD deposits. The absolute depth of the piston core was determined by core-to-seismic correlation.

3.2 MULTIPROXY SEDIMENT CORE ANALYSES

Cores were split, described and photographed with a 400 d.p.i. resolution digital camera, before u-channels were sampled in the middle of the cores in order to study geochemical and physical properties at very high resolution in undisturbed sediments. CAT-Scan (computerized axial tomography) analyses were carried out at INRS-ETE (Québec City) with a pixel resolution of 1 mm for the extraction of CT number profiles. The CT number primarily reflects bulk density variations (St-Onge et al., 2007). Micro-fluorescence-X (XRF) analyses were performed with an ITRAX core scanner (Croudace et al., 2006) with a downcore resolution of 300 μm for BRA03-1 and 100 μm for BRA03. BRA03-1 was irradiated during 1 s whereas BRA03 was irradiated during 10 s. The radiographs obtained were transformed in negative X-ray images. Finally, in short core BRA03-1 and piston core BRA03, sediment grain size distribution was determined at ISMER (Rimouski) using a Beckman-Coulter LS-13320 (0.04 to 2000 μm) laser sizer with a sampling interval ranging from 0.5 to 1 cm in the rapidly deposited layers. Data was processed with Gradistat software (Blott and Pye, 2001).

4. Results and discussion

4.1 AD 1881 EARTHQUAKE-TRIGGERED SLUMP

Based on varve counting, the top of the MWD affecting most of the deep basin (Fig. 2) is dated at AD 1886 ± 5 and can thus be associated to the nearby AD 1881 Allemond earthquake (Fig. 1) (Guyard et al., subm.). In addition, this type of mass movements creates failure scars on the subaqueous slopes. Such scars were detected just below the lake floor and suggest thus that unstable sediments were remobilized by earthquake ground accelerations in various simultaneous mass flows, a criteria used previously to single out earthquake shaking as trigger mechanism (Schnellmann et al., 2002). The base of the MWD is observed at 217 cm, just above deformed laminae deposited during a normal sedimentation period (Fig. 3). The deformed background sediments likely result from the rapid deposition of an important quantity of sediments (1.10 m at the coring site). Sediments incorporated in the slump have a lighter colour with regards to the rest of the core and are remoulded, laminated and highly deformed in specific intervals. The base of core BRA03-1 shows that the MWD is composed of locally folded and remoulded sediments and of reworked sediment blocks with their original laminations as seen on the Rx images. In addition, a homogenous layer is observed at the base of BRA03-1 at 80-71 cm, whereas a sand layer is recorded at 65-68 cm. The mechanism responsible for the deposit of the sediments between 46-217 cm thus involved unequivocally a mass wasting event.

This earthquake-induced mass movement is an example of a rotational sediment slide (Locat & Lee, 2002), corresponding to a slump following the classification of Mulder & Cochonat (1996). Moreover, the skewness *vs.* sorting diagram (Fig. 3b) does not illustrate a specific energy evolution such as for a classical turbidite or for a hyperpycnal deposit (see fig. 4), indicating a slumping mechanism.

Below the slump, a fining upward sequence beginning with an erosive contact is detected between 227-223 cm (Fig. 3) just above a thin olive-green layer, likely corresponding to the past sediment/water interface that was rapidly recovered by an instantaneous deposit. Coarser sediments at the base of this sequence are also indicated by the higher CT number values and the lighter X-ray grey scale, whereas mud clasts are observed on the Rx images from 226 to 224 cm. In addition, the observed increase in Fe and decrease in Rb contents within that interval are also suggesting a coarser base. Indeed, a Fe enrichment was previously interpreted as an indicator of the base of turbidites, whereas variations of the Rb content were related to fluctuations in the amount of detrital clays (Rothwell et al., 2006). Similarly, Fe in Lake Bramant sediments is concentrated in the coarse fraction especially at the base of turbidites, whereas Rb reflects the finer sediment fluctuations associated with the "glacial flour" resulting from glacial erosion. Based on this sharp contact, the normal grading and the presence of mud clasts, this sequence likely corresponds to an incomplete Bouma turbidite (e.g., Bouma, 1962; Mulder et al., 2001b), where only facies Ta-Tb were recorded. This turbidite could have been triggered by an earthquake, but could have also been generated by a storm event or an exceptional flood.

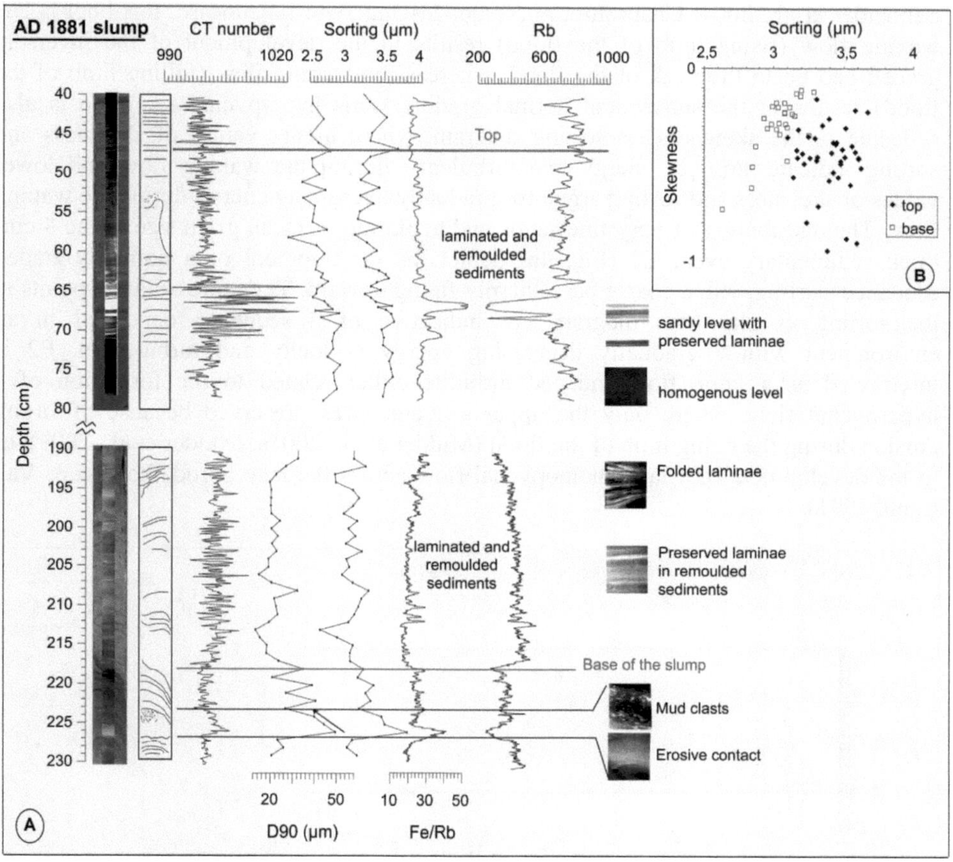

Figure 3: (A) Physical, sedimentological and geochemical properties of the base (190-217 cm) and the top (46-80 cm) of the slump triggered by the AD 1881 Allemond earthquake (MSK VII). Digital photographs, Rx images and chemical composition are obtained from the ITRAX measurements; VFSa: very fine sand percent (from 63 to 125 µm). The interval between 80-190 cm was not retrieved due to coring strategy. Just below the slump, an incomplete Bouma turbidite sequence is deposited. (B) Skewness *vs.* sorting diagram for the slump deposit.

4.2 FLOOD-INDUCED TURBIDITES

Rapidly deposited layers, labelled E2 and E3, are also recorded at 25-28 cm and 31-35 cm, respectively (Fig 4).

In the 4 cm-thick sedimentary event E3 (Fig. 4b), the basal sequence is characterized by an inverse grading (coarsening upward). The middle of the deposit (around 33 cm) is characterized by a thin layer of very fine sand, notably highlighted by a lighter grey scale and by a peak in the CT number, indicating a higher bulk density. Above the sand layer, the different proxies are reflecting a normal grading (fining upward). This grain size evolution is typical of hyperpycnal flows generated by large flood events in marine or lacustrine environments (e.g., Mulder et al., 2001a; 2003; St Onge et al., 2004;

Schneider et al., 2004; Chapron et al., 2006; St-Onge and Lajeunesse, this book). The waxing flow (rising limb of the flood) results in the development of the inversely graded bed up to the peak of the flood, whereas the waning flow (falling limb of the flood) results in the subsequent normal grading. This hyperpycnal sequence is also reflected in the skewness *vs.* sorting diagram, where higher values of skewness and sorting indicate stronger energy and turbulence during the waxing flow and lower values of skewness and sorting argue for gradually decreasing energy during the waning flow. The distribution of very fine sands and evolution of mean grain size in the 4 cm-thick sedimentary event E2 (Fig. 4a) depict the development of a normally-graded sequence starting with a coarse base sharply fining upward. The representative points in the sorting vs skewness diagram are indicative of a sequence deposited in an environment with a gradually decreasing energy (velocity and turbulence). E2 is interpreted as a large flood-induced turbidite either related to the formation of a hyperpycnal flow, where only the upper sequence was preserved because of strong erosion during the rising limb of the flood (Mulder et al., 2001a; Mulder et al., 2003), or to the development of a large homopycnal flow across the lake (Brodzikowski & Van Loon, 1991).

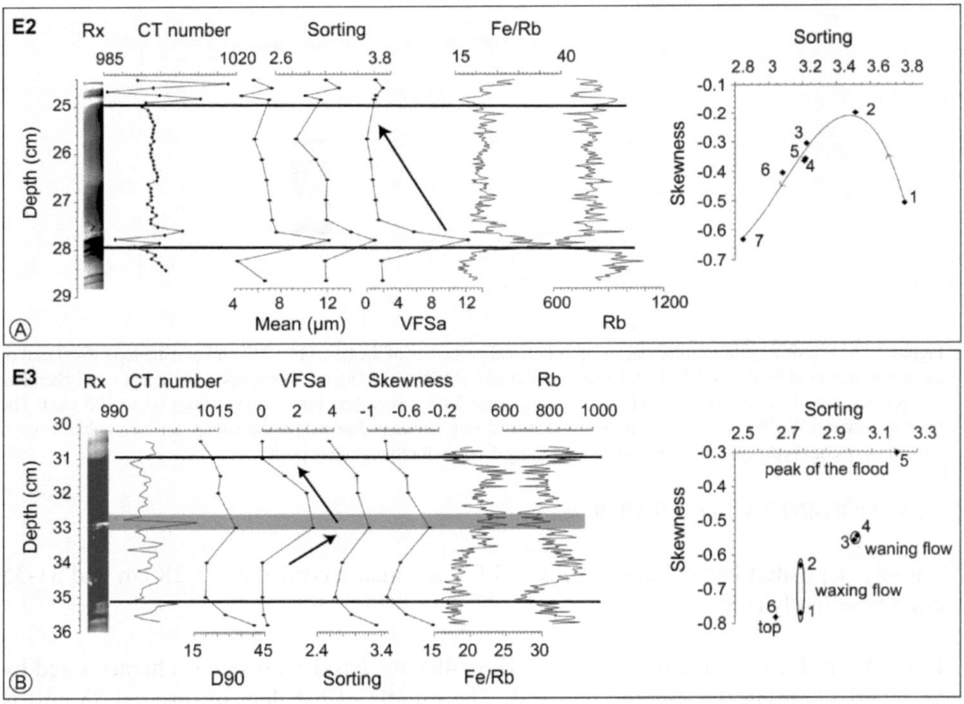

Figure 4: Physical, sedimentological and geochemical properties of sedimentary events E2 (A) and E3 (B). Rx images are obtained from the ITRAX measurements; VFSa: very fine sand percent (from 63 to 125 µm). Skewness *vs.* sorting diagrams for both layers are also displayed.

The formation of such exceptional flood deposits in Lake Bramant may result from the catastrophic drainage (i.e., outburst) of a temporary ice-contact lake or even a subglacial lake. The exceptional flood events E2 and E3, dated by varve counting (Guyard et al., subm) to AD 1904-1905 and AD 1908-1910, respectively, occurred when the St. Sorlin glacier was already retreating from its last advance following the end of the "Little Ice Age" (LIA), whereas the precise dating of the lowermost turbidite is still ongoing. Exceptional flood events in Lake Bramant could thus be linked to climatic oscillations through the outbursts of temporary ice contact lakes or subglacial lakes during warmer periods.

5. Conclusions

Various recent rapidly deposited layers were identified in high-altitude Lake Bramant sediments using several continuous high-resolution methods. The energy evolution of these different deposits is well distinguished by grain size statistical parameters and their trigger mechanisms were thus determined. The AD 1881 earthquake has triggered a slide that evolved into a slump. The recent hyperpycnal deposits (E2 and E3) are likely related to significant fluctuations of the St. Sorlin glacier following the end of the LIA. Nevertheless, as previously noted by Mulder et al. (2003) it can be difficult to distinguish flood-induced turbidites from hyperpycnites because erosion of the basal reverse grading sequence can occur during the rising limb of the flood or because the discharge and velocity during a high-magnitude flood are simply too important to allow sediment deposition. Ongoing complementary studies of Lake Blanc Bramant sediments, the second proglacial lake of the chain and the main tributary of Lake Bramant, will allow a better understanding of the relationship between flood-induced turbidites and the fluctuations of the St. Sorlin glacier.

6. Acknowledgements

We thank Michel Vallon (LGGE) for fruitful discussions and Urs Gerber (ETH Zürich) for the core photographs. Thanks are also due to Fabien Arnaud and Olivier Magand for the radionuclide dating. Atle Nesje, Oyvind Lie and all the coring team from Bergen University (Norway) are sincerely acknowledged for the collection of piston core BRA03 and for the logistic support. This project was funded by the French Mountain Institute. This constitutes GEOTOP-UQAM-McGill contribution n° 2007-0017.

7. References

Arnaud, F., Lignier, V., Revel, M., Desmet, M., Beck, C., Pourchet, M., Charlet, F., Trenteseaux, A. & Tribovillard, N., 2002. *Dating sediments from an alpine lake (Lake Anterne, NW Alps): influence of flood-events and gravity reworking on 210Pb vertical profile.* Terra Nova, 14, 225-232.

Blott, S.J. & Pye, K., 2001. *Gradistat: a grain size distribution and statistics package for the analysis of unconsolided sediments.* Earth Surf. Process. Landforms, 26, 1237-1248.

Bouma, 1962. *Sedimentology of Some Flysch Deposits*. Elsevier, Amsterdam.

Brodzikowski, K. & Van Loon, A.J., 1991. *Review of glacigenic sediments.* Development in Sedimentology, 49, 688 p.

Chapron, E., Beck, C., Pourchet, M. & Deconinck, J-F., 1999. *1822 earthquake-triggered homogenite in Lake Le Bourget (NW Alps).* Terra Nova, 11, 86-92.

Chapron, E., Ariztegui, D., Mulsow, S., Villarosa, G., Pino, M., Outes, V., Juvignié, E. & Crivelli, E., 2006. *Impact of the 1960 major subduction earthquake in Northern Patagonia.* Quaternary International, 158, 58-71 .

Croudace, I.W., Rindby, A. & Rothwell, R.G., 2006. *ITRAX: description and evaluation of a new X-Ray core scanner.* In R. G. Rothwell (Ed.), New Techniques in Sediment Core Analaysis. Geol. Soc. Spec. Publ. 267, pp. 51-63.

Doig, R., 1986. *A method for determining the frequency of large-magnitude earthquake using lake sediments.* Canadian Journal of Earth Science, 23, 930-937.

Guyard, H., Chapron, E, St-Onge, G., Anselmetti, F. S., Arnaud, F., Magand, O., Francus, P. & Mélières, M.A., in review. *High-altitude varve records of abrupt environmental changes and mining activity over the last 4000 cal yr BP in the Western French Alps (Lake Bramant, Grandes Rousses Massif)*. Quaternary Science Reviews.

Locat, J. & Lee, H. J., 2002. *Submarine landslides: advances and challenges*. Canadian Geotechnical Journal, 39, 193-212.

Monecke, K., Anselmetti, F., Becker, A., Sturm, M. & Giardini, D., 2004. *The record of historic earthquakes in lake sediments of Central Switzerland.* Tectonophysics, 394, 21-40.

Mulder, T. & Cochonat, P., 1996. *Classification of offshore mass movements.* Journal of Sedimentary research, 66, 43-57.

Mulder, T., Migeon S., Savoye B. & Faugères J.C., 2001a. *Inversely graded turbidite sequences in the deep Mediterranean: a record of deposits from flood-generated turbidity currents?* Geo-Marine Letters, 21, 86-93.

Mulder, T., Weber, O., Anschutz, P., Jorissen, F. J. & Jouanneau, J.-M., 2001b. *A few months-old storm generated turbidite deposited in the Capbreton Canyon (Bay of Biscay, SW France).* Geo-Marine Letters, 21: 149-156.

Mulder, T., Syvitski, J.P.M., Migeon, S., Faugères, J-C. & Savoye, B., 2003. *Marine hyperpycnal flows : initiation, behavior and related deposits. A review.* Marine and Petroleum Geology, 20, 861-882.

Nomade, J., Chapron, E., Desmet, M., Reyss, J-L., Arnaud, F. & Ligner, V., 2005. *Reconstructing historical seismicity from lake sediments (Lake Laffrey, Western Alps, France).* Terra Nova, 17, 350-357.

Rothwell, R. G., Hoogakker, B., Thomson, J., Croudace, I. W. & Frenz, M., 2006. *Turbidite emplacement on the southern Balearic Abyssal Plain (western Mediterranean Sea) during Marine Isotope Stages 1-3: an application of ITRAX XRF scanning of sediment cores to lithostratigraphic analysis.* From Rothwell, R. G: New techniques in sediment Core Analysis, pp. 79-98. Geological Society, London.

Schnellmann, M., Anselmetti, F., Giardini, D., McKenzie, J.A. & Ward, S.N., 2002. *Prehistoric earthquakes history revealed by lacustrine slump deposits.* Geology, 30, 1131-1134.

Schneider, J-L., Pollet, N., Chapron, E., Wessels, M. & Wassmer, P., 2004. *Signature of Rhine Valley sturzstrom dam failures in Holocene sediments of Lake Constance, Germany.* Sedimentary Geology, 169, 75-91.

Sims, J. D., 1975. *Determining earthquake recurrence intervals from deformational structures in young lacustrine sediments.* Tectonophysics, 29, 141-152.

St-Onge, G. & Lajeunesse, P., 2007. *Flood-induced turbidites from northern Hudson Bay and Western Hudson strait: a two pulse record of Lake Agassiz final outburst flood?* (this book)

St-Onge, G., Mulder, T., Piper, D.J.W., Hillaire-Marcel, C. & Stoner, J.S., 2004. *Earthquake and flood induced turbidites on the Saguenay Fjord (Québec): a Holocene paleoseismicity record.* Quaternary Science Reviews, 23, 283-294.

St-Onge, G., Mulder, T. & Francus, P, 2007. *Continuous physical properties of cored marine sediments.* In C. Hillaire-Marcel and A. de Vernal (Eds.), Proxies in Late Cenozoic Paleoceanography (Developments in Marine Geology), pp. 65-101.

Strasser, M., Anselmetti, F.S., Fäh, D., Giardini, D. & Schnellmann, M., 2006, *Magnitudes and source areas of large prehistoric northern Alpine earthquakes revealed by slope failures in lakes.* Geology 34, 1005-1008.

MORPHOSEDIMENTOLOGY OF SUBMARINE MASS-MOVEMENTS AND GRAVITY FLOWS OFFSHORE SEPT-ÎLES, NW GULF OF ST. LAWRENCE (QUÉBEC, CANADA)

P. LAJEUNESSE[1], G. ST-ONGE[2], G. LABBÉ[3] AND J. LOCAT[4]

[1] *Centre d'études nordiques & Département de géographie, Université Laval, Québec, QC, Canada G1K 7P4*

[2] *Institut des sciences de la mer de Rimouski (ISMER), Université du Québec à Rimouski, Rimouski, QC, Canada, G5L 3A1*

[3] *Département de géographie, Université Laval, Québec, QC, Canada G1K 7P4*

[4] *Département de géologie et génie géologique, Université Laval, Québec, QC, Canada G1K 7P4*

Abstract

Recent multibeam sonar and acoustic subbottom profiler surveys and sediment coring offshore the city of Sept-Îles (NW Gulf of St. Lawrence) reveal different types of submarine mass-movements and gravity flows in glaciomarine, paraglacial and postglacial deposits. These mass-movement and gravity flow features are slumps, gullies and channel-levee systems and fans. The key results of this study include: 1) slumps involve the entire deglacial and postglacial sequence, indicating their recent triggering; 2) identification of a 57-cm thick turbidite and several sand layers in sediment cores collected in a deep and isolated basin unaffected by fluvial inputs, along with ^{210}Pb measurements, indicate the recent activity of mass wasting events derived from slope instabilities; 3) important volumes of sediments are being transported from the coastal to the deeper marine environment by gravity flows processes on the prodelta of the Moisie River. Hypotheses for explaining the widespread occurrence of recent mass-movements due to slope instabilities in the area possibly include their possible triggering by the AD 1663 (M~7) or another large earthquake.

Keywords: Submarine mass-movements, multibeam bathymetry, marine geomorphology, glaciated margin, Gulf of St. Lawrence

1. Introduction

Multibeam bathymetry is now recognized as a valuable tool for studying postglacial geomorphic and sedimentary processes on formerly glaciated coastal zones and continental margins (e.g., Weaver *et al.*, 2000; Locat, 2001; Urgeles *et al.*, 2002). Digital Terrain Models (DTM) produced from multibeam sonar surveys allow a three dimensional visualisation of the seafloor and offer a greater detail and resolution of the extent, geometry and morphology of submarine features than datasets provided by conventional geophysical instruments. This paper reports and describes various submarine mass-movements and gravity flows offshore Sept-Îles (NW Gulf of St. Lawrence) (Figures 1, 2) from the analysis of a multibeam bathymetric dataset coupled with subbottom profiler data and sediment cores. The Sept-Îles area offers a great opportunity to study submarine mass-movements and gravity flows in an environment characterized by: 1) rapid land emersion following deglaciation (Dredge, 1983); 2) rapid deposition of a large volume of sediments during deglaciation (Hein *et al.*, 1993);

V. Lykousis, D. Sakellariou and J. Locat (eds.), Submarine Mass Movements and Their Consequences, 287–296.

3) important sediment inputs from rivers since ice retreat (Hein *et al.*, 1993); 4) pervasive modern coastal erosion (Savard, 2006) that possibly transferred sediments from the coast to the basin floor through the longshore drift; and 5) its proximity to an active seismic zone (Lamontagne *et al.*, 2003).

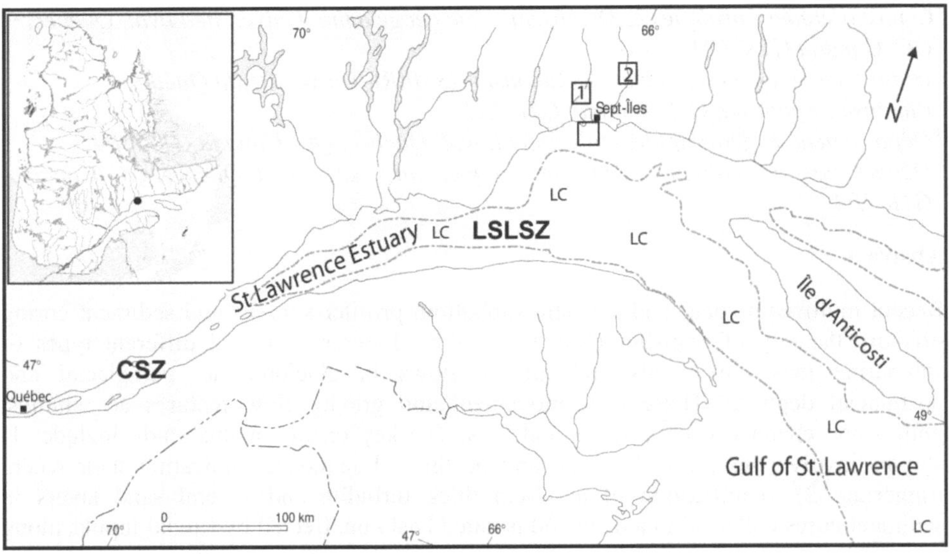

Figure 1. Location map of the Sept-Îles area, Québec, Canada (NW Gulf of St. Lawrence); (1) Ste-Marguerite and (2) Moisie rivers; (LC) Laurentian Channel; (CSZ) Charlevoix Seismic Zone; (LSLSZ) Lower St. Lawrence Seismic Zone. The 200 m bathymetric contour is outlined by the dashed grey line.

2. Study area and physical settings

The study area is located on the north shore of the Gulf of St. Lawrence, south of the city of Sept-Îles, Québec (Figure 1). It extends over 350 km^2, with depths ranging between 5 and 213 m. Beyond the study area, depths gradually increase to <340 m in the center of the Laurentian channel (Figure 1), a long U-shaped submerged glacial valley. The study area is located between two major deltaic systems: the Moisie River delta to the east and the Sainte-Marguerite River delta to the west (Figure 1). Sediments deposited by these rivers during the Holocene consist of thick paraglacial and postglacial units that completely drape the underlying glacial landforms and glaciomarine deposits near their mouth (Hein *et al.*, 1993). Two geological provinces are present offshore Sept-Îles (Faessler, 1942): the gently inclined Palaeozoic sedimentary rocks of the St. Lawrence Platform in the south that form cuestas and the highly deformed Precambrian crystalline rocks of the Canadian Shield (Grenville Province) in the north. The Sept-Îles archipelago consists of seven small islands separated from each other by series of longitudinal and transverse faults, forming fault-blocks. The islands represent the uplifted terrain resulting from the formation of these faults. Faulting is probably post-Ordovician because it affects some sedimentary rock layers of Trenton age (Faessler, 1942).

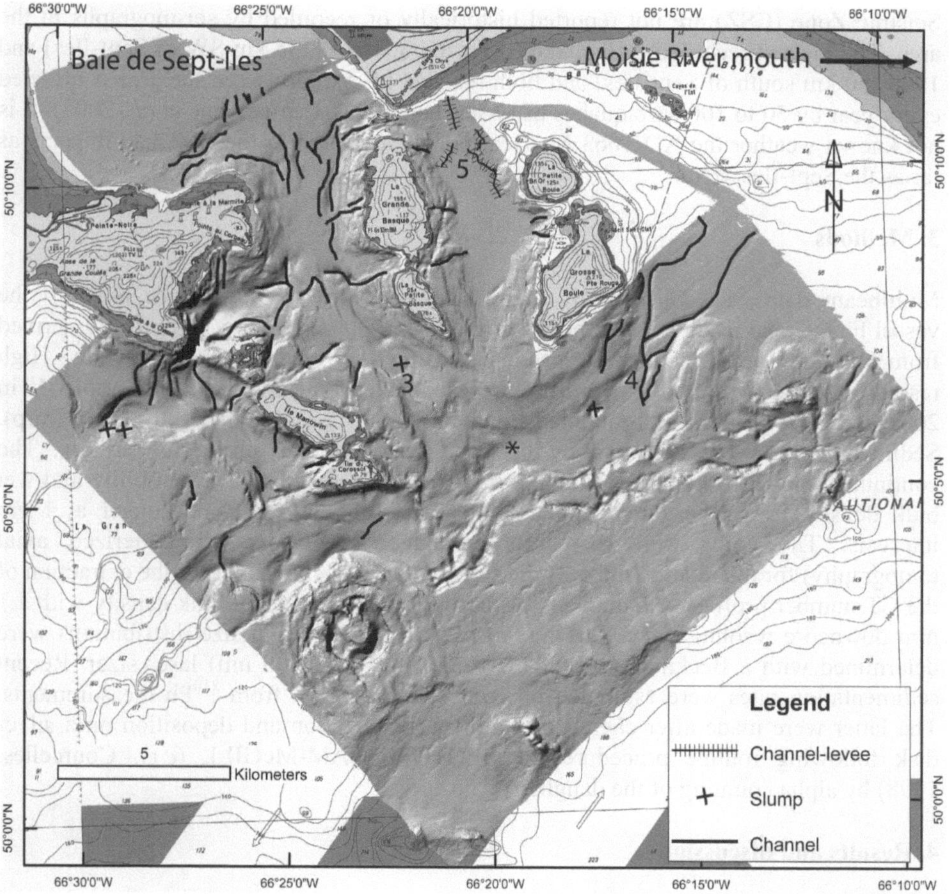

Figure 2. Shaded multibeam bathymetry of the study area with location of channels, slumps and channel-levees. Location of Cores 12BC and 12P: (*); Figure 3: (3); Figure 4: (4); Figure 5: (5).

In accordance with previous studies (Syvitski and Praeg, 1989; Hein *et al.*, 1993), five stratigraphic units were identified on Chirp profiles in the area: till (Unit 1), ice-proximal glaciomarine (Unit 2), ice-distal glaciomarine (Unit 3), paraglacial (Unit 4) and postglacial (Unit 5). However, the complete vertical sequence of these five units is rarely observed on acoustic profiles (Lajeunesse *et al.*, in preparation). These units record the stages of deglaciation, glacio-isostatic rebound and the establishment of modern conditions. The Goldthwait Sea, the marine transgression that occurred during and after deglaciation in the Estuary and Gulf of St. Lawrence (Elson, 1969; Dionne, 1977), reached ~130 m above sea level (asl) in the Sept-Îles area (Dredge, 1983). Emergence was at its peak prior to 8 kyr BP when it underwent a temporary stabilization, as indicated by a terrace found at 60 m asl (Dredge, 1983). During emergence, large paraglacial deltaic systems (Unit 4) were constructed along the Québec North Shore, including the Moisie and the Ste-Marguerite river deltas near Sept-Îles.

The Sept-Îles area is located within the Lower St. Lawrence Seismic Zone (LSLSZ) (Figure 1). Evidence for large earthquakes as important as those of the Charlevoix

Seismic Zone (CSZ) are not reported historically or recorded by seismographs in the area. The highest recorded earthquakes occurred in 1944 (135 km SW of Sept-Îles) and 1999 (60 km south of Sept-Îles) and both reached M=5.1 However, the area is affected every year by 50 to 100 earthquakes of lesser magnitude (Lamontagne *et al.*, 2003). It is not known weather the AD 1663 M~7 earthquake centered in the CSZ had impacts as far as the Sept-Îles sector.

3. Methods

Multibeam data was collected by the Canadian Hydrographic Service onboard the vessel F.G. Creed using a Kongsberg Simrad EM-1000 system (95 kHz). DTM derived from the multibeam data were produced using the Fledermaus® software. High resolution acoustic subbottom profiles were collected onboard the R/V Coriolis II in 2005 and 2006 using a hull-mounted Edgetech X-Star 2.1 system (2-20 kHz) (Chirp). Sediment cores were collected by box and piston corers during the cruises. The continuous physical properties (volumetric whole core magnetic susceptibility and wet bulk density) were measured with a GEOTEK Multi Sensor Core Logger at 1 cm intervals. The cores were also run through a CAT-scan (computerized axial tomography) for the identification of the sedimentary structures and for the extraction of the CT number profiles. CT numbers primarily reflect changes in bulk density with a 1 mm downcore resolution (see St-Onge *et al.*, 2007). The grain size distributions were determined with a Beckman-Coulter LS-13320 (0.04 to 2000 μm) laser sizer. Recent sedimentation rates were estimated on core SEPT08-12BC from ^{210}Pb measurements. The latter were made after chemical treatment, purification and deposition on a silver disk following routine procedures at GEOTOP-UQAM-McGILL (*e.g.,* Courcelles, 1998) by alpha counting of the daughter ^{210}Po.

4. Results and discussion

Four morphological features observed on the Sept-Îles bathymetric data can be associated with submarine mass-movements and gravity flows: slumps, gullies, channel-levee systems and fans. Slumps occur at three sites located on the sidewalls of submarine valleys where slopes are >7°. They are distant from the two river mouths and are not influenced by modern fluvial processes. The slumps remobilized a thick sedimentary package involving the glaciomarine, paraglacial and postglacial units. The most important and clearest slump occurs on the relatively steep slope (7.5°) of a submarine glacial valley located 8 km south of Sept-Îles (Figures 2, 3). The multibeam grid shows that this slump left behind a typical semi-circular headwall and a convex deposit in the submarine valley floor. On the Chirp profile (Figure 3), the slump cuts through the glaciomarine units and the entire deglacial-postglacial sequence, thus indicating a postglacial origin.

Evidence of strong activity of turbidity currents is provided by the presence of two feature types: a) the gullies found along the relatively steep slopes of submarine valley walls; and b) the channels occurring on more gentle sedimentary slopes. The gullies that cut through valley slopes and fans are particularly well developed east along the slopes of the islands located offshore Sept-Îles. These gullies are incised up to 20 m deep and are >150 m wide. They are carved in bedrock, occasionally draped by thin Quaternary sediments, as indicated by the high backscatter values on the multibeam data. They can reach lengths of 900 m and occur on slopes that are generally >5°. Small-scale submarine aprons are in many cases discernible at the lower reaches of these gullies. They indicate recent activity of turbidity currents because they have not been completely buried by postglacial sediments. Channels are found on sediment dominated gentle slopes (<2°). Many channels are observed west of Moisie delta river mouth (Figures 2, 4) and at the entrance of Baie de Sept-Îles (Figure 2). The low multibeam backscatter values observed where these channels occur indicate that they are carved in fine-grained sediments within the thicker Quaternary sequence. These channels are incised up to 10 m deep and are <200 m wide. They have lengths that can reach >3 km and are generally much longer than the gullies. Their fans are also much larger than those of the gullies and are not easily distinguishable on the multibeam data because they fill and smooth the deep basins.

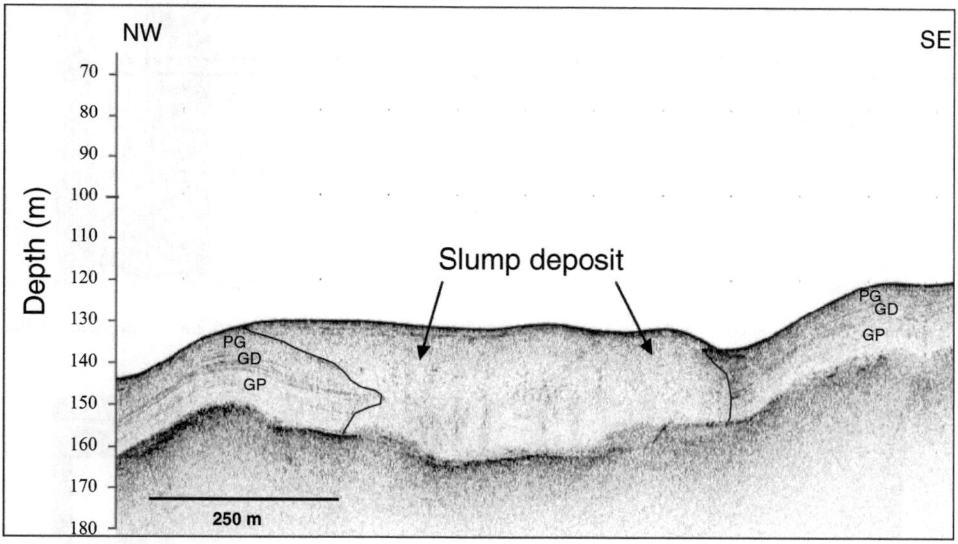

Figure 3. Chirp profile through a slump deposit (see Figure 2 for location on multibeam grid); (GP): Glaciomarine ice-proximal deposits (Unit 2); (GD) Glaciomarine ice-distal deposits (Unit 3); (PG) Postglacial deposits (Unit 5).

Large near-shore sand bodies are present ~1 km SE of Sept-Îles and ~19 km from the Moisie River mouth. These deposits show very low acoustic penetrations on the Chirp profiles. They form a very irregular surface that evolves downslope into a gently dipping slope and, later, into a flat sediment surface (Figures 2, 5). The slope section shows submarine channel-levee systems with a bird-foot geometry. The origin of these deposits cannot be clearly determined here, but they are most probably associated with downslope sediment transport by turbidity currents along the frontal slope of the Moisie River delta. A large part of these sediments could also be the product of coastal erosion, which is very active in the Sept-Îles area (Savard, 2006). Eroded sediments would then be remobilized westward by the longshore current and transported and deposited down the frontal gentle slopes of the Moisie River delta by density currents.

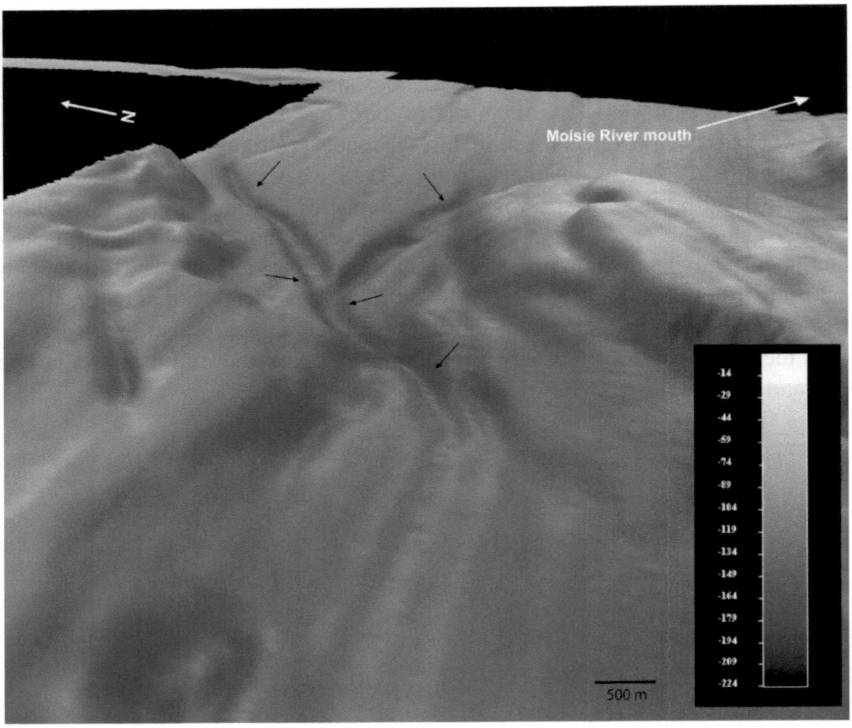

Figure 4. Channels (shown by black arrows) incised in Late-Quaternary sediments on the prodelta of the Moisie River (see Figure 2 for location on multibeam grid).

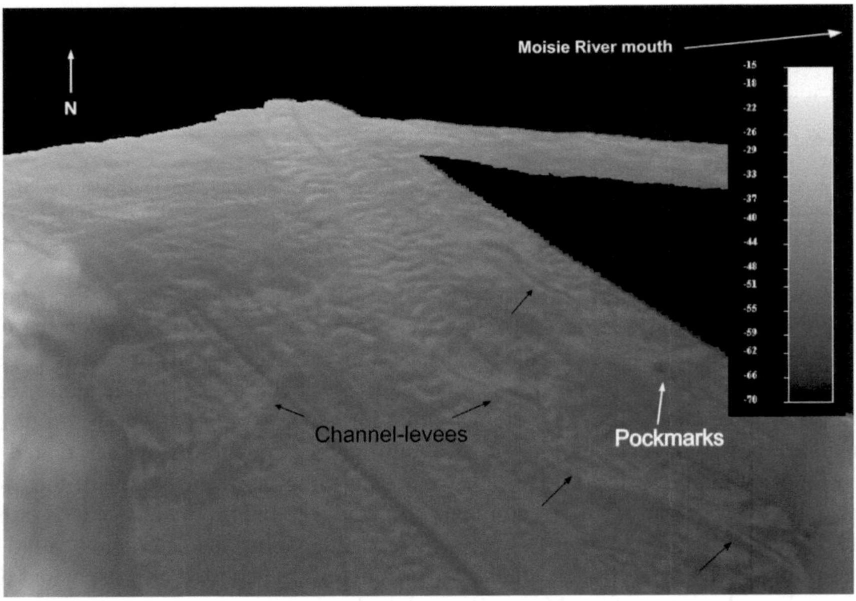

Figure 5. Sand bodies forming submarine channel-levees with a bird-foot geometry (19 km SE of the Moisie River mouth). Pockmarks are visible at the termini of the easternmost levee (see Figure 2 for location on multibeam grid).

On the Chirp data, discontinuities of internal reflectors in ice-distal glaciomarine deposits located near the frontal escarpment of some cuestas indicate the occurrence of sediment failures due to strong slope gradients. The slumped sediments accumulated at the bottom of the central elongated trough where the cores were collected. Core 12PC and 12BC (Figure 6) collected in the deeper part of the trough show distinctive turbidite layers that can be associated with these slope failure events. For example, grain size measurements of core SEPT08-12BC reveal at least 4 coarse sand pulses, whereas a 57 cm-thick classical turbidite is observed in core SEPT08-12PC from 343 to 286 cm (Figure 6), highlighting the importance of turbidity currents in the area. This turbidite has a coarse base, is normally graded and shows parallel and sub-horizontal laminations.

The basin in which these cores were collected is not exposed to sedimentation associated with the submarine channels of the Moisie River delta because it is separated by an important bathymetric high from the basin where these channels terminate (Figure 2). It is also located away from the shoreline, thus reducing sediment inputs by the longshore drift. Therefore, sedimentation by turbidity currents in this basin is more likely to have been caused by slope failures than by other sediment delivery mechanisms. The presence of a strong ^{210}Pb activity in the box cored sediments indicates that the sediments of core SEPT08-12BC were recently deposited. The average sedimentation rate for that core is estimated by the ^{210}Pb analyses at 0.14 cm/yr (Figure 7), a rate of comparable scale with the ones previously determined in the Gulf of St. Lawrence (*e.g.*, Smith and Schafer, 1999; Muzuka and Hillaire-Marcel, 1999). Slumps that remobilized glaciomarine, paraglacial and postglacial deposits as well as turbidites of postglacial age show the importance of recent submarine mass-movements and gravity flows in the Sept-Îles area.

Although some slumps could have been produced during the early phase of deglaciation, when glacio-isostatic rebound was very rapid, the observed slumps were triggered much later because they integrate acoustic units 2-5 —although Unit 4 is not always present on some profiles due to their relative distance from paraglacial deltas. As these features occur on slopes that have little influence from river inputs (which could constitute prone settings for the occurrence of mass-movements) and involve sediment layers that remained stable for a relatively long period of time, it is probable that they were triggered by a postglacial seismic event.

Earthquakes are a likely triggering mechanism because the area is located near the presently active LSLSZ. Even if only M≤5.1 earthquakes have been recorded in this region, is it not impossible that stronger events occurred prior to instrumental records and historical archives. Further dating of these deposits should provide information on the timing of the mass-movement events and their possible link with the AD 1663 (M~7) event or other large earthquakes. For example, Cauchon-Voyer *et al.* (this volume) have dated turbidites and associated mass wasting deposits from the Betsiamites area (Lower St. Lawrence Estuary) with large earthquakes that most likely occurred in AD 1870 (M~6.5) or AD 1860 (M~6), AD 1663 (M~7) and 7250 cal BP.

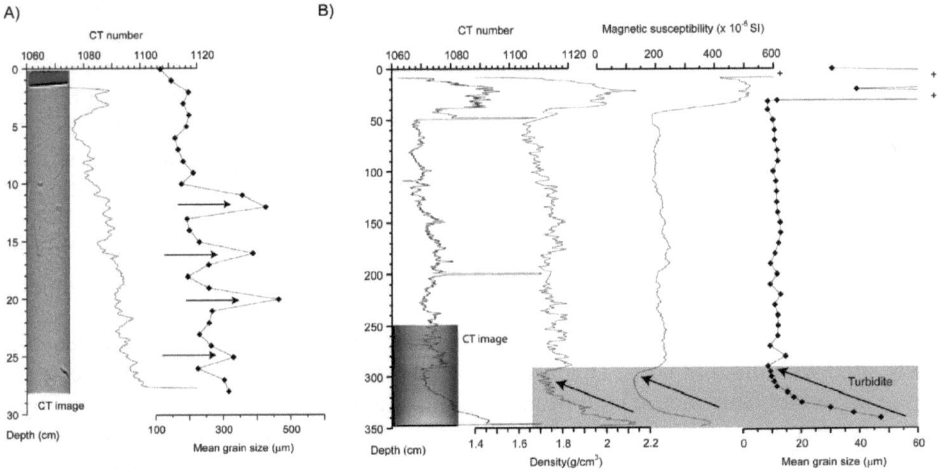

Figure 6. (a) CAT-Scan image and CT number values of core 12BC outlining four sand layers; (b) CAT-Scan image, CT number values, magnetic susceptibility and grain size of core 12 PC showing a normal graded turbidite.

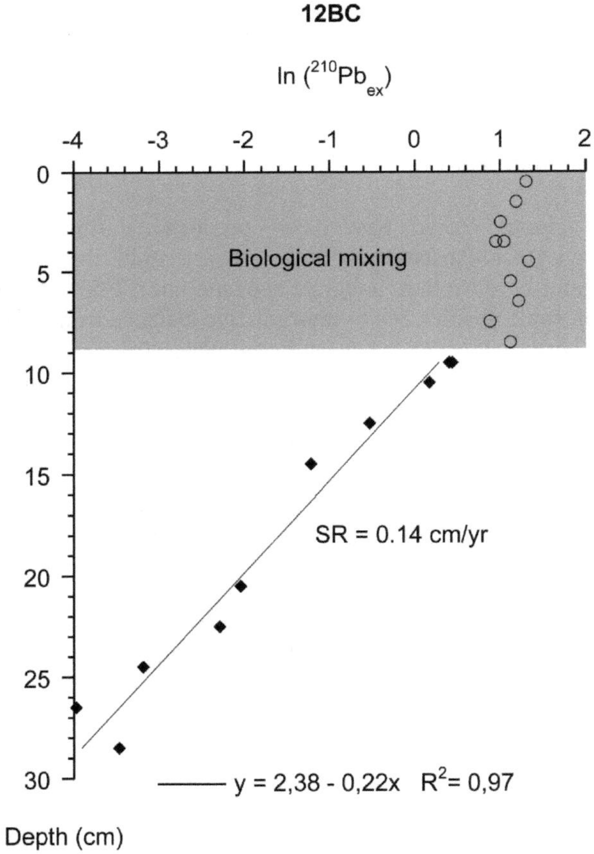

Figure 7. ^{210}Pb in Core 12BC showing recent sedimentation rates of 0.14 cm yr^{-1}.

5. Conclusions

The multibeam, Chirp and sediment core data presented here reveal the postglacial and recent activity of slumps and gravity flows in the Sept-Îles area. Sumps were likely triggered by a (late?) postglacial seismic event, possibly the AD 1663 (M~7) or another large earthquake. Some turbidity current gullies and channels and their related submarine fans are either related to fluviodeltaic and/or longshore current transport processes or to slope instabilities. The presence of submarine levees and channels offshore the Moisie River delta indicates important sediment transfer from the coast to the deeper marine environment by turbidity currents related to hyperpycnal flow during high river discharge. The identification of a 57-cm thick turbidite and of several sand layers in sediment cores collected in a deep and isolated basin unaffected by fluvial and coastal inputs, along with ^{210}Pb measurements, indicate recently active mass wasting events in the area. These instabilities could also be related to recent seismic events. Further studies should provide more information on the history of seismic events in the region and their impacts on submarine mass-movements.

6. Acknowledgements

This research was supported by the Natural Sciences and Engineering Research Council of Canada (NSERC) through Discovery grants to P.L., G.S.O. and J.L. and ship-time grants in 2005 and 2006. We thank the Captain, crew and scientific participants of the R/V Coriolis II for their help during the research cruises. The multibeam data was gratefully provided by the Canadian Hydrographic Service in Mont-Joli (Québec). We also thank R. Urgeles and M. Allard for their formal reviews that helped to improve the quality of the manuscript.

7. References

Cauchon-Voyer, G., Locat, J. and St-Onge (in press). *Submarine mass movements in the Betsiamites area, Lower St. Lawrence Estuary, Québec, Canada.* Advances in Natural and Technological Hazards Research, Springer.

Courcelles, M. 1998. *Enregistrement sédimentaire des flux récents de métaux lourds (Pb, Hg) et d'isotopes à courte période (^{210}Pb, ^{137}Cs et ^{228}Th) dans un lac sub-arctique à faible vitesse de sédimentation (Lac Jobert, Québec).* Ph.D. Thesis, Université du Québec à Montréal, Montréal, Québec.

Dionne, J.-C., 1977. *La Mer de Goldthwait au Québec.* Géographie physique et Quaternaire, 31: 61-80.

Dredge, L. 1983. *Surficial geology of Sept-Îles area, Québec North Shore.* Geological Survey of Canada, Memoir 408.

Elson, J.A., 1969. *Late Quaternary marine submergence of Québec.* Revue de géographie de Montréal, 23: 247-258

Faessler, C. 1942. *Sept-Îles area, Saguenay County.* Québec Bureau of Mines, Geological Report 11, 26 pp. Map. No.503

Hein, F.J., Syvitski, J.P.M., Dredge, L.A. and Long, B.F. 1993. *Quaternary sedimentation and marine placers along the North Shore, Gulf of St-Lawrence.* Canadian Journal of Earth Sciences, 30(3): 553-574

Lajeunesse, P., St-Onge, G. Labbé, G., Locat, J. and Sanfaçon, R. (in preparation). *Marine geomorphology of a formerly glaciated coastal zone, Sept-Îles area, NW Gulf of St. Lawrence.*

Lamontagne, M., Keating, P. and Perreault, S. (2003) *Seismotectonic characteristics of the Lower St. Lawrence Seismic Zone, Québec: insights from geology, magnetics, gravity, and seismics* Canadian Journal of Earth Sciences 40(2): 317-336.

Locat, J. 2001. *Instabilities along ocean margins: a geomorphological and geotechnical perspective.* Marine and Petroleum Geology, 18(4): 503-512.

Muzuka, A.N.N. and Hillaire-Marcel, C. 1999. *Organic matter burial rates along the eastern Canadian Margin and stable isotope constraints on its origin and diagenetic evolution.* Marine Geology, 160: 251-270.

Savard, J.P. 2006. *Érosion des berges dans le golfe du Saint-Laurent.* 74e Congrès de l'ACFAS, Université McGill, Québec.

Smith, J.N. and Schafer, C.T. 1999. *Sedimentation, bioturbation, and Hg uptake in the sediments of the estuary and Gulf of St. Lawrence.* Limnology and Oceanography, 44: 207–219.

St-Onge, G., Mulder, T., Francus, P., Long., B. 2007. *Continuous physical properties of cored marine sediments.* In: C. Hillaire-Marcel and A. de Vernal (Eds.), Proxies in Late Cenozoic Paleoceanography, Elsevier: 65-101.

Syvistki, J.P.M. and Praeg, D.B. 1989. *Quaternary sedimentation in the St. Lawrence estuary and adjoining areas, eastern Canada: an overview based on high resolution seismo-stratigraphy.* Géographie physique et Quaternaire, 43: 291-310.

Urgeles, R., Locat, J., Lee, H.J. and Martin, F.M. 2002. *The Saguenay Fjord, Québec, Canada: integrating marine geotechnical and geophysical data for spatial seismic slope stability and hazard assessment.* Marine Geology, Volume 185, Issues 3-4, 30 June 2002: 319-340

Weaver, P.P.E., Wynn, R.B., Kenyon, N.H. and Evans, J. 2000. *Continental margin sedimentation with special reference to the Northeast Atlantic margin,* Sedimentology, v.47 (Supplement 1): 239-256.

SEDIMENT FAILURE PROCESSES IN ACTIVE GRABENS: THE WESTERN GULF OF CORINTH (GREECE)

V. LYKOUSIS, D. SAKELLARIOU, G. ROUSAKIS, S. ALEXANDRI, H. KABERI, P. NOMIKOU, P. GEORGIOU AND D. BALAS

Hellenic Centre for Marine Research, 46,5 km Athens-Sounio Ave, 19013, Anavyssos, Greece, e-mail: vlikou@ath.hcmr.gr

Abstract

On steep (2-6°) offshore fan deltas of Western Gulf of Corinth medium to small scale (10^{6}-10^{7} m^{3}) debris flows and avalanches are the prominent slope features. Loose sands, gravels and pebbles are observed in the lower fan while silty/sandy turbidites/tsunamites detected in cores, implying sediment dissociation during failure and downslope transport. These failures are associated with significant upslope retrogression that has caused coastal retreat with important human and economic impact. All the events were estimated to have occured during the late 4-5 kyr after or during the deposition (progradation) of the HST fan delta. Recent (late 0.1-0.15 kyr) prodelta failures in the Gulf of Corinth are evidenced by the destruction of telecommunication cables, coastal collapse and the initiation of destructive tsunami waves (i.e. 1963AD). The frequency of major failure events in the Western Gulf of Corinth is estimated to 2-3 events/0.1 kyr, usually associated with strong earthquakes and tsunamis (i.e. 1817, 1861, 1917?/,1963 1995AD events)

Keywords: Offshore sediment failure, Fan delta, geotechnical properties, active graben, Gulf of Corinth, NE Mediterranean

1. Introduction

The western Gulf of Corinth is one of the more seismically active areas in Europe due to its proximity to the Hellenic trench (Jackson et al., 1982; Armijo et al., 1996; Clarke et al., 1997; Davies et al., 1997; Moretti et al., 2003) (Fig. 1). The greater area includes the Gulf of Corinth (graben) and the Ionian margins with high river run-off and terrigenous sediment supply. Intense shallow earthquake motions particularly pronounced in Western Gulf of Corinth are expected to generate peak ground accelerations of 25-35%g (Makropoulos and Burton, 1985; Papoulia et al., 1998). Within this region frequent sediment mass failures during the late century have been responsible for numerous cable breaks (Heezen et al., 1966) and catastrophic tsunamis (Papadopoulos and Chalkis, 1984).

From the late Pleistocene stage, the Gulf of Corinth basin was filled up by a thick sequence of gravity deposits (turbidites, debris flows, mudflows etc.) (Heezen et al., 1966; Brooks and Ferentinos, 1984; Poulos et al., 1995; Lykousis et al., 2007a). Fan delta prograding deposits during Pleistocene sea-level changes coupled by extensive slumping, are the predominant sedimentary processes in the steep flanks of the gulf (Ferentinos et al., 1988; Lykousis, 1990; Lykousis et al., 1995; Perissoratis et al., 2000; Lykousis et al., 2007b). These processes are mostly related to seismically triggered mass-gravity failures that initiated in the coastal zone and/or the uppermost slope

V. Lykousis, D. Sakellariou and J. Locat (eds.), Submarine Mass Movements and Their Consequences, 297–305.

(Perissoratis et al., 1984; Lykousis, 1990; Piper et al., 1990; Ferentinos et al., 1988; Papatheodorou and Ferentinos, 1997). Slope stability calculated using the Normalised-Soil-Parameter (NSP) method (Lee and Edwards, 1986) indicates that instabilities could be induced by earthquake ground accelerations of 26.6-29.6 %g (Lykousis et al., 2007b). Since the expected ground accelerations over the next 100 yrs are 25-35% g (Papoulia et al., 1998) the fan delta slopes of the Western Gulf of Corinth are potentially unstable.

The purpose of this paper is to study the mass failures and associated processes, related to the prograding sedimentary sequences in the steeper fan delta slopes (mean gradient 2^0-6^0) of the Western Gulf of Corinth.

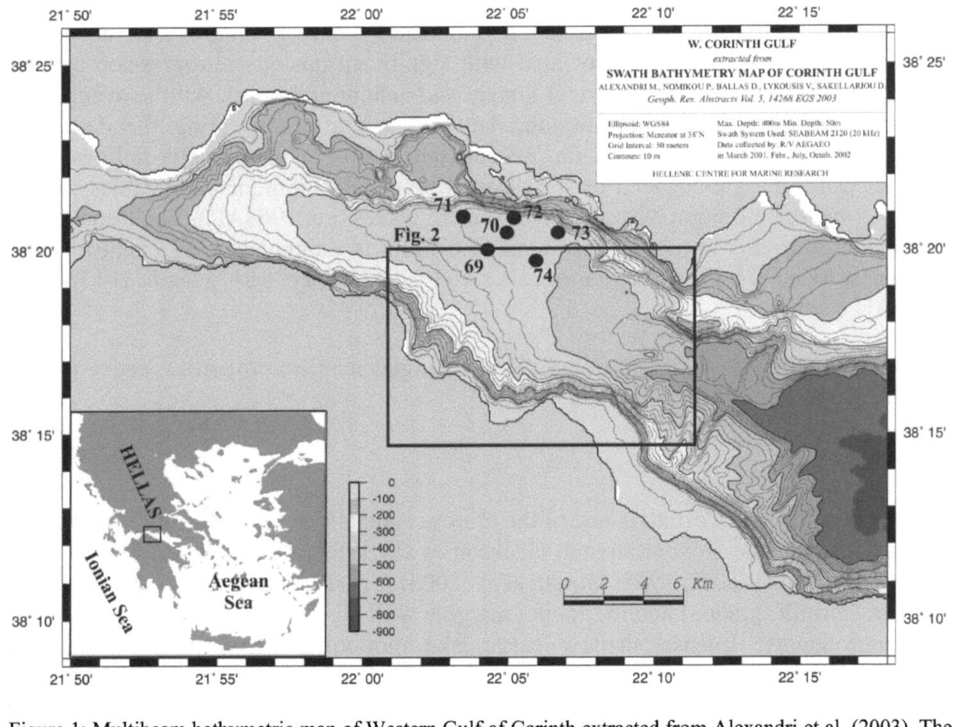

Figure 1: Multibeam bathymetric map of Western Gulf of Corinth extracted from Alexandri et al. (2003). The black box indicates the location of the 3D view shown in Fig. 2. Approximate location of box-core samples is indicated by full circles.

2. Methods

High resolution seismic reflection profiles taken from various surveys during the last decade (3.5 kHz and Air Gun) with the R/V AEGAEO were used to identify the sediment sequences and the associated instabilities. Sediment cores (2 - 5m long) were recovered using a BENTHOS INSTR. gravity corer that provides relatively undisturbed samples. Grain-size analysis was performed with the Sedigraph laser technique (Micrometrics 5100) at 5 cm intervals along the split cores. The multibeam mapping was conducted using the SEABEAM 2120 and 2200 models operating at 20 kHz and

200 kHz respectively. Recent sedimentation rates (late 0.1 ka) were calculated by measuring the ^{210}Pb activity via its α-particle-emitting granddaughter ^{210}Po, and assuming secular equilibrium with ^{210}Pb (Appleby and Oldfield, 1978) and following the methodology described by Sanchez-Cabeza et al. (1998) and Radakovitch (1995).

3. Fan delta slope failures

Extensive Gilbert –type fan delta deposits were developed along the Gulf of Corinth steep, faulted slopes. The offshore fan deltas are well pronounced along the southern slopes which are characterised by high sediment fluxes and rapid tectonic uplift (Brooks and Ferentinos 1984; Ferentinos et al. 1988; Papatheodorou and Ferentinos 1997; Lykousis et al. 2007a). The dip of the slope ranges between 3^0-6^0 and is locally higher, close to fault or slump scarps.

Small-medium size sediment failures with volumes of 10^6-10^7 m^3 have been imaged with the high resolution (200 kHz) multibeam mapping of the Western Gulf of Corinth particularly along its southern margin (Fig. 2).

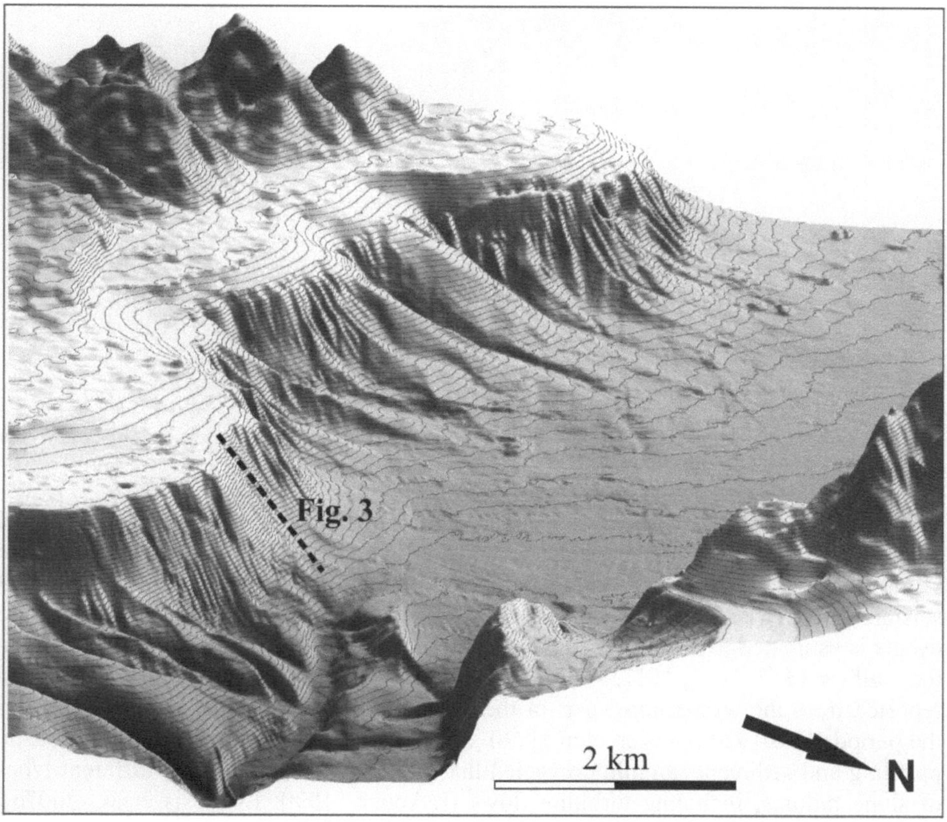

Figure 2: 3D view of fan delta failures (debris flows) along the southern flanks of Western Gulf of Corinth (imaged from ENE). Note the "floated" sediment block debris in the lowermost fan. For approximate location see figure 1. Location of figure 3 is also indicated.

The majority of the observed failures are downslope debris flows/avalanches (Fig 3). Slab (translational) slides occur also (locally) along gently dipping slopes (2^0-3^0 HST prodeltas (i.e. Mornos delta) (Lykousis, 1990). The failure scarps are shallow (10-50m depth), located close to (100-200 m) the coastline and are usually associated with coastal zone retreat and submergence, due to upslope (cross shore) retrogression. Although more difficult to locate in offshore fan deltas, failure planes are expected to be basal muddy layers (flooding surface-MFS??) and/or biogenic methane gas charged sediment horizons (Lykousis et al., 2007b).

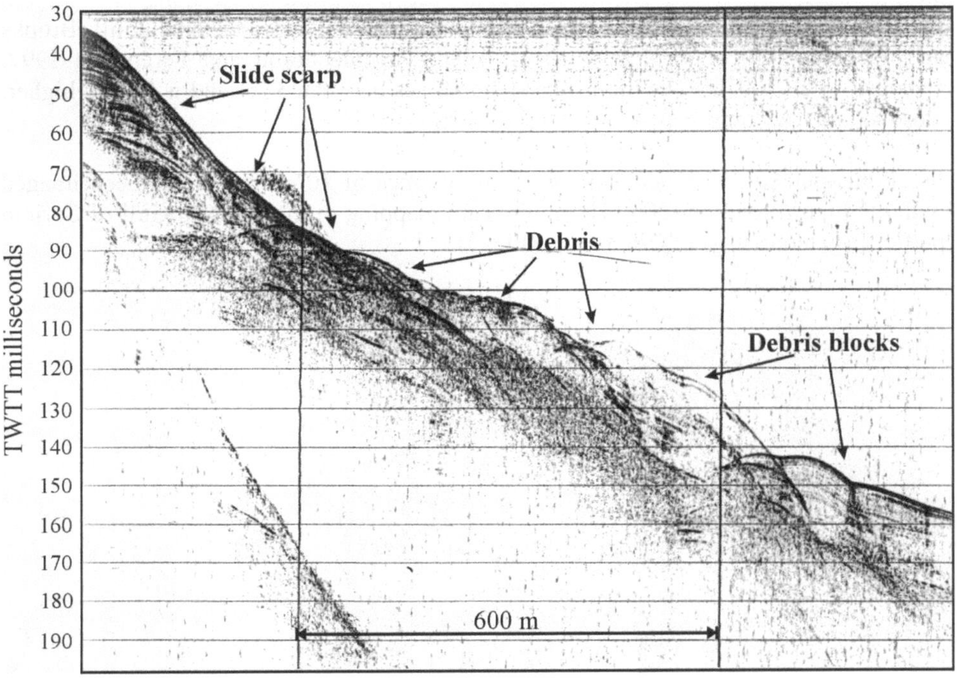

Figure 3: 3.5 kHz subbottom profile showing debris flows off the fan deltas of Western Gulf of Corinth. For approximate location see figure 2.

Since the fan deltas have been developed (prograded) during the late high sea level stand (HST fan deltas) all the events are younger than 4-5 ka BP. The majority of these failures are associated with strong earthquakes, although major failures initiated without seismic activity (Papadopoulos and Chalkis, 1984). The repetition rate of major failure events is estimated to 2-3events/100yr and is coincident with occurrence of tsunamis in the Gulf (ie 1817, 1861, 1917, 1963, 1995AD etc). At least five cable brakes have been reported from the westernmost part of the Western Gulf of Corinth (Mornos delta) for the period 1890-1920 (Heezen et al 1966). Successive surveys with subbottom acoustic profiling and sediment sampling revealed that cable brakes were due to different types of slope failures, including turbidite flows (Lykousis, 1990; Lykousis et al., 2007b). Repeated dives with submersible and ROV deployments upslope the evacuated zones displayed loose sands, gravels and pebbles dispersed in the lower fan and numerous downslope small channels (0.5m wide and 0.4 m deep) indicating sediment (turbidity?)

flows. Towards the steeper scarp zone repeated arcuate (crown) cracks with 10-20 cm offset are evidence of the frequent small scale failures that promotes the upslope erosion and the oversteepening of the uppermost slope (Figure 4).

In a long term basis, this process play important role in the upslope and inshore retrogressive failures, with ultimate result the costal failure and subsidence of significant coastal zones. Shallow gravity normal faults recorded on 3.5 kHz profiles between the failure scarp and the coastline support this interpretation. Consequently, smaller scale events occur very often as indicated from indirect approaches (cable brakes, small tsunamis without seismic shock, local ground tremors and sounds etc) but also from repeated direct ROV/submersible observations. At least one medium-big failure event (2-3 km^3 in volume) (Lykousis et al., 1995) has been determined in the Western Gulf of Corinth and may be associated with the regional coastal submergence (drowning of ancient Helike and destruction of ancient Voura) and the giant tsunami which followed the 363 BC earthquake.

.

Figure 4: Arcuate (crown) crack with small offset, observed on the southern slope of the Western Gulf of Corinth, is interpreted as a precursor of a forthcoming small failure. Approximate scale of the frame 2m x 2m.

4. Basin turbidites/tsunamites?

A series of sediment cores (box and gravity) were recovered from the basin of the Western Gulf of Corinth between Trizonia Island and the south slope. All the cores systematically display two distinct coarse grained layers with sharp upper and lower contacts and comparable thickness among the sediment cores (Figure 5). The predominant layer, that occurs basin-wide, ranges in thickness from 6-10 cm (mean 8 cm) and the total volume of this coarse-grained deposit is estimated to 0,5 x 106 m^3.

This is comparable to a single failure event in the Western Gulf of Corinth based on the estimation from the detailed multibeam bathymetry. Microscopic analysis showed that it consists almost exclusively of relatively well rounded terrigenous coarse silt-fine sand. The grains are mostly feldspars, quartz and minor amount of terrigenous calcite. There was a lack of shell debris and foraminifera. These characteristics, the basin wide spread and the uniform thickness probably implies tsunamigenic derivation rather than gravity flow (turbidite) deposition (i.e. distal turbidites from debris flow along transport dissociation), although, this late assumption should not be excluded.

The mean accumulation rates for the last 100-120 yrs obtained by the ^{210}Pb radiometric method from two short sediment box cores are shown in Figure 6. The obtained ^{210}Pb activity profile versus depth below sea floor displayed, irregularities which can be directly related to episodic deposition of gravity driven mud flows or mud turbidites on the seafloor. Apart from the internal structure of the core, the calculated mean accumulation rates for the last 100-120 yrs, according to the downcore distribution of the excess ^{210}Pb activity, were 20-30cm 100 yr^{-1}. This also indicates that the turbidites have been deposited at the margin before 100-120 yrs or earlier (about 150 yrs). The region experienced two very intense seismic events the late 100-200 yrs followed by destructive tsunamis and extensive coastal collapse that indicated significant offshore failures (1817AD and 1861 AD). We conclude that the coarse grained layers were deposited during the 1861 AD stronger and more recent event either as tsunami deposit or as a distal end member of debris flow.

5. Summary and conclusions

Multibeam, seismic profiling, submersible dives and sedimentological data (cores, x-ray images, AMS and Pb^{210} dating) provide new information regarding the mass failure processes in one of the most active rifts among the European margins. The Gulf of Corinth, Greece is characterized by a complex set of natural hazards, (earthquakes, slumps, tsunamis, fluid escape etc) mainly associated with its strong neotectonic and seismic activity. The seismicity, the steep slopes of the basin, together with the large amount of sediments deposited by the several rivers (sedimentation rates about 3 m/kyr) is the key factors of sediment failure. The latter explains why the sliding masses were found offshore the major delta plains of the western part of the Gulf of Corinth. The submarine slope failures are characterized by relatively small-medium volumes (up to $9x10^7$ m^3) but of high number of events and aerial extension on the slopes of the Western Gulf of Corinth. The sediment slides are initiated from very shallow depths (20-40 m) and sometimes from the coastal zone (i.e. 1963 AD slump and associated tsunami). During the failure most of the slumped sediment masses are disintegrated to debris flows (muddy sands with gravels) in the mid-lower slope and probably sandy/silt turbidites in the proximal basin. These coarse-grained layers may also have been derived from deposition of material resuspended by tsunami (tsunami deposits).

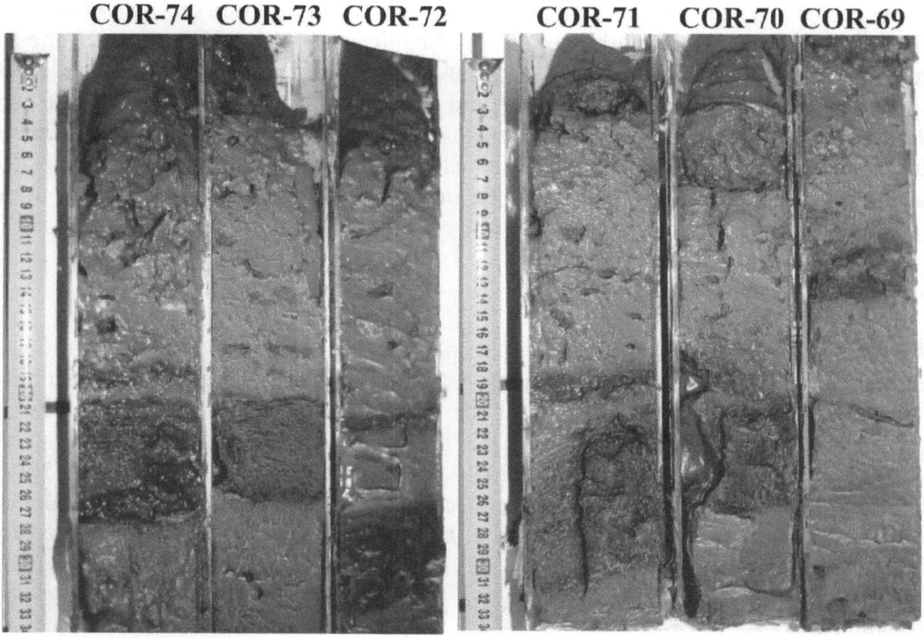

Figure 5: An 8cm-thick coarse-grained layer (turbidite/tsunamite?) occurs at ca. 20-30 cm depth below the seafloor of the Western Gulf of Corinth basin.

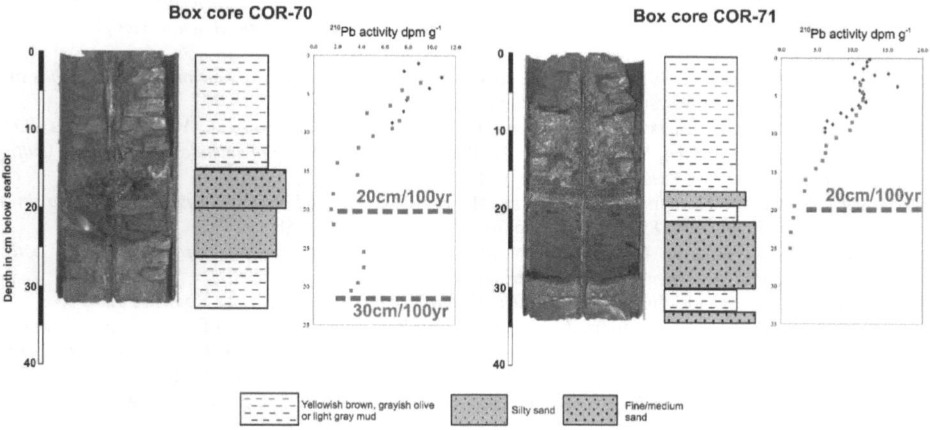

Figure 6: Photographs and sedimentological description of two box cores (COR-70, COR-71) from the Western Gulf of Corinth. The ^{210}Pb downcore activity profiles and the calculated actual sedimentation rates are also shown (from Sakellariou et al., 2004).

The upper slope and the scarp of the failure zone undergo active retrogression as indicated by the upslope (up to the coastal zone) shallow gravity faults and the arcuate (crown) cracks that have been observed by submersible dives. This process is responsible for the coastal zone subsidence during major, mass failure events. Estimated reoccurrence of major failure event is at least 5 events/100yr (based on cable brakes and Pb^{210} dating of successive turbidite events). Smaller and minor failure events, mostly due to retrogression and gravity faults activation, are often and continuous as evidenced by submersible observations, local ground vibrations and sounds due to offshore slumping. One of the greater, recent, extensive offshore and coastal failure occurred during and after the 1861AD strong earthquake. The sandy-silt turbidite layer deposited on the flat basin of the Western Gulf of Corinth because of this event is 8 cm thick, extends basin wide and display total volume of about 0,5 x 10^6 m^3.

6. Acknowledgements

We thank the officers and the crew of the R/V AEGAEO for their significant assistance in the field work during various cruises. The authors gratefully acknowledge Mr. P. Mantopoulos and Mr. I. Pampidis for assisting and operating the subbottom profiling systems and the gravity and box coring equipment. This work is partly funded by ASSEM and 3HAZ EC projects (EVK3-2001-00038/2002-2004 and SSPI 004043 contract numbers respectively) and National projects funded by the Greek government. The referees J. Locat and T. Alves are acknowledged for their upgrading reviews.

7. References

Alexandri, S., Nomikou, P., Ballas, D., Lykousis, V. and Sakellariou, D. 2003. *Swath bathymetry map of Corinth Gulf.* Geoph. Res. Abstracts. V.5, 14268, EGS, Nice.

Appleby, P.G. and Oldfield, F. 1978. *The calculation of lead-210 dates assuming a constant rate of supply of unsupported ^{210}Pb to the sediment.* Catena, **5**, 1-8.

Armijo, R., Meyer, B., King, G.C.P., Rigo, A. and Papanastassiou, D. 1996. *Quarternary evolution of the Corinth rift and its implications for the late Cenozoic evolution of the Aegean.* Geoph J Int 126: 11-53.

Brooks, N. and Ferentinos, G. 1984. *Tectonics and sedimentology in the Gulf of Corinth and Zakynthos and Kefallinia chanels, western Greece.* Tectonophysics 101: 25-54.

Clarke, P.J., Davies, R.R., England, P.C., Parsons, B.E., Billiris, H., Paradissis, D., Veis, G., Denys, P.H., Cross, P.A., Ashkenazi, V. and Bingley, R. 1997. *Geodetic estimate of seismic hazard in the Gulf of Korinthos.* Geophys. Res. Lett. 24, 1303-1306.

Davies, R.R., England, P.C., Parson, B.E., Billiris, H., Paradissis, D. and Veis, G. 1997. *Geodetic strain of Greece in the interval 1892-1992.* J. Geophys. Res. 102, 24, 571-588.

Ferentinos, G., Papatheodorou, G. and Collins, M.B. 1988. *Sediment transport processes on an active submarine fault escarpement: Gulf of Corinth, Greece.* Mar Geol. 83:43-61.

Heezen, B.C., Ewing, M. and Johnson, G. 1966. *The Gulf of Corinth floor.* Deep Sea Res 13: 381-411.

Jackson, J.A., Gagnepain, J., Houseman, G., King, G.C.P., Papadimitriou, P., Soufleris, C. and Virieux, J. 1982. *Seismicity, normal faulting and the geomorphological development of the Gulf of Corinth (Greece): the Corinth earthquakes of February and March 1981.* Earth & Plan Sc Lett. 57: 377-397.

Lee, H.J. and Edwards, B.D. 1986. *Regional method to assess offshore slope stability.* J Geot Eng 112(5): 489-509.

Lykousis, V. 1990. *Prodelta sediments: seismic stratigraphy, sedimentology, slope stability* Ph.D. thesis, University of Patras, Greece, pp 280.

Lykousis, V., Sakellariou, D., Papanikolaou, D., Chronis, G., Papoulia, I., Rousakis, G. and Georgiou, P., 1995. *Contribution of Neotectonic and Marine Geology-Geophysics to the Discovery of Ancient Heliki.* 2nd Congr Ancient Helike & Aegialia, Proceedings pp. 451-469.

Lykousis, V., Sakellariou, D., Moretti, I. & Kaberi, H., 2007a. *Late quaternary basin evolution of the Gulf of Corinth: sequence stratigraphy, sedimentation, fault-slip and subsidence rates.* Tectonophysics, in press.

Lykousis, V., Sakellariou, D. and Rousakis, G., 2007b. *Offshore slope stability analysis of prograding sediment sequences at active margins, W. Greece, NE Mediterranean.* J. Earth Sci. (accepted for publication).

Makropoulos, K. and Burton, P. 1985. *Seismic hazard in Greece, II: ground acceleration.* Tectonophysics. 117:259-294.

Moretti, I., Sakellariou, D., Lykousis, V. and Micarelli, L. 2003. *The Gulf of Corinth: an active half graben?* J Geod 36:323-340.

Papadopoulos, G.A. and Chalkis, B.J. 1984. *Tsunamis observed in Greece and the surrounding area from Antiquity up to the present times.* Mar Geol. 56: 309-317.

Papatheodorou, G. and Ferentinos, G. 1997. *Submarine and coastal sediment failure triggered by the 1995, M_s=6.1 R Aegion earthquake, Gulf of Corinth, Greece.* Mar Geol. 137: 287-304.

Papoulia, J., Lykousis, V. and Sakellariou, D. 1998. *Neotectonic activity and seismic hazard in central Creece.* Boll Geof Teort Applicata 39:113-124.

Perissoratis, C., Mitropoulos, D. and Angelopoulos, I. 1984. *The role of earthquakes in inducing sediment mass movements in the eastern Corinth Gulf: an example from the February 24 - March 4 activity.* Mar Geol. 55: 35-45.

Perissoratis C., Piper D.J.W. and Lykousis V., 2000. *Alternating marine and lacustrine sedimentation during late Quaternary in the Gulf of Corinth rift basin, central Greece.* Mar. Geol. 167: 391-411.

Piper, D.J.W, Kontopoulos, N., Anagnostou C., Chronis, G. and Panagos, A.G., 1990. *Modern fan deltas in the Western Gulf of Corinth, Greece.* Geo-Mar Letters 10: 5-12.

Radakovitch, O., 1995. *Étude du transfert et du dépôt du matériel particulaire par le ^{210}Po et le ^{210}Pb. Application aux marges continentales du Golfe de Gascogne (NE Atlantique) et du Golfe du Lion (NW Méditerranée).* PhD. Thesis Univ. of Perpignan.

Sakellariou, D., Kaberi, H. and Lykousis, V. 2004. *The impact of active tectonics on the actual sedimentation in the Gulf of Corinth basin.* 10th Int. Cong. Geol. Soc. Greece, Abstracts, p. 232-233, Thessaloniki (in Greek).

Sanchez-Cabeza, J.A., Masqué, P. and Ani-Ragolta, I. 1998. *^{210}Pb and ^{210}Po analysis in sediments and soils by microwave acid digestion.* Journal of Radioactivity and Nuclear Chemistry 227 (1-2): 19-22.

Section 6 - Submarine landslides in volcanic island settings

HIGH FREQUENCY SEDIMENT FAILURES IN A SUBMARINE VOLCANIC ENVIRONMENT: THE SANTORINI (THERA) BASIN IN THE AEGEAN SEA

T. HASIOTIS

Dept. of Marine Sciences, University of the Aegean, University Hill, 81100 Mytilene, Lesvos, Greece, tel: +30-2251036829, (e-mail: hasiotis@marine.aegean.gr)

G. PAPATHEODOROU, M. CHARALAMPAKIS

Laboratory of Marine Geology & Physical Oceanography, Dept. of Geology, University of Patras, Patras, Rio, 26500, Greece

A. STEFATOS

Rocksource ASA, Olav Kyrres gate 22, N-5808 Bergen, Norway

G. FERENTINOS

Laboratory of Marine Geology & Physical Oceanography, Dept. of Geology, University of Patras, Patras, Rio, 26500, Greece

Abstract

Seismic data have been used to evaluate the extent, characteristics and importance of the sediment failures in the Santorini basin. The failures are small but abundant, occupying almost half of the areal extent of the study area, and modify the relief of the basin and the surrounding slopes. The fact that surficial mass flow deposits are the source area of younger sediment failures is evident of the high intensity and frequency of the sediment instabilities. The major factors which are responsible for the observed instabilities are seismic activity and seismicity related to modern volcanic activity, steep slopes and the open sediment structure due to the specific texture of the volcanic material. Sediment failures are believed to compose a big part of the deeper sedimentary column.

Keywords: Santorini, sediment failures, seismic activity, volcanic activity, volcanic sediments

1. Introduction – General Setting

The Santorini basin is located between the Ios and Santorini islands and covers a small part of the Hellenic Volcanic Arc (HVA in the following) (Fig. 1a). The arc was formed due to the subduction of the African plate beneath the Aegean microplate. Volcanic activity in the HVA began approximately 3–4 Ma ago and the area is considered as a region of extensive Quaternary volcanism. The Santorini island is the most striking geomorphological feature of the HVA. Activity of the Santorini complex started 600 ka B.P. and the volcano is well known for its very large eruption around 1640 BC, which also formed the present caldera. Historic activity has resulted in the present-day islands of Palea and Nea Kameni. Approximately 7 km NE of Santorini, a new volcanic center last erupted in 1650 AD, referred to as the Columbo volcanic reef. The main fault trend in the wider area is SW-NE, which offset earlier faults of E-W and N-S trend (Piper and Perissoratis 2003; Piper et al., 2004).

V. Lykousis, D. Sakellariou and J. Locat (eds.), Submarine Mass Movements and Their Consequences, 309–316.

Seismic activity at the HVA is smaller compared to the forearc region and concentrated at the volcanic centers (mainly Milos, Santorini and Nisyros) and along a SW–NE trending zone of crustal weakness, the Santorini–Amorgos zone (Bohnhoff et al. 2006) (Fig. 1a). This area is characterised by strong microseismic activity of a small hypocentral depth (<15km). Within this zone, the submarine Columbo volcano exhibits a strong temporal variation of seismic activity, which has been interpreted to be directly linked to the migration of magma and fluids towards the surface. Small-scale seismic activity has also been reported NE of Columbo, probably representing local pathways of upward migrating fluids or even developing volcanic activity within a zone of crustal weakness (Bohnhoff et al. 2006). The Santorini-Amorgos zone has also hosted the two largest earthquakes (7.4R and 7.2R) in the entire south Aegean region the last century (in 1956).

The general stratigraphic and structural framework around Santorini island has been examined by Perissoratis (1995), whilst sediment distribution and origin in the Santorini basin is presented in the Santorini (Thera) sheet (I.G.M.E. 1995). However, the importance of sediment failures in the Santorini basin architecture has not been evaluated in detail. This paper examines the geomorphological and instability conditions of the surficial sedimentary cover in the western Santorini basin (between Thera and Ios) and gives indications for the basin filling processes during the late Quaternary, using a suite of high resolution 3.5 kHz subbottom profiles and side scan sonar images, and shallow sediment coring obtained in early 1998.

2. Data Analysis

2.1 GEOMORPHOLOGY

The Santorini basin (within the limits of the study area) is surrounded by the 300 m isobath, has an average width of 7.5 km and trends to the east with low gradients. The basin is bounded by the Ios slope to the north and the Santorini slope to the south. The Ios slope extends between the 80 and the 300 m isobaths. The slope gradients are steeper between the 100 and 200m (13°) than between 200 and 300m (8°), and increase to the east (Fig. 1). The Santorini slope develops between the 20 and 300 m isobaths and is characterized by very steep gradients between the 20 and 200 m (max. 17°). The lower slope has smaller gradients that range between 5 and 11°.

The seabed, especially in the Santorini basin, is characterized by abundant acoustic changes, which were attributed to the presence of cohesionless volcanic sediments that constitute a major part of the seafloor sediments and to the numerous instability phenomena. The abundance of the acoustic characters in the 3.5 kHz profiles made difficult the distinction of certain echo types. For that reason the seafloor was divided in four types (Fig. 1b and 2) depending on the acoustic response of the deformational characteristics of the recent sedimentary cover that were observed in the seismic records. The first type corresponds to an almost coherent seafloor that is interpreted as "undisturbed sediment islands", which have survived from the failures rather than transported in the form of failed sediment slabs. The second type corresponds to a medium degree of deformation, whilst the third type is characterized by a high deformational degree and concerning to the acoustic characteristics is associated with sediment failure deposits and channeled seafloor along the slopes.

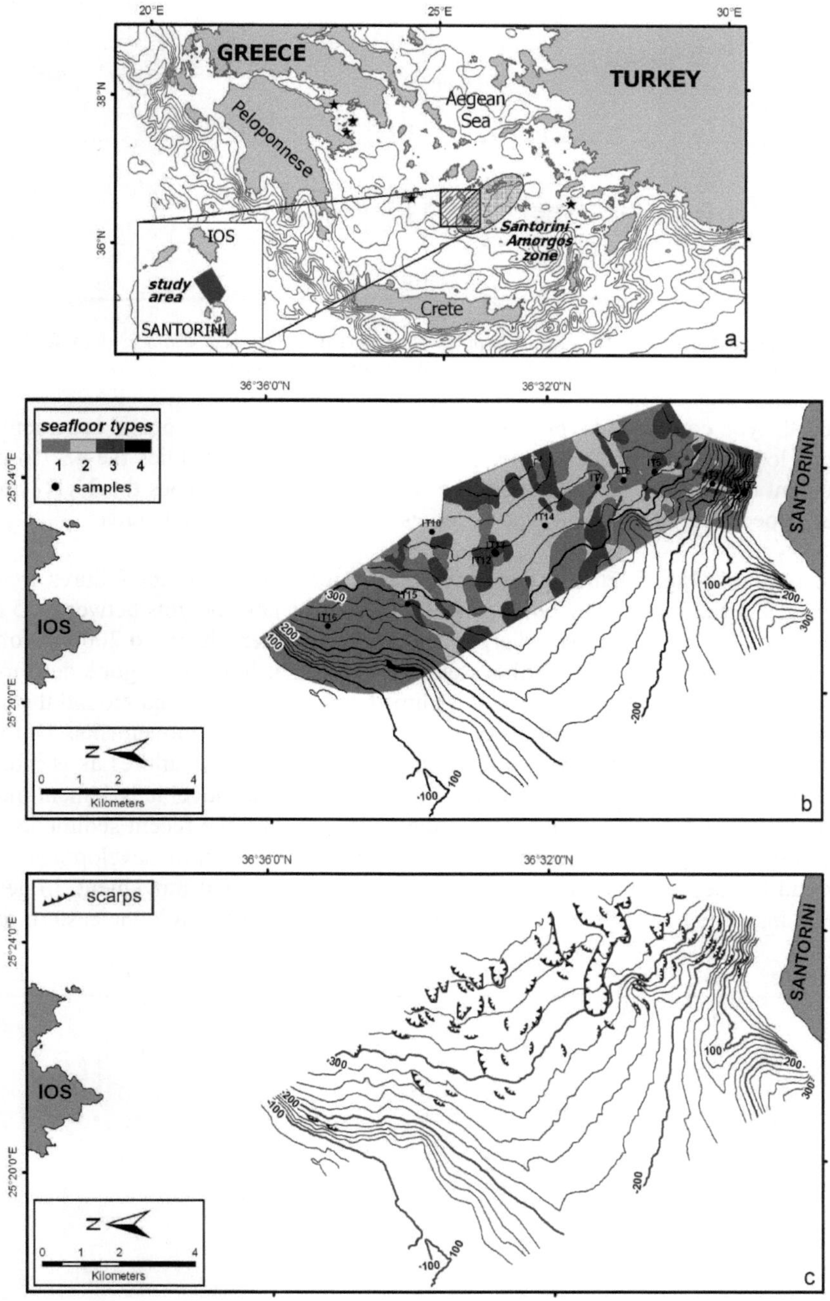

Figure 1. (a) General map of the southern Aegean sea showing the location of the study area, the Santorini-Amorgos zone and the main volcanic centers (asterisks from east: Nisyros, Santorini, Milos); (b) seafloor map of the study area showing the deformational degree of the surficial sediments (1: undisturbed, 2: medium deformational degree, 3: high deformational degree, 4: acoustic basement) and the sample locations; (c) bathymetric chart of the showing the main scarps in the study area. The subbottom profiler and side scan sonar survey extended only to the eastern part of the area under consideration.

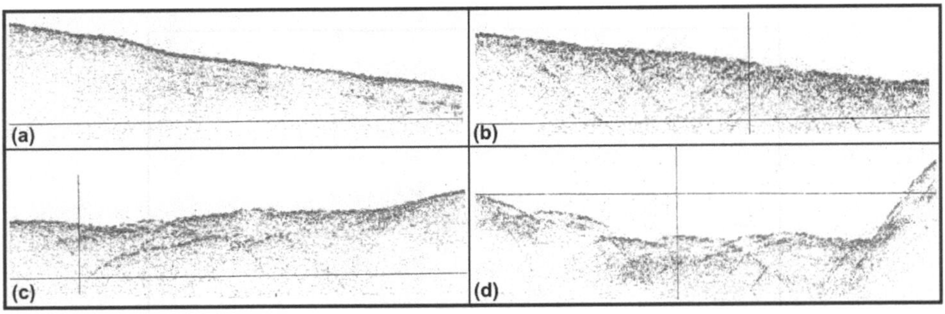

Figure 2. Representative 3.5 kHz seismic profiles showing (a) undisturbed seafloor, (b,c) seafloor of a medium deformational degree and (d) seafloor of a high deformational degree.

The fourth type corresponds, probably, to a bedrock exposure (acoustic basement) and develops locally along the Ios slope. The first seafloor type occupies the 48.7% of the areal extent of the study area, whereas the second and the third types the 35.1% and the 15.8%, respectively. The fourth type occupies only 0.4% of the study area.

The main morphologic elements that prove the existence of submarine failures are scarp areas (Fig. 1c) and hummocky relief (Fig. 3). The scarps have heights between 1.5 and 8 m and some of them retain their sharp characteristics for lengths up to 2000 m forming low-relief small valleys. These valleys enclose areas that have undergone deformation and transportation. The hummocky relief immediately downslope and around the scarps provides evidence of the young age of the observed mass movements. Deformed sediments and mass flow deposits are subsequently affected by failures as is indicated by the well-developed scarps and the hummocky relief. Volcanogenic basement mounds near Santorini base of slope, that pre-existed or intrude into the recent sediments, were locally found to control the sediment transport paths. The overall development of the scarps and the depositional areas indicate an eastward sediment movement. In general, sediment instability appears to be more intense and frequent towards the eastern part of the basin.

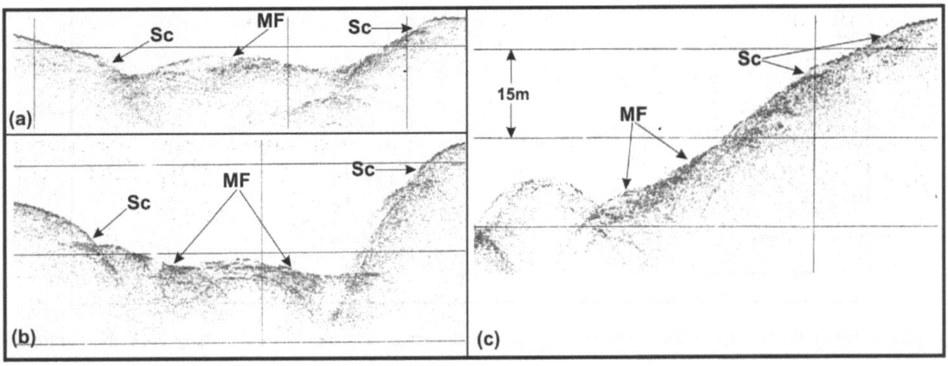

Figure 3. Representative 3.5 kHz seismic profiles in the Santorini basin showing scarps (Sc) and mass flow deposits (MF) (the vertical scale is the same in all the profiles).

The Santorini slope is covered by a thin surficial sedimentary unit (<3m) that is characterized by well-developed sediment detachment surfaces and rotational slides (Fig. 4). The detachment surfaces are about 2 m in height, arrange almost parallel to the isobaths and are usually narrow and elongated. Some of them bound distinct areas, which correspond to sliding planes. The absence of discrete transported sediment blocks indicates that the detached surficial sediments were immediately transformed into flows or they were gradually disintegrated during transport to mass flows. Disintegration is also favored due to the non-cohesive nature of the volcanic material. The lower slope (deeper than the 200 m isobath) is locally characterized by steep gradients and larger sediment evacuation zones.

The Ios slope is covered by a thin sedimentary unit, which is affected by small in thickness slides and few and small detachment surfaces of smaller importance (Fig. 5).

2.2 SEDIMENTS

Eighteen sampling attempts (gravity cores and grab samples) were made in areas of different deformational characteristics and thirteen samples were finally retrieved. The failed attempts were made in areas of a high deformational degree. Due to the coarse material of the seafloor the gravity cores were very short (max. 50 cm). The surficial sediments were found to consist of different grain sizes (from muds to gravels) (Table 1). The coarse-grained components (sands and gravels) are mainly of volcanic origin (ash and pumice).

The short cores consist, in general, of less than 50 cm of loose coarse-grained material. Coarse sand and gravels are scattered within the sediment matrix and appear to gradually increase towards the bottom of the cores. Discrete layers of sandy material with erosional base contacts within a muddy matrix as well as alternations of muddy /sandy layers, which were found in some cores of the Santorini slope and basin, are probably of turbiditic origin.

The sediments in Santorini slope are principally of a sandy nature, whilst in Ios slope and Santorini basin are muddy in nature. Locally, sampling from areas of a medium and a high deformational degree, wherever possible, appear to have a major gravel component.

3. Discussion - Conclusions

Giant slides, debris flows and avalanches and turbidity currents are common gravitative processes around volcanic islands (Gee et al. 2001; Ollier et al. 1998). The Santorini basin is a marine volcanic environment that is controlled by recent failures which affect the surficial sediments. The failures develop mainly along the Santorini slope but also influence the Santorini basin sediments and move towards the east. The observed instabilities are small, but very frequent, occupy half of the areal extent of the study area and seem to be a dominant process in valley architecture. The development of scarps in regions of already deformed sediments or over mass flow deposits indicates the high intensity and the frequency of the failures as well as a relatively recent formation. There is no evidence of volcanic debris avalanche with large debris standing on the seafloor like in other similar environments (Gee et al. 2001).

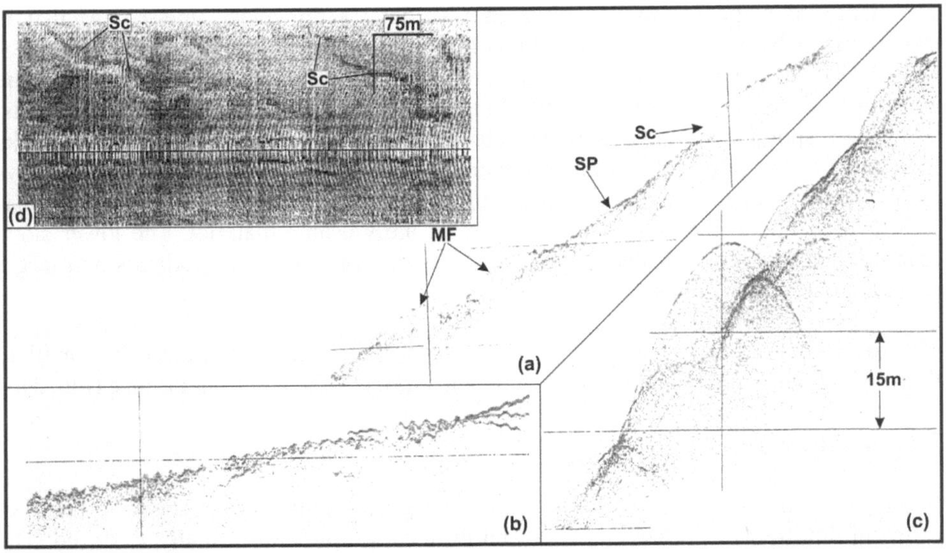

Figure 4. Representative 3.5 kHz seismic profiles (a,b,c) and side scan sonar image (d) along the Santorini slope showing scarps (Sc), sliding planes (SP), mass flow deposits (MF) and seabed of a medium (b) and of a high (c) deformational degree (the vertical scale is the same in all the profiles).

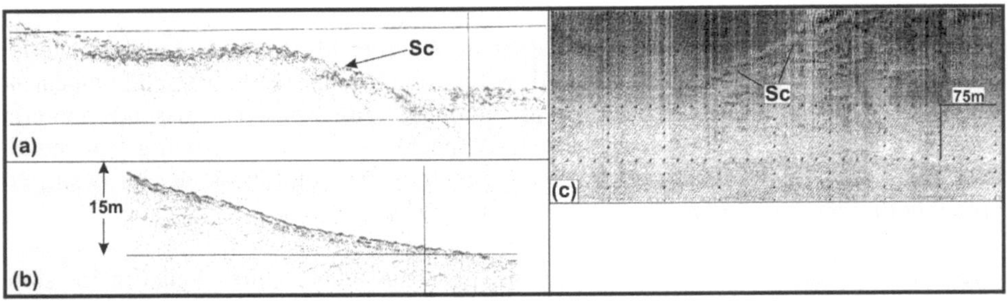

Figure 5. Representative 3.5 kHz seismic profiles (a,b) and side scan sonar (c) along the Ios slope showing the thin sedimentary cover, which is locally affected by low in height scarps (Sc) (the vertical scale is the same in all the profiles).

Factors that contribute to the triggering and to the frequent repetition of the observed sediment failures are the paroxysmic seismic activity and/or seismicity related to volcanism, the steep gradients of the Santorini slope and of the local volcanic intrusions and the specific texture of the surficial sedimentary cover (rich in non-cohesive volcanic sediments - pumice and ash) that contribute to a relatively open sediment structure. Hampton (1989) demonstrated that the ash-rich sediments of the Chiniak Trough (Gulf of Alaska) exhibit considerable cyclic strength degradation, being also highly susceptible to liquefaction type failure. The unstable nature of the volcanic material is also evident along the coastline, where subaerial slopes are constantly eroded by aeolian processes forming narrow (up to some tens of centimeters) transport chutes.

Table 1. Sediment sampling details and surficial sediment types.

Location	Sample	Type	Length (cm)	Deformational degree	Surficial Sediment Type
Ios slope	IT16	c	20	no	Sandy mud
Santorini basin	IT14	c	16	medium	Sandy mud
	IT15	c	27	no	Sandy mud
	IT10	c	21	medium	Mud
	IT13	c	16	high	Slightly gravelly sandy mud
	IT5	g		high	Gravelly mud
	IT6	g		medium	Muddy gravel
	IT7	g		high	Gravels
	IT12	g		high	Slightly gravelly sandy mud
Santorini slope	IT1	g		no	Gravelly muddy sand
	IT3	g		medium	Sandy gravels
	IT2	c	48	medium	Slightly gravelly muddy sand
	IT4	c	50	medium	Muddy sand

c: short gravity core, g: grab sample

The studied area lies (i) at the southwestern tip of the NE-SW striking Santorini-Amorgos zone that exhibits frequent seismic activity, locally as a result of upward magma and fluid migration and (ii) very close to the epicenter of the two largest earthquakes in the south Aegean sea, that occurred in 1956. Frequent earthquakes, even if of a small magnitude, may account for sediments to maintain a relative open structure, whereas big events, such as those in 1956, could have triggered numerous of the observed sediment failures.

The modern volcanic activity, since the Minoan explosion (~3.6ka B.P.), has been manifested by the formation of Columbus crater, the magma domes in northern and western Santorini Caldera basins (Perissoratis 1995) and the emergence of Kameni islands. This activity probably was accompanied by shallow seismic activity that promoted part of the detected sediment failures.

Although hydrothermal fluid venting is well-documented only within the Santorini caldera (Varnavas and Cronan 2005) and is assumed by Bohnhoff et al. (2006) in the wider area, it was not documented in the present study. However, it is expected to occur and to influence the stability of the surficial sedimentary cover.

The results of this survey complement previous studies by I.G.M.E. (1995), in which it was believed that mass movements affected only the NNE part of Santorini slope. Similar processes should have taken place since when Santorini volcano started its activity, thus abundant, small in thickness, sediment failure deposits are likely to constitute a major part of the deeper sedimentary sequence.

4. Acknowledgements

The authors would like to thank D. Piper and A. Velegrakis for their constructive comments while reviewing the manuscript.

5. References

Bohnhoff, M., Rische, M., Meier, T., Becker, D., Stavrakakis, G. and Harjes, H., 2006. *Microseismic activity in the Hellenic Volcanic Arc, Greece, with emphasis on the seismotectonic setting of the Santorini–Amorgos zone.* Tectonophysics, 423: 17-33.

Gee, M., Watts, A., Masson, D. and Mitchell, N., 2001. *Landslides and the evolution of El Hierro in the Canary islands.* Marine Geology, 177: 271-293.

Hampton, M.A., 1989. *Geotechnical properties of sediment on the Kodiak continental shelf and upper slope, Gulf of Alaska.* Marine Geotechnology, 8: 159-180.

I.G.M.E. (Institute of Geology and Mineral Exploration of Greece), 1995. Santorini (Thera) sheet. Scale 1:200000.

Ollier, G., Cochonat, P., Lenat, J. and Labazuy, P., 1998. *Deep-sea volcaniclastic sedimentary systems: an example from La Fournaise volcano, Reunion island, Indian ocean.* Sedimentology, 45: 293-330.

Perissoratis, C., 1995. *The Santorini volcanic complex and its relation to the stratigraphy and structure of the Aegean arc, Greece.* Marine Geology, 128: 37-58.

Piper, D. and Perissoratis, C., 2003. *Quaternary neotectonics of the South Aegean arc.* Marine Geology, 198: 259-288.

Piper, D., Pe-Piper, G., Perissoratis, C. and Anastasakis, G., 2004. *Submarine volcanic rocks around Santorini and their relationship to faulting.* 5th International Symposium on Eastern Mediterranean Geology, Thessaloniki, Greece, p. 873-876.

Varnavas, S. and Cronan, D., 2005. *Submarine hydrothermal activity off Santorini and Milos in the Central Hellenic Volcanic Arc: A synthesis.* Chemical Geology, 224: 40– 54.

SEDIMENT STABILITY CONDITIONS WEST OF MILOS ISLAND, WEST HELLENIC VOLCANIC ARC

T. HASIOTIS
Dept. of Marine Sciences, University of the Aegean, University Hill, 81100 Mytilene, Lesvos, Greece, tel: +30-2251036829, (e-mail: hasiotis@marine.aegean.gr)

G. PAPATHEODOROU, G. FERENTINOS
Laboratory of Marine Geology & Physical Oceanography, Dept. of Geology, University of Patras, Patras, Rio, 26500, Greece

Abstract

Examination of seismic profiles west of Milos Island (Aegean Sea) show that the north Milos slope is affected by extensive mass movements, whereas the south slope is generally stable. Conditions that promote sediment failures include late Quaternary volcanic activity and related seismic activity, tectonically oversteepened slopes and possibly hydrothermal fluid escape. Big and small depressions indicative of phreatic explosions and hydrothermal venting, respectively, have been observed in Milos shelf. Sediment coring revealed that the surficial sediments are relatively soft and thus prone to failure. The lack or presence of sapropel layers is also indicative of unstable or stable slope conditions of the surficial sedimentary cover, respectively.

Keywords: Milos Island, mass movements, seismic activity, volcanic activity, hydrothermal venting, sediment stability

1. Introduction

The Hellenic Volcanic Arc is one of the major geomorphological features of the Aegean Sea. It is located at the southern margin of the almost aseismic Cyclades Plateau and has resulted from the subduction of the African plate beneath the Aegean–Anatolian microplate. The main explosive volcanic centres of the Upper Quaternary are Milos, Santorini and Nisyros (Fig. 1a). Milos Island separates the Myrtoon basin to the north and the Cretan basin to the south. The late Neogene sedimentation and its association with volcanism around Milos has been examined by Anastasakis and Piper (2005), studying mainly airgun seismic profiles and multichannel seismic lines. This study focuses on the stability conditions of the surficial sedimentary cover west of Milos, along the south Myrtoon and the north Cretan basins bounding margins (Fig. 1b), during the late Quaternary. A suite of side scan sonar images, subbottom profiles and shallow sediment cores were examined for the purpose of this study.

2. Geological Backround

The western Hellenic Volcanic Arc, active since 3-4 Ma ago, extends from Crommyonia in western Attica through the island of Aegina and Methana peninsula to Milos (Fig. 1a). These volcanic centres are dominated by the presence of domes and lava flows with subordinate pyroclastic rocks. Milos Island has been the site of explosive rhyolitic volcanism during Plio-Quaternary times. As a result, volcanic rocks are widespread in submarine areas around the island (Anastasakis and Piper 2005). The

V. Lykousis, D. Sakellariou and J. Locat (eds.), Submarine Mass Movements and Their Consequences, 317–324.

small island of Antimilos and the islets of Ananes consist only of volcanic rocks (Fig. 1b). According to Anastasakis and Piper (2005), evidence of Quaternary volcanism offshore appears only south of Milos and at Antimilos (one age determination at 0.32 Ma). Submarine hydrothermal venting locations occur near the coast of Milos (Varnavas and Cronan 2005). No submarine volcanoes are known around Milos.

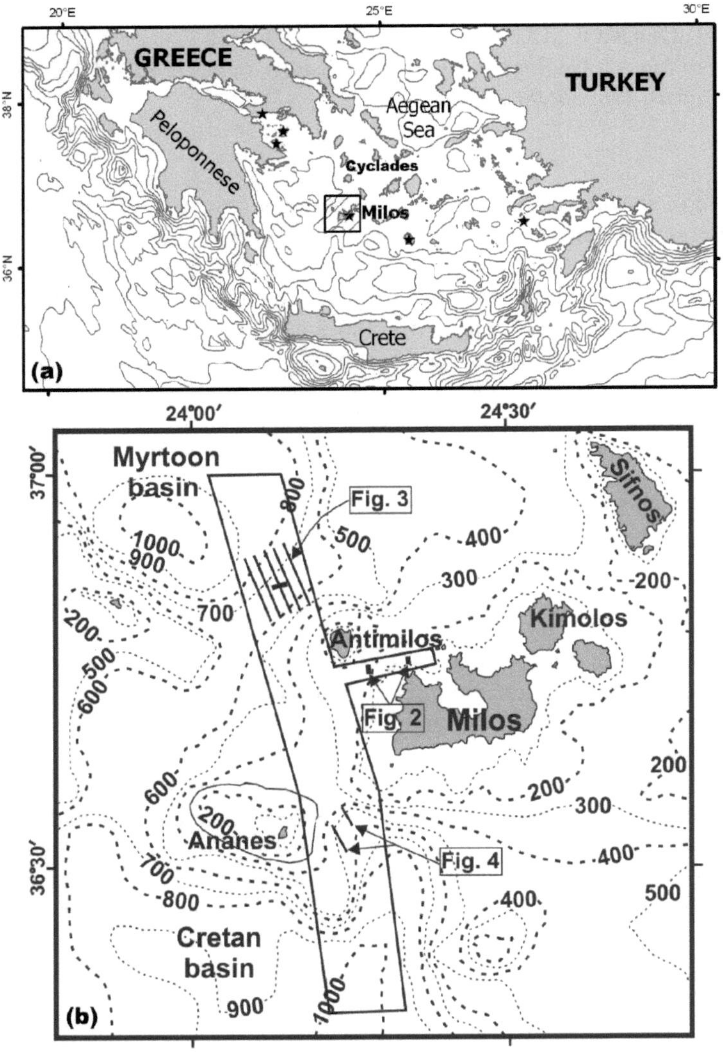

Figure 1. (a) General map of the southern Aegean Sea showing the location of the study area (box) and the main volcanic centers (asterisks from east: Nisyros, Santorini, Milos, Methana, Aegina, Crommyonia); (b) bathymetric chart around Milos Island showing the study area (frame) and the figure locations.

Young (mid to late Quaternary age) and older fault systems of N-S and E-W direction, respectively, as well as NE-trending faulting (Mascle and Martin 1990; Piper and Perissoratis 2003) is responsible for regional basin subsidence and for the formation of small deep-water basins (Fig. 1b). Locally, the sedimentary column is strongly offset by faults, with scarps up to 150 m high (Piper and Perissoratis 2003).

The study area is characterized by weak seismic activity as is indicated by (a) the earthquake hypocentre distribution for the time interval between 1964-1998 (Bohnhoff et al. 2006), (b) the last century National Observatory of Athens (NOA) earthquake catalogues and (c) the new Greek antiseismic regulation. A local temporary network, which was operated on Milos (Ochmann et al. 1989) for several months, identified a strong spatio-temporal hypocentral clustering below Milos, where time intervals with average seismic activity of 3–5 events per day were interrupted by single days with up to 600 events. The events were weak (M~2 R) and occurred within the uppermost 10 km (Bohnhoff et al. 2006). A 4.8 R earthquake in 1992 with an epicentre close to Milos, although small in magnitude, caused liquefaction features in Hivadolimni area in Milos.

3. Data Presentation

3.1 BATHYMETRY - MORPHOLOGY

The north Milos slope extends between the 250 and the 800-900 m isobaths, whilst the south slope extends between the 400 and 800 m isobaths (Fig. 1). They have an average gradient of about 3° with local maxima up to 26°. The south slope is in general smooth, whereas the north slope is characterised by multiple and abrupt changes in slope direction. Locally, the seabed has a step-like morphology of small relief due to the detachment and downslope movement of sediment slabs of relatively small thickness. Small channels and canyons appear near the base of the slope.

The Milos shelf has in some places an anomalous relief due to the presence of volcanic rock outcrops, extended mounds and troughs and depressions. The depressions are of two distinct types: either big, having diameters and depths of 500 m and 5 m respectively, or small with a maximum depth of 0.5 m and a diameter of 30 m (Fig. 2a,b).

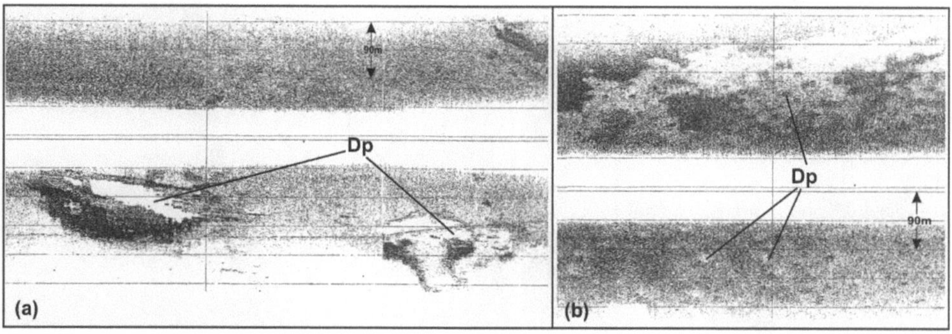

Figure 2. Side scan sonar images showing (a) the big and (b) the small depressions (Dp) in the Milos shelf.

3.2 SHALLOW GEOLOGICAL CHARACTERISTICS

The slopes around Milos are covered by parallel to sub-parallel sedimentary layers up to 90 m or more in thickness, which are influenced by active faulting. In the north Milos slope the surficial sediments have an average thickness of about 15 m and are affected by mass movements. The 3.5 kHz seismic profiles show truncations of the surficial layered sequence, outcropping sedimentary units, thinning and wedging out of the surficial layers, sudden slope gradient changes and relatively thin mass flow deposits (less than 7-8 m thick) that are all indicative of sediment failures (Fig. 3). Side scan sonar records show the existence of elongated less than 3 m scarps indicative of sediment detachment along bedding planes. The instabilities are small but abundant and involve less than 5-10 m of the upper surficial sediments. The base of slope is built by stacked, small (both in term of thickness and areal extent) deposits of a semi-transparent acoustic character, denoting multiple mass flow events. Successive, buried mass flow deposits have also been detected along the slope.

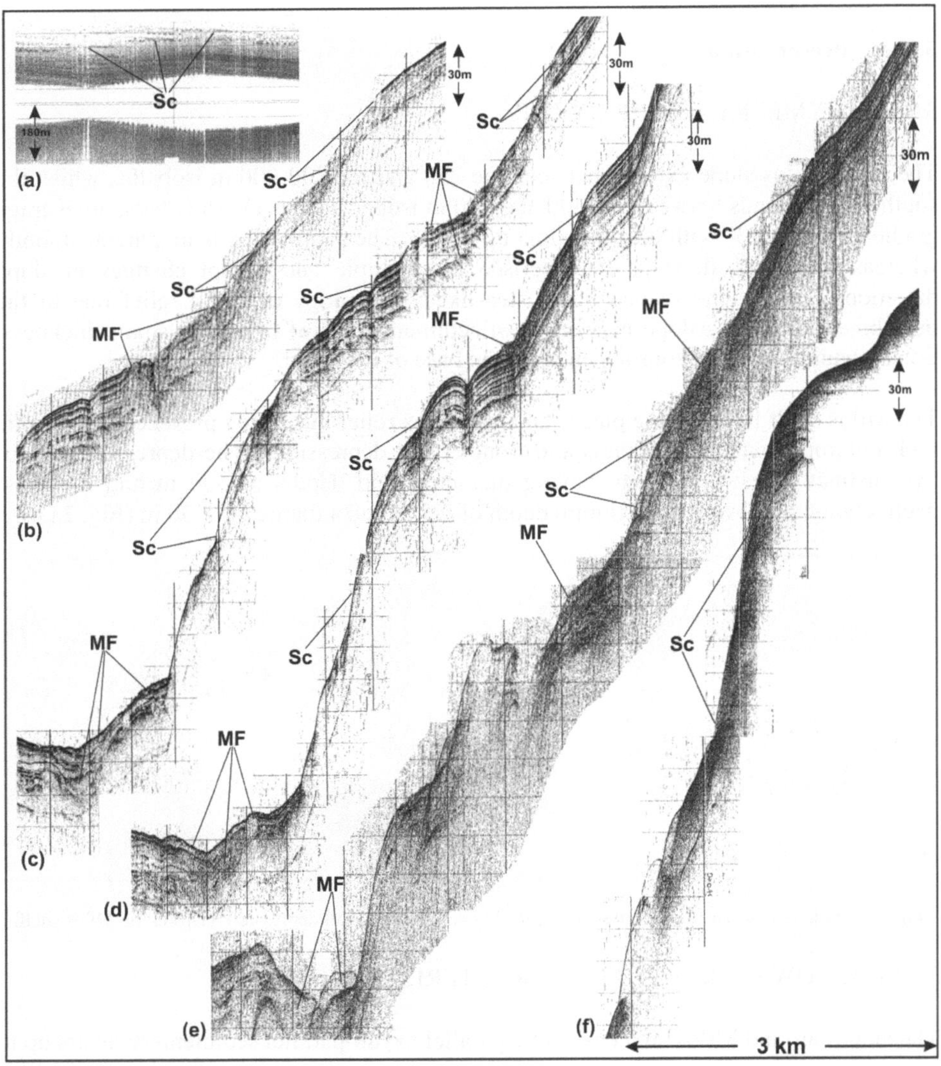

Figure 3. Representative side scan sonar image (a) and 3.5 kHz seismic profiles (b,c,d,e,f) along the north Milos slope showing extensive sediment failures. Sc: scarp, MF: mass flow deposit.

In the south Milos slope the surficial stratified sediments have a maximum thickness of 15 m and overlie unconformably slightly folded layers (Fig. 4). Near the base of the slope the sediments are deformed probably due to volcanic intrusions. In general, the seabed is considered stable since only a few relatively small instability features have been detected.

In the Milos shelf, the recent sediments have a variable thickness and overlie folded sediments or volcanic rocks. The observed big depressions are probably small craters due to old phreatic explosions, whereas the small depressions develop along areas covered by a thin sedimentary cover (< 15 m thick) and most likely resemble to small pockmarks caused by hydrothermal solution venting.

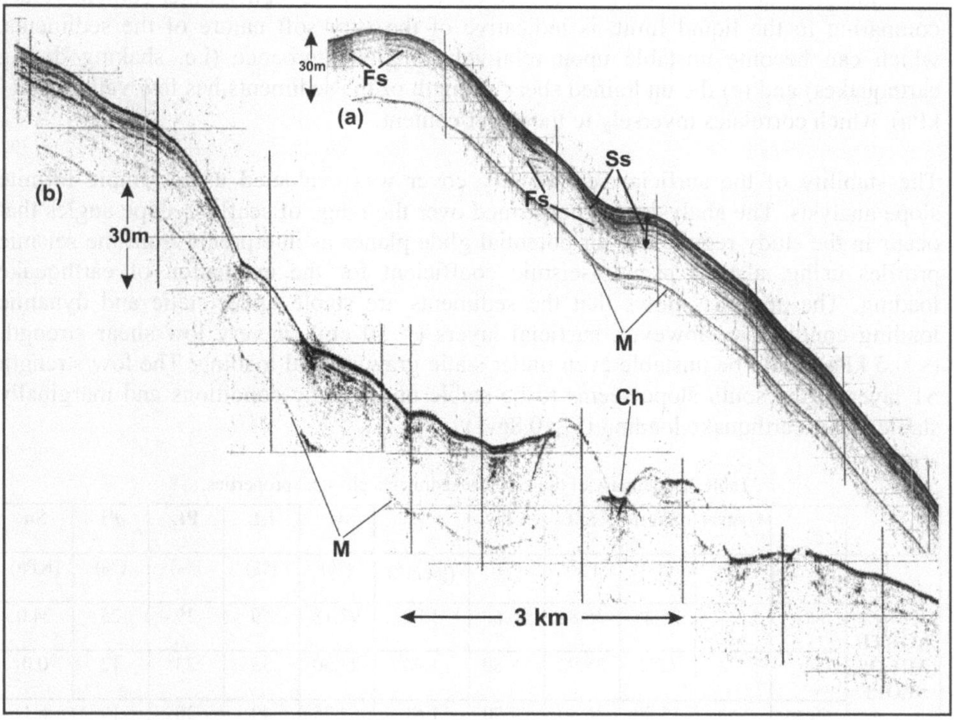

Figure 4. Representative 3.5 kHz seismic profiles showing (i) stratified sediments (Ss) overlying slightly folded sediments (Fs) and (ii) the smooth morphology of the south Milos slope, which is locally interrupted by channels (Ch). M: Multiple.

3.3 SEDIMENT PHYSICAL PROPERTIES AND STABILITY REGIME

Six cores (from the north and south slopes) and twelve grab sediment samples were collected from the study area. The surficial sediments are coarse-grained (sand to muddy sand) in the west Milos shelf whereas the slope sediments are fine-grained (sandy mud to mud) in nature. The main sediment types (lithofacies) of the surficial sedimentary column (upper 2 m) in water depths greater than the shelf-edge are: (i) calcareous mud with a small proportion of sand, (ii) sapropels (organic rich mud - S1 and locally S2) (Anastasakis and Stanley, 1984), (iii) hemipelagic mud and (iv) sand to sandy muds of turbiditic origin. Sand and silt-sized volcanic ash is present almost all along the sediment cores but does not form discrete ash layers. The most recent traces have been locally observed at about 15 cm below surface and probably correspond to the latest ash-layer in the Aegean Sea (due to the Santorini eruption, around 3.6ka B.P.). Sapropel layers have not been detected in the north slope, whereas they appear in the south Milos slope.

The vertical variation in the physical properties (Table 1) shows that: (a) the wet bulk density (1.39-1.82 gr/cm^3) and the water content (23.3-113.45%) values show great variability, and are mainly related to the sediment composition and secondary to the morphological unit where they have been collected, (b) the higher water content values comparing to the liquid limit, is indicative of the very soft nature of the sediments, which can become unstable upon relatively small disturbance (i.e. shaking during earthquakes) and (c) the undrained shear strength of the sediments has low values (0-24 kPa), which correlates inversely to the water content.

The stability of the surficial sedimentary cover was evaluated using simple infinite slope analysis. The analysis was performed over the range of seafloor slope angles that occur in the study region and for potential glide planes as interpreted from the seismic profiles using, also, a pseudo-seismic coefficient for the evaluation of earthquake loading. The analysis shows that the sediments are stable under static and dynamic loading conditions. However, surficial layers (< 50 cm) of very low shear strength (< 0.5 kPa) could be unstable even under static gravitational loading. The low strength S1 layer in the south slope seems to be stable under static conditions and marginally stable under earthquake loading (F_D: 0.86-1).

Table 1. Summary of the surficial sediments physical properties.

		Gravel	Sand	Silt	Clay	γ	w	LL	PL	PI	Su
		(%)	(%)	(%)	(%)	(g/cm^3)	(%)	(%)	(%)	(%)	(KPa)
NORTH MILOS SLOPE	max		25.28	36.88	56.60	1.82	92.18	59	39	25	24.0
	min		6.52	35.92	38.80	1.44	23.30	35	23	12	0.0
	mean		15.90	36.40	47.70	1.61	64.35	49	30	19	6.1
MILOS SHELF	max	0.77	75.83	41.60	30.85						
	min	0.11	40.63	13.94	7.86						
	mean	0.43	55.64	27.42	16.61						
SOUTH MILOS SLOPE	max		9.97	36.88	56.60	1.68	113.45	59	40	25	13.8
	min		6.52	35.78	54.25	1.39	49.47	35	23	11	0.5
	mean		8.25	36.33	55.43	1.60	66.69	45	29	17	5.7

γ: wet bulk density, w: water content, LL: liquid limit, PL: plastic limit, PI: plasticity index, Su: undrained vane shear strength

4. Discussion - Conclusions

According to Anastasakis and Piper (2005) the sedimentary column in the basins surrounding Milos island consists mainly of hemipelagic deposits interbedded, locally, with gravity flow deposits (< 20 m thick) stemming from (i) small collapses of the neighboring volcano flanks, (ii) pyroclastic flows and (iii) deeper-seated failure scarps in the hemipelagic sediments. Although the previous authors conclude that the basin margins have been generally stable, the results of this high resolution survey demonstrate that this is only true for the Cretan basin margin (E-SE of Ananes, within the surveyed corridor), since the north Milos slope (south Myrtoon margin) shows a variety of evidence of mass movements affecting the surficial sedimentary cover.

The major factors that are responsible for the sediment failures in the study area are the paroxysmic seismic activity and/or seismicity related to volcanism, neotectonic slope steepening and, probably, hydrothermal fluid venting and the presence of relative coarse-grained volcanic sediments (ash) throughout the sediment column that contribute to a relatively open sediment structure. Hampton (1989) reported considerable cyclic strength degradation in ash-rich sediments from the Gulf of Alaska and high vulnerability to liquefaction type failures. Storm waves as a mechanism triggering mass movements is considered of negligible importance since the observed failures are located in water depths that are not influenced by waves, even during times of lower sea level conditions.

Many of the north Milos slope failures could be associated with the late Quaternary submarine volcanism of Antimilos (younger than 0.32 Ma), which possibly produced paroxysmic seismic activity and more intense hydrothermal fluid venting. The presence of big and small depressions in the Milos shelf gives evidence for old phreatic explosions and synchronous hydrothermal fluid venting. The fluid escape could destabilize the sediment framework, form an open structure and maintain this configuration even with slow seepage. These conditions could be also sustained due to the small but frequent seismic activity as reported by Bohnhoff et al. (2006).

The sapropel layers, which have periodically formed in the Mediterranean Sea, can be considered as trace horizons. The most recent sapropel S1 was formed (6900-9000) years B.P. and it is characterized by low shear strength and high water content values. Due to these geotechnical properties, the sapropel layer is considered as a horizon prone to failure. Its presence or absence in the sedimentary column gives evidences for the relative age of the most recent mass failures, sometimes independently from the study of the seismic profiles. Thus, the absence of the S1 locally from the upper north Milos slope indicates that failures took place the last 9000 years. In some cores, the presence of S1 relatively deep in the sedimentary column suggests that it might have been covered by thin gravity flows (skin flows). Insights into this process arise from the stability analysis, which shows that the S1 in the south Milos can be a potential glide plane. Along this level, surficial sediments could slide under relatively low seismic horizontal accelerations (5.3%g). Furthermore, the most recently deposited (~50 cm) soft and unconsolidated sediments can be locally unstable, even under their own weight, producing skin gravity flows.

5. Acknowledgements

The authors would like to thank R. Urgeles and D. Masson for their constructive comments while reviewing the manuscript.

6. References

Anastasakis, G. and Stanley, J., 1984. *Sapropels and organic-rich variants in the Mediterranean: sequence development and classification.* In: Fine-grained sediments: Deep-water processes and facies (Edit. D. Stow and D. Piper), Spec. Pap. Geol. Soc. London, 15: 497–510.

Anastasakis, G. and Piper, D., 2005. *Late Neogene evolution of the western South Aegean volcanic arc: sedimentary imprint of volcanicity around Milos.* Marine Geology, 215: 135–158.

Bohnhoff, M., Rische, M., Meier, T., Becker, D., Stavrakakis, G. and Harjes, H., 2006. *Microseismic activity in the Hellenic Volcanic Arc, Greece, with emphasis on the seismotectonic setting of the Santorini–Amorgos zone.* Tectonophysics, 423: 17–33.

Hampton, M.A., 1989. *Geotechnical properties of sediment on the Kodiak continental shelf and upper slope, Gulf of Alaska.* Marine Geotechnology, 8: 159–180.

Mascle, J. and Martin, L., 1990. *Shallow structure and recent evolution of the Aegean Sea: a synthesis based on continuous seismic reflection profiles.* Marine Geology, 94: 271–299.

Ochmann, N., Hollnack, D. and Wohlenberg, J., 1989. *Seismological exploration of the Milos geothermal reservoir, Greece.* Geothermics, 18(4): 563–577.

Piper, D. and Perissoratis, C., 2003. *Quaternary neotectonics of the South Aegean arc.* Marine Geology, 198: 259-288.

Varnavas, S. and Cronan, D., 2005. *Submarine hydrothermal activity off Santorini and Milos in the Central Hellenic Volcanic Arc: A synthesis.* Chemical Geology, 224: 40–54.

Section 7 - Submarine mass movements and tsunamis

MASS WASTING PROCESSES - OFFSHORE SUMATRA

D.R. TAPPIN
British Geological Survey, Nottingham, England

L.C. MCNEIL, T. HENSTOCK
National Oceanography Centre, Southampton, England

D. MOSHER
Bedford Institute of Oceanography, Natural Resources Canada, Challenger Dr., Dartmouth,, NS B2Y 4A Canada

Abstract

Earthquakes are a commonly cited mechanism for triggering submarine landslides that have the potential to generate locally damaging tsunamis. With measured runups of over 35 metres in northern Sumatra from the December 26th 2004 tsunami source, these runups might be expected to be due, in part, to local submarine landslides. Mapping of the convergent margin offshore of Sumatra using swath bathymetry, single channel seismic and seabed photography reveals that seabed failures are common, but mainly small-scale, and composed of blocky debris avalanches and sediment flows. These failures would have contributed little to local tsunami runups. Large landslides are usually formed where there is significant sediment input. In the instance of Sumatra, most sediment is derived from the oceanic plate, and there is little sediment entering the system from the adjacent land areas. Input from the oceanic source is limited because of the diversion of sediment entering the subduction system off of Sumatra, that is attributed to collision between the Ninetyeast ridge and the Sunda Trench at approximately 1.5 million years ago.

Keywords: Sumatra margin, tsunami, mass wasting, multibeam bathymetry, single channel seismic

1. Introduction

The December 26th 2004 earthquake in the Indian Ocean was the worlds largest for over 40 years and created the most devastating tsunami ever recorded, with fatalities around the Indian Ocean of over 220,000. The Sumatran subduction-zone system is subject to great (Mw 8–9) earthquakes, (e.g., Newcomb and McCann, 1987; Ortiz and Bilham, 2003). Earthquakes are a commonly cited mechanism for triggering submarine landslides (e.g. Hampton et al., 1996; Lee et al., 2007). However, it has only recently been recognized that they can generate locally damaging tsunamis (e.g. Papua New Guinea 1998, Seward 1964, Grand Banks 1929, and Storegga 8,200BP). Thus runups of over 35 metres reported from Banda Aceh in northern Sumatra, close to the tsunami source, might have been enhanced by the failure of local submarine landslides. In 2004, knowledge of offshore bathymetry for the region off Sumatra was generally poor because of the sparse coverage of single-beam echo soundings. However, in the Aceh Basin there was some evidence for sediment failures identified from arcuate features on GEBCO bathymetric maps. In 1998 a submarine slump located in an arcuate shaped feature off of northern Papua New Guinea caused a tsunami that led to over 2,000

V. Lykousis, D. Sakellariou and J. Locat (eds.), Submarine Mass Movements and Their Consequences, 327–336.

deaths (Tappin et al, 2001). In January 2005 multibeam bathymetry was acquired offshore of northern Sumatra (Figure 1) (Henstock et al, 2006). This was the first seafloor survey immediately after a great subduction-zone earthquake; and an ideal opportunity to identify coseismic deformation features in soft sediment. During a second marine survey in August, 2005 additional single channel seismic (SCS) data were acquired in the same region (Moran and Tappin, 2006). Based on these datasets, we here report on the active mass wasting processes taking place in the mapped region and their regional controls. We describe the seabed morphology based on the multibeam data and the related subseabed structure interpreted from the SCS. By this means we determine the 3D architecture of the primary mechanisms through which mass wasting is taking place in the region.

Figure 1. Map of the margin off Sumatra and the location of the HMS Scott bathymetry data. Inset shows of the December 26th rupture (in red) and the earthquake epicentre (red hexagon).

2. Methodology

The swath bathymetric data were acquired by the Royal Navy hydrographic vessel HMS Scott, in January- February 2005, using a 12 kHz SASS-IV system with 361 beams and a 120° swath width. The theoretical beam width (horizontal resolution) is ~25m directly beneath the ship to 100 m for the outer beams (50-60°) (flat seabed, 4500m depth). The ideal depth precision (vertical resolution) is ~5m directly beneath the ship, assuming no position or attitude errors and using Rayleigh's criterion; smaller features might be identifiable due to their coherence over several pings. Minor roll artifacts and noise may reduce the vertical precision to 10-15m for outer beams. Bathythermographs (XBTs) were taken at <3 h intervals to constrain water sound velocity. Available pre-existing bathymetry data in the area are sparse, single beam, and subject to navigation errors. The seismic data and seabed images were acquired aboard the RV Performer in August 2005. The seismic reflection profiling system consisted of a pneumatic sound source, a Seismic Systems Inc. Generator Injector (GI) gun (or a two

GI gun array) and a hydrophone, the Teledyne model 28420 streamer, 61 m (200 ft) in length.

3. Seabed morphology and structure

The two main areas mapped were the Aceh Forearc Basin, lying offshore of northern Sumatra and a ~550 km long section of the central Sunda convergent margin, including the outer-arc high fault system (Figures 1 and 2). It is in the southern part of this region that the December 2004 earthquake epicenter is located.

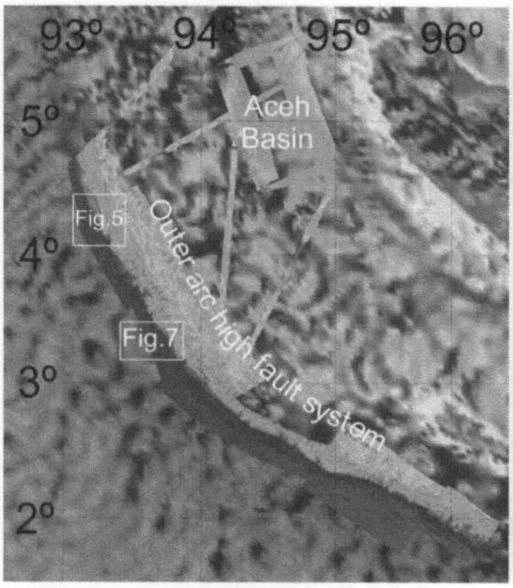

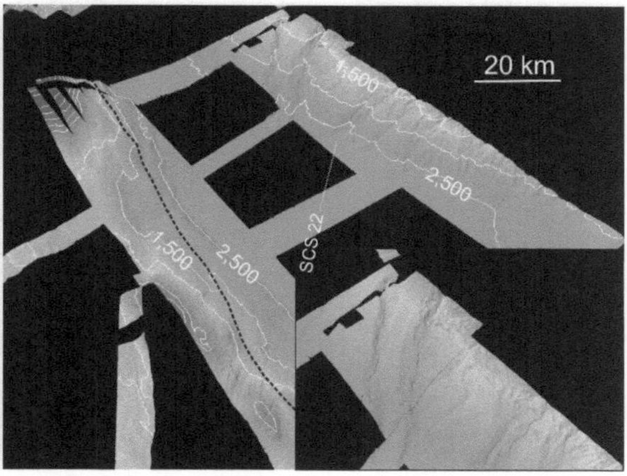

Figure 2 (left). Overview of the HMS Scott bathymetry and the locations of Figures 5 and 7.
Figure 3 (right). Morphology and bathymetry of the Aceh Basin. Bathymetric contours in metres, black dashed line – the West Andaman Fault, white dotted SCS Line 22 shown in Figure 4. Inset shows enlargement of the channels on the northwest. (Location in Figure 2.)

3.1 ACEH BASIN

The Aceh Basin is a forearc basin lying 45 km off of northern Sumatra (Figures 2 and 3). Its western boundary is the strike-slip, West Andaman Fault (Curray, 2005). In the east lies the shelf off of Sumatra. The basin trends NNW-SSE, the basin floor is planar, slightly sloping to the SSW and with water depths of ~2,500 m. Seabed gradients on the basin margins vary between 6° and 12°. The new multibeam data prove that the evidence for sediment failures on the eastern margin of the basin identified on the GEBCO bathymetry are artifacts probably due to the sparse coverage of single beam data. However, the data does show incised submarine channels, indicating several phases of downcutting. There is no evidence for the channels extending onto the basin floor, nor of any significant sediment build up, such as sediment fans, at their points of entry. There are no channels on the western margin of the basin along the Andaman Fault.

On the SCS data the basin infill mainly comprises parallel, moderate amplitude reflections with subordinate, semitransparent, chaotic reflections (Figure 4). The chaotic reflections form units that taper into the basin from the eastern margin. The seismic sequence is interpreted as representing dominantly hemipelagic sediment with subordinate sediment flows sourced from the eastern basin margin where they originate at the mouths of the channels identified on the multibeam bathymetry. None are recent. The flows are up to 50 m thick and extend up to 10 km into the basin, tapering out towards the basin centre. A major unconformity surface dips westward attaining a maximum depth of ~4 secs TWT (Figure 4). This indicates a significant episode of subsidence along the West Andaman Fault that may relate to the downcutting observed in the channels on the eastern basin margin.

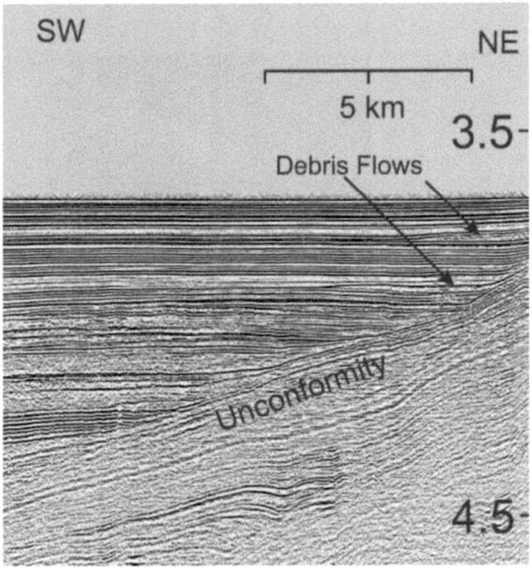

Figure 4. SCS Line 22 across the eastern Aceh Basin. Vertical scale in seconds two-way-time. See text for description. Location in Figure 3.

3.2 OUTER ACCRETIONARY PRISM

The multibeam data acquired along the plate margin cover the toe of the accretionary prism up to 75 km inboard (Figures 1 and 2) (Henstock et al., 2006). The lower part of the prism is defined by a rapid change in water depth from 4300–4900 m at the deformation front to ~1500 m on the broad plateau at the top of the slope. The steep lower slope is ~20 km wide, with mean slope gradients in excess of ~8°. There are two morphologies present; those sections with thrust folds (comprising 70% of the margin mapped) and those without. Two main types of mass failure are recognized, blocky debris avalanches and sediment flows, with the majority of failures small-scale.

3.2.1 Blocky debris avalanches

On the toe of the accretionary prism, located on the young thrust folds, there are planar erosional scars that on the bathymetry are usually associated with a hummocky seabed topography, with the hummocks interpreted as outrunner blocks. The failure scars are typically ellipsoidal although some exhibit linear side margins. These features represent sediment failures that are particularly common on the seaward limbs of the thrust folds. At some locations the associated slipped blocks lie outboard of the folds on the ocean basin, at others they are on the surface of upraised sections of the detached young thrust folds or lie between them and the main body of the accretionary prism. A prominent example, and probably the youngest, is found in the north of the mapped region (Figure

5). This is an 18 km wide, semi-elliptical slump scar on the outboard limb of a young fold. At its base, on the abyssal plain (the trench has little morphological surface manifestation at this location), the thrust fold lies at a depth of 4,400 m with a crest at 3,200 m. The outboard fold limb slopes at an angle of 11-12°, but at the crest the slope is up to 23°. The failed area corresponds to the steepest gradient and greatest elevation along the fold. The top of the headwall scarp (see Hampton et al 1996 for terminology) lies on a notable, sharply defined spine, on the culmination of the fold ridge. Within the scar there are three areas of mass failure (Figure 5) with a total thickness of sediment excavation of ~100 m. Measurement of the vertical steps at the boundaries between the three areas of failure indicate that the individual layer thicknesses vary between 20 and 35 m. The outrunner blocks occupy a triangular shaped area seaward of the fold with the furthest block outboard forming the apex. The largest outrunner block is ~100 m proud of the seabed and up to 2 km by 1 km in length/width; it lies furthest from the source, with its outboard face lying 10 km from the foot of the thrust fold.

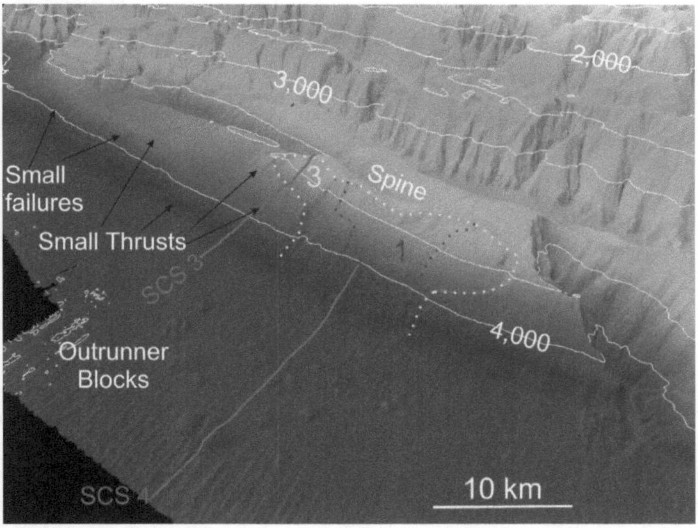

Figure 5. Morphology and bathymetry of the blocky debris avalanche in the north of the area. Bathymetric contours in metres, internal dotted lines and numbers within the failure represent the three internal subdivisions, purple lines are the tracks of the SCS lines. Note the large 'colliform' erosion feature to the right of the debris avalanche. (For location see Figure 2).

Two SCS lines cross the young failure, one in the centre (SCS 4) and one in the north (SCS 3) (Figure 6A and B). On northern SCS line 3 where seabed gradient is not so steep, the subseabed structure is well imaged and reveals a number of small thrusts on the outboard thrust fold limb (Figure 6a). These thrusts can be correlated with linear features on the multibeam data that may be traced along the fold limb into the failure scar (Figure 5). They generally correlate with the horizontal internal boundaries of the three failed areas. SCS Line 4 crosses the abyssal plain to seaward of the fold, revealing a complex relationship between the blocks identified on the bathymetry and the underlying substrate (Figure 6B). The blocks lying nearest to the toe of the accretionary prism appear to be part of a more continuous unit of chaotic reflections. The combination of bathymetry and seismic indicate the failure to be a blocky debris avalanche (see Normark et al., 2004). Towards the thrust fold the avalanche unit thickness tapers and its upper surface is slightly upraised, whereas the present seabed is not. Nearest the thrust fold the base of the avalanche unit is a moderate reflector but oceanward, midway between the outrunner blocks, this reflection fades and the unit thickens to a maximum of 100 ms (excluding the block heights), cutting down into the underlying sediment, indicating erosion during emplacment. The farthest outrunner block appears to sit on the underlying chaotic unit, which rapidly thins and tapers out beneath it. The blocky debris flow is overlain by a horizontal, seismically bedded unit (labelled 'Upper Unit' on 6B) that onlaps the margins of the outrunner blocks. It is thickest at 50 ms TWT between the thrust fold and the nearest block.

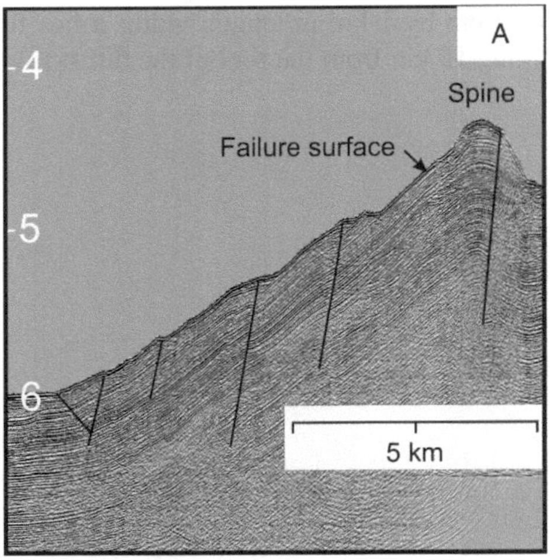

Figure 6A. SCS Line 3. The black lines are the thrusts. Vertical scale in seconds two-way-time. Location on Figure 5.

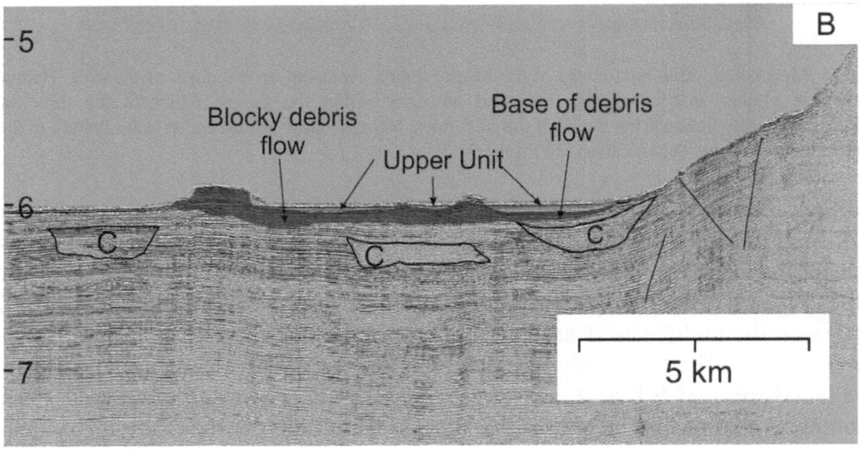

Figure 6B. SCS Line 4. See text for discussion. Vertical scale in seconds two-way-time. C labels are channels. Black lines are thrusts or faults. Location on Figure 5.

Seabed photographs reveal that the surface of the sediment forming the farthest block is not fresh (Moran and Tappin, 2006). By comparison with images of very recent seabed movement from farther south in the area, the formation of the block surface is apparently not recent. Photographs on the slide scar reveal fresh slumped blocks, and fissures but these are not as common as might be anticipated from a recent major failure event. In addition a plant, Umbellula *sp*. was located in the centre of the slide scar that was dated at ~20 years old (Moran and Tappin, 2006).

3.2.2 Sediment Flows

Along the toe of the accretionary prism where the young thrust folds are absent the prism toe rises abruptly from the abyssal plain and seabed gradients approach 30° (Figure 7). The outboard slopes are heavily incised by numerous gullies. Landward, the gullies cut through the older thrust folds and lead into arcuate, incised 'colliform' features (described because of their similarity to a cauliflower) that are similar in form to stream catchments in mountainous regions. Seaward, the gullies lead into channels on the abyssal plain, which are up to 100 m deep. The channel morphology varies; with some channels are linear, others are meandering. In some locations at the mouth of the gullies there are small (10 – 20 m high) sediment blocks. Meandering channels are commonly seen to have a number of episodes of activity. On the abyssal plain there are sediment waves, and at one location a sediment fan has been formed.

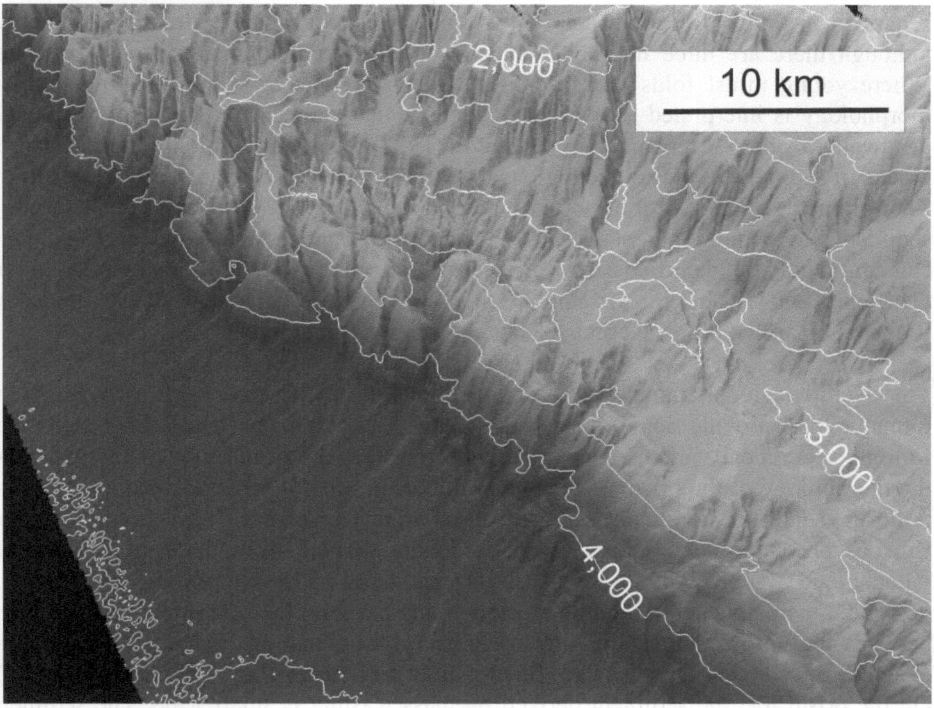

Figure 7. Toe of accretionary prism without thrust folds. Bathymetric contours in metres.

4. Discussion

Mapping of the convergent margin offshore of Sumatra using swath bathymetry, SCS and seabed photography reveals that seabed mass wasting takes place mainly through small-scale events. Within the Aceh forearc basin there is no evidence for large-scale landslides or slumps that may have contributed to local runups during tsunamis. There is no evidence to indicate that any sediment flow is very recent. Sediment is transported into the basin through channels, by small-scale sediment flows. Deposition in the basin is mainly hemipelagic.

On the outer part of the accretionary prism there are two main types of mass wasting process, blocky debris avalanches and sediment flows. On the seaward faces of young thrust folds the avalanches are interpreted to be formed in cohesive, but relatively unconsolidated, sediment. The mechanism of failure is interpreted to be due to a combination of factors, primarily tectonic oversteepening of thrust fold limbs, with the failures controlled by movement along small-scale thrusts activated during earthquakes. These thrusts we interpret to reflect larger, deeper seated seaward dipping thrusts (landward vergence) (Henstock et al 2006). There is no evidence of fluid expulsion. The thickness of the failed blocks is controlled by the bedding, inherited from the original depositional sediment character. Failures take place on the steepest slopes. Internal structures seen on the failure scars may not represent individual failures that are widely separated in time. In the instance of the blocky debris avalanche described in detail, although there are three internal boundaries, these represent one episode of failure. Where young thrust folds are absent, the deeply dissected, steeply sloping, gullied morphology is interpreted as a result of incremental sediment failure, mainly through headwall erosion. As recounted by Hampton (1972) we interpret initial failure of larger blocks of sediment in the source regions that break down during transport. The resulting finer-grained sediment is transported onto and deposited on the abyssal plain forming sediment waves and sediment sheets. Of the debris avalanches and sediment flows, none can confidently be identified as of very recent origin (for example formed on the 26th of December 2004). This throws into perspective a widely held view that earthquakes trigger submarine failures that may create destructive tsunamis.

The overall regional framework of sediment supply to the convergent margin off of Sumatra may provide some answers as to the lack of major landsliding. Sediment forming accretionary prisms may be derived from the land or the subducting plate, and the morphology of the prism off of Sumatra indicates that it is sediment starved. There is no evidence for significant sediment input from the landward direction. No major canyons cross the accretionary prism. Sediment derived from the land is of small volume and trapped within the Aceh forearc basin. The interior of the prism is degraded; there appears to be little erosion taking place on the fold limbs, the synclines between the uplifted thrust folds form elongate basins that are flat floored with little evidence of sediment fans, they are filled mainly with hemipelagic sediment (Figure 2). On the prism toe, the sediment derived by mass wasting is small scale. The sediment comprising the accretionary prism appears to be dominantly derived from the oceanic plate. On the prism toe, there is a southward decrease in the size of the thrust folds, and an increasing isolation from the prism toe, and an increase in their erosion. All these features indicate a decrease in sediment supply in a southward direction.

On the abyssal plain, the seismic data record a change in the sedimentation regime at some previous time (Figure 6B). An extensive system of channels, is present only at depth; there are none at the surface. Channel size and internal structure records a previous period of vigorous activity that is margin parallel. Consideration of sedimentation in the Indian Ocean suggests that there should be a considerable volume of sediment delivered to accretionary prism off Sumatra that is derived from the Bengal Fan (Curray, 2006). However, this does not seem to be the case. To the contrary, our data indicate a southward reduction in the sediment section entering the subduction system, that indicates a reduction in the volume of sediment delivered to the prism. We attribute this to the collision, during the early Quaternary, of the Ninetyeast ridge with the Sunda margin (Figure 1). This collision resulted in a cut off of supply as sediment was diverted away from the margin.

5. Conclusions

Although earthquakes trigger submarine landslides that may create destructive tsunamis, there are other factors, apart from earthquake magnitude and frequency, that require consideration when assessing the tsunami hazard from landslides along convergent margins. These include sediment supply and tectonism. The Sumatra margin is prone to large earthquakes, but is sediment starved. This is due to the small volume of sediment entering the margin, both from land and the ocean. Landward derived sediment is mainly trapped in the Aceh forearc basin. Although a potential location for large landslides, the restricted sediment input to the Aceh Basin, leads to small-scale sediment flows, widely spaced in time.

On the prism toe, there is a close relationship between sediment failure and local small-scale tectonics. The common blocky debris avalanches are mainly confined to the young thrust folds, with failure taking place on the steep outboard fold limbs in soft cohesive sediment, the thicknesses of the slumped blocks being controlled by bedding planes. Failure is in part determined by movement along small scale seaward dipping thrust planes. Although small in themselves the thrusts reflect underlying, large scale landward thrust vergence. Where young thrust folds are absent the presence of dominant sediment flows may reflect a different style of thrust vergence and/or more lithified sediment. The most likely trigger for the both styles of failures is considered to be earthquakes.

The volume of sediment derived from the oceanic plate is limited despite the proximity of the massively thick Bengal Fan. The explanation for this apparent contradiction is explained by collision in the early Quaternary along the subduction zone with the Ninetyeast Ridge, that resulted in the diversion of sediment away from the Sumatra region. This reduction in sediment input is regarded as contributory to the small scale nature of sediment failure observed along the Sumatra margin. In both the Aceh Basin and on the prism toe, the scale of sediment failure is considered a limited threat as a causative mechanism for a major locally generated tsunami.

6. Acknowledgements

We thank the ship's company of the HMS *Scott* a UK Royal Navy survey vessel, for acquiring the multibeam data during marine scientific research coordinated by the Joint Environment Directorate of Defence Intelligence. We also thank; The Royal Navy, British Geological Survey, National Oceanography Centre, United Kingdom Hydrographic Office, and the Government of Indonesia. The Government of Canada is thanked for funding the use of the seismic equipment provided by the Canadian Geological Survey. General thanks to the British Embassy Jakarta, British Broadcasting Corporation, Discovery Channel, Science Applications International Corporation, BP Marine Limited, Oceaneering International Corporation, and Darlow Smithson Productions. Many thanks to Homa Lee and Gary McMurtry for providing reviews at short notice that considerably improved the manuscript. David Tappin publishes by permission of the Executive Director of the British Geological Survey (NERC).

7. References

Curray JG. (2005) *Tectonics and history of the Andaman Sea region.* Journal of Asian Earth Sciences. 25: 187–232.

Goldfinger C Kulme LD McNeill LC Watts P (2000) *Super-scale failure of the southern Oregon Cascadia margin.* Pure and Applied Geophysics. 157: 1189-1226.

Hampton M (1972) *The role of subaqueous debris flow in generating turbidity currents.* Journal of Sedimentary Petrology. 42: 775-993.

Henstock T McNeill LC Tappin DR. (2006) *Seafloor morphology of the 26 December 2004 Indian Ocean earthquake rupture zone.* Geology. 34: 485–488. doi: 10.1130/22426.1.

Lee HJ Locat J Desgagnes P Parsons J McAdoo B Orange D Puig P Wong F Dartnell P Boulanger E (2007) *Submarine Mass Movements.* In C.A. Nittrouer (ed) Continental-Margin Sedimentation: Transport to Sequence. Blackwell/IAS.

McAdoo BG Capone MC Minder J (2004) *Seafloor geomorphology of convergent margins: implications for Cascadia seismic hazard.* Tectonics. 23: TC6008, doi:10.1029/2003TC001570

Moran K Tappin D (2006) *SEATOS 2005 Cruise Report: Sumatra Earthquake and Tsunami Off shore Survey (SEATOS).* pp92 [Online] available at http://ocean. oce.uri.edu/seatos.

Newcomb KR McCann WR (1987) *Seismic history and seismotectonics of the Sunda arc.* Journal of Geophysical Research. 92: 421–439.

Ortiz M Bilham R (2003) *Source area and rupture parameters of the 31 December 1881 Mw ~7.9 Car Nicobar earthquake estimated from tsunamis recorded in the Bay of Bengal.* Journal of Geophysical Research. 108: doi:10.1029/2002JB001941

Tappin DR Watts P McMurtry GM Lafoy Y Matsumoto T (2001) *The Sissano, Papua New Guinea tsunami of July 1998-offshore evidence on the source.* Marine Geology. 175: 1-23.

SLOPE FAILURES OF THE FLANKS OF THE SOUTHERN CAPE VERDE ISLANDS

T.P. LE BAS, D.G. MASSON, R.T. HOLTOM
National Oceanography Centre, Empress Dock, European Way, Southampton, SO14 3ZH. UK

I. GREVEMEYER
Leibniz-Institute of Marine Sciences, IFM-GEOMAR, Wischhofstr. 1-3, D-24148 Kiel, Germany

Abstract

The flanks of the Cape Verde Islands Fogo, Brava and Santiago show extensive evidence of past and potential slope failure. Analyses of bathymetric and subaerial datasets show debris avalanches, turbidite pathways and debris flows, with the transportation of large volumes of rock and sediment. Similarities are seen with many of the processes operating on the flanks of the Canary Islands. In this paper we show the use of different multibeam bathymetry systems for geomorphological description and quantitative analysis. Slope maps, profiles and backscatter analysis can be used to classify the bathymetric characteristics. These derivative processes delineate the size and shape of the debris avalanches and flows and identify channel systems, as well as areas of recent seafloor volcanic activity. Two distinct debris fields covering at least 2000 km^2 sourced from the east coast of Fogo are thought to contain up to 250km^3 of material. Large slope failures, such as those from Fogo, may result in damage to seafloor installations such as submarine cables. However, greater hazards may be posed by the consequent generation of tsunamis.

Keywords: Slumping, debris flows, debris avalanches, Cape Verde, landslides, tsunamis

1. Introduction

Slope failure is a common feature of mid-plate oceanic island evolution and plays a significant role in shaping the seafloor environment. Numerous marine geophysical and bathymetric studies have identified and constrained the characteristics and dimensions of slope failures located on the flanks of the volcanic oceanic island of Réunion (Oehler et al., 2004) as well as those in the Hawaiian (Moore et al., 1994), and the Canary island chains (Masson et al., 2002). These slope failures are now considered to be widespread on the submerged flanks of ocean island volcanoes (Holcomb & Searle 1991). Furthermore, catastrophic large-scale landslides of infrequent occurrence have been suggested as a fundamental process in the evolution of Cape Verde Islands (Ali & Watts, 2003). Although infrequent at all known locations worldwide, it is clear that huge destructive potential exists in any one failure. Debris avalanches, slumps, turbidity currents and debris flows have the potential to transport vast volumes (hundred to thousands of km^3) of rock and sediment over distances of hundreds to a thousand or more kilometres (Hampton, Lee & Locat, 1996). Catastrophic slope failures may result in damage to seafloor installations, although greater hazards may be posed by the threat associated with tsunamis (Ward & Day., 2001).

V. Lykousis, D. Sakellariou and J. Locat (eds.), Submarine Mass Movements and Their Consequences, 337–345.

Slope failures on the flanks of volcanic oceanic islands can typically be divided into two main types: debris avalanches and slumps (Moore et al., 1989). Debris avalanches are relatively thin failures, typically 1 km thick, that affect the superficial volcanic sequences of the island flanks. They are believed to be emplaced rapidly and produce a field of blocky rock debris covering the lower island slopes and adjacent seafloor. Slumps affect a much greater thickness of the island sequences (up to 10 km in the Hawaiian case), and involve slow or intermittent movement of a coherent part of the island flank. Debris flows, cohesive flows of sedimentary material, can also affect island slopes, either when slope sediments are remobilised as part of a larger flank failure, or where volcaniclastic sediments deposited on the island slopes fail due to sediment or earthquake loading. All forms of landsliding can trigger turbidity currents, which can travel great distances (> 1000 km) from the source.

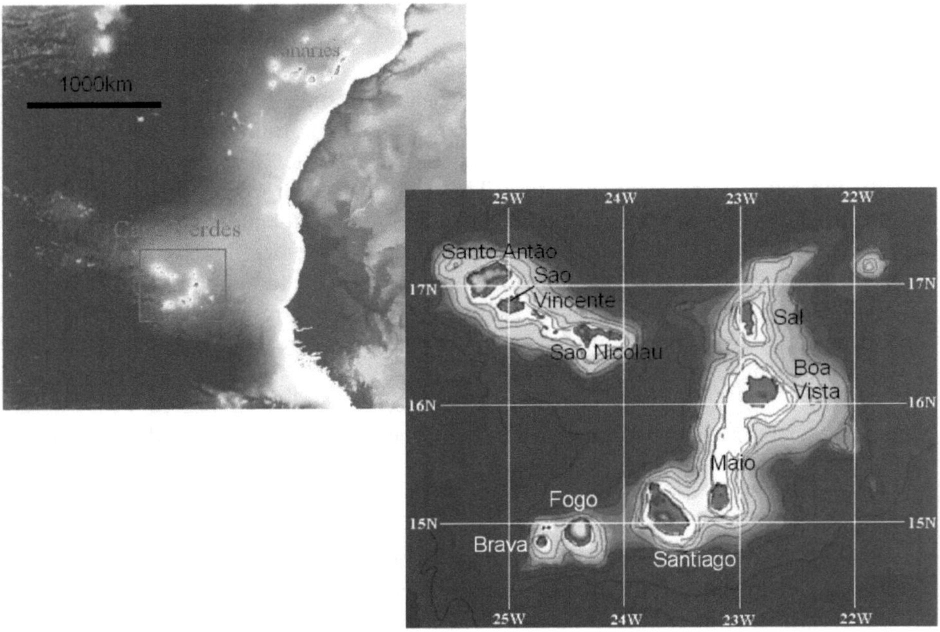

Figure 1: Location maps of the Cape Verde rise off the west coast of Africa and a 500m contour map showing the islands.

Until recently most bathymetric and seismic studies around the Cape Verde Islands had focussed on the tectonic attributes of the Cape Verde Rise as a whole. Seismic investigations around the islands, showing evidence for superficial slumping and debris avalanches, were carried out by Dash et al. (1976) and McNutt (1988). The combination of steep slopes and rugged geomorphology make the Cape Verde Islands a likely candidate for flank instability leading to catastrophic lateral collapses. The steep sided slopes of islands such as Fogo (Day et al., 1999) are comparable to those of volcanoes in the Canary Islands where a significant number of giant lateral collapse structures have been identified and well documented (Masson et al., 2002; Mitchell et al., 2002).

2. Geological Setting

The Cape Verde Islands are located on the southwest flank of the Cape Verde Rise off the west coast of Africa (Figure 1). The Cape Verde Rise is an elevated region of the present ocean floor approximately 1200km in width and encompassing an area >$3x10^5$ km^2 enclosed by the 3500m isobath. The islands do not form a linear chain as observed in other hot spot regions such as the Hawaiian Islands. The origin of the island chains is believed to relate to volcanic activity from an underlying mantle plume. It has been surmised that the volcanoes are of late Tertiary origin, resting upon a Mesozoic-aged seafloor, with most of the islands having formed between 5 and 15 Ma on lithospheric of approximately 130 Ma age (McNutt, 1988).

Fogo and Brava are the youngest and most seismically active in the archipelago and are inferred to be on the upwelling axis of the mantle plume, their characteristics being consistent with those of other volcanic islands associated with drifting hotspot models (Detrick & Crough, 1978). From a morphological aspect, there seems to be an overall progression from older volcanism in the East to younger in the West in both the northern and southern chains of islands. However knowledge of ages of volcanic activity exists for only a few islands (Maio, Santo Antão and Fogo) and is too sparse for determining a definitive age progression (Mitchell et al., 1983; Holm et al., 2006).

3. Methods

A compilation of data consisting of Simrad EM12 swath bathymetry and backscatter data collected on the UK vessel RRS Charles Darwin during cruise CD168 in February 2005 and Hydrosweep bathymetry collected on the German vessel RV Meteor from cruise 62/3 in September 2004 was created. A digital elevation model (DEM) of the Cape Verdes land surface was combined with the bathymetric data (Figure 2). This subaerial topography of the islands is based on the Shuttle Radar Topography Mission (SRTM) dataset (Rabus et al., 2003) which has similar resolution to multibeam bathymetry data acquired here.

Backscatter data from multibeam bathymetry data is often overlooked. However, with optimum processing such data can provide insight to geomorphological shapes seen in bathymetry as well as pick out lithological variations in sediments in a way similar to sidescan sonar. The EM12 backscatter data and beam intensity values from the Hydrosweep data (an approximate equivalent of backscatter) were processed using the PRISM (Processing of Remotely-sensed Imagery for Seafloor Mapping) software (Le Bas, 2005). The EM12 backscatter and Hydrosweep intensity values were then combined to produce a backscatter map (see Figure 3). High backscatter on this map can be interpreted as rock outcrop, coarse-grained sediment or rougher surface whereas low backscatter is associated with fine-grained sediments.

4. Slope failures

The east-facing landslide scar that dominates the island of Fogo, and the large field of blocky debris immediately to the east of the island, provide spectacular evidence for failure of the east flank of Fogo (Figures 2-6).

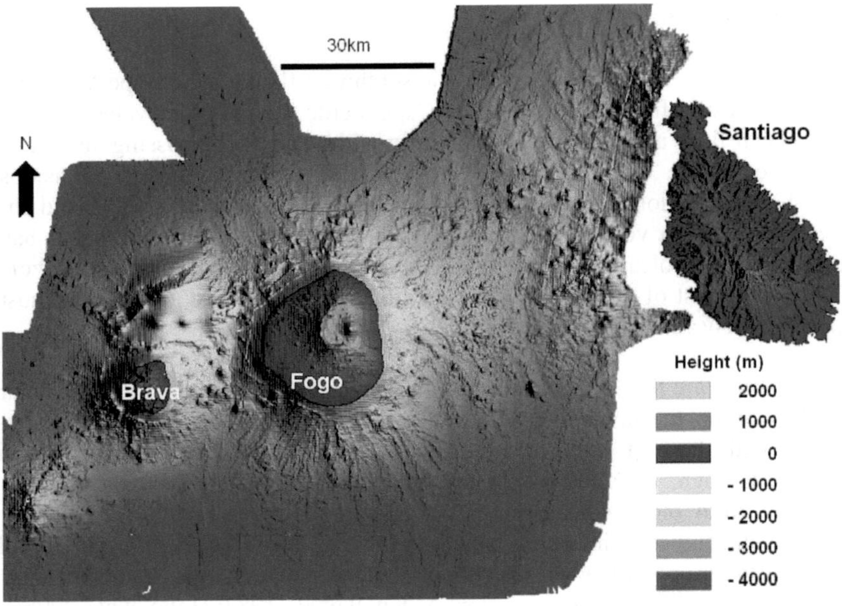

Figure 2: Topography of, and bathymetry around, the islands of Brava, Fogo and Santiago. Data compiled from Simrad EM12 and Atlas Hydrosweep bathymetry together with Shuttle Radar Topography Mission topography. The image is shown in shaded relief and shows many channels running off Fogo into deeper water (from example to the south) and some large blocks and volcanic edifices between Santiago and Fogo and between Fogo and Brava. A new volcanic seamount is also seen about 30km south-west of Brava.

Figure 3: Multibeam backscatter imagery from the Simrad EM12 and Atlas Hydrosweep systems around the islands of Brava, Fogo and Santiago. Low backscatter is shown as dark and high as white. Level of backscatter is used as a measure of seafloor roughness and grainsize. A combination of morphology and backscatter imagery is used to interpret process related features, such as individual debris avalanches or debris flows.

Backscatter imagery, slope data and geomorphology all show the debris field to be heterogeneous and more irregular than the fields of volcanic cones that occur elsewhere around Fogo (see below). Lineations in the slope analysis (Figure 4) and backscatter imagery (Figure 3) suggest the debris flowed east initially and then spread southeast into the deeper water, coving an area of 750 km^2. Blocks are most numerous in the centre of the debris field and decrease in both size and abundance towards the margins (Figures 4, 6). In this respect, the debris avalanche resembles the El Golfo avalanche on the island of El Hierro in the Canaries (Masson, 1996). However, the rim of large blocks seen around the margins of several Canaries avalanches is absent in the Fogo avalanche.

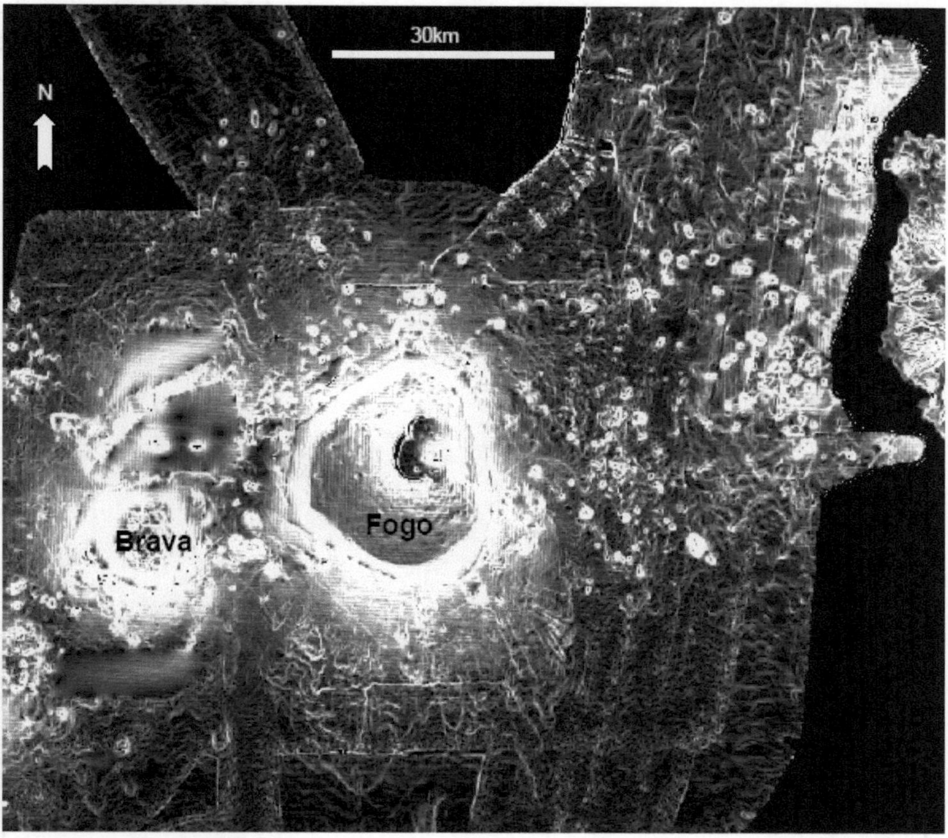

Figure 4: Slope map derived from DEM compiled from the multibeam bathymetry and ShuttleRadar Topography Mission data. Low slopes are shown as dark and high slopes as white. Theadvantage of this type of imagery is the better definition of blocks and features that may be presenton slopes. The sediment waves to the south of Fogo are an example of this and cannot be seen inFigures 2 or 3.

The margins of the most recent debris avalanche deposit east of Fogo are clearly defined in the bathymetry data (Figure 6). Backscatter data, however, show that the typical 'speckled' signature of the debris avalanche extends outside the area of topographic expression, particularly to the south of the surficial avalanche deposit (Figure 3) into an area of flat seafloor (Figures 4, 6). We interpret this as an older debris avalanche, buried beneath a thin layer of younger turbidites.

The area to the west of the island of Santiago is marked by large blocks or edifices up to 5 km^2 in area. It is evident that several of these features have a cone-like form, with the presence of a small central summit. They can be distinguished using their backscatter values which are greater than the surrounding sediments (Figure 3), and on the slope map due to their regular, steep slopes (Figure 4). They are fewer in number, more regularly spaced, and have a more regular conical profile and spherical outline compared to the more angular blocks observed within the debris avalanche deposits seen east of Fogo. They are interpreted as small volcanic cones. Similar small cones are present between Brava and Fogo. These latter cones coincide with recent small earthquake foci (Heleno et al., 2005), though no associated lava flows can be identified on the backscatter imagery.

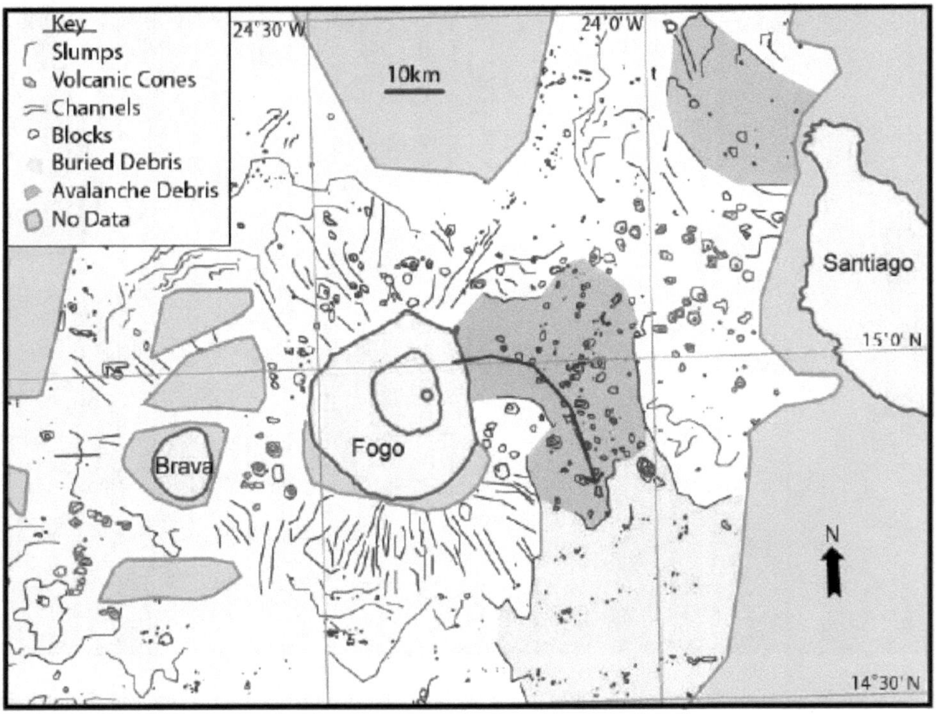

Figure 5: Final slope failure interpretation map for the Southern Cape Verde Islands.

As far as can be discerned from geomorphological evidence, only the east flank of Fogo has been affected by large-scale flank collapse and debris avalanching. However, other flanks, notably to the south and northwest, show complex topography related to the formation of small submarine canyons and small-scale landsliding within the sedimentary cover of the island slopes (Figures 5, 6). Some canyons can be traced upslope to the limit of the bathymetry data, suggesting that they may relate to debris or turbidity current flow directly associated with volcaniclastic sediments or lava flows entering the sea at the coast. The slope map (Figure 3) shows subtle contour parallel features (perpendicular to the canyons) in the canyoned areas. These can be interpreted either as turbidity current sediment waves or as evidence for slumping.

5. Discussion and Conclusions

Geomorphological data show that, in common with many volcanic oceanic islands worldwide, Fogo Island in the Cape Verdes is affected by large but infrequent flank failures. Debris avalanches are the principal landslide mechanism. The age of the most recent debris avalanche is not known, although indirect evidence suggests that it is pre-Holocene (Day et al., 1999). Fogo is an extremely volcanically active island, with at least 25 major eruptions in the last 500 years (Torres et al., 1998). Eruptions since the 18th century have occurred from north-south oriented fissures on the floor of the old landslide scar, suggesting that east-west extension plays a part in island evolution. While extension and volcanic loading will inevitably lead to a new flank failure, the present-day volcanic edifice is much smaller than that which failed previously, and there is no evidence that such a failure is likely in the foreseeable future

Figure 6: Three-dimensional model showing the seafloor to the east of the Island of Fogo. It shows the debris avalanche deposits, volcanic cones and debris flow channels interpreted from the bathymetry, backscatter and slope analysis data.

Much has been made of the possible tsunami hazard posed by flank collapse of volcanic oceanic islands, with particular reference to La Palma in the Canary Islands (e.g. Ward and Day, 2001). The volume of the most recent Fogo avalanche has been estimated at 200-300 km^3 (Day et al., 1999) and is comparable with a typical Canary Island avalanche (Masson et al., 2002). A volume of around 250 km^3 is broadly compatible with the topographic anomaly associated with the debris avalanche deposits offshore Fogo, corresponding to a maximum thickness of 250 m in the deposit centre and slopes to the edges of about 1°. Bathymetry profiles show that this is acceptable. However, this assumes that the entire topographic anomaly was generated by a single event, rather than a series of events, such as seen off Tenerife in the Canary Islands (Watts and Masson, 1995). The homogeneous characteristics of the oceanic island submarine debris fields appear to suggest a single failure event. However, this has been disputed by Wynn and Masson (2003) who have shown that apparently single debris avalanche deposits around the Canary Islands can be correlated with up to nine turbidity current pulses further offshore, indicating that the apparently catastrophic debris avalanche is made up of a series of closely linked but smaller slope failures. This should reduce the magnitude of any associated tsunami. Nevertheless, the tsunami threat from any future Fogo flank collapse should not be underestimated. The island of Santiago, with the Cape Verde capital city of Praia at its southern tip, lies only 80 km east of Fogo. Even a relatively small landslide could produce significant tsunami amplitudes that close to source. Furthermore, the densely populated and low lying West African coast lies only a further 700 km to the east, and could be vulnerable to a larger event.

6. Acknowledgments

We would like to thank the ship personnel of RRS Charles Darwin and RV Meteor for their assistance in data acquisition. Critical reading and suggestions by Dr Russell Wynn was much appreciated. This work was partly funded by Natural Environment Research Council.

7. References

Ali, M.Y., Watts, A.B., & Hill, I., 2003, *A seismic reflection profile study of lithosphere flexure in the vicinity of the Cape Verde Islands:* Journal of Geophysical Research, v. 108 (B5), 24pp.

Dash, B.P., Ball, M.M., King, G.A., Butler, L.W., & Rona, P.A., 1976, *Geophysical investigation of the Cape Verde Archipelago:* Journal of Geophysical Research, v. 81 (29|), p. 5249-5259.

Day, S.J., Heleno da Silva, S.I.N., & Fonseca, J.F.B.D., 1999, *A past giant lateral collapse and present-day flank instability of Fogo, Cape Verde Islands*: Journal of Volcanology and Geothermal Research, v. 94, p. 191-218.

Detrick, R.S., & Crough, S.T., 1978, *Island subsidence, Hot spots, and Lithospheric thinning:* Journal of Geophysical Research, v. 83 (B3), p. 1236-1244.

Hampton, M.A., Lee, H.J., & Locat, J., 1996, *Submarine landslides*: Reviews of Geophysics, v. 34, p. 33-59.

Heleno da Silva, S.I.N., Faria, B., Bandamo, Z., & Fonseca, J., 2005, *Observations of high-frequency harmonic tremor in Fogo Island, Cape Verde.* SAL2005 Workshop on Ocean Island Volcanism, Abstract volume p. 61.

Holocomb, R.T., & Searle, R.C., 1991, *Large landslides from oceanic volcanoes*: Marine Geotechnology, v. 10, p. 19-32.

Holm, P.M., Wilson, J.R., Christensen, B.P., Hansen, L., Hansen, S.L., Hein, K.M., Mortensen, A.K., Pedeersen, R., Plesner, S., & Runge., M.K., 2006, *Sampling the Cape Verde Mantle Plume: Evolution of Melt Compositions on Santo Antão , Cape Verde Islands*: Journal of Petrology, v. 47 (1), p. 145-189.

Le Bas T.P., 2005, *PRISM - Processing of Remotely-sensed Imagery for Seafloor Mapping - Version 4.0.* National Oceanography Centre Southampton report. Unpublished manuscript. 192pp.

Masson, D.G., 1996, *Catastrophic collapse of the flank of El Hierro about 15,000 years ago, and the history of large flank collapses in the Canary Islands*: Geology, v. 25, p. 231-234.

Masson, D.G., Watts, A.B., Gee, M.J.R., Urgeles, R., Mitchell, N.C., LeBas, T.P., & Canals, M., 2002, *Slope failures on the flanks of the western Canary Islands*: Earth-Science Reviews, v. 57, p. 1-35.

McNutt, M., 1988, *Thermal and Mechanical Properties of the Cape Verde Rise*: Journal of Geophysical Research, v. 93, p. 2784-2794.

Mitchell, J.G., Le Bas, M.J., Zielonka, J., & Furnes, H., 1983, *On the magmatism on Maio, Cape Verde Islands:* Earth and Planetary Science Letters, v. 64, p. 61-76.

Mitchell, N.C., Masson, D.G., Watts, A.B., Gee, M.J.R., & Urgeles, R., 2002, *The morphology of the submarine flanks of volcanic islands. A comparative study of the Canary and Hawaiian hotspot islands:* Journal of Volcanology and Geothermal Research, v. 115, p. 83-107.

Moore, J.G., Clague, D.A., Holcomb, R.T., Lipman, P.W., Normark, W.R., & Torresan, M.E., 1989, *Prodigious submarine landslides on the Hawaiian Ridge*: Journal of Geophysical Research, v. 94, (B12), p. 17465-17484.

Moore, J.G., Normark, W.R., & Holcomb, R.T., 1994, *Giant Hawaiian Landslides*: Annual Review of Earth Planetary Science, v. 22, p. 119-144.

Oehler, J-K., Labazuy, P., & Lénat, J-F., 2004, *Recurrence of major flank landslide during the last 2-Ma history of Reunion Island:* Bulletin of Volcanology, v. 66, p. 585-598.

Rabus, B., Eineder, Roth, A., & Bamler, R., 2003, *The shuttle radar topography mission- a new class of digital elevation models acquired by spaceborne radar*: Photogrammetry Remote Sensing, v. 57, p. 241-262.

Torres, P.C., Madeira, J., Silva, L.C., Brum da Silveira, A., Serralheiro, A., & Mota Gomes, A., 1998, *Carta Geologica da Ilha do Fogo (Republica de Cabo Verde):* Lisbon, Departmento de Geologia da Faculdade de Ciencas da Universidade de Lisboa.

Ward, S. N. & Day, S. J. 2001. *Cumbre Vieja volcano – potential collapse and tsunami at La Palma, Canary Islands.* Geophysical Research Letters 28, 3397-3400.

Watts, A.B. & Masson, D.G. 1995. *A giant landslide on the north flank of Tenerife, Canary Islands*. Journal of Geophysical Research 100(B12) pp 24487-24498.

Wynn, R.B. & Masson, D.G., 2003, *Canary Islands landslides and tsunami generation: Can we use turbidite deposits to interpret landslide processes?,* p. 325-332. In: Submarine mass movements and their consequences. J. Locat and J. Mienert Eds., Advances in Natural and Technological Hazards Research Series, 2003, Kluwer.

TRIGGERING FACTORS AND TSUNAMIGENIC POTENTIAL OF A LARGE SUBMARINE MASS FAILURE ON THE WESTERN NILE MARGIN (ROSETTA AREA, EGYPT)

S. GARZIGLIA, M. IOULALEN, S. MIGEON, E. DUCASSOU, J. MASCLE, O. SARDOU AND L. BROSOLO
Géosciences Azur, BP 48, 06235 Villefranche-sur-Mer, France

Abstract

A large-scale mass-transport deposit (MTD) called Sl6 was recognized on the upper slope of the western Nile margin, downslope from of a 30 km-long scarp located along the outer shelf. Regional mapping indicated that this MTD extends on nearly 505 km^2 and involved about 14 km^3 of Pleistocene-Holocene sediment. Sl6 was triggered between 10 and 9 kyr BP, during the Holocene sea-level rise and coeval pluvial period (increased river flow). The consequent enhanced sediment supply on the upper slope and the outer shelf area caused local overburdening. This factor combined with the potential accumulation of gas in the sediment and earthquake activity is thought to have been the main factor triggering the Sl6 MTD. From the estimated volume of the MTD, a potential slide-generated tsunami was numerically simulated using the GEOWAVE software. The results indicate that the ~80 km wide Egyptian continental shelf protects the main part of the coastline from a slide-induced tsunami coming from the Rosetta area. An exception is the part of the coastline around Alexandria because focussing and shoaling processes can be simulated very close to the coast.

Keywords: submarine mass failure, triggering factors, tsunamogenic potential, numerical simulation

1. Introduction

Climate-driven environmental changes are commonly considered to affect the slope susceptibility to failure (*Owen et al., 2007*). Studies from submarine river-fed depositional systems increasingly demonstrate that Quaternary submarine mass failures have been influenced by climate through a combination of sedimentary processes associated with relative sea-level oscillations (*Owen et al., 2007*). A review of the architecture, distribution and age of a large mass-transport deposit (MTD) called Sl6 on the western Nile margin fed by the Rosetta branch of the Nile delta, provided an additional example of the interaction between climate and slope failures. Moreover, it reinforces the strong evidences of sediment failures triggered during the late-Quaternary sea-level rise. Despite these relationships, the role of gas release from active gas chimneys and of earthquakes in the complex geodynamic setting of the Eastern Mediterranean is also evaluated to explain slope failure. Additionally, in view of the considerable economic and societal importance of Egyptian margin and coastline, the tsunamogenic potential of such a large-scale event is discussed using numerical modelling.

V. Lykousis, D. Sakellariou and J. Locat (eds.), Submarine Mass Movements and Their Consequences, 347–355.

2. Methods

This paper is based on the interpretation of 1) a bathymetric map with a spatial resolution of 100 m (combined from EM12, EM300-Dual multibeam data and data derived from a 3D seismic block provided by BP-Egypt), 2) a grid of high-resolution seismic profiles with a ~2 km regular spacing provided by BP-Egypt, 3) one Kullenberg core collected in the upper part of Sl6 MTD. Geochronological control of the triggering of Sl6 was performed using a detailed ecostratigraphic scheme based on planktonic foraminifera distribution, constrained in time with oxygen isotope records, tephrochronology and radiocarbon data developed for the Nile deep-sea turbidite system (NDSTS) by Ducassou et al. (in press).

Based on estimated volume of Sl6 MTD, a landslide-induced tsunami was modelled to evaluate its impact on the nowadays Egyptian coastal area. The tsunami generation is estimated using GEOWAVE which combines TOPICS software to approximate landslide tsunami source, with the fully non-linear Boussinesq water-wave model FUNWAVE (*Watts and Grilli., 2003*).

3. Geological setting

The Rosetta area, off the Rosetta feeding branch of the Nile delta, lies on the western Nile margin, which is characterized by a flat continental shelf about 200 m deep and 80 km wide. The continental slope extends over more than 130 km with typical angles of 1° to 2.5°. The Rosetta canyon cuts the upper continental slope and the shelf (Fig. 1). The western Nile margin is located in the vicinity of three known active plate boundaries namely: the Red-Sea, the Gulf of Aqaba and the Hellenic Arc (*Mesherf, 1990*). Consequently, the Rosetta area is affected both by frequent small to moderate local earthquakes (M_s = 6.7) and large remote earthquakes (M_s = 7.8) (*El-Sayed et al., 2004*). The main tectonic element identified, at depth, in the vicinity of the study area is the NE-SW Rosetta fault trend (*Aal et al., 2000*). That fault trend appears to have been active mostly during the early Cretaceous (*Camera, 2006*), but these pre-existing fractures acted as conduits for fluid ascent that generated North Alex and Horus mud volcanoes on the upper slope of the Rosetta area (Fig. 1) (*Loncke et al., 2004*).

During the Quaternary, turbidite sand, clastic mud related to Nile flooding periods and hemipelagic mud dominate the sedimentation over the NDSTS (*Ducassou et al., subm*). Major sedimentary transfers and formations occurred especially during pluvial periods, in particular during sea level low stands, but also during rising and high sea-level stands. The whole Rosetta area recorded the highest sedimentation rates between 12 and 8 kyr BP (~150 cm/kyr) during a period of enhanced pluvial conditions over the catchment area of the Nile.

4. Results

4.1 CHARACTERISTICS AND DISTRIBUTION

East of the Rosetta canyon head, slope failures have significantly impacted the seafloor morphology in the form of wide imbricated scars affecting the outer shelf (Fig. 1).

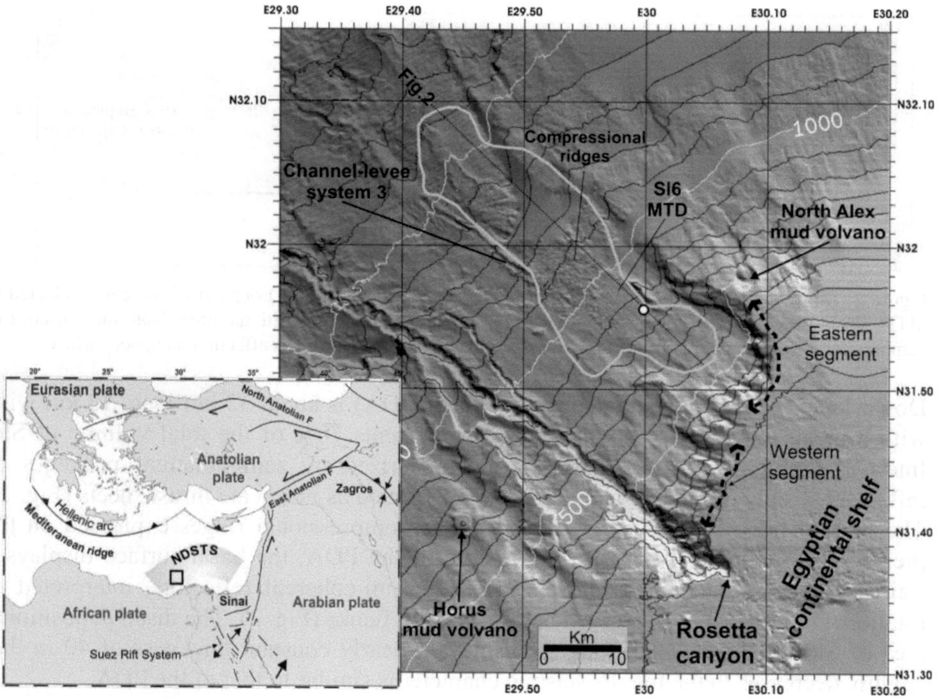

Figure 1: Interpreted bathymetric map of the upslope part of the Rosetta area (western Nile margin). Location of the two segments of the scarp affecting the outer shelf is shown. The area affected by the MTD Sl6 is bounded by the light grey line. Note the uneven seafloor topography interpreted as compressional ridges within the intermediate depositional area of Sl6. The black dotted lines indicate the location of the seismic profile shown on Figure 2. The white dot shows location of the Kullenberg core reaching the upper part of Sl6 MTDs. Inset shows geodynamical sketch of the Eastern Mediterranean and the black box shows the location of the study area in the western province of the Nile deep-sea turbidite system (NDSTS).

The cumulated width of scars is about 30 km, for a mean height of 200 m. They typically extend from 200 to 400 m water depth and display two segments of strongly differing dips (Fig. 1). The eastern segment displays the highest slope values (between 11° and 35°), and may correspond to the source area of a relatively recent failure event.

Downslope from the eastern segment, MTD Sl6 has still a significant bathymetric signature. It induced the sharp disruption of the channel-levee system 3 (Fig. 1) at water depth ranging between 400 and 1100 m. Its surface displays compressional ridges between 800 and 1200 m water depth (Fig. 1). On seismic profiles, MTD Sl6 is dominated by chaotic internal configuration. It is bounded at its base and laterally by high amplitude reflections corresponding to channel-levee deposits (Fig. 2). MTD Sl6 exhibits a lens-shaped cross section. It extends 50 km in a downslope direction, about 10 km across the slope and covers a total area of 505 km^2. With a mean thickness of about 70 m, Sl6 estimated volume is about 14 km^3.

Sl6 is composed of three depositional areas. The proximal depositional area (PDA) of Sl6 is located about 6 km away from its headscarp and displays rather homogeneous thickness of about 30 m over a length of nearly 9 km. The PDA is characterized by low-amplitude and highly chaotic internal configuration.

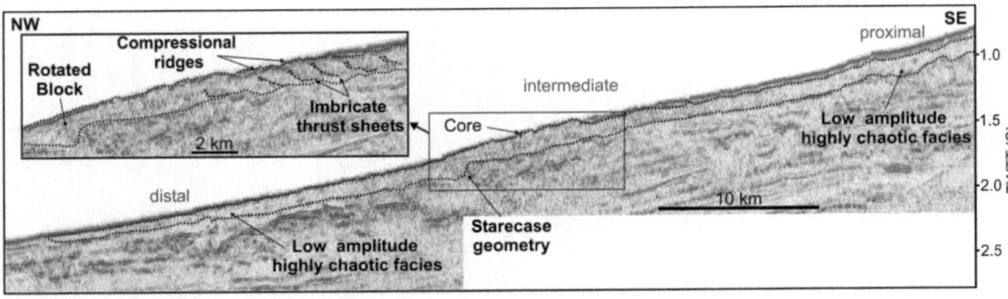

Figure 2: Dip-oriented seismic-reflection profile (see Fig. 1 for location). Location of the core collected in MTD Sl6 is shown. Inset shows a zoom on MTD Sl6 intermediate depositional area. Note the presence of compressional features within the MTD and associated compressional ridges affecting its upper surface.

Downslope, the intermediate depositional area (IDA) is about 32 km long, 13 km wide with a mean thickness of about 80 m. It represents 70% of the total volume of Sl6. Internal configuration of the IDA is characterized by semi-coherent packages of reflectors steeply dipping upslope that are interpreted as imbricate thrust sheets (Fig. 2). The tops of these packages correspond to the compressional ridges expressed on the present day seafloor (Fig. 1). In the middle of the PDA, the basal surface displays a staircase-like arrangement while packages of semi-coherent reflectors, interpreted as rotational blocks, can be observed below its steep flanks (Fig. 2). The distal depositional area of Sl6 (DDA) is 7 km long. It displays a nearly constant thickness of 40 m that slightly decreases to the toe. Its seismic character is similar to that of the PDA.

4.2 SMALL-SCALE STRUCTURES AND AGE OF MTD S16

Visual descriptions of core sections from the IDA of Sl6 (see location on Fig. 2) reveal that tens of centimetre-thick intervals of alternating clayey and silty layers are well-preserved (Fig. 3). However, these intervals are tilted by more than 30° and disconnected from each others by discrete shear surfaces (Fig. 3). If many intervals exhibit preserved structures, many of them also contain deformed strata affected by core-scale faults (Fig. 3). Some intervals exhibit core-scale folds and contorted laminae.

Stratigraphic analyses reveal that the upper part of Sl6 deposits is located at about 10 cm below the sapropel S1 (base at 8940 ± 30 yr cal. BP; *Ducassou et al., in press*) and contains planktonic fauna typical of the early Holocene (~10-9 kyr BP; *Ducassou et al., in press*; Fig. 3). The age of MTD Sl6 is thus estimated between 10 and 9 kyr.

5. Triggering factors

Sl6 event was triggered from the shelf break, between 10 and 9 kyr which correlates with a period of enhanced sediment supply (> 150 cm/kyr), resulting from frequent and high-magnitude floods (*Ducassou et al., 2006*). That period also corresponds to the end of a rapid relative sea-level rise (Fig. 4), from -120 to -35 m below the present-day sea-level that induced a short landward shift of sediment accumulation.

Rapid sediment accumulation, on the upper slope and outer shelf, might have induced gradual but significant excess pore-pressure generation in fine-grained sediment and hence, reductions in effective stress and a consequent reduction of their shear strength.

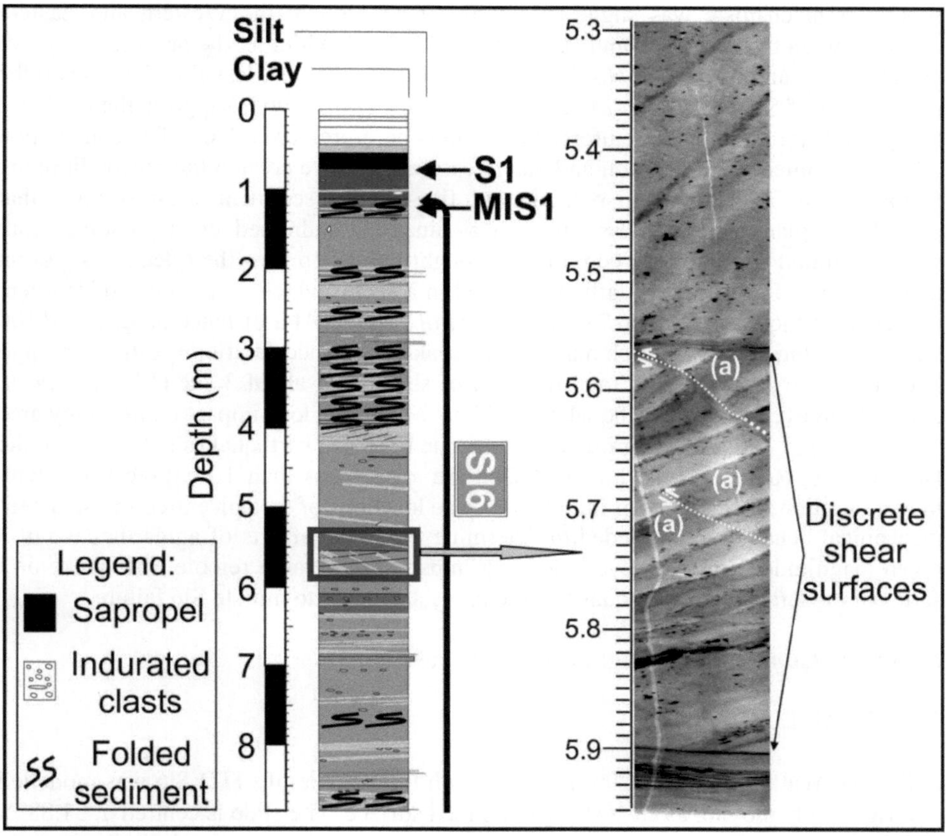

Figure 3: Lithological log of the Kullenberg core collected in the upper part of Sl6. Sapropel layers S1 is shown in black color. Base of marine isotopic stage 1 (MIS-1) at 10 kyr BP is also labelled. The X-ray image right to Sl6 log shows three intervals with differing dipping separated by discrete slide surfaces. In the upper one, clayey and silty beds are tilted by 45°. In the middle one, clayey and silty beds are less tilted but affected by inverse faults. The top of the lowest interval is characterized by sub-horizontal silty beds.

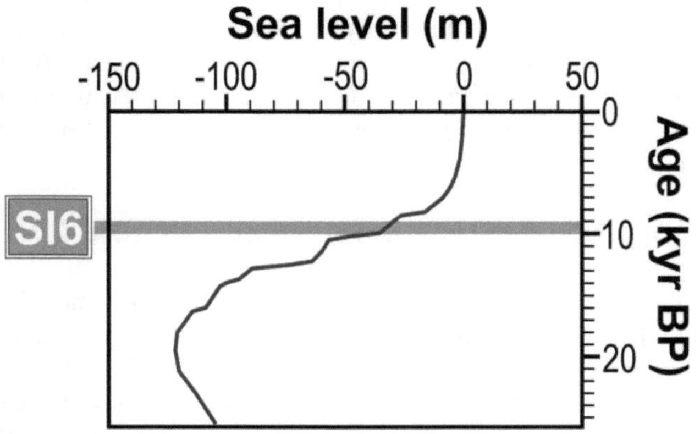

Figure 4: Timing of MTD Sl6 formation relative to Mediterranean sea level oscillations (Ducassou, 2006).

A similar mechanism was suggested for the generation of the Western and Eastern debris flow on the Amazon Fan (*Maslin et al., 2005*). In addition, the presence of active gas chimneys and associated North Alex mud volcano (Fig. 1), less than 3 km from the source area of S16 (*Loncke et al., 2004; Prinzhofer et al., 2005*) supports the evidence that gas release might have reinforced the excess pore-pressure effects. The conjunction of rapid sediment accumulation and gas release might have created the preconditioning factor of S16 failure through reduction of fine-grained sediment shear strength that provided a plane of weakness. In such a situation, undrained cyclic loading from seismic-ground accelerations could have brought rapidly to zero the effective stress on that potential slip plane. S16 failure occurred in a sector which is the locus of frequent moderate earthquakes ($M_s = 6.7$) (*El-Sayed et al., 2004*). At a distance of less than 100 km from earthquakes with such magnitudes, peak-ground acceleration is estimated up to 300 cm/sec^2 for an expected total duration of shaking of about 3 sec (*El-Sayed et al. 2004*). Moreover, remote earthquakes could have a tremendous impact on the study area (*El-Sayed et al., 2004*). The ground motions due to strong earthquakes ($M_s > 7.8$) in the Hellenic arc, Red Sea and Gulf of Aqaba are small (less than 10 cm/sec^2), but with longer duration, in the order of minutes. Precise locations of the epicentres and sediment mechanical behaviour are needed to determine the likely effects of an earthquake of a given magnitude. However, cyclic loading imparted by a large remote earthquake or a series of moderate local earthquakes were likely sufficient to initiate S16 failure.

6. Derived tsunami modelling: preliminary results

6.1 INITIAL ELEVATION

Following Watts and Grilli (2003), a slide with the volume of MTD S16 was modelled as a rigid body moving along a straight inclined surface. The slide is centred at 31.88°N and 30.10°E, at 420 m water depth and it is oriented 52°N along a 1.5° dipping slip plane. The slide block is 11 km long, 8.5 km wide, 150 m thick and 14 km^3 in volume (Fig. 5). The sliding characteristics are estimated through the TOPICS software that uses semi-empirical formulas deduced from laboratory experiments and potential flow numerical experiments (*Watts and Grilli, 2003*). In TOPICS, an initial particles velocity is also prescribed (at depth 0.53 × local depth). It is derived from the linear theory of progressive waves on water of arbitrary depth. Considering the level of accuracy of the slide representation, the following results must be taken cautiously. A run-out length of 21 km was obtained during 14 minutes. The derived initial elevation of the free surface exhibits a 9 m-deep depression oriented towards the coast and a 3 m-high crest oriented offshore with an initial wavelength of 53 km (Fig. 5).

6.2 PRELIMINARY RESULTS OF THE TSUNAMI PROPAGATION

We simulated the tsunami propagation using the FUNWAVE model, based on fully non-linear and dispersive Boussinesq equations, including bottom friction and turbulence sub-gridding and a moving shoreline algorithm (*Chen et al., 2000; Kennedy et al., 2000*). The numerical simulation indicates that the local tsunami propagating towards the coast exhibits amplitude maxima on the continental shelf (Fig. 6). The shoaling process occurs on the continental slope and then is inhibited on the continental shelf because of the low bathymetric gradient. As imaged on figure 5, in some areas, the seafloor morphology of the continental slope up to the shelf break exhibits a landward concave shape. Consequently, refraction processes focuses the tsunami waves offshore and far from the coastline (Figs. 5 and 6) where energy and wave height can be locally amplified. In contrast, in the vicinity of the city of Alexandria, where the bathymetry exhibits a concave-shaped pattern, the wave focuses at the coastal area. The numerical simulation showed therefore that a wave height of about 4 m could reach the coast around Alexandria (Fig. 6). These results must be taken as order of magnitude only because of the strong assumptions made on the slide representation.

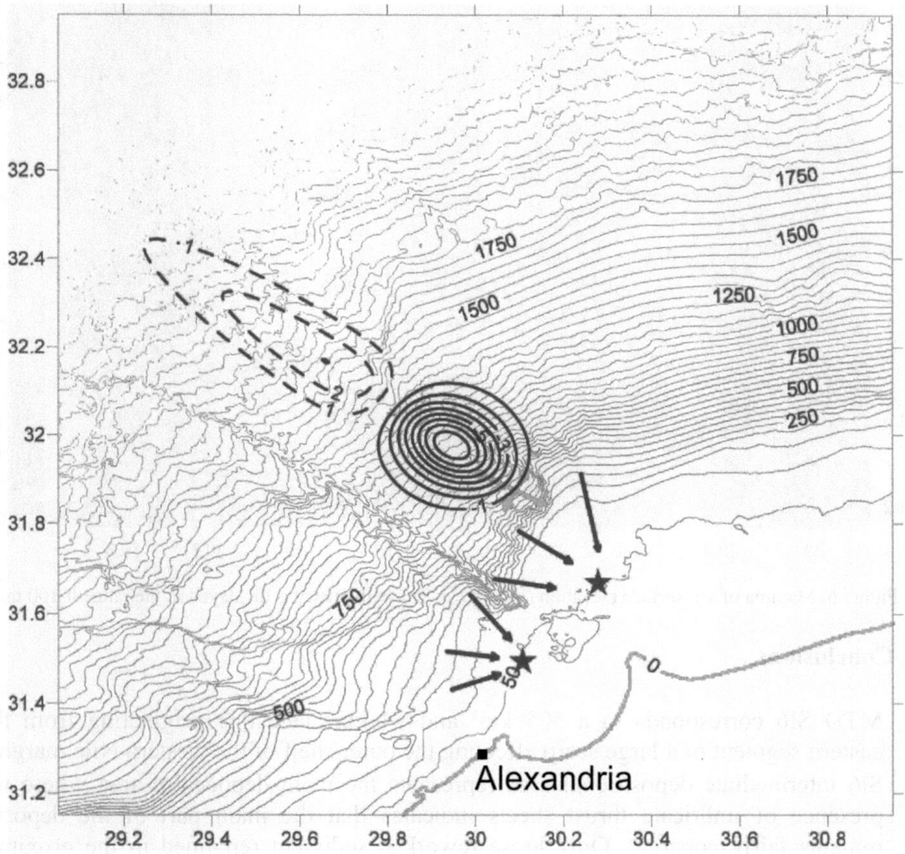

Figure 5: Initial surface elevation: dashed contours for crest and filled line for depression. The bathymetry is displayed with 50 m contours. The green cross indicates the initial location of the slided volume. Arrows indicate the normal to the wave direction for concave-shape bathymetry (refraction). Stars indicate potential convergence areas (focussing).

As a preliminary result, the largest part of the western Egyptian coastline can be considered as weakly vulnerable to tsunamis because (i) there is no shoaling effect and (ii) the wave focussing occurs offshore. However, the coastline around Alexandria might be considered as a vulnerable area because the focussing process could happen really close to the coast.

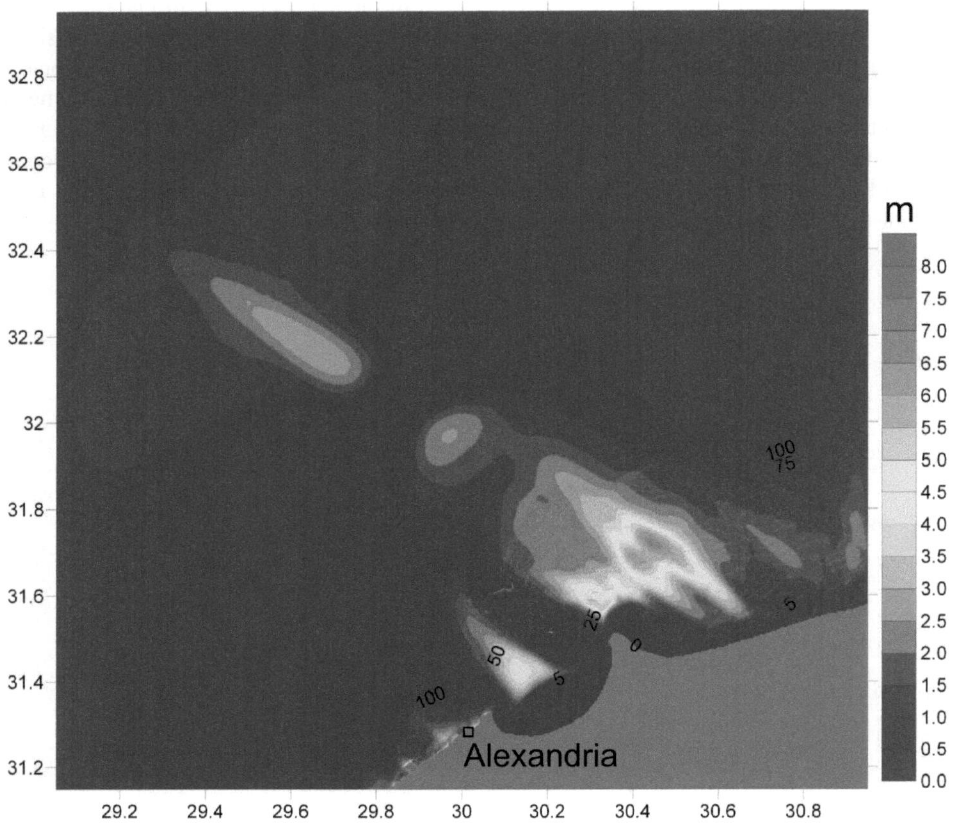

Figure 6: Maxima of sea surface elevation (in metres). The bathymetry is displayed in the range 0-100 m.

8. Conclusions

- MTD Sl6 corresponds to a 505 km^2 and 14 km^3 landslide originating from the eastern segment of a large scarp affecting the outer shelf of the western Nile margin.
- Sl6 intermediate depositional area represents the main depocentre area where the presence of imbricate thrust sheets indicates that the main part of the deposits remains fairly coherent. Only loose reworked sediment remained in the proximal depositional area or reached the distal depositional area.
- Age estimate for Sl6 formation (10-9 kyr BP) points out a close relationship between climate and submarine mass failure through a combination of eustatic sea-level changes and sedimentary processes.
- Effects of pore pressure, which increased within fine-grained sediment because of rapid sediment accumulation and likely by gas release, acted as a preconditioning factor for Sl6 failure.
- The final trigger is likely a large remote earthquake ($M_s > 7.8$) or a series of moderate local earthquakes ($M_s > 6.5$).
- Even if Sl6 appears large enough to produce a tsunami of notable size, preliminary results of numerical modelling show that the extended Egyptian continental shelf

(~80 km) protects the larger part of the coast because the focussing happens offshore and there is not anymore shoaling. On the contrary, the coastline around Alexandria is more vulnerable because the focusing and shoaling happen directly on the coast.

7. Acknowledgments

We thank the French Margin program (GRD Marges) and the former CNRS/INSU Ocean group for their financial support during completion of this work. A. Cattaneo for critical review of this paper. We are indebted to V. Felt and P. Boucher (BP-Egypt) for the access to some BP unpublished data.

8. References

Aal A, El Barkooky A, Gerrits M, Meyer H, Schwander M, Zaki H. 2000. *Tectonic evolution of the Eastern Mediterranean Basin and its significance for hydrocarbon prospectivity in the ultradeep water of the Nile Delta.* The Leading Edge 19(10): 1086-1102.

Camera L. 2006. *Structure de la marge passive egyptienne au large du Nil: contribution des données de sismique multitrace et de gravimètrie.* PhD thesis UPMC, Villefranche/Mer, pp 161.

Chen Q, Kirby JT, Dalrymple RA, Kennedy AB, Chawla A. 2000. *Boussinesq modelling of wave transformation, breaking and run-up.* II:2D. J. Wtrwy, Port, Coast, and Oc. Engrg 126(1): 48-56.

Ducassou E, Migeon S, Mulder T, Gonthier E, Murat A, Bernasconi S, Duprat J, Capotondi L and Mascle J. 2006. *The Nile deep-sea turbidite system: characterization and evolution of sedimentary processes during Late Quaternary*, External controls on deep-water depositional systems: climate, sea-level and sediment flux Conference. SEPM-Geological Society, London.

Ducassou E, Capotondi L, Murat A, Bernasconi S, Mulder T, Gonthier E, Migeon S, Duprat J, Giraudeau J, Mascle J. (in press). *Multiproxy Late Quaternary stratigraphy of the Nile deep-sea turbidite system – Towards a chronology of deep-sea terrigeneous systems.* Sedimentary Geology.

Ducassou E, Mulder T, Migeon S, Gonthier E, Murat A, Capotondi L, Bernasconi, S. and Mascle, J. (submitted). *Nile floods recorded in deep Mediterranean sediments.* Quaternary Research.

El-Sayed A, Korrat I, and Hussein HM, 2004. *Seismicity and seismic hazard in Alexandria (Egypt) and its surroundings.* Pure and Applied Geophysics 161(5-6): 1003-1019.

Loncke L, Mascle J, and Fanil Scientific Parties, 2004. *Mud volcanoes, gas chimneys, pockmarks and mounds in the Nile deep-sea fan (Eastern Mediterranean): geophysical evidences.* Marine and Petroleum Geology 21(6): 669-689.

Maslin M, Vilela C, Mikkelsen N and Grootes P. 2005. *Causes of catastrophic sediment failures of the Amazon Fan.* Quaternary Science Reviews 24: 2180-2193.

Mesherf WM. 1990. *Tectonic framework of the northern Egypt and the eastern Mediterranean region.* The Geology of Egypt (R. Said, ed.).

Owen M, Day S, Maslin M. 2007. *Late Pleistocene submarine mass movements: occurrence and causes.* Quaternary Science Reviews.

Prinzhofer A, Deville E, Foucher J-P, Caprais J-C and Mascle J. 2005. *Geochemistry of Gas Emanations from Mud Volcanoes and Pockmarks from Offshore Egypt.* AAPG International Conference, Paris, France.

Kennedy AB, Chen Q, Kirby JT, Dalrymple RA. 2000. *Boussinesq modelling of wave transformation, breaking and run-up.* I:1D. J. Wtrwy, Port, Coast, and Oc. Engrg 126(1): 39-47.

Watts P and Grilli ST. 2003. *Underwater landslide shape, motion, deformation, and tsunami generation.* Proc. of the 13th Offshore and Polar Engrg. Conf., Honolulu, Hawaii 3: 364-371.

REASSESSMENT OF SEISMICALLY INDUCED, TSUNAMIGENIC SUBMARINE SLOPE FAILURES IN PORT VALDEZ, ALASKA, USA

H.J. LEE, H. RYAN
U.S. Geological Survey, 345 Middlefield Road, Menlo Park, CA 94025 USA

P.J. HAEUSSLER
U.S. Geological Survey, 4200 University Drive, Anchorage, AK 99508 USA

R.E. KAYEN, M.A. HAMPTON
U.S. Geological Survey, 345 Middlefield Road, Menlo Park, CA 94025 USA

J. LOCAT
Dept of Geology and Geological Engineering, Universite Laval, Quebec, Canada, GIK 7P4

E. SULEIMANI
Geophysical Institute, University of Alaska, 903 Koyukuk Dr., Fairbanks, AK 99775 USA

C.R. ALEXANDER
Skidaway Institute of Oceanography, 10 Ocean Sciences Circle, Savannah, GA 31411 USA

Abstract

The M9.2 Alaska earthquake of 1964 caused major damage to the port facilities and town of Valdez, most of it the result of submarine landslide and the consequent tsunamis. Recent bathymetric multibeam surveys, high-resolution subbottom profiles, and dated sediment cores in Port Valdez supply new information about the morphology and character of the landslide deposits. A comparison of pre- and post-earthquake bathymetry provides an estimate of the net volume of landslide debris deposited in the basin and the volume of sediment removed from the source region. Landslide features include (1) large blocks (up to 40-m high) near the location of the greatest tsunami-wave runup (~50 m), (2) two debris lobes associated with the blocks, (3) a series of gullies, channels and talus, near the fjord-head delta and badly damaged old town of Valdez, and (4) the front of a debris lobe that flowed half-way down the fjord from the east end. A transparent unit, with a maximum thickness of 10 m, above the debris flow deposits in the deepest part of the fjord, likely represents a lower density sediment flow related to the other failure deposits. Calculations from the bathymetric difference map suggest a total net volume of displaced sediment on the order of 0.4 km^3. However, an integration of the volume of debris flow deposits (mapped according to their acoustic signature) indicates a gross volume of about 1 km^3, showing that the landslides incorporated significant additional sediment from the fjord floor into the debris flow as it moved. Despite the large volume of sediment failures in the eastern part of the fjord, smaller, but more coherent block failures in the western part appear to be the primary cause of the largest tsunamis that impacted the shorelines.

Keywords: Submarine landslide, tsunami, Alaska, earthquake

V. Lykousis, D. Sakellariou and J. Locat (eds.), Submarine Mass Movements and Their Consequences, 357–365.

1. Introduction and Background

The Great Alaska Earthquake of 1964 (Mw = 9.2), the largest earthquake ever recorded in North America, caused major damage to all coastal communities in south-central Alaska. Valdez, a small town on the shores of the glacial fjord known as Port Valdez (Fig. 1), was particularly hard hit (Coulter and Migliaccio, 1966). A total of 32 people died in Valdez, most as the result of tsunamis with runups of up to 50m that were induced by submarine landslides that also destroyed the waterfront. Following the earthquake, extensive investigations were conducted of the effects of the earthquake on Alaska's communities and landscape (Coulter and Migliaccio, 1966; Plafker et al., 1969; Hampton et al., 1993). These studies showed dramatic changes in bathymetry off the fjord-head

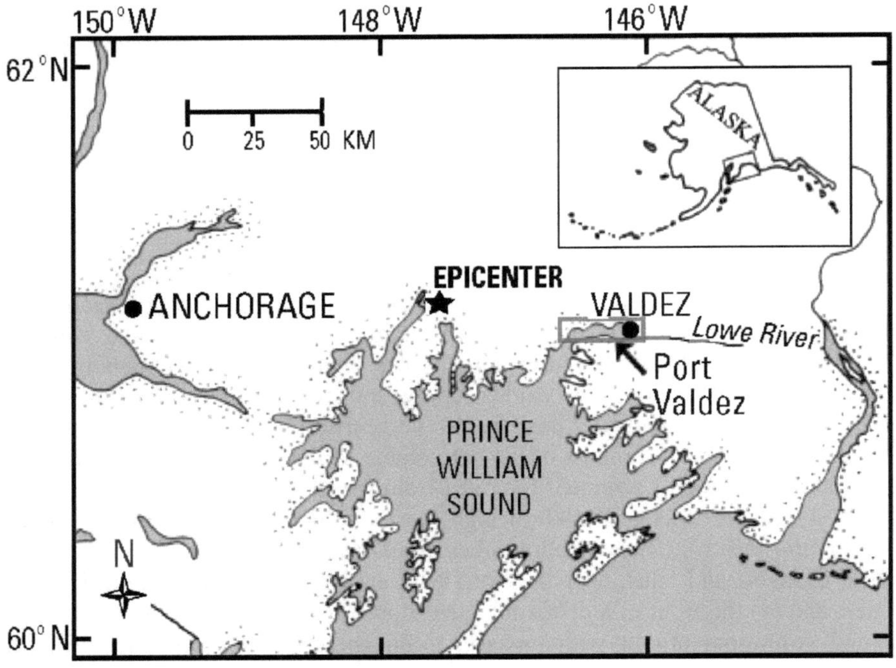

Figure 1. Location map showing Valdez, Alaska, and the epicenter of the 1964 Mw = 9.2 earthquake. Red box identifies study area shown in Figs. 2-4, 7 and 8.

delta near the town of Valdez (up to 100 m of deepening) and the highest tsunami wave runup heights measured anywhere in Alaska during the event. Tsunami waves at the waterfront of Valdez were estimated to have been 10-12 m (Coulter and Migliaccio, 1966), but the greatest tsunami wave runup heights (52 m, Fig. 2) were in the western part of the fjord, at the opposite end from the community (Plafker et al., 1969). Using a comparison of bathymetric data (within 1 km of the Valdez waterfront), obtained in 1959 and 1964, Coulter and Migliaccio (1966) estimated that about 75×10^6 m^3 of material failed during the earthquake. Based on an engineering analysis of the fjord-head delta (Shannon and Wilson, 1964), a decision was made to move the entire town of Valdez to a more stable site 5.5 km to the west, and the move was completed by 1967.

Recent multibeam maps of the fjord, prepared by the United States National Oceanic and Atmospheric Administration (NOAA) (Fig. 2), disclose apparent landslide-related features, prompting the US Geological Survey (USGS) to initiate a research effort to better understand the 1964 submarine landslide-generated tsunami events. These efforts include a reassessment of pre- and post-earthquake bathymetry and a field study to obtain high-resolution subbottom profiles and several gravity cores for radioisotope dating.

Figure 2. Multibeam imagery of Port Valdez showing landslide features discussed in the text. Labels around margin of the fjord show the estimated tsunami wave heights (runups) resulting from the 1964 earthquake (Plafker et al., 1969).

Preliminary findings from these data sources (Lee et al., 2006) note the presence of several large blocks (40-m high and 300-m across) on the basin floor near a moraine deposit in the western part of the fjord (front of Shoup Bay, Fig. 2) and a complex series of gullies and chutes in the area off the old community where depth increased by over 100 m during the earthquake. We speculated that the large blocks resulted from a failure of the moraine front during the earthquake and that their motion contributed to the particularly high waves observed immediately onshore from them. We also noted a subtle north-south trending step near the middle of the fjord, which we speculated was the front of a large debris lobe that had flowed westward from the fjord-head delta (Fig. 2). In the present paper, we provide improved evaluations of these data, present an inventory of sources and sinks of landslide material, and evaluate the scale of post earthquake sedimentation within Port Valdez.

2. Methods

Multibeam data from Port Valdez, Alaska, were obtained using high-resolution MBES systems (Reson 8101, 8125, differential GPS navigation, no backscatter) aboard the NOAA ship *Rainier*. These systems are capable of producing geodetic-quality bathymetry with spatial errors of less than 1 m and vertical errors of 0.5% or less of the water depth (see Hughes Clarke et al., 1996, for a discussion of the technology). In addition to the multibeam data, two hydrographic surveys conducted by the U.S. Coast and Geodetic Survey in 1901 and 1966 were available for Port Valdez, spanning the 1964 earthquake. The 1901 gridded data set was created by digitizing soundings of the Port Valdez, AK smooth sheet. The 1966 data set, which contained considerably more soundings than the 1902 set, was created by combining surveys conducted at a 1:20,000 scale over the entire fjord with a more detailed survey near the head of the fjord at a scale of 1:5000. Because there was an overall subsidence of the Valdez area of 1 m as a result of the earthquake (Coast and Geodetic Survey, 1966), the 1966 grid was adjusted by adding 1 m to the grid before comparison with the 1902 grid. Soundings from the steep north, south, and west walls of the fjord were not included in the analyses as small location errors could result in large bathymetric changes, and these had the potential for introducing large errors into the difference map.

High-resolution subbottom profile data were acquired using an Edgetech 512i Chirp System (tracklines shown in Fig. 3). Frequencies were typically swept over the range of 500 to 7200 Hz during a 30 ms period. Chirp technology uses multiple sound frequencies as opposed to older, single-frequency systems. Resolution (ability to image thin layers) is improved without compromising significantly in terms of penetration. Vertical resolution ranged from 0.25 to 0.5 m; and depth penetration, from 10 to 50 m, depending upon sediment type

Sediment cores were taken with a conventional 350-kg gravity corer that recovered samples up to 2 m in length (locations shown in Fig. 3). Cores were analyzed for short-half-life radioisotopes using the methods of Alexander and Venherm (2003). The most useful measured isotope was ^{137}Cs, an impulse tracer produced from atmospheric nuclear tests, which was first introduced into the environment in significant amounts in 1954, and had peak input in 1964 (Kuehl et al., 1986). Because the peak ^{137}Cs signal coincides with the year of the earthquake, the thickness of sediment lying above the ^{137}Cs peak identifies the amount of post-landslide sedimentation.

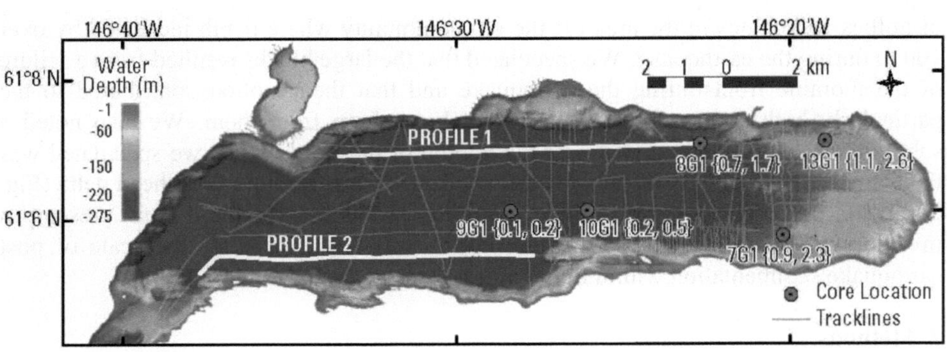

Figure 3. Locations of cores analyzed for ^{137}Cs dating and track lines for chirp profiles. Locations of profiles shown in Figs. 5 and 6 are identified. Numbers next to core station locations identify thickness (m) of sediment younger than 1964 (^{137}Cs peak) and sediment accumulation rate for the last 41 yrs in cm/yr.

3. Results

3.1 BATHYMETRIC CHANGES, 1901-1966

Differences in bathymetry between 1901 and 1966 show areas as much as 40 m shoaler southeast of Shoup Bay and as much as 60 m deeper near the fjord head. Overall, the bathymetric difference map (Fig. 4) shows a shoaling over 60% of the surface area, with a deepening over 40% of the fjord. Volumetrically, our calculations show 4.3×10^8 m^3 removed from the area that deepened and an increase in volume of 3.8×10^8 m^3 over the area that shoaled. Because we focused on areas with gentle gradients, we expect that the overall errors in our volume calculations are low, perhaps in the range of 10 to 20%. Accordingly, the sediment budget roughly balances and shows significantly more sediment movement than reported by previous investigators (75×10^6 m^3, Coulter and Migliaccio, 1966).

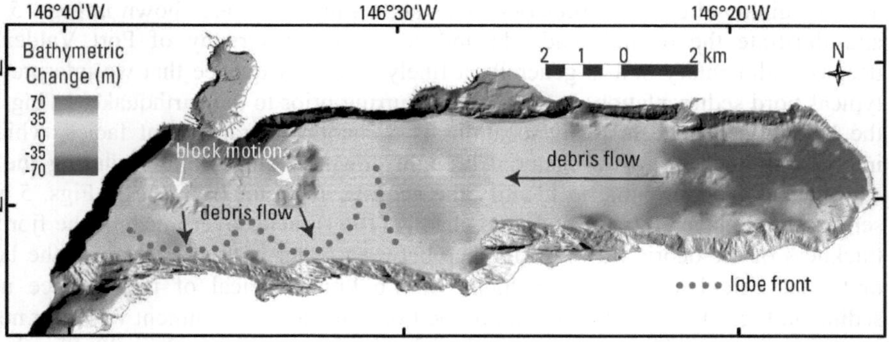

Figure 4. Map showing changes in bathymetric depth between surveys run in 1901 and 1966, spanning the 1964 earthquake.. Our interpretation of the fronts of three large debris flows produced by slope failures generated in Port Valdez during the 1964 earthquake and their presumed directions of motion are also shown (see below for discussion).

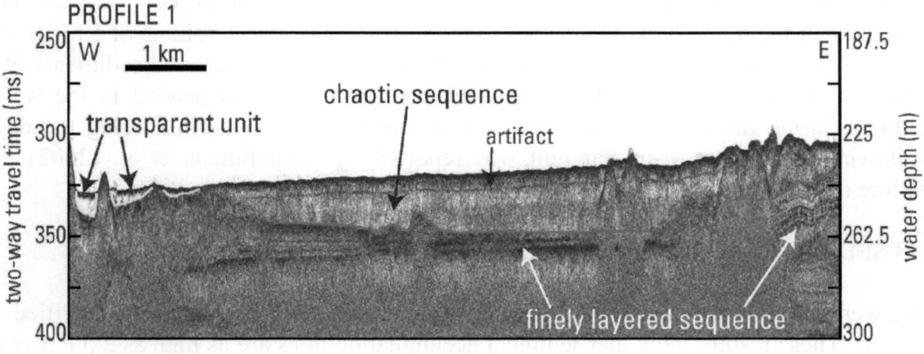

Figure 5. E-W chirp profile line across northern part of Port Valdez (see Fig. 3 for location). Acoustic facies features discussed in text are identified. Red line shows boundary between layered and chaotic facies. Yellow line shows base of transparent unit.

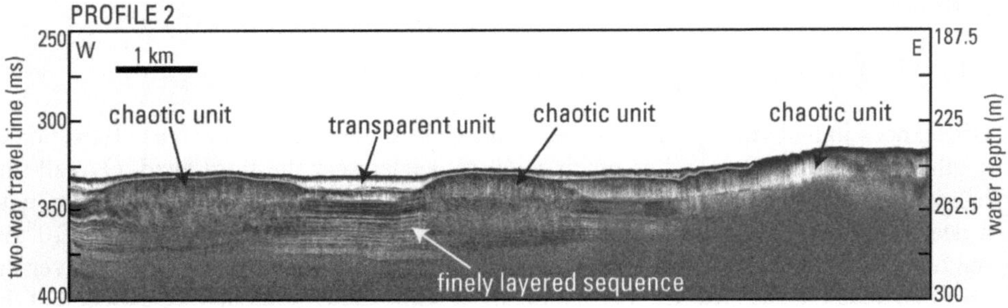

Figure 6. E-W chirp profile line across the southern part of Port Valdez (see Fig. 3 for location). Three different acoustically defined chaotic debris lobes are identified (mapped in Fig. 4). Colored lines represent the same boundaries as in fig. 5.

3.2 SEDIMENT VOLUMES BASED ON ACOUSTIC STRATIGRAPHY

Two longitudinal chirp profiles (locations shown in Fig. 3) are shown in Figs. 5 and 6 and illustrate the overall landslide and sediment stratigraphy of Port Valdez. The deepest sedimentary unit is generally a finely layered sequence that we interpret to be typical fjord sedimentation, such as was occurring prior to the earthquake. Lying above the layered sequence in many locations is a chaotic to transparent facies, which we interpret to be debris flow material brought down by slope failures during the 1964 earthquake. We were able to identify the acoustic reflector (red line in Figs. 5 and 6) separating the layered (pre-1964) and chaotic (1964) facies over much of the fjord. The thickness of the debris flows (difference between water bottom and top of the layered unit) was calculated using a velocity of 1.6 km/s (typical of near-surface marine sediment, Fig. 7). At the eastern end of the fjord, the layered sediment was at or near the sea floor; at the middle to western end of the fjord, it was as much as 30-40 m beneath the sea floor. The total calculated volume of material above the surface of layered sediment is 9.8×10^{8} m^3.

In the deepest part of the fjord and lying above the chaotic facies is an acoustically transparent unit that fills depressions in the seabed of western Port Valdez. The thickness of the transparent unit ranges from 0-12 m, with a total volume of 5.8×10^{7} m^3 (about 6% of the total volume of material above the surface of layered sediment). We interpret this facies to be redeposited sediment that became suspended in the water column during the failure events and flowed into the deepest basins during the days following the event. A similar unit was reported by Schnellmann et al. (2002) for failure deposits in Swiss lakes.

3.3 SEDIMENT ACCUMULATION RATES FROM CORE ANALYSIS

We were able to recognize the ^{137}Cs peak in five cores from Port Valdez (identified in Fig. 3). These results show that sediment accumulation rates are as high as 2.5 cm/yr in the eastern part of Port Valdez, off the fjord-head delta of the Lowe River. These rates decrease almost linearly with distance from the delta to about 0.2 cm/yr in the central fjord. The thickness of post earthquake sediment is slightly over 1 m in the eastern fjord most standards of marine sedimentation, but they produce a post earthquake sedimentary unit that is generally too thin to be resolvable in chirp profiles. Accordingly, almost all reflectors observed in the chirp profiles were deposited either during or prior to the 1964 earthquake.

4. Discussion

According to the bathymetric difference map (Fig. 4), the floor of the eastern 40% of Port Valdez deepened during the 1964 earthquake. This area includes the fjord-head delta where massive failures were directly observed at the waterfront of old Valdez (Coulter and Migliaccio, 1966). However, the area of sediment removal we delineate from bathymetric differences (Fig. 4) also includes the low seafloor gradient region west of the fjord-head delta. We now estimate that $4.3\text{x}10^8$ m^3 of sediment were removed, 6 times the original estimate. Because so much of this area of removal has a low slope (~2°), we believe that it did not fail directly but rather was scoured by very erosive debris flows moving westward from the delta head.

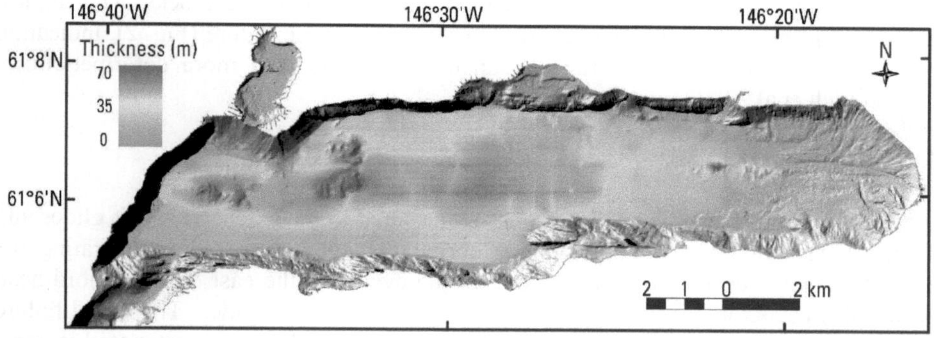

Figure 7. Map showing the thickness of landslide related sediment overlying the surface of the layered acoustic facies assumed to be pre-1964 sediment.

As discussed above, by integrating the thickness of sediment lying above the surface of layered sediment (Fig. 7), we estimated the volume of landslide debris to be $9.8\text{x}10^8$ m^3 (nearly 1 km^3). This is in stark contrast to the value calculated from bathymetric differences over the area that shoaled, which is $3.8\text{x}10^8$ m^3. That is, more than twice as much material is deposited as landslide debris than is represented by shoaling of the seafloor. We interpret these values as suggesting that the debris flows were very erosive when they entered the zone that would ultimately become depositional. They eroded some of this area, incorporated the sediment into the flow and essentially doubled the volume of flowing material. A similar phenomenon was observed in the tsunamigenic submarine debris flow at Kitimat, British Columbia in 1975, where more than half of the debris flow mass was derived from incorporated fjord-bottom sediment (Prior et al., 1984). The entire deposit was ultimately emplaced in the deep basin with a landslide debris thickness of as much as 30 m. Note in Fig. 5 that the landslide deposit is thickest toward the eastern part of the depositional zone and thins toward the west. This likely indicates that the ability of the flow to scour decreased as it moved westward, gradually leaving thinner deposits. Scouring by debris flows has also been reported from marine settings (Canals et al., 2004; Lastras et al., 2004).

The debris flow that began at the fjord-head delta and moved west was clearly the most significant failure component produced in Port Valdez by the 1964 earthquake. However, there were also failures contained in the western part, including large blocks generated by the failure of the Shoup Bay moraine (Fig. 2). The blocks form two clusters, one to the southeast of the moraine (smooth, pointed promontory south of Shoup Bay itself) and one to the southwest. Behind (north of) each set of blocks is a zone of sediment removal shown by the bathymetric difference map (Fig. 4), likely indicating gouging or eroding of the seafloor as the blocks translated across the bottom. In front of the blocks (to the SE and SW of the block clusters, respectively), are smaller debris lobes, which can be seen in the western (left) part of Fig. 6. Using this and other profile lines in the area, we mapped the fronts of these lobes (Fig. 4), showing arcuate patterns surrounding the block clusters. A plausible scenario is that the moraine failed catastrophically on both flanks setting in motion a series of large blocks, which in turn gouged the seabed. The gouged bottom material, as well as part of the failed moraine, mobilized into two debris flows that spread out in front of the blocks. The highest tsunamis for the earthquake were recorded near this moraine failure (Fig. 2), indicating that intact blocks are more efficient tsunami generators than more deformed flows (Rabinovich et al., 2003).

5. Conclusions

We conclude that two major failure zones contributed to the extensive block glides and debris flows that impacted Port Valdez during the 1964 earthquake, generating the highest tsunamis observed anywhere during the event. At the east end, the fjord-head delta failed catastrophically, destroying the waterfront of old Valdez. The initial failure may have been comparatively small (on the order of 75×10^6 m^3 as estimated in early surveys), but the resulting debris flow increased in volume as it proceeded westward, scouring the seabed and incorporating this scoured sediment into the flow. A zone of sediment removal extended 40% of the length of the fjord. Ultimately the flow deposit volume was in the range of 1 km^3.

In the western end, the moraine at the head of Shoup Bay also failed catastrophically but did not disintegrate as thoroughly as the failure at the east end. The resulting blocks (up to 40 m high and several hundred meters across) were efficient tsunami generators, based on the greater tsunami heights in the western part of the fjord (although the source of these higher waves cannot be fully constrained). Parts of the moraine, and material gouged by the blocks in their translation, mobilized into two arcuate debris flows (Figs. 4 and 6).

6. Acknowledgements

We acknowledge help from the crew of the *R/V Montague* and support from the U.S. Geological Survey, the National Oceanic and Atmospheric Administration and the Geophysical Institute of the University of Alaska. We thank Miguel Canals, Brian

Bornhold, and Brain Collins for helpful reviews that greatly improved the manuscript and Lori Hibbeler for technical assistance.

7. References

Alexander, C.R. and Venherm, C., 2003. *Modern sedimentary processes in the Santa Monica, California continental margin: sediment accumulation, mixing and budget*, Marine Environmental Research, 56: 177-204.

Canals, M., Lastras, G., Urgeles, R., Casamor, J.L.., Mienert, J., Cattaneo, A., De Batist, M., Haflidason, H., Imbo, Y., Laberg, J.S., Locat, J., Long, D., Longva, O., Masson, D.G., Sultan, N., Trincardi, F. and Bryn, P., 2004. *Slope failure dynamics and impacts from seafloor and shallow sub-seafloor geophysical data: Case studies from the COSTA project.* Marine Geology 213 (1-4), 9-72.

Coast and Geodetic Survey, 1966. *Descriptive report to accompany Hydrographic Survey*, Registry Number: H-8900, Field Number: HO-20-1-66, 24 pp.

Coulter, H.W. and Migliaccio, R.R., 1966. *Effects of the earthquake of March 27, 1964 at Valdez, Alaska.* United States Geological Survey Professional Paper, 542-C, 36 pp.

Hampton, M.A., Lemke, R.W. and Coulter, H.W., 1993: *Submarine landslides that had a significant impact on man and his activities: Seward and Valdez, Alaska.* in Schwab. W.C., Lee, H.J., and Twichell, D.C. (Eds.), Submarine Landslides: Selected Studies in the U.S. EEZ: United States Geological Survey Bulletin 2002, p. 123-142.

Hughes Clarke, J.E., Mayer, L.A. and Wells, D.E., 1996. *Shallow-water imaging multibeam sonars: a new tool for investigation of seafloor processes in the coastal zone and on the continental shelf.* Marine Geophysical Research 18, 607-629.

Kuehl, S.A., Nittrouer, C.A., and DeMaster, D.J., 1986. *Nature of sediment accumulation on the Amazon continental shelf*, Continental Shelf Research, 6: 209-225.

Lastras, G., Canals, M., Urgeles, R., De Batist, M., Calafat, A.M. and Casamor, J.L., 2004. *Characterisation of the recent BIG'95 debris flow deposit on the Ebro margin,Western Mediterranean Sea, after a variety of seismic reflection data.* Marine Geology 213 (1-4), 235-255.

Lee, H.J., Ryan, H., Kayen, R.E., Haeussler, P.J., Dartnell, P. and Hampton, M.A., 2006. *Varieties of submarine failure morphologies of seismically-induced landslides in Alaskan Fjords.* Norwegian Journal of Geology, 86: 221-230.

Plafker, G., Kachadoorian, R., Eckel, E.B. and Mayo, L.R., 1969. *Effects of the earthquake of March 27, 1964 on various communities.* United States Geological. Survey Professional Paper 542-G, 50 pp.

Prior, D.B., Bornhold, B.D. and Johns, 1984. *Depositional characteristics of a submarine debris flow*: Journal of Geology, 92: 707-727.

Rabinovich, A.B., Thomson, R.E., Bornhold, B.D., Fine, I.V. and Kulikov, E.A., 2003. *Numerical estimation of tsunami risk associated with landslide-generated tsunamis with application to the Strait of Georgia, British Columbia.* Pure and Applied Geophysics, 160, 1273-1313.

Schnellmann, M, Anselmetti, F.S., Giardini, D., McKenzie, J.A. and Ward, S.N., 2002. *Prehistoric earthquake history revealed by lacustrine slump deposits,* Geology, 30: 1131-1134.

Shannon and Wilson, Inc., 1964. *Report on subsurface investigation for Mineral Creek townsite, City of Valdez, Alaska.* Shannon and Wilson, Inc., Seattle, Wash., 12 pp.

TOWARDS THE MITIGATION OF THE TSUNAMI RISK BY SUBMARINE MASS FAILURES IN THE GULF OF CORINTH: THE XYLOCASTRO RESORT TOWN CASE STUDY

M. CHARALAMPAKIS,
Laboratory of Marine Geology & Physical Oceanography, Geology Department, University of Patras, Patras 26500, Greece, email: cmarinos@upatras.gr

A. STEFATOS,
Rocksource ASA, Olav Kyrres gate 22, N-5808 Bergen, Norway

K. MPOURDOPOULOS & G. FERENTINOS
Laboratory of Marine Geology & Physical Oceanography, Geology Department, University of Patras, Patras 26500, Greece

Abstract

Submarine-mass-failure-generated tsunamis pose a significant threat to the coastal communities around the Corinth Gulf. An effort was made towards the mitigation of a potential tsunami generated in the eastern part of the gulf, due to a submarine landslide. The impact of the tsunami was assessed along the coastal segment of the summer resort town of Xylocastro. The analysis (study) of the data within a Geographical Information System revealed that a 4 m tsunami run-up will flood 12% of the town's district. One fifth of the permanent residents are expected to be affected, while one fourth of the infrastructure is likely to undergo damages. Although alarm and reaction times to a possible local tsunami are short, there is just enough time for evacuation. Therefore a more detailed hazard assessment and an emergency management plan should be undertaken, not only for Xylocastro, but also for other coastal regions in the Corinth Gulf, where extensive development is taking place.

Keywords: tsunami, run-up, inundation zone, submarine mass failures, natural hazard, GIS, Corinth Gulf, Greece

1. Introduction

Tsunamis are among the most devastating natural hazards affecting the littoral zone. Even though infrequent, they can cause severe damage to coastal infrastructures, destruction of properties and even loss of human life (UNESCO-IOC, 2006). All the above can violently affect the socio-economic structure of a populated area, particularly in our days, where the density of occupation and utilization of the coastal zone has been significantly increased. Since the majority of damages are expected within the flooding zone, the inundation zone extent is the most critical tsunami hazard that must be estimated, in order to assess tsunami risk along coastal segments. Although, tsunamis are difficult to predict, their expected inundation zone along the coastline can be mapped as the maximum run-up height on shore.

This paper focuses on the 5 km coastal segment of Xylocastro summer resort town, in order to examine the impact of a local, submarine-landslide-generated tsunami, to the coastal zone (Figure 1). Utilizing published work on existing submarine landslides, with

V. Lykousis, D. Sakellariou and J. Locat (eds.), Submarine Mass Movements and Their Consequences, 367–375.

a tsunami potential in the Gulf of Corinth, the paper examines the effects of a tsunami generated offshore the Perachora peninsula to the east (Papatheodorou & Ferentinos, 1993; Stefatos et al., 2006). Potential areas for tsunami flooding are identified and damages to the town's infrastructure are estimated. Moreover, possible human casualties are estimated, while potential escape methods are suggested.

2. Historical record of Tsunamis in the Gulf of Corinth

The Gulf of Corinth is the region with the highest potential for tsunami occurrence in the Mediterranean Sea (Papadopoulos, 2003). Thisisnot only revealedin the published

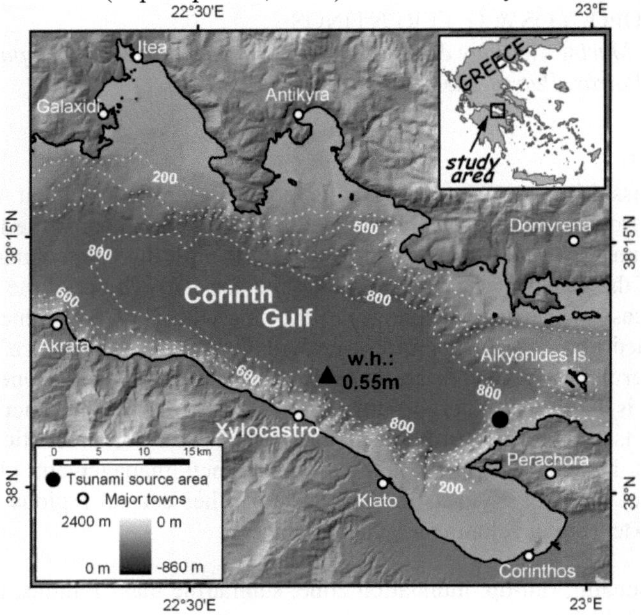

Figure 1. Location of potential tsunami generation area utilized in this study. Estimated deep water tsunami wave height (w.h.) offshore Xylocastro town.

historical records (Heck, 1947; Galanopoulos, 1960; Ambraseys, 1962; Antonopoulos, 1980; Papadopoulos and Chalkis, 1984; Papazachos et al., 1986; Soloviev, 1990; Papazachos and Papazachou, 1997; Papadopoulos, 2000; Soloviev et al., 2000; Papadopoulos, 2003), but also in the results of recent studies which suggest that the Gulf could experience relatively destructive tsunami events in the near future, posing a significant threat to coastal communities (Stefatos et al., 2005; Stefatos et al., 2006). Detailed marine geophysical surveys have shown that the steep offshore morphology, the presence of under-consolidated alluvial sediments along the shelf and the increased seismicity of the area, favor the development of submarine landslides (Ferentinos et al., 1988; Lykousis et al., 2003). Numerous such submarine slides have been reported in literature, while recently more detailed studies demonstrated the potential of such slides for destructive tsunamigenesis (Stefatos et al., 2006).

3. Submarine landslide generated tsunami and wave run-up

For the purpose of this paper a potential submarine-mass-failure-generated tsunami offshore Perachora peninsula to the east of the Corinth Gulf, has been used. The tsunami generation area lies 20 km, northeast of Xylocastro, which is a characteristic average distance considering the size of the Gulf and the distribution of towns around it (Figure 1). Stefatos et al. (2005, 2006), have calculated that existing submarine mass failures in the area could generate a tsunami with up to 4.04 m wave height and 6270 m wavelength. After simulating the tsunami propagation in the Gulf of Corinth with TUNAMI-N2 model (Imamura, 1995), they reported that tsunami wave height offshore Xylocastro, in the deep waters (i.e. 800m), reaches a maximum of 0.55 m (Stefatos et al., 2005, their figure 2, Figure 1).

In order to identify areas of potential tsunami flooding, the corresponding run-up (R) was calculated using the equation that Pelinovsky & Mazova (1992) have proposed:

$$R/H_0 = 2\pi \, 2L/\lambda_0$$

Where H_0 is the wave height and λ_0 the wave length, at distance L from the shoreline (their Figure 9).

The above formula is exact for predicting maximum run-up for both linear and non-linear theory (Synolakis 1991; Pelinovsky & Mazova, 1992) provided that the tsunami wave will not break while climbing up the coastline. Athough 75% of the tsunamis don't break as they climb up the coastline (Pelinovsky et al., 1989; Pelinovsky, 1989), a check was performed if the breaking criterion is violated in our case. For this reason the breaking criterion proposed by Synolakis (1991) was used:

$$H_0 > 0.818(\cot\beta)^{-10/9}$$

Where β is the angle of the slope.

Considering that the slope offshore Xylocastro is about 13°, it is clear that the tsunami wave is not expected to break while reaching the shore.

By solving the run-up equation we estimate a maximum run-up of 3.65 m (Figure 2). This value falls within the reported range of historical tsunami wave heights for the Gulf of Corinth (0.3 - 15m) (Papadopoulos, 2000) and therefore, the proposed vulnerability model assumes an estimated tsunami run-up height of 4 m.

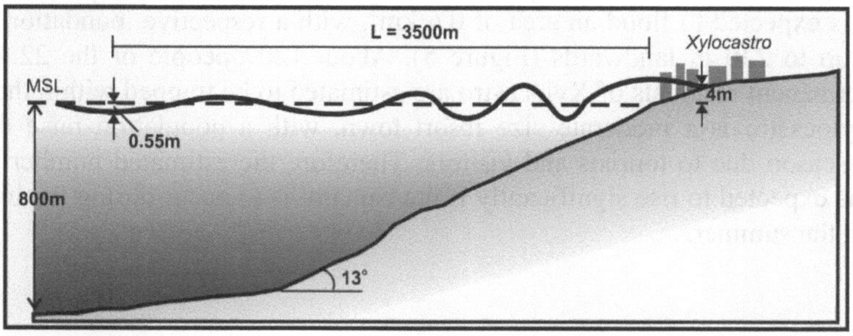

Figure 2. Schematic illustration of the estimated tsunami wave height and run-up. (Sketch not in scale, MSL: mean sea level).

4. Methodology – Presentation of data

Analysis of the data and presentation of the results were carried out within a Geographic Information System (GIS), which is an optimal decision making tool for hazard assessment and risk management. A geographical database was developed, integrating onshore and offshore morphological data, urban plan data (buildings, road and railway network), with the estimated tsunami run-up. Since morphology is one of the most imperative factors that affects the spatial extent of flooding, a detailed digital elevation model (DEM) for the study area was constructed (Figure 3). Topographic maps (Hellenic Military Geographical Service) scaled 1:5.000 was used for the reconstruction of the relief on land, while offshore, detailed bathymetric maps were used (Charalampakis et al., 2005). Data regarding the resident population and the urban plan of the town was provided by the General Secretariat of the National Statistical Service of Greece (Census of December 2000 – 2001). The urban plan was updated using supplementary data by the Municipality of Xylocastro and in situ observations. The road and railroad network were specifically digitized from topographic maps scaled 1:5.000.

A series of thematic maps were constructed and a representative map, containing the most important features is presented herein (Figure 4). The main spatial features that were employed in this study were the town blocks, the buildings, the road network (including the national road) and the railroad (Figure 4). Attribute data-table were established and linked to each feature accordingly. Town blocks were linked to the population, while a series of attributes, like age of construction, material of construction, number of floors, usage etc., were linked to the individual buildings within each block. On top of the thematic maps the flooding area was plotted. The mapped flooding area corresponds to the estimated inundation zone, which is considered to extend between the shoreline and the 4m contour that equals the estimated tsunami run-up height (Figure 5).

In order to estimate the extent of the destruction, specific queries were raised. Such queries quantify and illustrate the results of the tsunami flood in regard to the mapped features (e.g. buildings, road network).

5. Results from tsunami impact and discussion

According to the analysis of the produced thematic maps, within the town limits the tsunami is expected to flood an area of 0.6 km^2, with a respective inundation zone that extends up to 230 m landwards (Figure 5). About 1207 people or the 22.4% of the actual permanent residents of Xylocastro are estimated to be trapped within the flooding zone. Xylocastro is a moderate size resort town, with a population raise during the summer season due to tourists and visitors. Therefore the estimated number of people trapped is expected to rise significantly if the tsunami is to occur during the high tourist season in the summer.

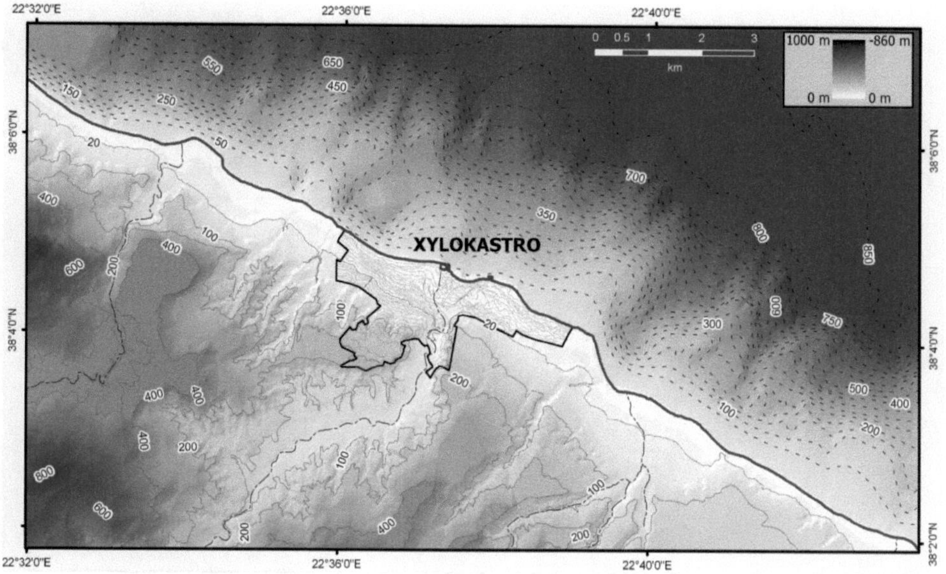

Figure 3. Shaded relief map of the study area. The detailed topography of Xylocastro district is also illustrated.

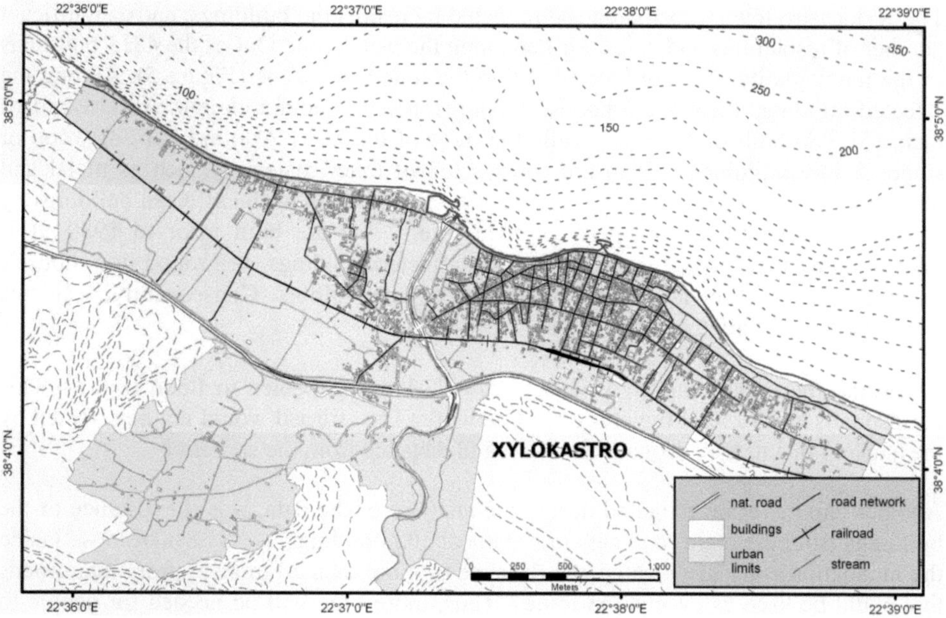

Figure 4. Xylocastro urban plan, displaying blocks, buildings, road and railroad network.

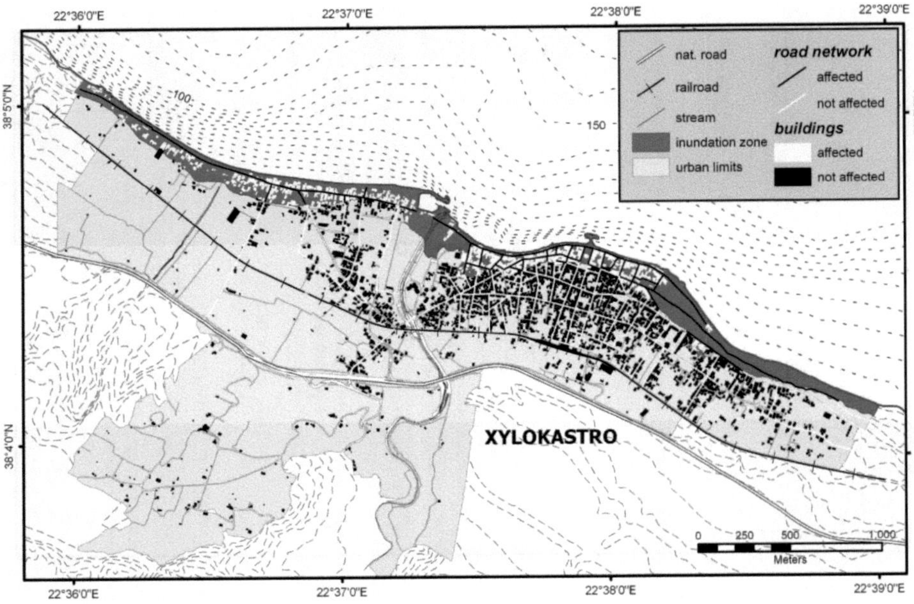

Figure 5. Thematic map illustrating the extent of tsunami flooding and the distribution of the affected buildings and roads.

The Xylocastro town, urban area is dominated by residential buildings, and a significant number of restaurants and cafes, located along the waterfront. Out of the 4813 buildings of the municipality, 983 are located within the inundation zone (Figure 5). Among the affected buildings there are 6 hotels, 31 stores, a school and a church. Almost half of them (52.3%) built of concrete, while the most of the rest (46%) are built of wood or stone. A few buildings (1.7%) are constructed of other materials, such as metal and other unclassified materials. According to the classification of the affected buildings by age of construction, the majority of them (i.e. 77%) are built prior to 1980. It is interesting to note the fact that 38.6% of the affected buildings are ground floor houses, meaning that there in no immediate escape to upper floors. Consequently, residents would be more easily trapped.

Regarding the urban road network, a total of 7 km is expected to flood (fig. 5). This corresponds to 28.2% of the entire road network. The railroad would not be affected by the tsunami due to the adequate elevation and distance from the shoreline.

Another parameter investigated during the analysis of the data, was the distance of the buildings from the main road network. Almost 300 buildings, i.e. 31.8% of those inside the inundation zone, are located at a distance of more than 20 m away from main roads that could be used as evacuation routes. Thus, more time will be needed for people to evacuate those districts.

Generally, the average distance that a person needs to cover in order to get out of the inundation zone ranges from 100 to 200m (Figure 6). Therefore, the required time for evacuation to safe ground varies from 1.5 to 3 minutes, taking into account an average walking speed of 0.948 m/s, which corresponds to the walking speed for a walking elderly person or an adult person with a child (Japanese Institute for Fire Safety & Disaster Preparedness, 1987; Sugimoto, 2003). Since the time for the tsunami to travel from its source area to the coastline of Xylocastro equals to almost 4 minutes (Stefatos et al., 2005), residents have just enough time to reach the evacuation areas.

The results of the analysis reveal that within a short distance, there are at least three locations that are suitable to serve as evacuation areas for the population to seek refuge (Figure 6). These areas have easy access from the main roads; they are situated on flat and open ground with elevation more than 8m above sea level (i.e. twice the run-up height) and they are close to the national road, where the rescue teams are expected to reach the town.

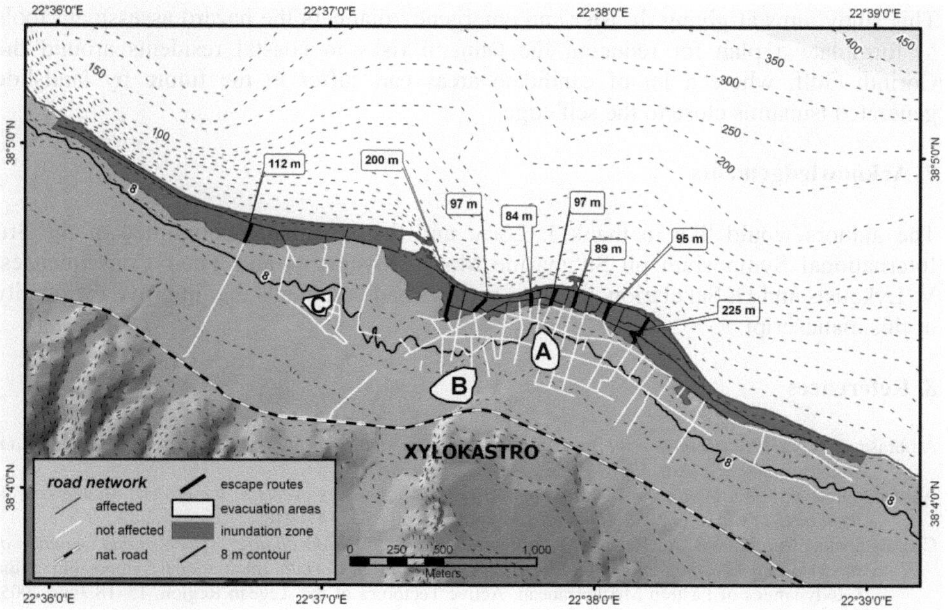

Figure 6. Thematic map showing indicative escape routes and suggested evacuation areas. The required distances to reach safe ground are annotated on the map.

6. Concluding remarks

This survey estimates the impact of a local, submarine-mass-failure-generated tsunami, on the coastal zone of Xylocastro, a typical summer resort town in the Gulf of Corinth. The data analysis indicates that about 1/5 of the resident population, not including tourists and visitors in the summer season could be affected by the tsunami, while almost one fourth of the town's infrastructure is expected to suffer severe damages. The expected high damage ratio is attributed to the fact that half of the buildings lying within the inundation zone are highly vulnerable since they are old and made of either stone or wood. Furthermore, 33% of the affected buildings are single storey buildings and preclude vertical evacuation to higher floor. Approximately 32% of them don't have easy access to main roads, thus increasing the required time for evacuation and therefore the risk of human casualties. Special care should be taken for the school and the six hotels that are situated inside the flooding zone.

Three locations were identified, within short distance from the affected region that are suitable as evacuation areas for the population to seek refuge. For this case study, the required time for the residents to seek refuge equals the time for the tsunami to travel from its source area to the coastline. This suggests that despite the fact that alarm and reaction times to a possible local submarine-landslide-generated tsunami are short, there is marginal time for evacuation and detailed hazard assessment plans should be undertaken in order to minimize the destructive effects of a potential tsunami in the area.

This study aims at giving disaster and emergency planners the hazard assessment tools to formulate a plan for reducing the tsunami risks to coastal residents around the Corinth Gulf, where a lot of candidate areas can suffer in the future by landslide generated tsunamis close to the self edge.

7. Acknowledgements

The authors would like to thank J. Locat and the organizing committee of the 3rd International Symposium on Submarine Mass Movements and Their Consequences, V. Lykousis and D. Sakelariou, for their reviews and comments that improve the quality of this manuscript.

8. References

Ambraseys, N.N., 1962. *Data for the investigation of the seismic seawaves in the eastern Mediterranean.* Bull. Seismol. Soc. Am. 52: 895–913.

Antonopoulos, J., 1980. *Data from investigation on seismic sea-waves events in the eastern Mediterranean from birth of Christ to 1980 AD (6 parts).* Annal. Geofis. 33: 141–248.

Charalampakis, M., Stefatos, A., Hasiotis, T. & Ferentinos, G., 2005. *Morphology, structure and evolution of the Xylocastro basin bounding fault margin, central Gulf of Corinth.* International Symposium on the Geodynamics of Eastern Mediterranean: Active Tectonics of the Aegean Region, 15–18 June 2005, Istanbul, Turkey.

Galanopoulos, A.G., 1960. *Tsunamis observed on the coasts of Greece from antiquity to present time.* Ann. Geofis. 13: 369–386.

Ferentinos, G., Papatheodorou, G. & Collins, M.B., 1988. *Sediment transport processes on an active submarine fault escarpment: Gulf of Corinth, Greece.* Marine Geology 83: 43-61.

Heck, N., 1947. List of seismic sea-waves. Bull. Seismol. Soc. Am. 37: 270.

Imamura, F., 1995. *Review of tsunami simulation with a finite difference method, long wave runup models.* World Scientific, 25-42.

Lykousis, V., Sakellariou, D. & Roussakis, G., 2003. *Prodelta slope stability and associated coastal hazards in tectonically active margins: Gulf of Corinth (NE Mediterranean).* In J. Locat and J. Mienert (eds). Submarine Mass Movements and their Consequences: Dordrecht, Kluwer.Academic Publishers, pp. 433-440.

Papadopoulos, G.A., 2000. *A new tsunami catalogue of the Corinth Rift: 373 B.C.–A.D. 2000.* In: Papadopoulos, G.A. (Ed.), Historical Earthquakes and Tsunamis in the Corinth Rift, Central Greece. Inst. Geodynamics, Natural Observatory, Athens, pp. 121–126.

Papadopoulos, G.A., 2003. *Tsunami hazard in the eastern Mediterranean: strong earthquakes and tsunamis in the Corinth Gulf Central Greece.* Natural Hazards 29: 437–464.

Papadopoulos, G.A. & Chalkis, B., 1984. *Tsunamis observed in Greece and the surrounding area from antiquity up to the present times.* Marine Geology 56: 309–317.

Papatheodorou, G. & Ferentinos, G., 1993. *Sedimentation processes and basin-filling depositional architecture in an active asymmetric graben: Strava graben, Gulf of Corinth,* Greece. Basin Res. 5: 235-253.

Papazachos, B., Koutitas, Ch., Hatzidimitriou, P., Karakostas, B. & Papaioannou, C., 1986. *Tsunami hazard in Greece and the surrounding area.* Ann. Geophys. 4: 79–90.

Papazachos, B. & Papazachou, C., 1997. *Earthquakes in Greece,* Ziti Publications, Thessaloniki.

Pelinovsky, E.N., 1989. *Tsunami climbing a beach and tsunami zonation.* J. Tsunami Soc. 7(2): 118-122.

Pelinofsky, E.N., Golinko, V.I. & Mazova, R.K., 1989. *Tsunami wave runup on a beach; Exact analytical results,* preprint No. 232, Inst. Appl. Phys., U.S.S.R. Academy of Sciences, Gorky (in English).

Pelinovsky, E. & Mazova, R., 1992. *Exact analytical solutions of non linear problems of tsunami wave run-up on slopes with different profiles.* Natural Hazards: 6, 227-249.

Regional Data for Disaster Prevention, Regional Evacuation Edition: 1987, Institute for Fire Safety & Disaster Preparedness, pp. 91–96. (in Japanese).

Soloviev, S.L., 1990. *Tsunamigenic zones in the Mediterranean sea.* Natural Hazards 3: 183–202.

Soloviev, S.L., Solovieva, O.N., Go, C.N., Kim, K.S. & Shchetnikov, N.A., 2000. *Tsunamis in the Mediterranean Sea—2000 B.C.–2000 A.D.* Dordrecht, Kluwer Academic Publishers, pp. 237.

Stefatos, A., Charalambakis, M., Papatheodorou, G., Ghionis, G. & Ferentinos, G., 2005. *Potential tsunami hazard from submarine landslides in the Corinth Gulf, Greece.* 22nd International Tsunami Symposium, Chania, Greece, 27-29 June, 2005.

Stefatos, A., Charalambakis, M., Papatheodorou, G. & Ferentinos, G., 2006. *Tsunamigenic sources in an active European half-graben (Gulf of Corinth, Central Greece).* Marine Geology, 232: 35–47.

Sugimoto, T., Murakami, H., Kozuki, Y. & Nishikawa, K., 2003. *A Human Damage Prediction Method for Tsunami Disasters Incorporating Evacuation Activities.* Natural Hazards, 29: 585-600.

Synolakis, C., 1991. *Tsunami runup on steep slopes: How good linear theory really is.* Natural Hazards, 4: 221-234.

UNESCO-IOC, 2006. *Tsunami glossary.* IOC Information document No. 1221. Paris, UNESCO.

PROBABILISTIC SMF TSUNAMI HAZARD ASSESSMENT FOR THE UPPER EAST COAST OF THE UNITED STATES

S. MARETZKI, S. GRILLI AND C.D.P. BAXTER
Departments of Ocean and Civil and Environmental Engineering, University of Rhode Island, Narragansett, RI, USA

Abstract

The level of tsunami hazard to the east coast of the United States is not well understood. This information is critical for the population, emergency services, and industry of the region. Assessing this hazard is particularly difficult because of the lack of tsunamis in the historical record and the uncertainty regarding the return periods of large-scale events that have been proposed, such as a large transoceanic tsunami possibly caused by a collapse of the Cumbre Vieja volcano in the Canary Islands, or a large co-seismic tsunami initiated in the Puerto Rican subduction zone. The most significant tsunami hazard in this region, however, may be due to local submarine mass failures (SMF), which could cause concentrated damage in coastal communities located near the failures. This paper presents results of a probabilistic analysis that estimates the hazard, expressed in terms of runup (at a given probability of occurrence), of SMF tsunamis triggered by earthquakes, on the upper northeast coast of the United States. A Monte Carlo approach is employed, in which distributions of relevant parameters (seismicity, sediment properties, type and location of slide, volume of slide, water depth, etc.) are used to perform large numbers of stochastic stability analyses of underwater slopes, based on standard geotechnical methods. When slope failure occurs, initial tsunami characteristic height and runup are estimated, based on earlier numerical work, for specified return periods of seismic events. The overall hazard associated with SMF tsunamis along the coast is found to be quite low at most locations as compared, e.g., to the typical 100 year hurricane storm surge in the region (5 m). Two sites, however, located off of Long Island, New York and Atlantic City, New Jersey, show an elevated risk of higher tsunami runup (5.0-7.5 m). These two sites should be the focus of more detailed studies.

Keywords: Submarine mass failures, tsunami triggering; probabilistic analysis; tsunami hazard; runup analysis

1. Introduction

The level of hazard posed by tsunamis to the upper east coast of the United States is not well understood. This information is critical for the populations, emergency services, and industry of the region. Assessing this hazard is particularly difficult because of the lack of tsunami observations made in the historical records of the region and the uncertainty regarding the return periods of large-scale events that have been identified in the literature. Events that could cause a large transoceanic tsunami include a volcano collapse of the Cumbre Vieja volcano in the Canary Islands (Ward and Day, 2001; Elsworth and Day, 1999; Hildenbrand et al., 2003), or a large co-seismic tsunami triggered by a magnitude 9 earthquake, initiated in the Puerto Rican subduction zone (W. Banks, United States Geological Survey, Personal Communication, 2006). A significant tsunami hazard in the east coast region may also be due to submarine mass

V. Lykousis, D. Sakellariou and J. Locat (eds.), Submarine Mass Movements and Their Consequences, 377–385.

failures (SMF) that could be triggered on the continental slope by local earthquakes of moderate magnitude (moment magnitude > 6 or so), and cause concentrated damage to coastal communities located near the failures. Landslide tsunamis can be triggered much closer to shore than co-seismic tsunamis, so despite their much lower energy release, the runup and flooding they induce may result in significant local destruction along small concentrated sections of the nearby coastline. Warning times for SMF tsunamis may only be minutes rather than hours, which may increase the human toll. One such case is the July 1998 tsunami in Papua New Guinea. The tsunami was generated by an underwater slump, triggered by a 7.1 magnitude earthquake, and caused up to 16 m runup on a nearby low lying barrier island, leading to 2,000 fatalities (e.g., Tappin et al., 2001, 2002, 2007).

For large but low probability single transoceanic events, once a tsunami source scenario has been identified and parameterized, based on *ad hoc* seismological/geological hypotheses, direct modeling with a long wave propagation model can be performed. Provided accurate ocean and coastal bottom bathymetry and topography are available, and a fine enough coastal grid is used in the model, estimates of runup distributions and flooding areas can be made. There are numerous examples of this approach in the literature, including for the catastrophic December 26, 2004 Indian Ocean tsunami (Grilli et al., 2007; Ioualalen et al., 2007). The same methodology has also been applied to known landslide tsunami case studies to provide a better understanding of the events (e.g., Watts et al., 2003; Day et al., 2005).

There is considerable geologic evidence of submarine mass failures along the northeast coast of the United States (e.g. Booth et al., 1985; Booth et al., 2002; Piper et al., 2003, and others). Because of this evidence, it may be assumed that potentially tsunamigenic SMFs can be triggered by seismic activity at thousands of sites on the continental shelf and slope, and appear in a variety of sizes and mechanisms. Hence, the first task is to identify and quantify these numerous SMF tsunami sources, and attach a probability to the coastal runup each of these could cause. Considering the many parameters, uncertainties, and sheer size of this investigation, in a first phase, we precluded direct modeling of each potential SMF tsunami event (both slope stability and tsunami generation/propagation/runup) and instead performed a probabilistic analysis aimed at estimating the coastal runup of seismically triggered SMF tsunamis, at given level of probability and return period. To this effect, a Monte Carlo Simulation (MCS) model was developed (e.g., Watts, 2004), in which distributions of relevant parameters affecting SMFs (seismicity, sediment properties, type and location of slide/slump, volume, water depth, etc.) are used to perform a large number of stochastic stability analyses of underwater slopes for many transects drawn across the continental shelf and slope in a selected section of the US East Coast. Slope stability analyses are performed in the MCS model using limit equilibrium methods (Maretzki, 2006). When slope failure is found to occur for a specified return period of seismic events (e.g., 100, 500 years), to which a local ground acceleration is associated, initial tsunami characteristic height and coastal runup are estimated based on earlier numerical work (Grilli and Watts, 2005; Watts et al., 2005).

In this paper, we report on the development and validation of the MCS model, and present results for 100 year and 500 year SMF tsunami runup, at a 95% confidence level.

2. Development and validation of MCS model

There are no major subduction zones in the upper North Atlantic Ocean, however, a 600 km long deep fault can be seen north of Puerto Rico (18 deg. N Lat. In Fig. 1a) in which a potentially large earthquake could occur. Along the upper east coast of the United States is a wide continental shelf that is supplied with sediment at several locations by large rivers (Fig. 1b). In this region, historical earthquakes have occurred mostly on land (Fig. 2), however, in 1929, the largest earthquake on record for Atlantic Canada, with a 7.1 magnitude, triggered a series of large underwater landslides on the slope of the Grand Banks that generated a tsunami that caused 27 fatalities and extensive damage in Newfoundland.

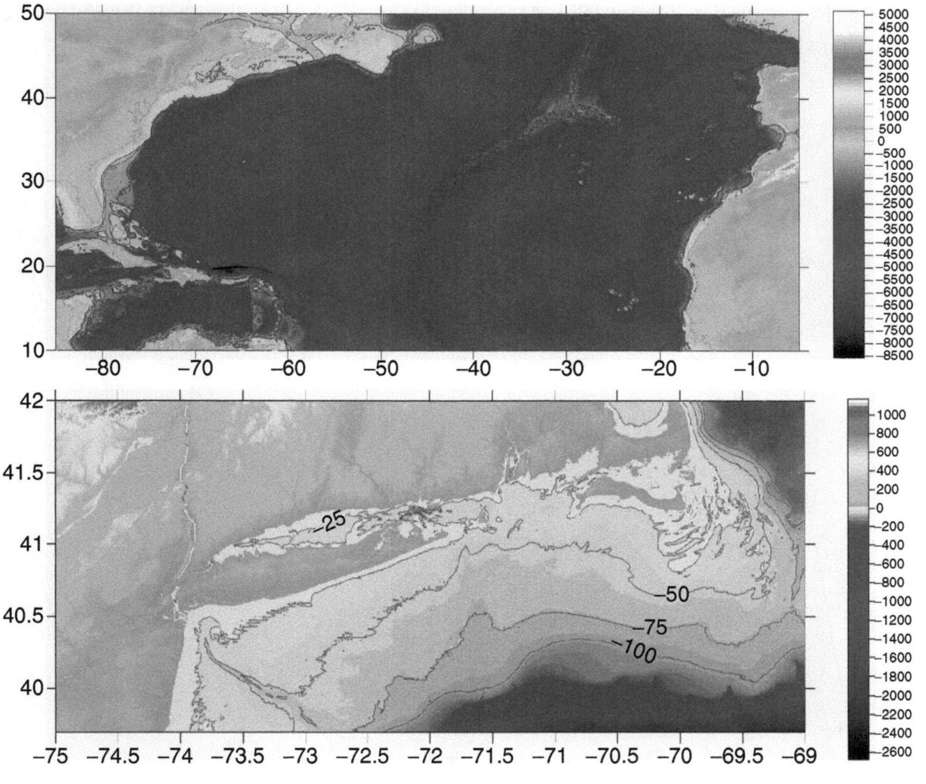

Figure 1: Bathymetry and topography in the North Atlantic Ocean (USGS data).

The MCS model was developed to investigate landslide tsunami risk in the region depicted in Fig. 1b, based on geographical, geological, sedimentary, and seismic data. For the latter, a historical (seismological) analysis of earthquakes (epicenter location, magnitude and extent, return period) was first performed for the region, based on publicly available United States Geological Survey (USGS) hazard maps (Fig. 2). Probability distributions were developed for peak horizontal accelerations in the bedrock as a function of the earthquake return period and the specified location on the continental shelf (Fig. 3). The choice of sediment properties provided the greatest uncertainty in the analysis due to the lack of information with depth. Information about surficial sediment types was obtained from the Continental Margin Mapping (CONMAP) database (Poppe et al., 2005). Based on these classifications, effective stress friction angles (ϕ') ranging from 28° to 44° were used for coarse-grained sediments and undrained shear strength ratios (S_u/σ'_v) ranging from 0.17 to 0.24 were used for fine-grained sediments in the MCS. The increase in bulk density with depth was assumed based on data from an Ocean Drilling Program Leg off the coast of New Jersey. Due to the high level of uncertainty in the types and thicknesses of sediment, no amplification or de-amplification of ground motions was considered. The sediment data, together with bathymetric and topographic data, was then used to create a number of GIS layers in ArcGIS 9 (Fig. 4). The bathymetric data was used to compute a series of transects (45) across the continental shelf and slope (Fig. 5), which, together with seismic (Fig. 3) and sediment (Fig. 4) data, were used as input for geotechnical models predicting slope stability in the MCS model.

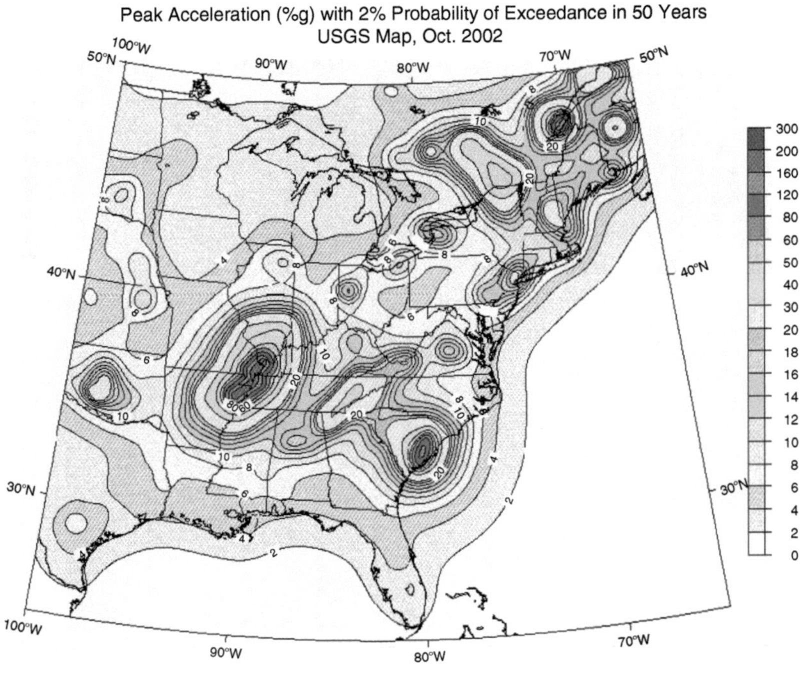

Figure 2: Seismicity of the Eastern continental United States (USGS, 2002).

Different limit equilibrium slope stability models were used depending on the type of sediment found along each transect. Failures in sandy sediments tend to be shallow and translational (i.e., in non-cohesive sediments), while clayey sediments tend to have deep rotational failures (i.e., slumps in cohesive sediments; Silva et al. 2004; Baxter et al. 2003; Soltau 2003). For selected cases the results of the MCS model have been verified using a commercial slope stability software package (Slope/W). (Details can be found in Maretzki (2006).)

In the MCS model, when slope failure is detected along one transect, given randomly selected slide/slump geometry, density and depth, an initial tsunami elevation is calculated, based on semi-empirical equations developed on the basis of numerical simulations and experimentally validated, for slides or slumps (Grilli and Watts, 2005; Watts et al., 2005; Enet and Grilli, 2007). The correspondence principle (which states that, all things considered, maximum runup caused by a SMF tsunami source on a nearby continental slope, is approximately equal to the initial tsunami characteristic amplitude; Watts et al., 2005) is then used to estimate runup, based on the maximum initial tsunami depression. Maximum runup for this particular event is assumed to occur in a primary direction of tsunami propagation, selected based on transect orientation (to which some randomness is associated), and runup is then modulated along-coast, assuming a normal variation and some angular spreading from the location of maximum runup. A large number of MCS simulations were performed for each selected return period and transect, and 800 points were used to calculate the runup probability distributions along the coast, based on MCS prediction for the triggered SMF events.

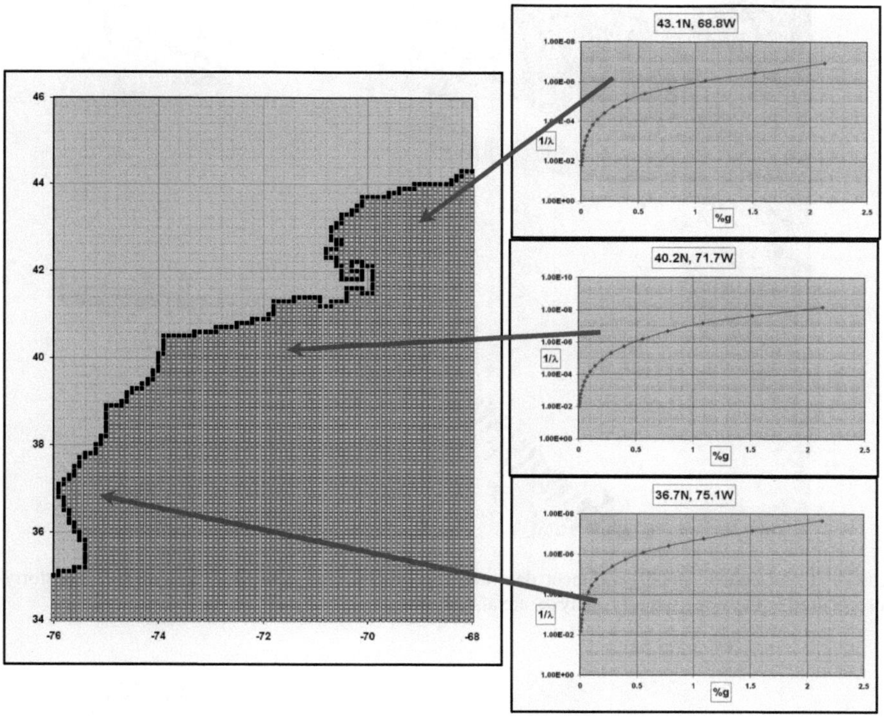

Figure 3: Typical annual probability of exceedance (1 x 10^{-2} to 1 x 10^{-10}) for peak horizontal acceleration (0 to 2.5g) at various locations in the study area. This data is developed from the USGS hazard map shown in Fig. 2.

Numerous validations of the MCS model output were performed to verify that statistical results satisfied expected probability distributions for both governing and output parameters; these were performed up to very long return periods, such as 10,000 years, using at least 10,000 MCS runs for each year and transect. In particular, the dimensions and volumes of estimated failures matched approximately distributions presented by Booth et al. (2002). All of these validations successfully confirmed the relevance of our MCS results. (Details can be found in Maretzki (2006).) We then concentrated on producing both detailed and accurate (i.e. converged) results for two return periods of interest : 100 and 500 years, for which we used 50,000 MCS runs for each transect. Along-coast runup results were then statistically analyzed.

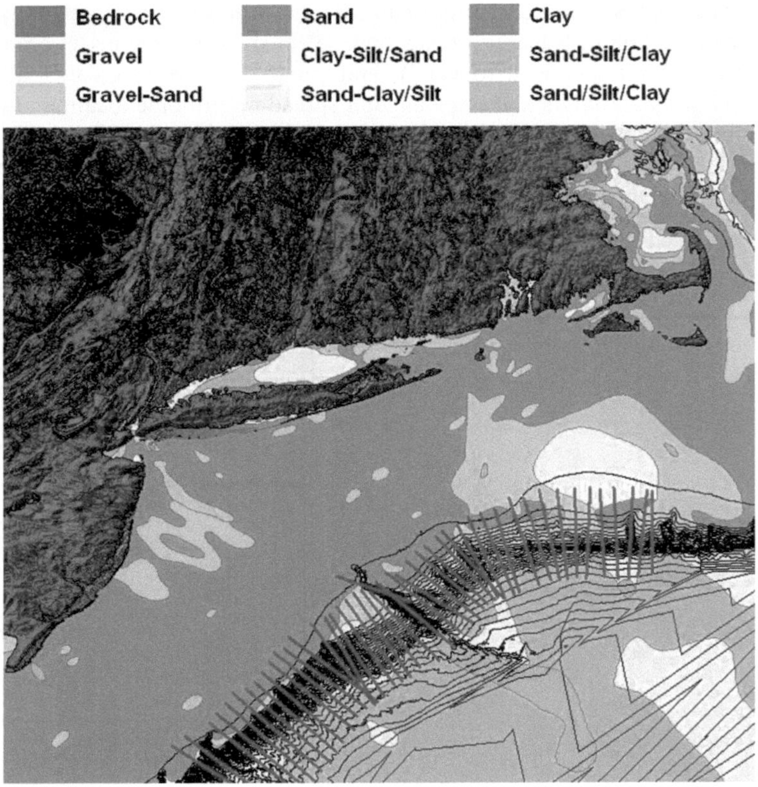

Figure 4: Color-coded ArcGIS sediment data and topography, and contoured bathymetry. The forty-five transects used in the slope stability analyses are also shown.

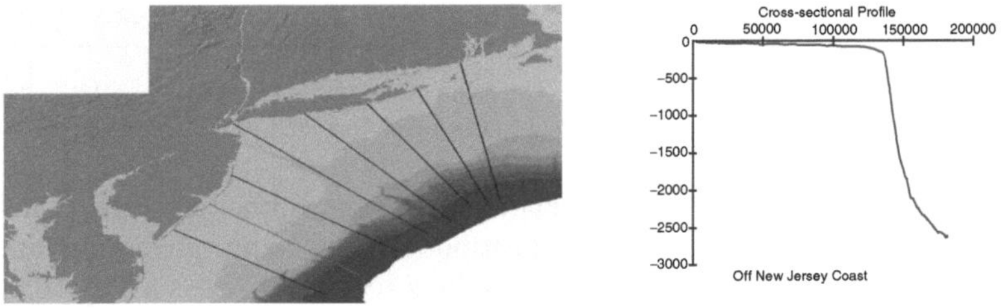

Figure 5: ArcGIS color coded bathymetry and one example of transect.

3. Results of MCS model

Coastal runups predicted for a given seismic return period were found to satisfy a log-normal probability distribution. Based on these predictions, confidence limits (e.g., upper bound of 95% or 97.5% confidence intervals) could be calculated using standard Gauss (or Student t for less than 30 individual runup results) statistical distributions. Figure 6 shows results of MCSs, for 100 and 500 year return periods, of SMF tsunami runup, at 95%, 97.5% levels, along the selected section of US coastline. In Figure 6a, which shows results for the standard 95% level used in probabilistic engineering design, the overall hazard associated with SMF tsunamis along the coast is found to be quite low at most locations, as compared to the typical 100 year hurricane storm surge in the region (5 m). Two sites, however, located near Long Island, New York (coastal points no. 500-520) and Atlantic City, New Jersey (coastal points no. 660-760) show an elevated risk of higher tsunami runup in the range 5-7.5 m. Figure 6b further shows that these values would range between 5 and 11.8 m, at a 97.5% confidence level, a level that could be selected, e.g., for designing more important structures (e.g., for energy production or homeland security).

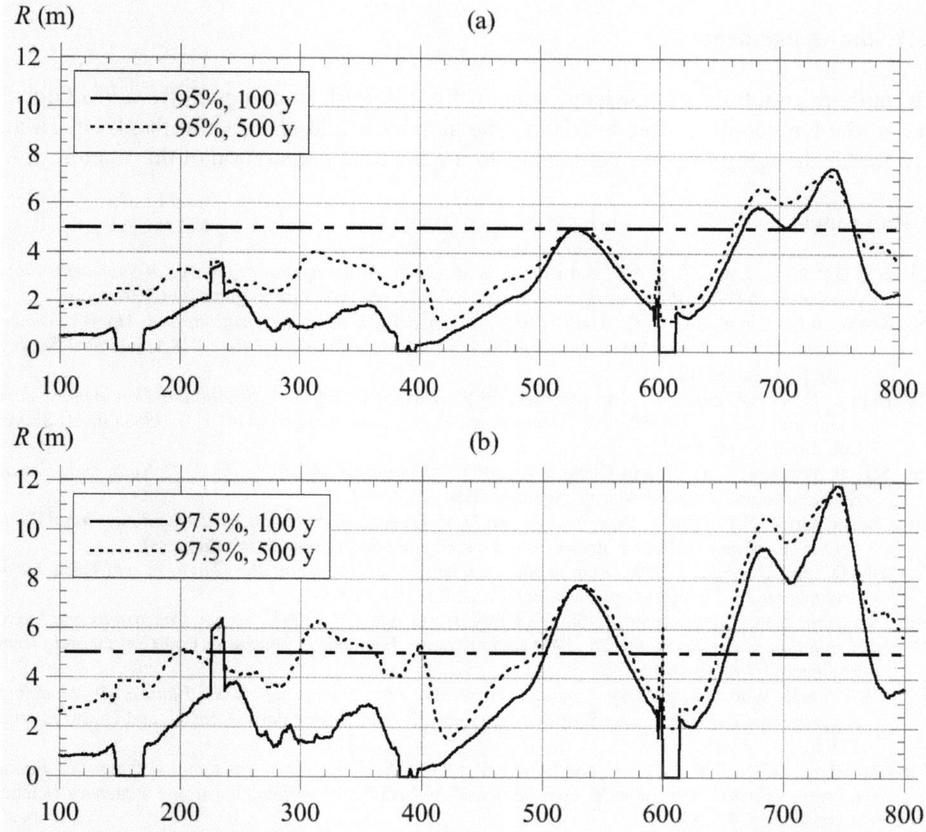

Figure 6: MCS Runup at a: (a) 95%; (b) 97.5%, confidence levels (upper bounds), caused by SMF tsunamis, triggered by 100 or 500 year seismic events (x-axis is index of studied coastal points, numbered E-W, as detailed in Maretzki, 2006).

4. Conclusions

We developed a MCS model and applied it to assess SMF tsunami risk, in terms of runup, along the upper US East Coast. Considering the many simplifications and assumptions inherent to this analysis, it is understood that it is only aimed at identifying coastal locations with a higher probability for large runup, for which more detailed tsunami modeling studies should be performed in order to better assess tsunami hazard.

We thus identified two sections of the US East coast with increased risk of tsunami runup, larger than the typical 100 year storm surge (5 m), at 95% and 97.5% confidence levels. These areas will be the focus, in future work, of more detailed studies, including direct modeling of SMF triggering, tsunami generation, and propagation/runup, over detailed coastal bathymetry and topography. Of critical importance will be reducing the uncertainty in the assumed geotechnical properties and stratigraphy.

Flooding area estimates could be inferred from runup distributions predicted along the coast, as a function of the coastal topography, but these will also be obtained in the more accurate direct simulations that will be performed in future work.

5. Acknowledgements

The authors gratefully acknowledge support for this work from a grant from FM Global, Inc. to the University of Rhode Island. The authors would also like to thank Dr. Homa Lee for his thoughtful review comments during the final preparation of this manuscript.

6. References

Baxter, C.D.P, King, J.W., Silva, A.J., and Bryant, W.R. (2003). *Investigation of the Age of Submarine Slope Failures in the Gulf of Mexico.* Proc. Intl. Soc. of Offshore and Polar Engng. Conf., Honolulu.

J.S. Booth, R.C. Circe, and A.G. Dahl. (1985). *Geotechnical Characterization and Mass-Movement Potential of the United States North Atlantic Continental Slope and Rise*, US Geological Survey, Tech. Rep., pp. 85-123.

J.S. Booth, D.W. O'Leary, P. Popenoe, and W.W. Danforth (2002). *U.S. Atlantic Continental Slope Landslides: Their Distribution, General Attributes, and Implications*, U.S. Geological Survey Bulletin 2002, pp. 14–22.

Day, S.J., P. Watts, S. T. Grilli and Kirby, J.T. (2005). *Mechanical Models of the 1975 Kalapana, Hawaii Earthquake and Tsunami.* Marine Geology, 215(1-2), 59-92.

Enet, F. and Grilli, S.T. (2007). *Experimental Study of Tsunami Generation by Three-dimensional Rigid Underwater Landslides*. J. Waterway Port Coastal and Ocean Engineering (in press).

Elsworth, D., and Day, S.J. (1999). *Flank collapse triggered by intrusion: the Canarian and Cape Verde Archipelagoes.* J. Volcanology and Geothermal Res., 94, 323-340.

Grilli, S.T., Ioualalen, M, Asavanant, J., Shi, F., Kirby, J. and Watts, P. (2007.) *Source Constraints and Model Simulation of the December 26, 2004 Indian Ocean Tsunami.* J. Waterway Port Coastal and Ocean Engineering (in press).

Grilli, S.T., and Watts, P. (2005). *Tsunami Generation by Submarine Mass Failure. I: Modeling, Experimental Validation, and Sensitivity Analyses.* J. Waterways, Ports, Coastal, and Ocean Engng., 131(6), pp. 283-297.

Hildenbrand, A., Gillot, P.Y , Soler, V. and Lahitte, P. (2003). *Evidence for a persistent uplifting of La Palma (Canary Island), inferred from morphological and radiometric data.* Earth and Planetary Science Letters, 210, 277-289.

Ioualalen, M. , Asavanant, J., Kaewbanjak, N., Grilli, S.T., Kirby, J.T. and Watts, P. (2007). *Modeling the 26th December 2004 Indian Ocean tsunami: Case study of impact in Thailand.* J. Geophys. Res. (in press).

Maretzki, S. (2006). *Numerical Simulation of Tsunami Hazard Maps for the US East Coast.* Masters Thesis. Department of Ocean Engineering, University of Rhode Island, 187 pp.

Piper, D.J.W., Mosher, D.C., Gauley, B.-J., Jenner, K., and Campbell, D.C. (2003). *The Chronology and Recurrence of Submarine Mass Movements on the Continental Slope off Southeastern Canada,* Submarine Mass Movements and their Consequences, Kluwer Academic Publishers, pp. 299-306.

Poppe, L.J., Williams, S.J., and Paskevich, V.F. (2005). *U.S. Geological Survey East-Coast Sediment Analysis: Procedures, Database, and GIS Data,* US Geological Survey, Tech. Rep. 2005-1001, 2005.

Silva, A.J., Baxter, C.D.P., LaRosa, P.T., and Bryant, W.R. (2004). *Investigation of Mass Wasting on the Continental Slope and Rise.* Marine Geology, 203 (3-4), 355-366.

Soltau, B. (2003). *Evaluation of Paleoseismicity of Eastern North America from Subaqueous Slope Failures,* Masters Thesis, Department of Civil and Environmental Engineering, University of Rhode Island, 152 pp.

Tappin, D.R., Watts, P., McMurtry, G.M., Lafoy, Y., and Matsumoto, T. (2001). *The Sissano, Papua New Guinea tsunami of July 1998 – Offshore evidence on the source mechanism.* Marine Geology, 175, 1-23.

Tappin, D.R., Watts, P., McMurtry, G.M., Lafoy, Y., and Matsumoto, T. (2002). *Prediction of slump generated tsunamis: The July 17th 1998 Papua New Guinea event.* Sci. Tsunami Hazards, 20(4), 222-238.

Tappin, D.R., Watts, P., and Grilli, S.T. (2007). *The Papua New Guinea tsunami of 1998: anatomy of a catastrophic event.* Natural Hazards and Earth System Sciences. (submitted).

United States Geological Survey (2002). *Conterminous States Probabilistic Maps & Data,* http://earthquake.usgs.gov/research/hazmaps/products_data/48_States/index.php (April 19, 2007).

Ward, S.N. and Day, S. (2001). *Cumbre Vieja Volcano—potential collapse and tsunami at La Palma, Canary Islands.* Geophys. Res. Lett., 28, 397-400.

Watts, P., Grilli, S.T., Kirby, J.T., Fryer, G.J. and D.R. Tappin, 2003. *Landslide tsunami case studies using a Boussinesq model and a fully nonlinear tsunami generation model.* Natural Hazards and Earth System Sciences, 3, 391-402.

Watts, P. (2004). *Probabilistic Predictions of Landslide Tsunamis off Southern California.* Marine Geology, 203, 281-301.

Watts, P., Grilli, S.T., Tappin, D.R., and Fryer, G.J. (2005). *Tsunami Generation by Submarine Mass Failure. II: Predictive Equations and Case Studies.* J. Waterways, Ports, Coastal, and Ocean Engng., 131(6), pp 298-310.

ROLE OF SOIL BEHAVIOR ON THE INITIAL KINEMATICS OF TSUNAMIGENIC SLIDES

A.S. BRADSHAW
Dept. of Civil Engineering, Merrimack College, North Andover, MA, 01845 USA

C.D.P. BAXTER, O.-D.S. TAYLOR AND S. GRILLI
Depts. of Ocean and Civil and Environmental Engineering, University of Rhode Island, Narragansett, RI, 02882 USA

Abstract

Recent investigations on tsunami generation from submarine mass failures show that one of the most important factors influencing the source characteristics of the wave is the initial acceleration of the failure itself. In a number of these studies, a translational slide is modeled as a rigid body sliding down an inclined plane and basal resistance is neglected. In this paper, a similar rigid body model is proposed that incorporates basal resistance, which is related to the shear strength of the soil. Initial slide kinematics were investigated under two triggering mechanisms including overpressures at depth and rapid sedimentation. The model results show that soil behavior significantly influences the acceleration time history as well as the magnitude of the peak acceleration. The slide kinematics depend largely on the initial stress state and on the undrained residual shear strength of the soil along a potential failure surface, which highlights the importance of performing detailed geotechnical site investigations when assessing these geohazards. More research is needed to determine the influence of using more realistic basal friction models on the initial wave heights generated by submarine mass failures.

Keywords: Tsunamis, landslides, clay, overpressures, rapid sedimentation

1. Introduction

Tsunamigenic mass failures are of major concern for proper risk analysis related to the development of offshore and coastal structures, seafloor resources and for the protection of coastal communities (Locat et al. 2001). There are a variety of types of submarine failures (Locat and Lee 2000); however many tsunamigenic slope failures can be classified into two basic categories (e.g. Grilli and Watts 2005): (1) *slides*, which are defined as thin translational failures with long runout distances, and (2) *slumps*, which are thick rotational failures. The focus of this paper is on submarine *slides* in normally consolidated clay.

In order to investigate the generation of tsunamis from submarine mass failures and its sensitivity to governing parameters, Watts and Grilli (2003) and Grilli and Watts (2005) modeled a translational slide as a semi-elliptical or Gaussian shaped rigid body sliding down an inclined plane. A semi-elliptical body was shown to produce the largest (i.e. worst-case) initial tsunami. In their analyses, the basal resistance was assumed to be negligible as compared to hydrodynamic resistance, thereby limiting the number of parameters in the model. However, since the parameter of greatest influence on tsunami generation was shown to be the initial acceleration of the center of the failed mass

V. Lykousis, D. Sakellariou and J. Locat (eds.), Submarine Mass Movements and Their Consequences, 387–394.

(Haugen et al. 2005, Watts et al. 2005), removing the soil behavior may result in an acceleration time history that is not representative of actual slide motion.

This paper presents an analysis of the effect of basal friction on the initial acceleration of a submarine slide. A modified solid body model was developed, which includes the resistance to sliding due to the shear strength of the soil. First, a slide model developed by Grilli and Watts (2005) is described along with the governing equations of motion. The incorporation of a new basal resistance function into this model is described, and the impact of soil behavior on the slide kinematics is thus investigated.

2. Slide Model

The equation of center of mass motion for a 2-D semi-elliptical body moving down an inclined plane, as shown in Figure 1, is given by the following expression (Grilli and Watts 2005):

$$(\gamma + C_m)\ddot{s} = (\gamma - 1)(\sin\theta - C_n \cos\theta)g - C_d \frac{2}{\pi \cdot B}\dot{s}^2 \qquad [1]$$

where γ = ratio of the bulk density of the soil composing the slide to the density of water, g = gravitational acceleration, B = slide length, C_m = added mass coefficient, C_n = Coulomb friction coefficient, C_d = hydrodynamic drag coefficient, $\ddot{s}$ = slide acceleration, and $\dot{s}$ = slide velocity (the upper dots denote time derivatives).

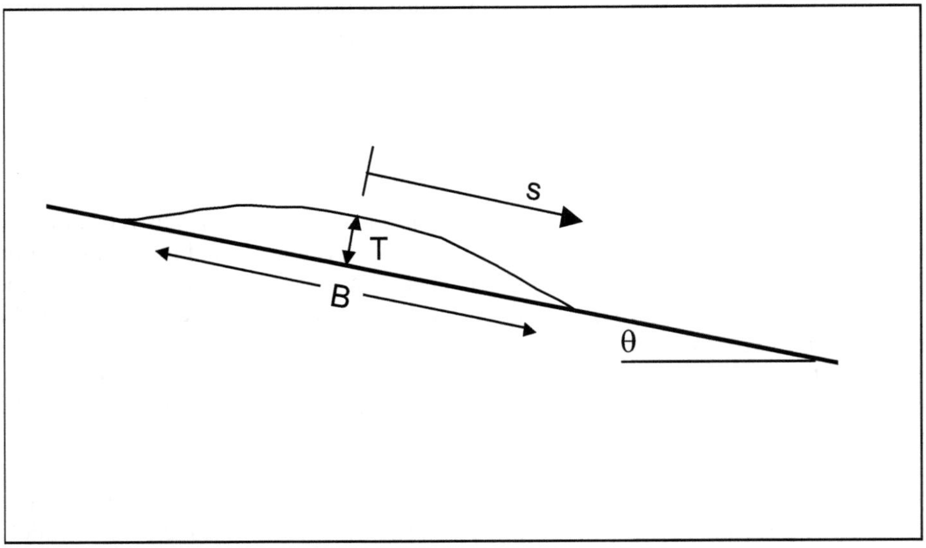

Figure 1. Rigid semi-elliptical body used in the underwater landslide model.

For translational slides, Grilli and Watts (2005) and Watts et al. (2005) assumed that C_n was approximately zero thus eliminating the basal resistance term from Equation 1. To account for more realistic soil behavior, a revised equation of motion is proposed:

$$(\gamma + C_m)\ddot{s} = (\gamma - 1)g \sin\theta - \frac{S(s,B)}{\rho_w \frac{\pi}{4} B \cdot T} - C_d \frac{2}{\pi \cdot B} \dot{s}^2 \quad [2]$$

where $S(s,B)$ = basal resistance function that depends on slide displacement (s) and slide length (B), ρ_w = density of water, and T = slide thickness.

3. Basal Resistance Function

A slope failure can occur when the applied shear stress exceeds the shear strength of the soil (e.g. rapid sedimentation), the shear strength is reduced (e.g. overpressures), or a combination of the two (e.g. earthquake loading). For this analysis, triggering due to both overpressures at depth and rapid sedimentation were considered. In order to develop a reasonable basal resistance function, it is important to first understand the stress paths that occur in the soil before, during, and after landslide triggering. Figure 2 illustrates the stress paths for both cases, where the shear stress (τ) on a potential failure surface is plotted against the effective normal stress (σ').

In the case of overpressures at depth (Figure 2a), an increase in pore pressure causes a decrease in effective stress with no change in the driving shear stress. From an initial stress state, the stress path moves horizontally to the left in the diagram, during which time the soil swells slightly from the decrease in effective stress. Some deformation occurs at this point but triggering is not yet initiated. Eventually, the stress path reaches the failure envelope (defined as $\sigma' \tan \phi_p'$) where the applied shear stress (τ_f) is equal to the shear strength of the soil. Any further reduction in effective stress initiates landslide motion, and the soil is sheared under undrained conditions. With continued shear displacement, the strength eventually reaches the residual undrained shear strength (S_{ur}) which is located on the residual strength envelope (defined by $\sigma' \tan \phi_r'$). Since S_{ur} is a steady-state strength, it remains constant at very large strains. It is important to note that landslide motion will only occur if the soil exhibits strain softening behavior (i.e. $S_{ur} < \tau_f$).

In the case of rapid sedimentation (Figure 2b), the thickness of the overburden soil increases thereby increasing the driving shear stresses within the slope. In low permeability soil layers (i.e. clays) the sedimentation may be so rapid that the soils do not have sufficient time to fully consolidate. Therefore, the effective stress will only increase slightly as the shear stress increases, causing the stress path to move up and to the right towards the failure envelope in the diagram. Similar to the first case, once the shear stress exceeds the strength of the soil, the motion of the slide is initiated and the clay is sheared undrained to a residual condition.

For purposes of modeling initial slide kinematics, only the conditions after triggering were considered. For the semi-elliptical body shown in Figure 1, incipient motion occurs when the driving shear stress equals the peak effective shear strength of the clay, given by the following expression:

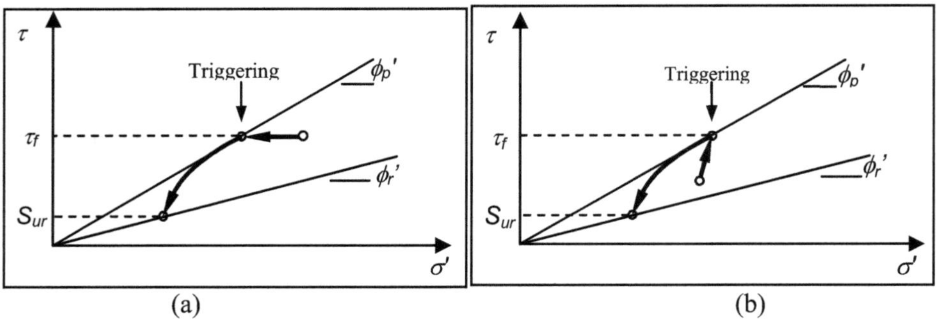

Figure 2. Stress paths showing the initiation of failure and reduction of shear strength during landslide triggering due to (a) overpressures at depth and (b) rapid sedimentation.

$$\tau_f = \frac{\pi}{4} T(\gamma - 1)\rho_w g \sin\theta = \sigma' \tan\phi_p' \qquad [3]$$

where θ = slope angle. Results of undrained ring shear tests performed on undisturbed samples of normally consolidated Drammen clay (Stark and Contreras 1996) were used as the basis for modeling the shear strength along the base of the slide post-triggering. These test results, shown in Figure 3, plot shear stress ratio versus shear displacement, where the stress ratio is defined as the applied shear stress normalized by the initial effective normal stress ($'_0$). Note that in this figure, the shear displacement is relative to the point at which the peak shear stress is mobilized. The test results show that the stress ratio decreases logarithmically with shear displacement from peak to residual where it remains constant. From these data, the undrained residual strength ratio (S_{ur}/σ'_0) is approximately 0.11 and is mobilized at a displacement of about 20 mm.

Large runout distances observed during past landslides also suggest that underwater landslides eventually reach a hydroplaning condition (e.g. Issler et al. 2003). Therefore, it was assumed that the basal resistance goes to zero for any portion of the slide that overrides the soils down slope of the initial slide location. Based on the results in Figure 3, the shear strength function was defined by the following set of equations

$$S(s,B) = 0.95\tau_f \cdot (B - s)\,;\, s < 0.001 \text{ m} \qquad [4]$$

$$S(s,B) = \left[S_{ur} + \frac{(0.95\tau_f - S_{ur})}{\log\left(\dfrac{s_r}{0.001}\right)} \cdot \log\left(\frac{s_r}{s}\right) \right] \cdot (B - s)\,;\, 0.001 \text{ m} \le s \le s_r \qquad [5]$$

$$S(s,B) = S_{ur} \cdot (B - s)\,;\; s > s_r \qquad [6]$$

where S_{ur} = undrained residual shear strength (in Pa), and s_r = displacement (in m) at which S_{ur} is fully mobilized. The soil model is shown along with the laboratory test data in Figure 3.

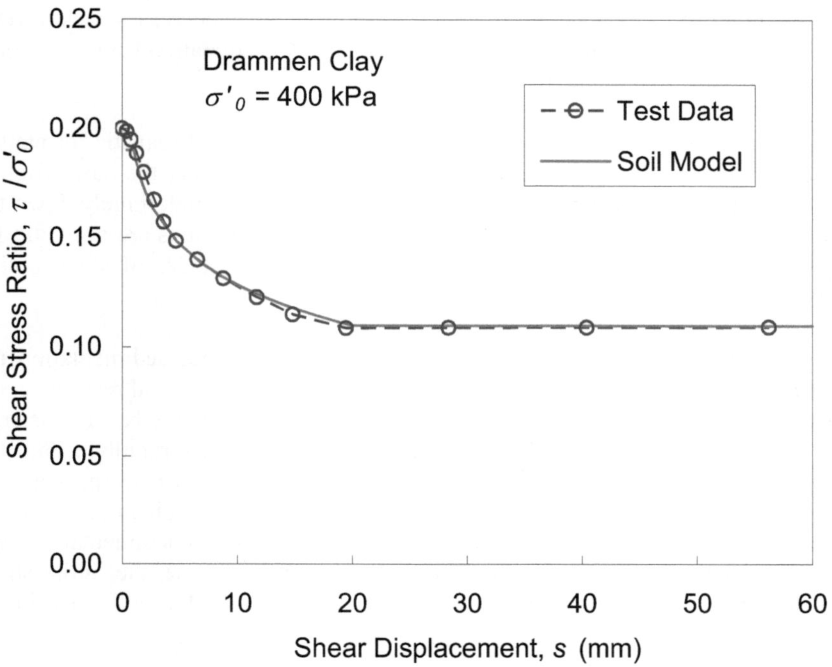

Figure 3. Ring shear test results obtained from an undisturbed sample of normally consolidated Drammen clay (after Stark and Contreras 1996). The soil model used in the slide modeling is also shown.

4. Initial Slide Kinematics

The equation of motion (Equation 2) was solved numerically using a finite difference approach with the following slide parameters: $T = 60$ m, $B = 4$ km, $\theta = 10°$, and $\gamma = 1.8$. These parameters were chosen to illustrate the impact of basal resistance, and represent a relatively short slide on a steep slope (Canals et al. 2004). The coefficients C_m and C_d were taken to be unity (Grilli and Watts 2005). The numerical results were first validated by comparing them to the analytical solutions of Grilli and Watts (2005) for a translational slide having zero basal resistance. Subsequent model runs were then performed for the two cases described above assuming $\phi_p' = 25°$, $S_{ur}/\sigma'_0 = 0.11$, and $s_r = 0.02$ m. The S_{ur} was estimated for each case from the strength ratio and the initial stress state.

Figure 4 plots the time history of slide velocity and acceleration for the case of overpressures at depth. Two plots are shown; one assuming zero basal resistance (i.e. $S(s,B) = 0$) and the other assuming $S(s,B)$ calculated using Equations 4 through 6. When basal resistance is neglected, the maximum acceleration (0.47 m/s^2) occurs at $t = 0$ and gradually decreases to zero as the slide reaches a terminal velocity of 90 m/s. When soil shear strength is included, the slide eventually reaches the same terminal velocity only the acceleration time history is very different. The acceleration increases from zero at

$t = 0$ to a maximum value of 0.35 m/s^2 at about $t = 200$ s and then decreases back to zero. Soil behavior in this case had the effect of shifting the peak acceleration later in the time history and decreasing magnitude by about 25%.

The high values of terminal velocity predicted by the model are not realistic for natural slides. They are the result of assumptions in the model, namely that the basal friction becomes zero at large displacements, the slide is allowed to travel indefinitely down the slope, and the slide remains a rigid body. Although these assumptions are not valid for very large slide displacements, they are realistic for the early stages of slide motion during which tsunami generation occurs.

The velocity and acceleration time histories for the case of rapid sedimentation are shown in Figure 5. The first curve shown in the figure assumes zero basal resistance and is identical to the curve in Figure 4, and the second curve includes the basal resistance function. When soil strength is included, the acceleration increases rapidly within the first second to a peak value of about 0.36 m/s^2. Similar to the case of overpressures at depth, the peak acceleration is about 25% less than the peak value obtained assuming zero basal resistance. However, the peak acceleration for the rapid sedimentation case occurs much earlier in the time history. This is because in this case, the initial stress state was lower which resulted in a lower estimated S_{ur} and thus a higher initial acceleration.

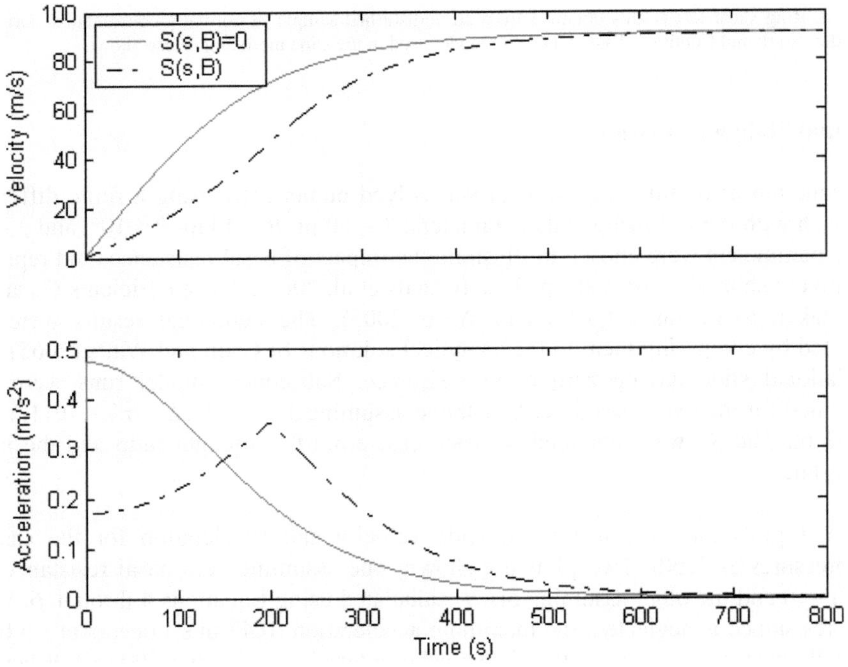

Figure 4. Time history of slide velocity and acceleration for the case of overpressures at depth (B=4 km, T=60 m, and $\theta = 10°$).

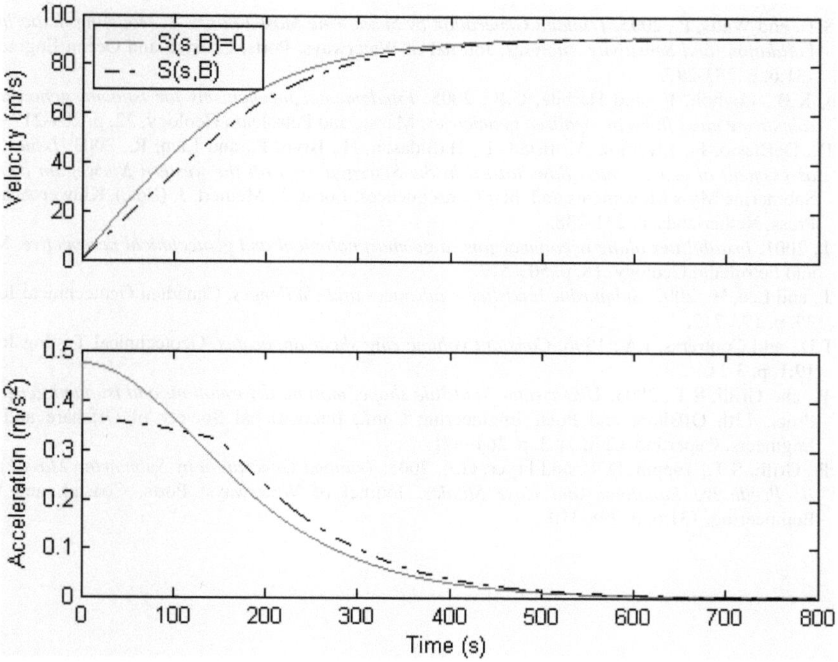

Figure 5. Time history of slide velocity and acceleration for the case of rapid sedimentation (B=4 km, T=60 m, and $\theta = 10°$).

5. Conclusions

A modified solid body slide model was proposed that includes basal resistance that is a function of the undrained residual strength of the soil. The equation of motion was solved to obtain the time history of velocity and acceleration for a relatively short translational slide on a steep slope. The model results indicate that the soil behavior had a significant influence on the acceleration time history of the slide. The effect of the modeled time histories on tsunami generation was not investigated in this paper. However, the model results do suggest the importance of characterizing both the initial stress state of the soils as well as their shear strength properties. Furthermore, identifying these properties in situ will ultimately allow for better prediction of actual landslide kinematics.

6. Acknowledgements

The authors would like to thank Dr. Jacques Locat for his thoughtful review comments during the final preparation of this manuscript.

7. References

Canals, M., Lastras, G., Urgeles, R., Casamor, J.L., Mienert, J., Cattaneo, A., De Batist, M., Haflidason, H., Imbo, Y., Laberg, J.S., Locat, J., Long, D., Longva, O., Masson, D.G., Sultan, N., and Bryn, P., 2004. *Slope failure dynamics and impacts from seafloor and shallow sub-seafloor geophysical data: case studies from the COSTA project.* Marine Geology, 213, p. 9-72.

Grilli, S.T., and Watts, P., 2005. *Tsunami Generation by Submarine Mass Failure. I: Modeling, Experimental Validation, and Sensitivity Analyses*. Journal of Waterways, Ports, Coastal, and Ocean Engineering. 131:6, p. 283-297.

Haugen, K.B., Løvholt, F., and Harbitz, C.B., 2005. *Fundamental mechanisms for tsunami generation by submarine mass flows in idealized geometries.* Marine and Petroleum Geology, 22, p. 209-217.

Issler, D., DeBlasio, F., Elverhoi, A., Ilstad, T., Haflidason, H., Bryn, P., and Lien, R., 2003. *Issues in the assessment of gravity mass flow hazard in the Storegga area off the western Norwegian coast.* In: Submarine Mass Movements and their Consequences. Locat, J., Meinert, J. (Eds.), Kluwer Academic Press, Netherlands, p. 231-238.

Locat, J., 2001. *Instabilities along ocean margins: a geomorphological and geotechnical perspective.* Marine and Petroleum Geology, 18, p. 503-512.

Locat, J., and Lee, H., 2002. *Submarine landslides: advances and challenges*. Canadian Geotechnical Journal, 39, p. 193-212.

Stark, T.D., and Contreras, I.A., 1996. *Constant volume ring shear apparatus.* Geotechnical Testing Journal, 19:1, p. 3-11.

Watts, P., and Grilli, S.T., 2003. *Underwater landslide shape, motion, deformation, and tsunami generation.* Proc., 13th Offshore and Polar Engineering Conf., International Society of Offshare and Polar Engineers, Cupertino, Calif., v.3, p. 364–371.

Watts, P., Grilli, S.T., Tappin, D.R., and Fryer, G.J., 2005. *Tsunami Generation by Submarine Mass Failure. II: Predictive Equations and Case Studies.* Journal of Waterways, Ports, Coastal, and Ocean Engineering. 131:6, p. 298-310.

REVISITING SUBMARINE MASS MOVEMENTS ALONG THE U.S. ATLANTIC CONTINENTAL MARGIN: IMPLICATIONS FOR TSUNAMI HAZARDS

J.D. CHAYTOR
Dept. of Geology and Geophysics, Woods Hole Oceanographic Institution, Woods Hole, MA, 02543 USA

D.C. TWICHELL, U.S. TEN BRINK, B.J. BUCZKOWSKI AND B.D. ANDREWS
U.S. Geological Survey, 384 Woods Hole Road, Woods Hole, MA 02543 USA

Abstract

Interest in the generation of tsunamis by submarine mass movements has warranted a reassessment of their distribution and the nature of submarine landslides offshore of the eastern U.S. The recent acquisition and analysis of multibeam bathymetric data over most of this continental slope and rise provides clearer view into the extent and style of mass movements on this margin. Debris flows appear to be the dominant type of mass movement, although some translational slides have also been identified. Areas affected by mass movements range in size from less than 9 km^2 to greater than 15,200 km^2 and reach measured thicknesses of up to 70 m. Failures are seen to originate on either the open-slope or in submarine canyons. Slope-sourced failures are larger than canyon-sourced failures, suggesting they have a higher potential for tsunami generation although the volume of material displaced during individual failure events still needs to be refined. The slope-sourced failures are most common offshore of the northern, glaciated part of the coast, but others are found downslope of shelf-edge deltas and near salt diapirs, suggesting that several geological conditions control their distribution.

Keywords: bathymetry, seismic reflection, side scan, sediments, Quaternary

1. Introduction and Background

Since the 1929 earthquake near the Grand Banks offshore of Nova Scotia, Canada (Heezen and Ewing, 1952), it has been realized that submarine landslides contribute to shaping passive continental margins. A full appreciation of the importance of this process was delayed several decades until bathymetric, sub-bottom profiling, and seafloor imaging techniques were sufficiently advanced to allow systematic mapping of large portions of continental margins. The importance of submarine mass movements was confirmed after completion of the systematic mapping of this margin by GLORIA in 1987. We have used available multibeam bathymetry, GLORIA sidescan sonar imagery, a regional grid of high-resolution seismic profiles, and published accounts of sediment cores to map the distribution and style of surficial landslides along the U.S. Atlantic margin between the eastern end of Georges Bank and the Blake Spur (Fig. 1). The near-complete coverage of the U.S. Atlantic continental slope and rise by multibeam bathymetry provides a uniform dataset and a more detailed and consistent view of the geomorphology of submarine landslides than had been available in the past. Here we review the distribution and style of submarine landslides along the continental margin, and speculate on the geologic conditions that have influenced their distribution.

V. Lykousis, D. Sakellariou and J. Locat (eds.), Submarine Mass Movements and Their Consequences, 395–403.

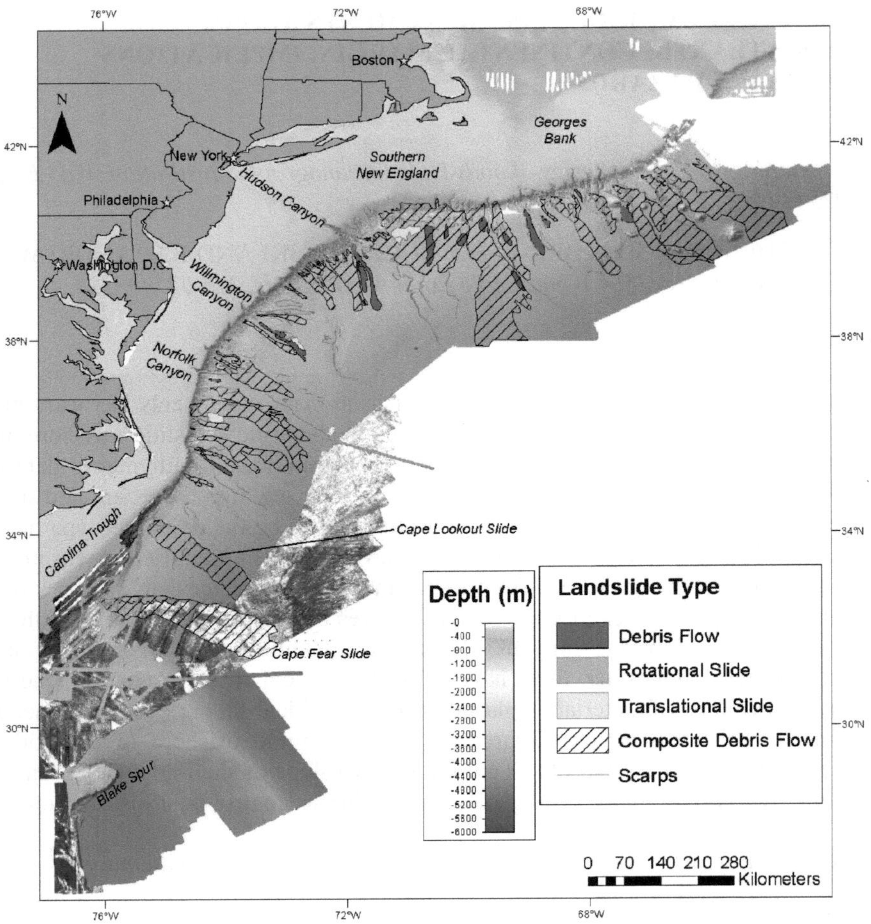

Figure 1. Distribution of different landslide types along the US Atlantic continental slope and rise between the eastern end of Georges Bank and the Blake Spur, overlain on the regional bathymetry compilation. GLORIA sidescan data (black and white) can be seen over areas where there is sparse or no multibeam bathymetry data available.

Studies of individual landslides along the U.S. Atlantic margin first appeared in 1960s (e.g., Uchupi, 1967) and continued through the 1980s (see Embley and Jacobi (1986) and Booth (1993) and numerous references therein). The first regional synthesis of landslide distribution was compiled by Embley (1980) using echo-sounder profiles and cores, with subsequent regional compilations benefiting from the collection of additional local (e.g., O'Leary, 1996; Popenoe and Dillon, 1996) and regional (Booth *et al.*, 1988; Pratson and Laine, 1989) data. Booth *et al.* (1993) provided a summary of landslide distribution and attributes and the first tabulated information on the dimensions of these features, characteristics of the source areas, and style of failure.

The U.S. Atlantic continental margin rifted asynchronously from south to north during the Mesozoic (Klitgord *et al.*, 1988). Salt deposition probably was extensive during early stages of continental margin formation, but only in the Carolina Trough offshore of North and South Carolina did salt domes form (Dillon *et al.*, 1982; Fig. 1). During the early Middle Jurassic a nearly continuous carbonate platform and barrier reef system formed that stretched northward from the Bahamas to the Canadian margin (Poag, 1991). Deposition during the Cenozoic was primarily siliciclastic sediments (Poag and Sevon, 1989) except during the Eocene when calcareous chalk was deposited along much of the margin (Weed *et al.*, 1974). The Quaternary saw large volumes of sediment eroded from the North American continent by glacial processes, with deposition by large river systems along the Georges Bank shelf edge (Schlee and Fritsch, 1982) and along the southern New England shelf. Beyond the extent of glaciers, the large river systems that underlie the present Hudson, Delaware, and Chesapeake estuaries extended across the shelf with shelf-edge deltas built off the Virginia and Delaware coasts, while the Hudson Canyon system transferred sediment to a deep-sea fan (Poag and Sevon, 1989).

2. Methods

2.1 BATHYMETRY

Data used in the compilation of the U.S. Atlantic margin bathymetry map were acquired from several sources and vary in age, sounding density, and positional accuracy. The primary data set was acquired by the University of New Hampshire (UNH) in support of the U.S. Law of the Sea Study (Gardner et al., 2006) and provides near continuous coverage of the U.S. Atlantic margin from the base of the continental slope down to the abyssal plain (~1500-m and 5000-m). Several additional multibeam data sets collected by ships from Woods Hole Oceanographic Institution, Lamont-Doherty Earth Observatory, and the National Oceanographic and Atmospheric Administration (NOAA) were used in several areas where there were no UNH data. In areas where multibeam soundings were not available sounding data from the National Ocean Service hydrographic database and the NOAA coastal relief model provided bathymetric coverage of the continental slope.

2.2 GLORIA SIDESCAN

In addition to the acoustic backscatter data from the UNH multibeam surveys, GLORIA sidescan sonar data collected in 1987 (EEZ-SCAN 87, 1991) were used to identify and map landslide features. These data provide near total coverage of the sea floor within the study area at a pixel resolution of 50-m, from the shelf edge out to 200 nm from shore.

2.3 SEISMIC REFLECTION PROFILES

Analogue records of 3.5 kHz seismic reflection profiles, co-acquired with the GLORIA sidescan data, helped determine location, geometry, and thickness of landslide features. Although other data sets are available, the acquisition parameters and quality of these data are consistent over the entire area of study, and they provide a relatively clear picture of the upper sedimentary section.

2.4 CORES

Over 1400 cores have been collected from the study area, and descriptions of the cores are available from the National Geophysical Data Center core repository database. Of these, approximately 1000 have been visually described, and 145 of them have had general ages assigned based on faunal content. While many of the descriptions are brief they provide a valuable summary of the overall lithology of the cores.

2.5 DATA ANALYSIS

The mapping of landslide affected areas was broken into several steps. The first step was to identify scarps around and within landslide source areas. Scarps showed clearly in shaded-relief and slope maps derived from the bathymetric data. Second, using shaded-relief bathymetry, backscatter mosaics, and GLORIA imagery, the areas affected by landsliding were outlined. Most landslide areas have a high-backscatter signature on the sidescan and multibeam imagery. In the southern part of the study area, offshore of the Carolinas, where the multibeam coverage was incomplete, the extent of the landslide areas was based on the GLORIA imagery alone (Fig. 1). Composite slides were differentiated from individual, single event slides based on the presence of more than one headwall scarp and multiple deposit lobes. Third, the thickness of landslide deposits were measured on sub-bottom profiles. In some cases the thickness of the deposit was clearly imaged, however in others a highly-reflective seafloor did not allow sub-bottom penetration, and in some cases the base of the deposits may have been deeper than could be penetrated by the profiling system. These data and interpretations were incorporated into a GIS where the volumes of landslide deposits were calculated based on estimates of the average thickness of the deposit and its areal extent, while the volumes of source areas of a subset of the mapped slides were calculated using the bathymetry and an interpolated smooth-surface technique reported by ten Brink *et al.* (2006).

3. Results

Fifty-five landslide areas were mapped between the eastern end of Georges Bank and the Blake Spur, from the shelf-slope break, down to the abyssal plain (Fig. 1). This number is considerably less than the 179 tabulated by Booth *et al.* (1988; 1993), which is due to our ability to better define the larger landslide complexes and resolve their composite nature with the new bathymetry. The types of landslides were interpreted from the morphology of the deposits as well as from their internal character, and we use the classification scheme presented by Locat and Lee (2002) to describe them. Rotational slides, translational slides, and debris flows were identified, and their distribution is shown in Figure 1. Debris flows were by far the most common type identified and originate from clearly defined headwall scarps and failure surfaces, extending as much as 200 km downslope. Many of the debris flows have several scarps in the source area suggesting they consist of multiple failures rather than a single event or retrograde slumping. Identified rotational slides predate a thick section of younger sediment, which onlaps the slide toes. Translational slides were identified on the multibeam bathymetry as slabs often with a toe that appears to have undergone some disintegration. Headwall scarps were clearly defined and, in the observed cases, indicate a short translation distance.

The area of landslides was measured using the GIS, and ranged from 9 to 15,241 km^2 with a mean of 1,880 km^2. The water depth of the source area for landslides was identified as the shallowest scarp upslope of the landslide area. Where the bathymetry coverage was adequate (33 of the 55 landslides), the depth of measured headwall scarps ranged from 92 to 3,263 m with a mean depth of 1,630 m; 50% of the scarps occurred on the middle and lower slope in 1,200-2,250 m water depths. The relief could be measured on 45 of the headwall scarps: 75% had less than 100 m relief. The toe of the landslide deposits occurred in water depths greater than 2,126 m, with a mean depth of 3,101 m. The thickness of landslide deposits, where they could be measured, ranged from 5 to 70 m (Fig. 2), with a mean thickness of approximately 20 m. The volumes of sediment contained within landslide deposits ranged from 0.05 to 392 km^3, with many of the large slides comprised of several smaller deposits. The largest landslide complexes occur off the southern New England region (190 and greater than 392 km^3) and in the Carolina Trough area (114 and 150 km^3). By contrast, most of the landslides between Hudson and Norfolk Canyons have deposit volumes less than 10 km^3 (Fig. 2).

The volumes 34 mapped landslides source excavation areas calculated from the bathymetry only were found to be between 0.06 km^3 and 179 km^3. Although only a subset of those currently identified was used in these calculations, they are a representative of the full range of landslide types in the region. The largest volumes calculated are associated with the extensive slope-sourced landslides off Georges Bank and southern New England, and south of Cape Hatteras. The smallest source volumes were those found within the canyon systems; both canyon head and canyon side-wall sources.

Although numerous core and surficial sediment samples have been recovered from shelf, slope, and rise of the U.S. Atlantic margin, little information is available on the age of the landslides described herein. Of the few reliable dates that are available, we see that most failures are older than 10,000 yr BP (Embley, 1980; Prior *et al.*, 1986; Popenoe *et al.*, 1993), with only one failure with a Holocene age (Embley, 1980). The Pliestocene age of the sediment comprising the landslide deposits indicates that the failure process removed only a relatively thin surficial skin rather than cutting deeply into older strata under the continental slope (Fig. 3).

4. Discussion

Compared with previous regional studies of the distribution and size of major submarine landslides along the U.S. Atlantic margin (e.g., Booth *et al.*, 1988; 1993; Hance, 2003), the availability of high-resolution multibeam bathymetry significantly improves resolution and interpretation of these features. As a result, although we report a smaller number of mass movement features, we have been able to better define the extent and thickness of individual and composite landslides, which in most cases incorporate a number of the singular features described by previous investigators. As Booth et al. (1988) first observed, we note two distinct source areas for landslides: submarine canyons (headwalls and sidewalls) and the open continental slope. The importance of the distinction between these two types is the difference in magnitude of the individual landslides, and hence their tsunami generating potential. Open-slope sourced landslides have deposits that can exceed 200 km^3, and although many are composite features, we see source area excavations that exceed 100 km^3. By contrast, the canyon-sourced landslide deposits rarely exceed 10 km^3, with source area excavations less than 1 km^3.

The spatial distribution of landslides along the U.S. Atlantic margin is, in part, controlled by the underlying geology. Landslides are the most common, tend to be largest and of the open slope-sourced nature, offshore of areas where Quaternary sediment is thickest on the outer shelf and upper slope (Fig. 3). Nearly 60% of the area affected by landslides occurs offshore of the thick Quaternary shelf deposits of the Georges Bank, southern New England, and Virginia areas. These three areas are also regions where the older strata underlying the slope dip sub-parallel to the gradient of the present slope (Uchupi, 1967; Uchupi and Emery, 1967; Rona, 1969; McGregor, 1981; O'Leary, 1986), and display failure surfaces suggestive of sliding along bedding planes. The two large landslide areas in the Carolina Trough are controlled by different geological processes. They are sourced near salt domes and both the tectonic activity of

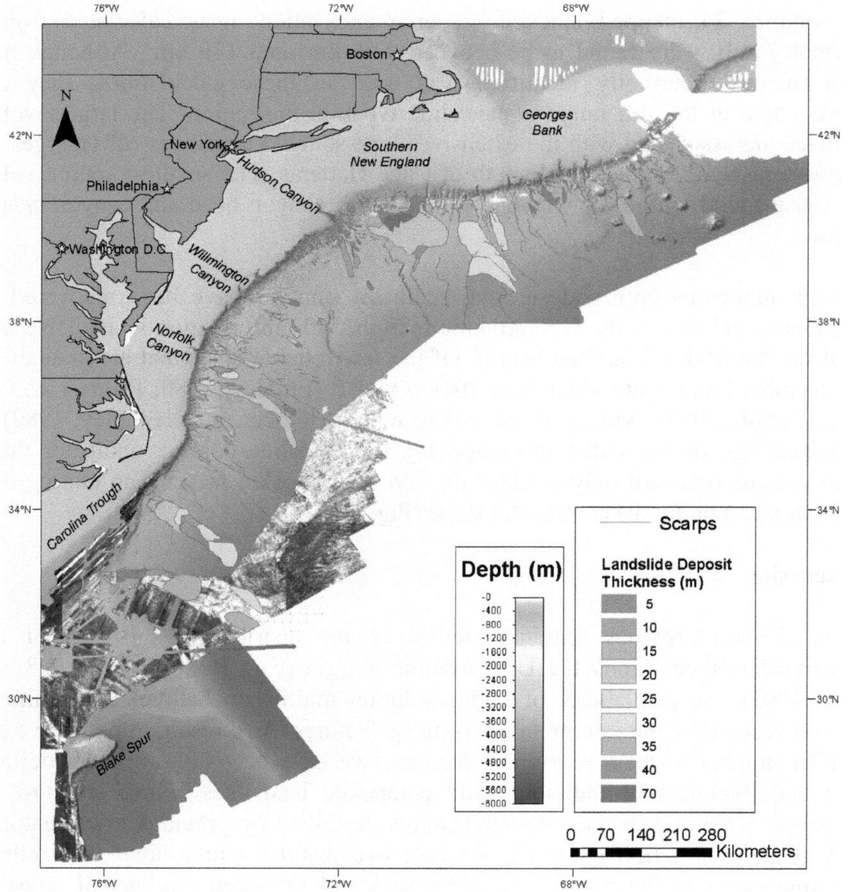

Figure 2. Thickness of landslide deposits mapped using 3.5-kHz profiles. Some were too thin to be resolved by these profiles and others had surface returns that attenuated the signal did not allow penetration to the base of the deposit.

the salt domes (Dillon *et al.*, 1982; Popenoe *et al.*, 1993) and hydrate destabilization (Carpenter, 1981) have been suggested as the triggering mechanisms for these failures. Landslides covering the remainder of the continental margin are off areas where Quaternary deposits at the shelf-edge are thin and older strata underlying the slope are nearly horizontal rather than having a dip sub-parallel to the seafloor and account for only 16% of the total landslide area of the margin. Therefore, other than localized tectonic mechanisms, rapid sediment accumulation rates during the Quaternary (Poag and Sevon, 1989) and the dipping nature of the subsurface strata may be the reason why sediments on parts of the margin are more prone to failure, given the appropriate triggering mechanism.

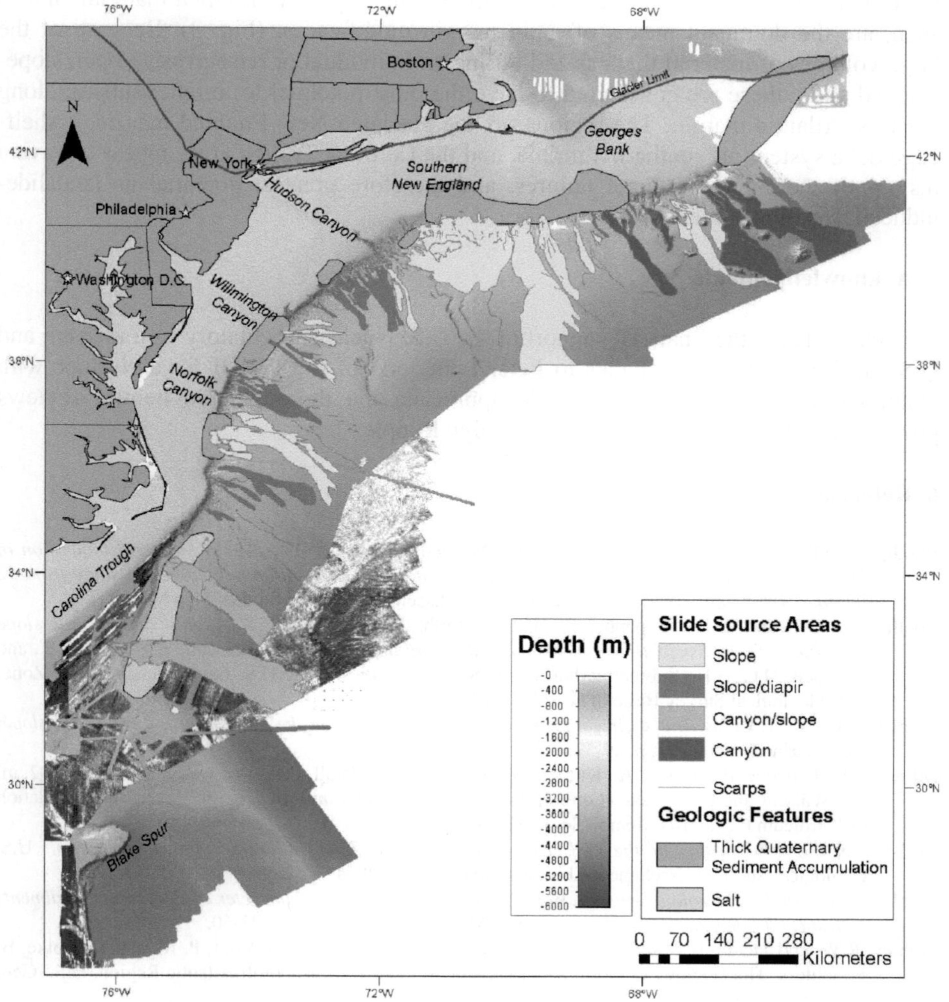

Figure 3. Map showing the distribution of landslides by source area and their relationship to regions of Quaternary sediment accumulation and salt diapirism. The largest landslides have open-slope source areas (in the Carolina Trough area they are associated with salt domes). The landslides off Georges Bank appear to have contributions from both canyon and open-slope sources.

Using the classification of Locat and Lee (2002), we find the dominant mass movement modes to be debris flows, translational and rotational slides, which in most cases are now found together as part of larger, multiphase composite deposits. In part, the reason for this may be because the bulk of the sediment that makes up the mass-wasting deposits was Quaternary in age and largely unconsolidated to semi-lithified, and could not be transported large distances without undergoing disintegration. The height of scarps in most landslide source areas have less than 75 m relief indicating that in most places only the Quaternary section (Poag and Sevon, 1989) is being removed.

Using the new bathymetry compilation we find that the open slope-sourced slides are larger both in the area of failure and overall volume of per-event failed material, and as such, are the dominant means of rapid margin modification (Fig. 3). Because of the large volumes of material that can fail during an individual or retrogressive open slope-sourced slide, these are considered to have the most potential to initiate tsunami along the U.S. Atlantic margin. The regions off the glaciated New England margin, a shelf-edge delta system off southern Virginia, and the Carolina Trough area, appear to have a history of these large volume failures, and therefore greatest potential as landslide-induced tsunami source zones.

5. Acknowledgements

We acknowledge the financial support of the U.S. Nuclear Regulatory Commission and U.S. Geological Survey. Thanks to Peter Lemmond of the WHOI for assistance with obtaining multibeam bathymetry. We appreciate the thorough and helpful reviews provided by Deborah Hutchinson and Carolyn Rupple.

6. References

Booth, J.S., O'Leary, D.W., Popenoe, P., Robb, J.M., and McGregor, B.A., 1988, *Map and tabulation of Quaternary mass movements along the United States-Canadian Atlantic continental slope from 32 00' to 47 00' N latitude*: US Geological Survey Miscellaneous Field Studies Map MF-2027.

Booth, J.S., O'Leary, D.W., Popenoe, P., and Danforth, W.W., 1993, *U.S. Atlantic continental slope landslides: Their distribution, general attributes, and implications,* in, Schwab, W.C., Lee, H.J., and Twichell, D.C., eds., Submarine landslides: Selected studies in the U.S. Exclusive Economic Zone: U.S. Geological Survey Bulletin no. 2002, p. 14-22.

Carpenter, G., 1981, *Coincident sediment slump/clathrate complexes on the U.S. Atlantic continental slope*: Geo-Marine Letters, v. 1, p. 29-32.

Dillon, W.P., Popenoe, P., Grow, J.A., Klitgord, K.D., Swift, B.A., Paull, C.K., and Cashman, K.V., 1982, in, J.S. Watkins and C.L. Drake (Editors), *Studies in continental margin geology*, American Association of Petroleum Geologists Memoir 34, p. 21-46.

EEZ-SCAN 87, 1991, *Atlas of the U.S. Exclusive Economic Zone, Atlantic continental margin*: U.S. Geological Survey Miscellaneous Investigations Series I-2054.

Embley, R.W., 1980, *The role of mass transport in the distribution and character of deep ocean sediments with special reference to the North Atlantic*: Marine Geology, v. 38, p. 23-50.

Embley, R.W. and Jacobi, R., 1986, *Mass wasting in the western Atlantic*, in, Vogt, P. R., and Tucholke, B. E., editors, The Geology of North America, Vol. M, The Western North Atlantic Region, U.S., Geol Soc. America, Boulder, CO, p. 479-490.

Gardner, J.V., Mayer, L.A., and Armstrong, A.A., 2006, *Mapping supports potential submission to U.N. Law of the Sea:* EOS Transactions, American Geophysical Union, v. 87, p. 157-159.

Hance, J.J., 2003. *Development of a Database and Assessment of Seafloor Slope Stability based on Published Literature.* M.S. Thesis, The University of Texas, Austin, p. 245.

Heezen, B.C. and Ewing, M., 1952, *Turbidity currents and submarine slumps, and the Grand Banks earthquake:* American Journal of Science, v. 250, p. 849-873.

Klitgord, K.D., Hutchinson, D.R., and Schouten, H., 1988, *U.S. Atlantic continental margin: structural and tectonic framework,* in, Sheridan, R.E. and Grow, J.A., editors, The Geology of North America, Vol. I-2, The Atlantic Continental Margin, U.S., Geol Soc. America, Boulder, CO, p. 19-55.

Locat, J., and Lee, H.J., 2002. *Submarine landslides: advances and challenges.* Canadian Geotechnical Journal, v. 39, p. 193-212.

McGregor, B.A., 1981, *Smooth seaward-dipping horizons – an important factor in sea-floor stability?:* Marine Geology, v. 39, p. M89-M98.

O'Leary, D.W., 1986, *The Munson-Nygren slide: A major lower-slope slide off Georges Bank*: Marine Geology, v. 72, p. 101-114.

O'Leary, D.W., 1996, *The timing and spatial relations of submarine canyon erosion and mass movement on the New England continental slope and rise*, in, Gardner, J.V., Field, M.E., and Twichell, D.C., eds., Geology of the United States Seafloor: The view from GLORIA, Cambridge Univ. Press, Cambridge UK, p. 47-58.

Poag, C.W., 1991, *Rise and demise of the Bahama-Grand Banks gigaplatform, northern margin of the Jurassic proto-Atlantic seaway*: Marine Geology, v. 102, p. 63-130.

Poag, C.W., and Sevon, W.D., 1989, *A record of Appalachian denudation in postrift Mesozoic and Cenozoic sedimentary deposits of the U.S. middle Atlantic continental margin*: Geomorphology, v. 2, p. 119-157.

Popenoe, P., Schmuck, E.A., and Dillon, W.P., 1993. *The Cape Fear Landslide: Slope failure associated with salt diapirism and gas hydrate decomposition.*, in, Schwab, W.C., Lee, H.J., and Twichell, D.C., eds., Submarine landslides: Selected studies in the U.S. Exclusive Economic Zone: US Geological Survey Bulletin no. 2002, p. 40-53.

Popenoe, P, and Dillon, W.P., 1996, *Characteristics of the continental slope and rise off North Carolina from GLORIA and seismic-reflection data: The interaction of downslope and contour current processes*, in, Gardner, J.V., Field, M.E., and Twichell, D.C., eds., Geology of the United States Seafloor: The view from GLORIA, Cambridge Univ. Press, Cambridge, UK, p. 59-79.

Pratson, L.F., and Laine, E.P., 1989, *The relative importance of gravity-induced versus current-controlled sedimentation during the Quaternary along the mideast US Outer continental margin revealed by 3.5 kHz echo character:* Marine Geology, v. 89, p. 87-126.

Prior, D.B., Doyle, E.H., and Neurauter, T., 1986, *The Currituck Slide, Mid-Atlantic continental slope; revisited:* Marine Geology, v. 73, p. 25-45.

Rona, P.A., 1969, *Middle Atlantic Continental Slope of United States: Deposition and erosion*: American Association of Petroleum Geologists Bulletin, v. 53, p. 1453-1465.

Schlee, J.S., and Fritsch, J., 1982, *Seismic stratigraphy of the Georges Bank basin complex, offshore New England,* in, Watkins, J.S., and Drake, C.L., eds., Studies in continental margin geology, AAPG Memoir, vol. 34, p. 223-251.

ten Brink U.S., Geist, E.L., Andrews, B.D., 2006, *Size distribution of submarine landslides and its implication to tsunami hazard in Puerto Rico.* Geophysical Research Letters, 33, L11307, doi:10.1029/2006GL026125.

Uchupi, E., 1967, *Slumping on the continental margin southeast of Long Island, New York*: Deep-Sea Research, v. 14, p. 635-639.

Uchupi, E. and Emery, K.O., 1967, *Structure of continental margin off Atlantic coast of United States*: American Association of Petroleum Geologists Bulletin, v. 51, p. 223-234.

Weed, E.G.A., Minard, J.P., Perry, W.J., Rhodehamel, E.C., and Robbins, E.I., 1974, *Generalized pre-Pleistocene geologic map of the northern United States Atlantic continental margin*: U.S. Geological Survey Miscellaneous Investigations Series I-861.

TSUNAMIGENIC LANDSLIDES IN THE WESTERN CORINTH GULF: NUMERICAL SCENARIOS

S. TINTI, F. ZANIBONI, A. ARMIGLIATO, G. PAGNONI, S. GALLAZZI, A. MANUCCI, B. BRIZUELA REYES, L. BRESSAN AND R. TONINI
Università di Bologna, Dipartimento di Fisica, Settore di Geofisica,
Email: filippo.zaniboni@unibo.it

Abstract

The western Corinth Gulf, central Greece, is characterized by steep slopes and large sediment river discharge, that are factors increasing the occurrence probability of underwater landslides. Thus the generation of tsunamis from submarine movements is expected to be frequent in this region, and this is confirmed in the historical tsunami catalogues, where reports of tsunamis related to landslides exist either triggered by earthquakes or by gravitational load. In this work we concentrate on the numerical simulation of submarine landslides and of the propagation of the ensuing tsunamis. We elaborate different scenarios basing on recent swath-bathymetry and seismic profiling surveys performed by the Hellenic Centre for Marine Research (HCMR). The most prominent potentially unstable bodies are found in three different regions: one is placed in the area off the city of Aigion, the second is located close to the Psathopyrgos fault, and the third occupies an elongated area off Eratini on the northern side of the gulf. All considered landslides are characterized by relatively small volumes (in the order ranging from 10^5-10^7 m^3). For each scenario, the slide motion is simulated by means of a Lagrangian block model, implemented in the numerical code UBO-BLOCK1, developed by the Tsunami Research Team (TRT) of the University of Bologna, Italy. The tsunami generation and propagation modelling is carried out through the finite-element code UBO-TSUFE, developed by the same research team, solving the Navier-Stokes equation in the shallow water approximation on a triangular-element mesh. We will show how landslide-induced sunamis propagate inside the western Corinth Gulf, the amplitude and period of the tsunami waves at some selected coastal points, and the spatial distribution of the extreme wave heights along the coast.

1. Introduction

The western Corinth Gulf, central Greece, is a well-known geologic structure: it is an active rift zone with a substantial N-S opening rate of about 1.5 cm/yr (De Martini et al., 2004; Cornet et al., 2004), that reflects into a well documented strong seismic activity (Papadopoulos, 2000). As a consequence, it is not surprising that the coasts of this basin have been affected in the past by a large number of destructive tsunamis of seismic origin (Papadopoulos, 2003). In addition, the morphology of the basin flanks is dominated by steep slopes, where large amounts of sediments are discharged by large rivers, creating favourable conditions for sediment failure under seismic or gravitational load (Lykousis et al., 2003). In fact, landslide tsunamis are here relatively more frequent than elsewhere in the Mediterranean. An important example is the 1963 tsunami that was generated by a submarine sediment slide, triggered by a small earthquake: the wave reached the maximum run-up height of about 6 m near the city of Aigion, with maximum horizontal inundation of 120 m (Papathoma and Dominey-Howes, 2003).

V. Lykousis, D. Sakellariou and J. Locat (eds.), Submarine Mass Movements and Their Consequences, 405–414.

Other significant cases are the tsunamigenic sediment failures in the Tolofonas delta, near Eratini in the northern coast, and in the Eliki delta in the southern coast of the gulf that were both the consequence of the 1995 M_S=6.1 Aigion earthquake (Papatheodorou and Ferentinos, 1997; Hasiotis et al, 2002). Therefore, in the western Gulf of Corinth, besides the hazard from seismic shocks, one has to count also the hazard due to tsunamis produced by co-seismic sea-floor displacement or to tsunamis generated by landslides deriving from earthquakes or simply from gravitational overload (Armigliato et al., 2006; Stefatos et al., 2006). In view of risk assessment and mitigation, studying tsunami generation and impact is crucial in this area where a high population density is concentrated within a narrow coastal belt especially in the peak tourist season.

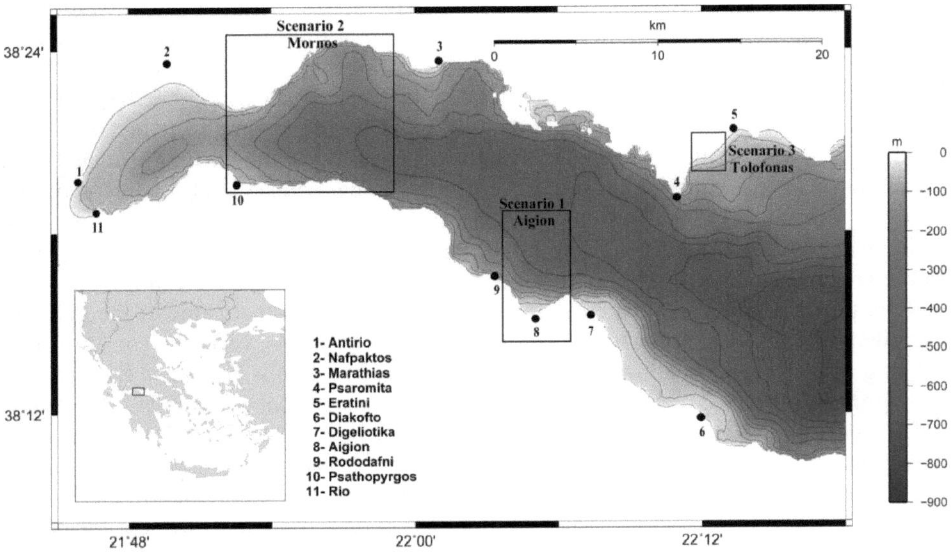

Figure 1. Map of the western Corinth Gulf, central Greece, showing the three areas where landslide-induced tsunami scenarios were developed.

This work, performed in the framework of the European project 3HAZ-Corinth, develops scenarios of tsunamis produced by underwater landslides in the western Gulf of Corinth. The sliding bodies are located in three different regions (Figure 1). The first one is in the Aigion area, where a number of slumps triggered by earthquakes associated with the Aigion fault (Koukouvelas, 1998; Pantosti et al., 2004) have been reported to occur (Papadopoulos, 2003). The second is located near the Psathopyrgos fault, in the western part of the Gulf, and includes also the Mornos river delta. Last, the third one is located in the Tolofonas-Eratini coast, again associated with a river mouth, where sedimentation rate and sediment accumulation are quite significant. In our approach, the simulation of the motion of the sliding bodies provides the input for the tsunami simulation code.

2. Numerical models

The numerical elaboration of the tsunami scenarios taken into account here implies a two-step model: first, modelling the landslide motion, and then modelling the ensuing tsunami. In principle, this excludes full coupling between the sliding body and the produced water waves: it accounts only for the one-way influence chain (the slide influences the tsunami), but not for the reverse one, with the exception of the resistance exerted by the water on the moving body. This approximation, however, is widely accepted, since 1) it has proven to be quite good, because the reverse process has limited effects on the overall dynamics, and 2) it has the further advantage of being quite efficient, because it simplifies calculations, substantially reducing the computational time.

2.1 THE LANDSLIDE MODEL

The simulation of the slide motion adopts a Lagrangian approach: the reference frame, which the equations are referred to, moves together with the mass, implying a significant simplification of the motion equations. The landslide body is partitioned into a set of interacting blocks, that conserve their volume and cannot separate nor penetrate each other during the motion. The centre of mass (CoM) dynamics of the interacting blocks is computed by evaluating the body forces (gravity, buoyancy) and the surface forces (basal friction, interaction of the exposed surfaces with environment) acting on each block, plus the interaction force, due to reciprocal pushes between adjacent blocks, causing stretching or contraction of the blocks. The equations of motion for all blocks are solved numerically. The computations stop when the mass exits the computational domain or when it comes to a rest. This Lagrangian block model has been implemented in the 1D and 2D versions in the numerical software packages UBO-BLOCK1 and UBO-BLOCK2 respectively, both developed by the Tsunami Research Team (TRT) of the University of Bologna (for a basic description of the method, see the paper by Tinti et al., 1997). The algorithms have been successfully applied in several cases to simulate the motion of tsunamigenic subaerial and submarine landslides (see e.g. Tinti et al., 2003; Tinti et al., 2006).

2.2 THE TSUNAMI MODEL

Tsunami generation and propagation is simulated by solving the hydrodynamical non-linear Navier-Stokes equations in the shallow water approximation, including a term that links the landslide model to the tsunami model. It represents the forcing at the sea surface due to the landslide motion and comes as a known output from the landslide simulations. Since the landslide and the tsunami models use different time-space grid, a grid-to-grid interpolation in the space and time domain is required to produce the right input to the tsunami model, which is carried out through the specifically developed code UBO-TSUIMP. The numerical computations are carried out through a finite-element code using a triangular mesh, with the dimension of each element proportional to the square of the local sea depth: the deeper the local bathymetry, the bigger is the triangle. The model is that one implemented by the code UBO-TSUFE, originally developed and subsequently improved by the same research group TRT (see Tinti et al., 1994).

3. Landslide simulations

As already mentioned, three sites have been selected for evaluating landslide generated tsunamis (Figure 1), starting from potentially unstable bodies mapped by recent swath-bathymetry and seismic profiling surveys performed by the Hellenic Centre for Marine

Research (HCMR). Notice, however, that the scenarios we developed are not meant to be reconstruction of specific past events, but rather we consider hypothetical, though realistic, sliding masses with the chief goal to explore the main tsunami features and to identify the coastal segments that are most exposed to tsunami attack, which will be a relevant contribution to the vulnerability estimation and risk assessment in the western Corinth Gulf. In this paper, all scenarios are treated by using the basic 1D code UBO-BLOCK1, that is quite apt to deal with elongated slump bodies, with longitudinal length prevailing on the transversal width. Notice that this model requires as input data the trajectories of the CoM's and the boundaries of the sliding area see Tinti et al., 1997).

3.1 AIGION LANDSLIDE SCENARIO

This scenario is intended to assess the risk connected to a coastal landslide in the Aigion bay, and is based on the 1995 landslide tsunami occurrence in the Aigion area. The sliding body has been hypothesized to detach from close to the shoreline (Figure 2). The geometry of the body has been conceived starting from the description by Papatheodorou and Ferentinos (1997) of the 1995 tsunamigenic slump, that involved sediments of the Meganitis fan delta. However, the thickness, and consequently also the total volume, are much higher: the assumed body is more than 50 m thick, with a total volume of about 28 10^6 m^3, one order of magnitude bigger than PF's estimate.

One reason for this is that previous simulations through the UBO-BLOCK1 and UBO-TSUFE codes performed by using PF volume (Armigliato et al., 2006) produced run-up heights (about 1 m) well below the ones observed, around 6 m (Papathoma and Dominey-Howes, 2003). A second reason is that volumes in the range of the one considered here seem to be quite reasonable in view of the geotechnical and morphological characteristics of the submarine slope (Lykousis, HCMR, personal communication). The simulation has been carried out using the value 0.05 for the friction coefficient, and taking the density of the sediment $\rho = 2000$ kg/m^3. The resulting run-out distance is about 3 km, the mass reaches more than 300 meters depth, and the mean (averaged over the blocks) slide velocity exceeds 15 m/s. Some blocks almost gain speeds of 20 m/s, about 50 s after slide initiation. The phase of the largest slide speed coincides with the phase of the highest tsunamigenic efficiency, when the slide touches the peak of the Froude-number curve. Consider that the closer is the slide Froude number to the critical value of 1, the more effective the slide is in generating tsunamis. After the peak phase, the slide decelerates slowly and stops its motion after about 265 s.

3.2 MORNOS LANDSLIDE SCENARIO

The second scenario assumes that a slide detaches in the Mornos river fan delta, which is characterized by a bathymetric profile not dissimilar from the Aigion bay (see Figure 3). In this case, we have assumed that the slide maximum thickness approaches 30 m and that the basal area is around 2 km^2, resulting in a total volume of about 9 10^6 m^3. Calculations of the tsunami are intended to evaluate the wave propagation in the westernmost part of the Corinth Gulf, with a particular attention to the village of Psathopyrgos, in the southern coast.

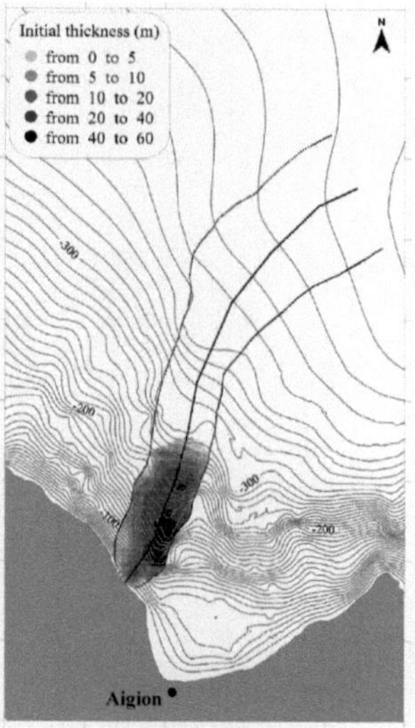

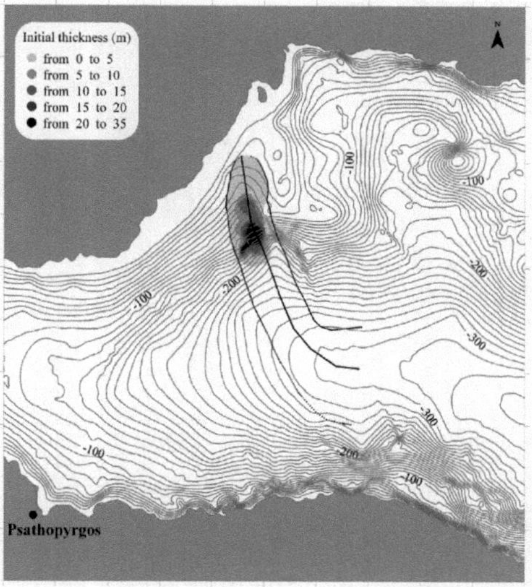

Figure 2 (left). Map of the Aigion landslide. The centre of mass path and the lateral boundaries required as input for the landslide simulation code UBO-BLOCK1 are indicated by the black lines.
Figure 3 (right). Map of Mornos scenario landslide, with centre of mass path and lateral boundaries required for the simulation code.

From Figure 3 we can observe that the runout distance is of about 2 km. The mean slide velocities are in the same range as for the Aigion case: about 15 m/s for the peak speed, attained around 50 s, with some blocks exceeding 20 m/s. The slide stops after 225 s at 250-300 m sea depth.

3.3 TOLOFONAS LANDSLIDE SCENARIO

The last scenario regards a site in the northern coast, more precisely in front of the Tolofonas river mouth, south-west of the city of Eratini. This landslide (Figure 4) has been positioned in the place where there is evidence of a sediment failure induced by the 1995 earthquake (Hasiotis et al., 2002).

The sliding body is considerably smaller than that assumed in the first two scenarios: the mobilized volume is around 2.5 10^5 m^3. Nonetheless, the study of such sliding mass and of the associated tsunami is quite interesting, since it gives a clue on the tsunamigenesis by small masses and allows one to compare the effects produced by masses spanning a range about two orders of magnitude.

The bathymetric profile here is not so steep and is almost linear. The total body displacement is about 500 m, with peak velocity below 4 m/s. The slide is too slow to produce large tsunami waves.

3.4 FROUDE NUMBER

The Froude number is generally used as an indicator of the tsunamigenic potential of a landslide: it is computed as the ratio between the mean horizontal velocity of a sliding mass and the mean velocity of the tsunami (mean values are calculated over the slide area).

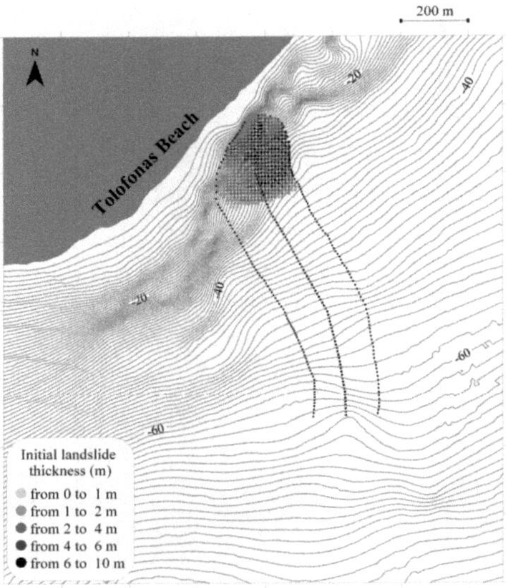

Figure 4. Tolofonas landslide, with centre of mass path and lateral boundaries required for used simulation code.

The closer this value is to 1, the more coupled are the sliding mass and the generated wave, meaning that the mass continuously "supplies" energy to the tsunami. For this reason the value 1 is considered as a critical value, and landslides exhibiting a value close to 1 are highly tsunamigenic. The Mornos case slide reaches the highest values of the Froude number, up to 0.35, and slightly lower values are obtained for the Aigion case with a 0.30 maximum. The Tolofonas landslide, probably due to its very low speed, has small Froude numbers (maximum around 0.15 at about 120 s).

4. Simulations of tsunamis

The landslide simulations provide the forcing term, which is used for the study of the wave propagation through the code UBO-TSUFE. As input, the program needs the computational grid, that is composed of irregular triangles. The grid is finer near the coast to describe adequately the complicated coastline of the gulf of Corinth as well as to represent properly the landslide detachment regions of the three scenarios, that are all located in the very shallow water zone.

4.1 ANALYSIS OF THE SIMULATIONS RESULTS

As an example of tsunami propagation, we show in Figure 5 the waves generated by the Aigion landslide by means of snapshots of the water elevation field taken in the first 8 minutes of tsunami radiation. The tsunami fills all of Aigion bay within 1-2 minutes, with waves reaching 3 m amplitude, and hits the opposite northern coast after 3 minutes, with waves higher than 1 m. Another relevant feature is the formation of a coastal wave train, particularly along the southern coast, proving that energy "trapping" takes place as the result of the wave interaction with the coast. As a general consideration, we can conclude that after 10 minutes all the western Corinth Gulf is affected by the tsunami, the first arrival being followed by a series of waves, either directly produced by the landslide or coming from reflection at the opposite coasts. The tide-gauge time-histories calculated in three cities (Aigion, Psathopyrgos, Eratini, that are each close to one of the simulated landslide source areas) for the three different landslide cases are shown in Figure 6.

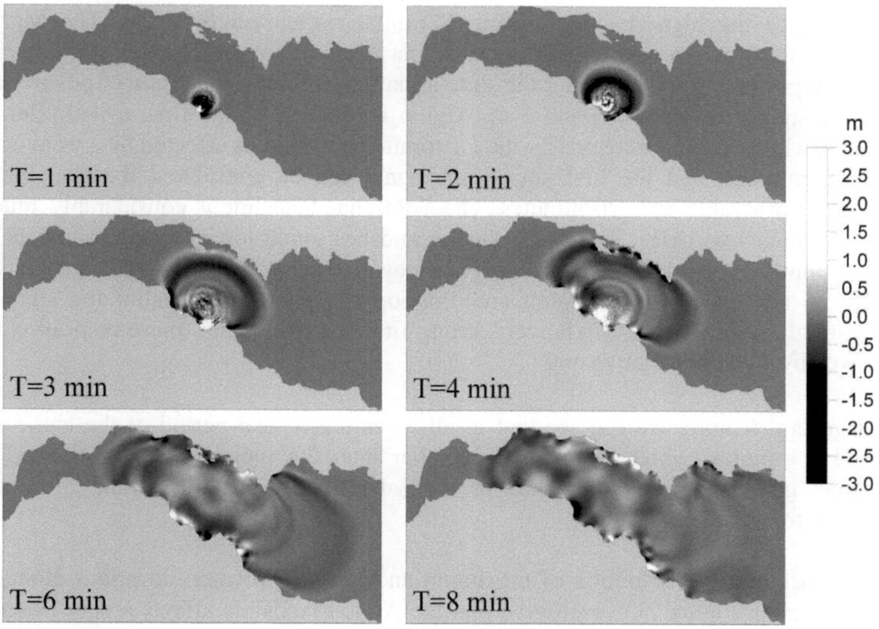

Figure 5. Propagation fields for the Aigion scenario. Darker areas denote negative elevation waves (troughs), and bright white areas denote positive waves (crests).

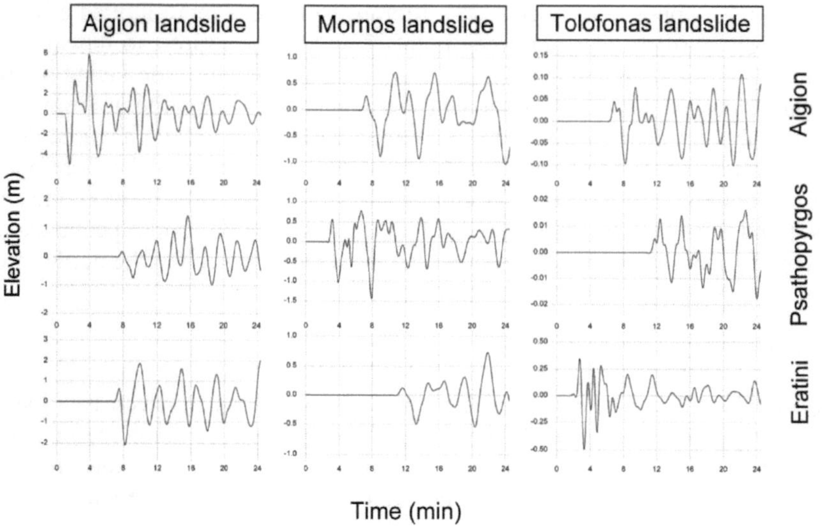

Figure 6. Synthetic tide-gauge 24-min long records of tsunamis for the three scenarios, calculated at three different localities: Aigion (8), Psathopyrgos (10), Eratini (5). See site location in Figure 1.

For the first slide, in Aigion (point 8 in Figure 1) we observe a strong negative first arrival, after about 1 minute, followed by positives and negatives: the highest wave is the second positive oscillation and reaches 6 m amplitude. After 24 minutes we still observe strong signals (>1m). In the other two localities, Psathopyrgos (point 10 in Figure 1) and Eratini (point 5), the tsunami arrives with about 8-minute delay, slightly positive. For Eratini, 2 m waves are observed already from the second arrival, while for Psathopyrgos the highest signals occur 8 minutes after the first one. The Mornos landslide, the westernmost case, induces a first positive arrival, reaching 0.5 meter in Psathopyrgos only 3 minutes after the slide initiation, followed by a train of positive and negative signals between 0.5 and 1 m. Aigion is reached by the first arrival after 6-7 minutes, while Eratini, protected by the Psaromita peninsula, is affected by a weak wave after 12 minutes, and the first significant signal can be considered the trough-crest sequence observable after 20 minutes. The Tolofonas landslide is considerably smaller than the other two cases, it is much slower and has much lower Froude numbers, all factors pointing to a very small tsunamigenic power. The synthetic marigrams indicate 5 to 10 cm sea elevation in Aigion and Psathopyrgos, while in Eratini the signal is interestingly not so small: a relatively strong first positive signal, more than 30 cm, is followed by a 50 cm negative one.

As a general feature, we can see that in all cases the typical period of the tsunami is around 2 minutes, which is distinctly smaller than the typical period estimated for tsunamis generated by earthquakes in the same region that is around 5 minutes (Armigliato et al., 2006).

If we consider the distribution of maximum and minimum water elevations along the coast (given in Figure 7), we observe that the Aigion landslide affects a very extended portion of coast with significant waves. Maxima and minima are expectedly located near the source area (points 7-9 correspond to the Aigion bay, see Figure 1). More than 3 m waves are also observed between points 3 and 4, corresponding to the northern portion of coast opposite to Aigion. The Mornos case has a more limited effect. In the portion of the northern coast between points 2 and 3 (the nearest to the source area), more than 2 m wave are computed, and also in the southern portion facing the Mornos fan delta a significant signal is observed (near point 10, Psathopyrgos). From point 3 to point 5, in the northern coast, we have a strong wave eastward attenuation, due to the screening effect of Trizonia island and of Psaromita peninsula. The Tolofonas slide shows a very narrow local peak, concentrated near the source area between the coastline points 4 and 5, with a signal exceeding 1 m. In this case Psaromita peninsula acts on wave coming from the opposite direction, i.e. westward, protecting the northern portion of coast between points 4 and 2.

5. Conclusions

The Corinth Gulf, characterized by a strong seismicity and by steep basin slopes, is particularly prone to sediment failures, triggered by seismic vibrations or simply by gravitational instability. The generation of landslide tsunamis is therefore frequent, and it is important to understand features and mechanisms of destructive waves generation and propagation, especially in view of risk assessment, considering further the high population density of the western Corinth Gulf coasts. Hence, the use of scenarios is a relevant tool, in order to determine which are the most affected areas and which measures can be taken to reduce the exposition to tsunamis.

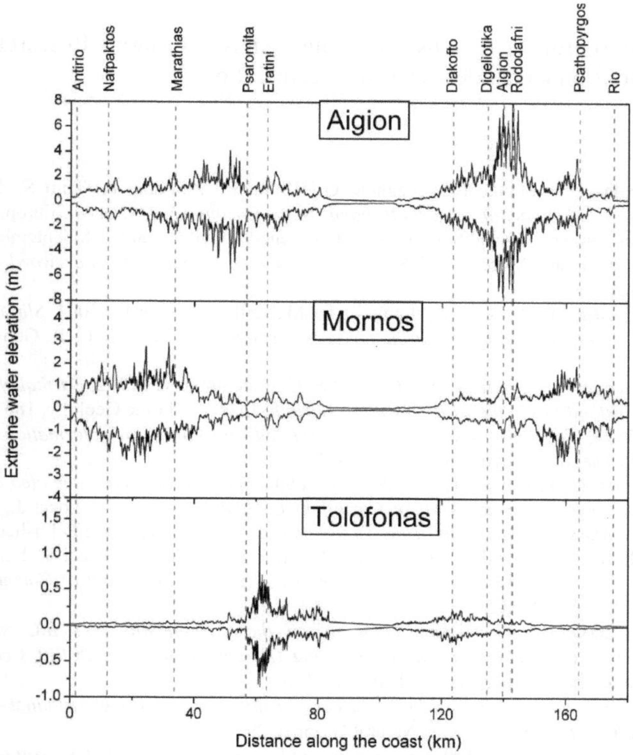

Figure 7. Maximum and minimum elevations of the sea level along the coast. The numbers refer to the localities reported in Figure 1. Note that the northern coast (1-5) is run west to east, while the southern coast of the gulf (6-10) is run from east to west.

In this work three scenarios have been taken into account: the first landslide has been positioned in Aigion Bay, near the Meganitis river mouth, with a volume of almost 30 10^6 m^3; the second near the Mornos fan delta, with a volume of 10 10^6 m^3; the third, considerably smaller, in front of Tolofonas river mouth, with a volume of 250 10^3 m^3 (two order of magnitude lower). The simulations have shown that landslides can reach velocities of 15-20 m/s within a very short time (less than 1 minute), corresponding to the most efficient tsunamigenic phase. It is interesting to stress that even the Tolofonas slide that is characterised by very limited mass and velocity is able to produce wave exceeding 1 m. Tsunamis produced on both sides of the gulf affect the nearest coast within 1 or 2 minutes and the opposite side within the first 8-10 minutes. The Aigion landslide generates a 1-2 m tsunami crossing the gulf, with maxima reaching 6 m (4 minutes after the landslide) in the Aigion bay and 4 m in the portion of coast facing it (Figure 7). The Mornos landslide causes more limited effects, but still relevant: 3 m wave maxima close to the source and 2 m maxima on the opposite coast. Further landslide-induced tsunamis seem to have dominant period around 2 minutes (Figure 6), definitely shorter than the typical periods of earthquake tsunamis that in the western Gulf of Corinth are estimated to be around 5 minutes.

6. Acknowledgments

We are grateful to Prof. V. Lykousis, Hellenic Centre for Marine Research (HCMR), for useful suggestions on landslide scenarios identification.

7. References

Armigliato A., Gallazzi S., Manucci A., Pagnoni G., Tonini R., Zaniboni F., Tinti S., 2006, *Generation, propagation and impact of scenario tsunamis in the Corinth Gulf (Greece)*, European Conference on Earthquake Engineering and Seismology 2006, Geneva (Switzerland), 3-8 September 2006.

Cornet, F.H., Bernard, P. and Moretti, I. (2004), *The Corinth Rift Laboratory – Foreword*, C. R. Geoscience, 336, 235-241.

De Martini P.M., Pantosti D., Palyvos N., Lemeille F., McNeill L., Collier R., 2004, *Slip rates of the Aigion and Eliki Faults from uplifted marine terraces, Corinth Gulf, Greece*, C. R. Geoscience, 336, 325-334.

Hasiotis T., Papatheodorou G., Bouckovalas G., Corbau C., Ferentonos G., 2002, *Earthquake-induced coastal sediment instabilities in the western Gulf of Corinth, Greece*, Marine Geology, 186, 319-335.

Koukouvelas I., 1998, *The Egion fault, earthquake-related and long-term deformation, Gulf of Corinth, Greece*, Journal of Geodynamics, 26, 501-513.

Lykousis V., Sakellariou D., Roussakis G., 2003, *Prodelta slope stability and associated coastal hazards in tectonically active margins: Gulf of Corinth (NE Mediterranean)*. In: Locat J., Mienert J. (eds.), *Submarine Mass Movements and Thier Consequences*, Kluwer Academic Publishers , 433-440.

Pantosti D., De Martini P.M., Koukouvelas I., Stamatopoulos L., Palyvos N., Pucci S., Lemeille F., Pavlides S., 2004, *Palaeoseismological investigations of the Aigion Fault (Gulf of Corinth, Greece)*, C. R. Geoscience, 336, 335-342.

Papadopoulos G., 2000, *A new tsunami catalogue of the Corinth Rift: 373 B.C.-A.D. 2000*. In: G. Papadopoulos (ed.), *Historical Earthquakes and Tsunamis in the Corinth Rift, Central Greece*, Inst. Geodynamics, Natl. Observatory Athens, 121-126.

Papadopoulos G., 2003, *Tsuanmi hazard in the Eastern Mediterranean: strong earthquakes and tsunamis in the Corinth Gulf, Central Greece*, Natural Hazards, 29, 437-464.

Papatheodorou G., Ferentinos G., 1997, *Submarine and coastal sediment failure triggered by the 1995, M_s=6.1 R Aegion earthquake, Gulf of Corinth, Greece*, Marine Geology, 137, 287-304.

Papathoma M., Dominey-Howes D., 2003, *Tsunami vulnerability assessment and its implications for coastal hazards analysis and disaster management planning, Gulf of Corinth, Greece*, Natural Hazards and Earth System Sciences, 3, 733-747.

Stefatos A., Charalambakis M., Papatheodorou G., Ferentinos G., 2006, *Tsunamigenic sources in an active European half-graben (Gulf of Corinth, Central Greece)*, Marine Geology, 232, 35-47.

Tinti S., Gavagni I., Piatanesi A., 1994, *A finite-element numerical approach for modeling tsunami*. Annals of Geophysics, 37, 1009-1026.

Tinti S., Bortolucci E., Vannini C., 1997, *A block-based theoretical model suited to gravitational sliding*, Natural Hazards, 16, 1-28.

Tinti S., Pagnoni G., Zaniboni F., Bortolucci E., 2003, *Tsunami generation in Stromboli and impact on the south-east Tyrrhenian coasts*, Natural Hazards and Earth System Sciences, 3, 299-309.

Tinti S., Pagnoni G., Zaniboni F., 2006, *The landslides and tsunamis of the 30th of December 2002 in Stromboli analysed through numerical simulations*, Bulletin of Volcanology, 68, 462-479.

TSUNAMIS GENERATED BY COASTAL AND SUBMARINE LANDSLIDES IN THE MEDITERRANEAN SEA

G.A. PAPADOPOULOS, E. DASKALAKI AND A. FOKAEFS
Institute of Geodynamics, National Observatory of Athens, 11810 Athens, Greece

Abstract

The main cause of tsunami generation in the Mediterranean Sea is tectonic activity associated with strong earthquakes. However, tsunami waves are also generated by landslides. From a compilation of 32 reliable cases of landslide tsunamis it comes out that most of them were caused by subaerial landslides or marine slides induced mainly by earthquakes and less frequently by volcanic eruptions. Others were caused by gravitative landslides or marine slides. The most frequent events were observed in the Corinth Gulf, Greece (11 out of 32 cases). In the volcanic Aeolian islands, Italy, 7 out of 32 cases were reported. In the Hellenic arc only 3 events are known, in contrast to the abundant, large-size tsunamis of seismotectonic origin historically documented. In Cyclades, South Aegean, only 2 but large landslide tsunamis were reported. Only few events have occurred in Marmara Sea, Cyprus, East Sicily, Liguria-Côte d'Azur and Algeria. Such a pattern of geographical distribution makes possible to assess the potential for landslide tsunami generation.

Keywords: landslide, tsunami generation, tsunami potential, Mediterranaen Sea

1. Introduction

Several types of landslides cause tsunamis with significant heights in near-source coasts that attenuate rapidly due to frequency dispersion. Due to the concentrated large wave heights, such tsunamis may result in catastrophic consequences. Therefore, their study is of practical interest to design engineers and decision makers. In the Mediterranean Sea region, tsunamis are mainly caused not only by earthquakes but also by several types of landslides. However, no systematic study of landslide tsunamis has been performed so far. We compile 32 cases of Mediterranean Sea landslide tsunamis. We summarize the tsunami characteristics and generation mechanisms, analyze the geographic distribution of tsunami sources and examine implications for tsunami potential. The term "landslide" is here used to include rock falls, subaerial earth slumps, settlements and marine slides, regardless the mechanism of cause.

2. The data

Cases of landslide tsunamis are summarized in Table 1 and analysed below:

[1] 373BC, winter, Helike, SW Corinth Gulf, Central Greece: Large shock caused subsidence of the coastal strip of the Helike city. A high tsunami was triggered. The city inundated while ten Spartan ships which happened to be at anchor close by were destroyed along with the city.
[2] c.1050AD, Sarkoy, NW Marmara Sea, Turkey: A tsunamigenic sediment layer was discovered in fluvio-alluvial sequences on the NW coast of Marmara Sea (Minoura et al. 2005). The layer consists of unsorted silty coarse sand including terrestrial molluscs and

V. Lykousis, D. Sakellariou and J. Locat (eds.), Submarine Mass Movements and Their Consequences, 415–422.

charcoal fragments. From AMS radiometric ages, Minoura et al. (2005) suggested that a tsunami occurred in the middle of 11th century and invaded the fluvial plains. The tsunami was generated by slope failure of coastal blocks due to fault movement. Possible historical earthquakes that occurred on AD990 or 1010 or 1050 are reported.

[3] 1650 09 30 (old style), Thera, Cyclades, South Aegean Sea: A large tsunami was caused during the eruption of the submarine volcanic edifice Columbo. On 30.09.1650 a pause was noted in volcanic activity. Then, the tsunami inundated the eastern coast of Thera and swept away churches, enclosures, boats, trees and agricultural land. Runup heights oup to 50 m were reported in nearby in nearby islands. Ships and fishing boats moored at Iraklio (Crete) were violently swept offshore, while the wave overtopped the city walls. The volcanic and seismic quiescence that prevailed before the tsunami struck, implies that it was generated by submarine collapse of the volcanic cone rather than by a strong shock or explosion (Dominey-Howes et al. 2000).

[4] 1783 02 06, Tyrrhenian Calabria, Italy: Very strong earthquake. Tsunami waves 6-9 m high were caused by earthquake-induced rockfall: a portion of the Monte Pacì, Scilla, collapsed suddenly into the sea. Many victims were counted in Scilla) and Messina.

[5] 1794 06 11 (old style), Galaxidi, NW Corinth Gulf, Central Greece: According to an anonymous manuscript written in c.1796, a tsunami was caused by coastal slump due to a strong earthquake near Galaxidi.

[6] 1861 12 26, Valimitika, SW Corinth Gulf, Central Greece: Large ($M \geq 6.6$), destructive shock caused surface-breaks 13 km long, soil liquefaction and a coastal subsidence in the area where same phenomena were historically reported on 373BC. Five tsunami waves, possibly caused by the coastal subsidence, were observed immediately after the shock with wave heights up to c.2 m in the south and north coasts of Corinth Gulf. Some damage in port facilities, cultivated land, merchantships and row-boats were reported from Aeghion, Valimitika, Galaxidi, Itea and Skala-Vitrinitza.

[7] 1888 09 09, Galaxidi, NW Corinth Gulf, Central Greece: Strong, destructive shock in Aeghion caused a moderate tsunami in Galaxidi. The tsunami possibly was caused by submarine slides that occurred offshore Aeghion.

[8] 1894 04 27, Atalanti, Greece: Large ($M=6.9$), shock caused coastal subsidence and a local tsunami 3 m high that penetrated inland for about 1 km near Kyparisi.

[9] 1898 06 02, East Corinth Gulf, Central Greece: Large, damaging shock caused tsunami that flooded the south coast of Corinth Gulf. The intermediate-depth focus favours tsunami generation rather by submarine slides than by co-seismic fault displacement.

[10] 1919 05 22, Aeolian Islands, Italy: During an eruption of Stromboli, a tsunami occurred carrying all the boats by more than 300 m in the neighbouring fields. The event looks like similar to the tsunami caused by a volcanogenic landslide on 30.12.2002.

[11] 1926 08 17, Aeolian Islands, Italy: Due to a shock near Salina island landslides occurred. An anomalous sea movement was observed in Salina.

[12] 1930 09 11, Aeolian Islands, Tyrrhenian Sea, Italy: During an explosion in Stromboli the sea retired c.100 m and penetrated the beach for 200 m. The run-up was about 2.5 m. It appears that this local tsunami was triggered by hot avalanches.

[13] 1944 08 20, Aeolian Islands, Tyrrhenian Sea, Italy: During a large explosion in Stromboli the sea penetrated at Punta Lena about 300 m inland, destroying one house. A lot of fish were found on the beach. This local tsunami was possibly triggered by hot avalanche at the Forcia Vecchia.

Table 1: Main characteristics of the tsunami events analyzed; data taken from Fokaefs and Papadopoulos (2007) for Cyprus, Papadopoulos (2003) for Corinth Gulf, Papadopoulos et al. (2007) for the East Hellenic Arc and Papadopoulos (2001) for rest Greece, Yalçiner et al. (2002) for Marmara Sea, Tinti et al. (2004) for Italy and Côte d'Azur, Soloviev et al. (2000) for west Mediterranean basin. Key: ER= submarine earthquake, EL=earthquake landslide, ES=earthquake marine slide, VO=volcanic eruption, VL=volcanic landslide, VS=volcanic marine slide, GL=gravitative landslide, GL=gravitative marine slide. Tsunami intensity has been estimated in the 6-grade Sieberg-Ambraseys scale and the 12-grade Papadopoulos and Imamura (2001) scale. Reliabilitiy (rel) is given in a 4-degree scale, where 1 and 4 correspond to improbable tsunami and definite tsunami, respectively.

no	year	month	day	latitude longitude	region	cause	intensity	rel
1	373BC	winter		38 11 22 09	W.Corinth Gulf	EL	5/9	4
2	c.1050			40 36 27 06	NW Marmara Sea	EL	?	3
3	1650	09	30	36 24 25 23	S.Aegean Sea	VL	6/10	4
4	1783	02	06	38 15 15 43	Tyrrhenian Calabria	EL	6/9	4
5	1794	06	11	38 23 22 23	W. Corinth Gulf	EL	3/5	3
6	1861	12	26	38 12 22 12	W. Corinth Gulf	EL	3(+)/5	4
7	1888	09	09	38 15 22 05	W. Corinth Gulf	ES	2/3	4
8	1894	04	27	38 42 23 00	Evoikos Gulf	EL	4/3	4
9	1898	06	02	38 05 22 38	E. Corinth Gulf	ES	2(+)/4	3
10	1919	05	22	38 48 15 12	Aeolian Islands	VO	3/5	4
11	1926	08	17	38 50 14 45	Aeolian Islands	EL	1/2	2
12	1930	09	11	38 48 15 12	Aeolian Islands	VO	3/4	4
13	1944	08	20	38 48 15 12	Aeolian Islands	VO	4/6	4
14	1947	10	06	36 54 22 00	SW Peloponnese	ES	2/3	3
15	1953	09	10	34 48 32 47	Cyprus	EL	2/3	4
16	1954	09	09	36 17 01 28	Algeria	ES	1/2	4
17	1956	07	09	36 38 25 58	S. Aegean Sea	ES	5/8	4
18	1963	02	07	38 12 22 12	W. Corinth Gulf	GS	4/7	4
19	1965	07	06	38 22 22 14	W. Corinth Gulf	EL	3/5	4
20	1968	04	18	44 05 08 00	Côte d' Azur	ER	2	4
21	1979	10	16	43 42 07 15	French Riviera	GS	3/4	4
22	1980	10	10	36 17 01 41	Algeria	ES	1/2	4
23	1981	02	24	38 04 23 00	E. Corinth Gulf	ER	2/3	4

24	1984	02	11	38 24 22 03	W. Corinth Gulf	EL	3/4	4
25	1988	04	20	38 24 14 58	Aeolian Islands	GL	2/3	4
26	1990	12	13	37 16 15 07	Eastern Sicily	ER	2/3	4
27	1995	06	15	38 22 22 14	W. Corinth Gulf	EL	2(+)/4	4
28	1996	01	01	38 15 22 07	W. Corinth Gulf	GS	3(+)/5	4
29	1999	08	17	40 45 29 52	E. Marmara Sea	EL	4/6	4
30	2000	04	05	35 21 25 09	Crete Island	ES	2/5	4
31	2002	03	24	36 27 28 12	Rhodes	GS	2/5	4
32	2002	12	30	38 48 15 12	Aeolian Islands	VS	4/7	4

[14] 1947 10 06, Pylia, SW Peloponnese, Greece: Large (M=7.0), destructive shock. A local tsunami, that was attributed to an offshore slide, advanced inland 15 m in Methoni.
[15] 1953 09 10, Paphos, Cyprus: Strong, destructive double shock (M=6.0, M=6.1) in SW Cyprus. In Paphos a small tsunami possibly triggered by coastal slumps.
[16] 1954 09 09, El-Asnam, Algeria: Strong (M=6.6), destructive shock. A weak tsunami, caused by underwater slides, was recorded by tide-gauges in Spanish and African coasts.
[17] 1956 07 09, Cyclades Islands, South Aegean Sea, Greece: Large (M=7.5), destructive shock caused a 15-20 m high tsunami in the Amorgos Basin. Tide-station records indicated that possibly it was triggered by earthquake-induced submarine slumps (Galanopoulos 1957). Bathymetric, shallow and intermediate penetration seismic profiles, and gravity cores provided evidence for sea-floor sediment instability and a large-scale, very recent sediment slump possibly triggered by the 1956 mainshock and/or its largest aftershock (Perissoratis and Papadopoulos 1999). It is supported that the tsunami was caused by the combined action of a reactivated normal fault and the sediment slump.
[18] 1963 02 07, West Corinth Gulf, Central Greece: Locally strong, destructive tsunami at both coasts of Corinth Gulf was generated by gravitative coastal and submarine sediment slump along the mouth of Salmenikos river. Two persons killed, twelve injured, while destruction was reported in houses, cultivated land, fishing-boats and vessels.
[19] 1965 07 06, Eratini, NW Corinth Gulf, Central Greece: Strong (M=6.5), damaging shock. Coastal landslide at Eratini caused a local tsunami as high as 3 m. One person drowned and some damage was caused.
[20] 1968 04 18, Liguria: Weak shock in the Ligurian coast. At Alassio coast, a small tsunami 3 m high was observed after the shock. The first sea motion was a withdrawal and then the water came back violently flooding the beach. The earthquake size is too small to generate a tsunami, that could have been caused by a triggered submarine slide.
[21] 1979 10 16, Nice-Antibes, French Riviera: A part of the Nice new harbour extension slumped into the sea during landfilling operations. The sea receded 300 m on shore between La Salis, south of Antibes and the vicinity of Nice, i.e. over a distance of 60 km. Then it returned with two crests 10 m high. Six people were killed and three were lost. Buildings were damaged, cars were smashed, boats were torn away and

carried off to sea. Sea oscillations continued for over 4 hrs along a 100 km stretch of coast. The tsunami was caused by the gravitative collapse of the embankment of c.10 millions m3, but mainly by a submarine slide of c.400 millions m3 (Assier-Rzadkiewicz 2000).

[22] 1980 10 10, El-Asnam, Algeria: Large (M=7.3), destructive shock. A weak tsunami, caused by one or more underwater slides, was recorded by tide-gauges in Spanish coasts.

[23] 1981 02 24, East Corinth Gulf, Greece: Large (M=6.7) shock. A high-frequency oscillation of amplitude 20-30 cm, recorded by a tide-station immediately after the shock and attenuated in about four days after the shock, possibly was caused by marine slump.

[24] 1984 02 11, Sergoula, NW Corinth Gulf, Central Greece: Moderate (M=5.5) shock. A local, moderate tsunami was caused in Sergoula beach because of a c.100 m long coastal landslide. Small boats moved ashore.

[25] 1988 04 20, Aeolian islands, Italy: A c.200,000 m^3 landslide occurred on La Fossa, Vulcano island, entered the sea and generated a tsunami 1-2 m high clearly observed in the harbour Porto di Levante. The first wave was positive, then other sea oscillations followed.

[26] 1990 12 13, Eastern Sicily, Italy: Strong (M=5.7), lethal shock with epicenter near Augusta coast where sailors observed an anomalous wave offshore. At Catania small submarine slides reported and at Agnone Bagni, close to Augusta, large submarine slides identified by bathymetric changes as large as 50 m.

[27] 1995 06 15, West Corinth Gulf, Central Greece: Strong (M=6.1), destructive shock in Aeghion. In Eratini, north coast, a coastal strip of c.100 m in length and 10 m in width along the local river mouth slumped in the sea as it occurred with the 1965 shock. A small tsunami was caused. The sea retreat was of 0.5-1 m, while wave amplitude of 40-50 cm was observed. In Aeghion, south coast, the wave was of c.1 m high.

[28] 1996 01 01, Aeghion, SW Corinth Gulf, Greece: During early hours of 01.01.1996 a coastal strip of width of c.20 m and length of several hundred meters submerged gravitatively causing strong sea-waves c.2 m high that attacked the coastline to the east of Aeghion, mainly the beach of Digeliotika. The sea disturbance continued for about two hours. The sea advanced inland for c.30 m, overtopped the coastal street and caused little damage to houses and cultivated land. A thin sand sediment was left behind.

[29] 1999 08 17, Izmit Bay, East Marmara Sea, Turkey: Large (M=7.4) earthquake ruptured the west section of the North Anatolian Fault. A tsunami was observed in the coast of Izmit Bay with runup heights of 1-3 m. In Değirmendere town, south coast, where a local slump occurred, eyewitnesses reported waves higher than 15 m. The tsunami was a the high waves seen in Değirmendere.

[30] 2000 04 05, Iraklion, Crete Island, Greece: Oscillations of amplitude c.50 cm and period of 10-12 min. were observed in Iraklion port from 09.00 to 20.00 local time. Small fishing-boats were moved ashore. The wave was observed c.1.5 hrs after a strong (M=5.7) shallow shock occurring (UTC 04:36:59; epicenter 34.220 N/25.690 E) offshore SSE Iraklion. Possibly a small-scale submarine slide, induced by the ground motion, caused the tsunami. It is unlikely that tectonic or meteorological causes triggered the tsunami since it was not reported from other observation points.

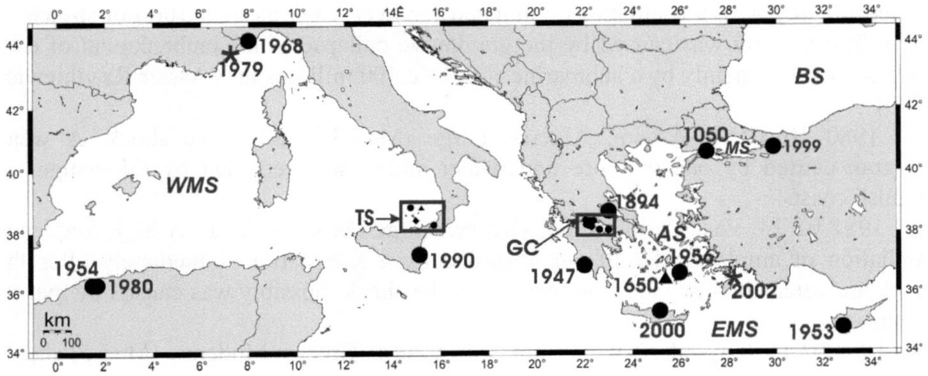

Fig. 1: Geographic distribution of the landslide tsunami sources analyzed. Key: EMS and WMS= East and West Mediterranean, respectively, MS=Marmara Sea, BS=Black Sea, AS=Aegean Sea, GC=Gulf of Corinth (see Fig. 2), TS=Thyrrenian Sea (see Fig. 3), solid circle= earthquake landslide, solid triangle=volcanic landslide, asterisk= gravitative landslide.

[31] 2002 03 24, Rhodes, Dodecanese Islands, Greece: Sea-waves 3-4 m high were reported along the Rhodes city coast (Papadopoulos et al. 2007). The waves inundated a coastal segment as long as 2 km. Some damage was noted in stores while small objects were drifted inland. No shock occurred before or after the tsunami. The wind intensity did not exceeded degree 6 in Beaufort scale. Possibly aseismic submarine slides caused the waves. Local fishermen reported that the days after the event they realized a significant increase of the sea depth at about 1 km offshore.

[32] 2002 12 30, Aeolian islands, Italy: During effusive eruption in Stromboli, a large subaerial landslide that moved downslope the Sciara del Fuoco, NW volcano cone, along with submarine slump, caused a local powerful tsunami. Runup heights up to 9 m were measured. The wave caused damage in villages of Stromboli but no victims were reported.

3. Results and discussion

The analyzed cases of landslide tsunamis indicate that a variety of causes are involved in their generation (Table 1). One may see in Table 2 that earthquake landslides (EL) along the coast and earthquake marine slides (ES) are the most frequent causes. Landslide is a possible but not verifiable generation cause for some tsunamis associated with submarine earthquakes (ER) and submarine eruptions (VO). Very few events are attributed to volcanic landslides (VL) or volcanic marine slides (VS). Also, there are cases of tsunamis caused no by geodynamic mechanisms but by gravitative marine slides (GS) and unfrequently by gravitative landslides (GL). However, the distinction between ES and EL as well as between GS and GL is not always possible but both mechanisms from each one of the two couples could be involved in a single case.

Table 2: Frequency distribution of the tsunami generation causes. Causes classification is explained in Table 1.

EL	ES	ER	VO	VL	VS	GS	GL
12	7	3	3	1	1	4	1

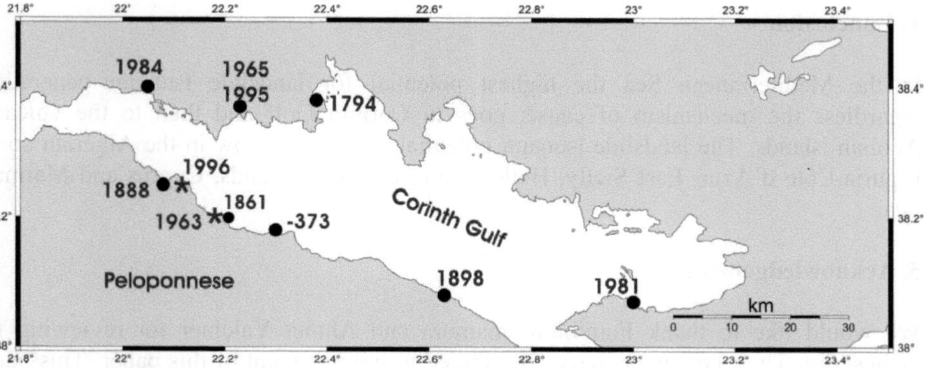

Fig. 2: Geographic distribution of landslide tsunami sources in the Gulf of Corinth. Symbol key as in Figure 1.

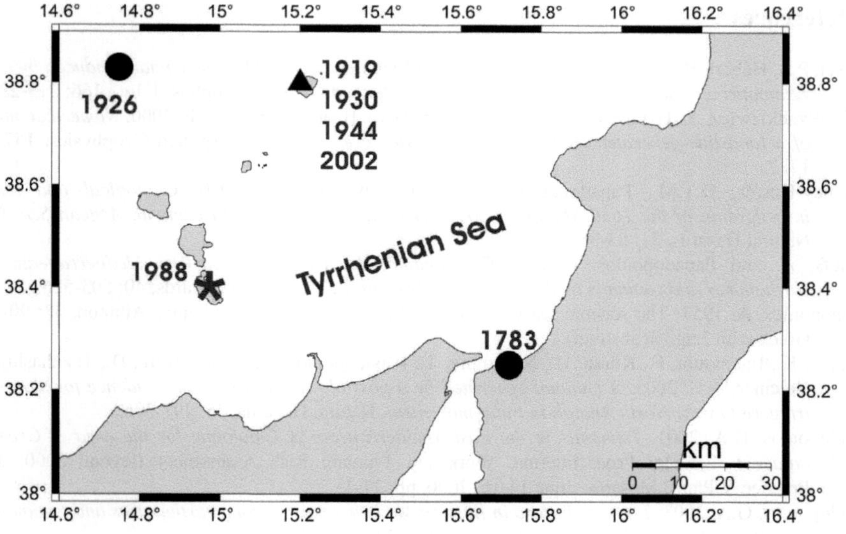

Fig. 3: Geographic distribution of landslide tsunami sources in the Thyrrenian Sea. Symbol key as in Figure 1.

From geographical point of view, the most frequent landslide tsunamis are observed in the Corinth Gulf where 11 out of 32 cases were reported. In the South Tyrrhenian Sea 7 cases were reported. In Algeria, marine slides triggered by the strong shocks caused moderate tsunamis. However, there is no evidence that the tsunami triggered by the 21.05.2003 shock was caused by landslides. Rather, numerical modelling results favour tectonic origin of it (Alasset et al. 2006). In the Hellenic arc only 3 events were reported, in contrast to the abundant, large-size tsunamis of seismotectonic origin historically documented. In Cyclades, only 2 but large events, were reported. In Marmara Sea, Cyprus, Sicily and Liguria-Côte d'Azur few events are known.

4. Conclusion

In the Mediterranean Sea the highest potential for landslide tsunami generation, regardless the mechanism of cause, goes to Corinth Gulf and then to the volcanic Aeolian islands. The landslide tsunami potential is relatively low in the Algerian coast, Liguria-Côte d'Azur, East Sicily, Hellenic arc, Cyclades islands, Cyprus and Marmara Sea.

5. Acknowledgements

We would like to thank Fumihiko Imamura and Ahmet Yalciner for reviewing the manuscript. Their comments have contributed to improvement of this paper. This paper is part of the project TRANSFER (Tsunami Risk ANd Strategies For the European Region) supported by the CEC, contract n. 037058, FP6-2005-Global-4.

6. References

Alasset, P.J., Hébert, H., Maouche, S., Calbibi, V. and Meghraoui, M. 2006. *The tsunami induced by the 2003 Zemmouri earthquake (M_w= 6.9, Algeria): modelling and results.* Geophys. J. Int., 166: 213-226.

Assier-Rzadkiewicz, S., Heinrich, P., Sabatier, P.C., Savoye, B. and Bourillet, J.F. 2000. *Numerical modelling of a landslide-generated tsunami: the 1979 Nice event.* Pure and Applied Geophysics, 157: 1707-1727.

Dominey-Howes, D.T.M., Papadopoulos, G.A. and Dawson, A.G. 2000. *Geological and historical investigation of the 1650 Mt. Columbo (Thera Island) eruption and tsunami, Aegean Sea, Greece.* Natural Hazards, 21: 83-96.

Fokaefs, A. and Papadopoulos, G.A. 2007. *Tsunami Hazard in the Eastern Mediterranean: Strong earthquakes and tsunamis in Cyprus and the Levantine Sea.* Natural Hazards, 40: 503-526.

Galanopoulos, A. 1957. The seismic sea wave of July 9, 1956. Praktika Academias Athinon, 32: 90-101 (in Greek with English abstract).

Minoura, K., Imamoura, F., Kuran, U., Nakamura, T., Papadopoulos, G.A., Sugawara, D., Takahashi, T. and Yalciner, A.C. 2005. *A tsunami generated by a possible submarine slide: evidence for slope failure triggered by the North Anatolian Fault movement.* Natural Hazards, 36: 297-306.

Papadopoulos, G.A. 2001. *Tsunamis in the East Mediterranean: A Catalogue for the area of Greece and adjacent seas.* In: Proc. Internat. Workshop Tsunami Risk Assessment Beyond 2000: Theory, Practice & Plans, Moscow, June 14-16, 2000, pp. 34-43.

Papadopoulos, G.A. 2003. *Tsunami hazard in Eastern Mediterranean: strong earthquakes and tsunamis in the Corinth Gulf, Central Greece.* Natural Hazards, 29:437-464.

Papadopoulos, G.A., Daskalaki, E., Fokaefs, A. and Giraleas, N. 2007. *Tsunami Hazard in the Eastern Mediterranean: Strong Earthquakes and Tsunamis in the East Hellenic Arc and Trench System.* Natural Hazards & Earth System Science, 7:57–64.

Papadopoulos, G.A. and Imamura, F. 2001. *Proposal for a new tsunami intensity scale.* In: Proc. 20th Internat. Tsunami Conference, Seattle, Aug. 7–9 2001, pp. 569-577.

Perissoratis, C. and Papadopoulos, G.A. 1999. *Sediment instability and slumping in the southern Aegean Sea and the case history of the 1956 tsunami.* Mar. Geol., 161: 287-305.

Soloviev, S.L., Solovieva, O., Go, C., Kim, K. and Shchetnikov, A. 2000. *Tsunamis in the Mediterranean Sea -2000 B.C. - 2000 A.D.*, Kluwer Acad. Publ., pp. 239.

Tinti, S., Maramai, A. and Graziani, L. 2004. *The new catalogue of Italian tsunamis.* Natural Hazards, 33:439–465.

Tinti, S, Armigliato, A., Manucci, A., Pagnoni, G., Zaniboni, F., Yalçiner, A.C. and Altinok, Y. 2006. *The generating mechanisms of the August 17, 1999 Izmit bay tsunami: Regional (tectonic) and local (mass instabilities) causes.* Mar. Geo., 225: 311-330.

Yalciner, A.C., Alpar, B., Altinok, Y., Ozbay, I. & Imamura, F. 2002. *Tsunamis in the Sea of Marmara: Historical Documents for the Past, Models for Future.* Mar. Geol., 190: 445-463.

AUTHORS INDEX

MIX
Papier aus verantwortungsvollen Quellen
Paper from responsible sources
FSC® C105338

FSC
www.fsc.org

If you have any concerns about our products,
you can contact us on
ProductSafety@springernature.com

In case Publisher is established outside the EU,
the EU authorized representative is:
Springer Nature Customer Service Center GmbH
Europaplatz 3, 69115 Heidelberg, Germany

Printed by Libri Plureos GmbH
in Hamburg, Germany